COMMON CORE EDITION

COURSE

3

CORE-PLUS
MATHEMATICS
Contemporary Mathematics in Context

Christian R. Hirsch · James T. Fey · Eric W. Hart

Harold L. Schoen · Ann E. Watkins

with

Beth E. Ritsema · Rebecca K. Walker · Brin A. Keller

Robin Marcus · Arthur F. Coxford

Cover (t)©Image State Royalty Free/Alamy Images, (tc)Getty Images, (bc)Echo/Getty Images, (b)©Brand X/JupiterImages

mheonline.com

This material is based upon work supported, in part, by the National Science Foundation under grant no. ESI 0137718. Opinions expressed are those of the authors and not necessarily those of the Foundation.

Send all inquiries to:
McGraw-Hill Education
STEM Learning Solutions Center
8787 Orion Place
Columbus, OH 43240

ISBN: 978-0-07-665796-4
MHID: 0-07-665796-5

Core-Plus Mathematics
Contemporary Mathematics in Context
Course 3 Student Edition

Printed in the United States of America.

2 3 4 5 6 7 8 9 DOW 18 17 16 15 14

Common Core State Standards © Copyright 2010. National Governors Association Center for Best Practices and Council of Chief State School Officers. All rights reserved.

McGraw-Hill is committed to providing instructional materials in Science, Technology, Engineering, and Mathematics (STEM) that give all students a solid foundation, one that prepares them for college and careers in the 21st century.

Core-Plus Mathematics Development Team

Senior Curriculum Developers

Christian R. Hirsch (Director)
Western Michigan University

James T. Fey (Emeritus)
University of Maryland

Eric W. Hart
Maharishi University of Management

Harold L. Schoen (Emeritus)
University of Iowa

Ann E. Watkins
California State University, Northridge

Contributing Curriculum Developers

Beth E. Ritsema
Western Michigan University

Rebecca K. Walker
Grand Valley State University

Brin A. Keller
Michigan State University

Robin Marcus
University of Maryland

Arthur F. Coxford (deceased)
University of Michigan

Principal Evaluator

Steven W. Ziebarth
Western Michigan University

Advisory Board

Diane Briars (formerly)
Pittsburgh Public Schools

Jeremy Kilpatrick
University of Georgia

Robert E. Megginson
University of Michigan

Kenneth Ruthven
University of Cambridge

David A. Smith
Duke University

Mathematical Consultants

Deborah Hughes-Hallett
University of Arizona

Stephen B. Maurer
Swarthmore College

William McCallum
University of Arizona

Doris Schattschneider
Moravian College

Richard Scheaffer
University of Florida

Evaluation Consultant

Norman L. Webb
University of Wisconsin-Madison

Collaborating Teachers

Mary Jo Messenger
Howard County Public Schools, Maryland

Valerie Mills
Oakland County Schools, Michigan

Jacqueline Stewart
Okemos, Michigan

Technical and Production Coordinator

James Laser
Western Michigan University

Support Staff

Angela Reiter
Hope Smith
Matthew Tuley
Teresa Ziebarth
Western Michigan University

Graduate Assistants

Allison BrckaLorenz
Christopher Hlas
University of Iowa

Madeline Ahearn
Geoffrey Birky
Kyle Cochran
Michael Conklin
Brandon Cunningham
Tim Fukawa-Connelly
University of Maryland

Dana Cox
AJ Edson
Nicole L. Fonger
Dana Grosser
Anna Kruizenga
Diane Moore
Western Michigan University

Undergraduate Assistants

Cassie Durgin
University of Maryland

Rachael Lund
Jessica Tucker
Western Michigan University

Core-Plus Mathematics, CCSS Edition Field-Test Sites

The CCSS Edition of *Core-Plus Mathematics* builds on the strengths of the 1st and 2nd editions, which were shaped by multi-year field tests in 49 schools in Alaska, California, Colorado, Georgia, Idaho, Iowa, Kentucky, Michigan, Missouri, Ohio, South Carolina, Texas and Wisconsin. Each text is the product of a three-year cycle of research and development, pilot testing and refinement, and field testing and further refinement. Special thanks are extended to the following teachers and their students who participated in the most recent testing and evaluation of Course 3.

Hickman High School
Columbia, Missouri
Sandra Baker
Lindsay Carlson
Melissa Hundley
Stephanie Krawczyk
Tiffany McCracken
Dana Meyer
Ryan Pingrey

Holland Christian High School
Holland, Michigan
Jeff Goorhouse
Brian Lemmen
Betsi Roelofs
Mike Verkaik

Malcolm Price Lab School
Cedar Falls, Iowa
Megan Balong
James Maltas

Riverside University High School
Milwaukee, Wisconsin
Cheryl Brenner
Scott Hanson
Alice Lanphier

Rock Bridge High School
Columbia, Missouri
Cynthia Francisco
Donna Lillard
Linda Shumate

Sauk Prairie High School
Prairie du Sac, Wisconsin
Joan Quenan
Mary Walz

Washington High School
Milwaukee, Wisconsin
Anthony Amoroso

Development and evaluation of the student text materials, teacher support materials, assessments, and computer software for *Core-Plus Mathematics* was funded through a series of grants from the National Science Foundation to the Core-Plus Mathematics Project (CPMP). We express our appreciation to NSF and, in particular, to our program officer John Bradley for his long-term trust, support, and input.

We are also grateful to Texas Instruments and, in particular, Dave Santucci for collaborating with us by providing classroom sets of graphing calculators to field-test schools.

As seen on page iii, CPMP has been a collaborative effort that has drawn on the talents and energies of teams of mathematics educators at several institutions. This diversity of experiences and ideas has been a particular strength of the project. Special thanks is owed to the exceptionally capable support staff at these institutions, particularly to Angela Reiter, Hope Smith, Matthew Tuley, and Teresa Ziebarth at Western Michigan University.

We are grateful to our Advisory Board, Diane Briars (formerly Pittsburgh Public Schools), Jeremy Kilpatrick (University of Georgia), Robert E. Megginson (University of Michigan), Kenneth Ruthven (University of Cambridge), and David A. Smith (Duke University) for their ongoing guidance and advice. We also acknowledge and thank Norman L. Webb (University of Wisconsin-Madison) for his advice on the design and conduct of our field-test evaluations.

Special thanks are owed to the following mathematicians: Deborah Hughes-Hallett (University of Arizona), Stephen B. Maurer (Swarthmore College), William McCallum (University of Arizona), Doris Schattschneider (Moravian College), and to statistician Richard Scheaffer (University of Florida) who reviewed and commented on units as they were being developed, tested, and refined.

Our gratitude is expressed to the teachers and students in the evaluation sites listed on page iv. Their experiences using the revised *Core-Plus Mathematics* units provided constructive feedback and suggested improvements that were immensely helpful.

Finally, we want to acknowledge Catherine Donaldson, Angela Wimberly, Justin Moyer, Michael Kaple, Karen Corliss, and their colleagues at McGraw-Hill Education who contributed to the design, editing, and publication of this program.

The first three courses in *Core-Plus Mathematics* provide a significant common core of broadly useful mathematics aligned with the Common Core State Standards for Mathematics (CCSS) and intended for all students. They were developed to prepare students for success in college, in careers, and in daily life in contemporary society.

Course 4: *Preparation for Calculus* continues the preparation of STEM-oriented (science, technology, engineering, and mathematics) students for success in college mathematics, especially calculus.

A separate alternative fourth-year capstone course, *Transition to College Mathematics and Statistics*, is intended for the large number of students planning to major in college programs that do *not* require calculus.

Core-Plus Mathematics is a problem-based, inquiry-oriented program that builds upon the theme of mathematics as reasoning and sense-making. Through investigations of real-life contexts, students develop a rich understanding of important mathematics that makes sense to them and which, in turn, enables them to make sense out of new situations and problems.

Each of the first three courses in *Core-Plus Mathematics* shares the following mathematical and instructional features.

- **Integrated Content** Each year the curriculum advances students' understanding of mathematics along interwoven strands of algebra and functions, statistics and probability, geometry and trigonometry, and discrete mathematics. These strands are unified by fundamental themes, by common topics, by mathematical practices, and by mathematical habits of mind. Developing mathematics each year along multiple strands helps students develop a connected understanding of mathematics and nurtures their differing strengths and talents.

- **Mathematical Modeling** The problem-based curriculum emphasizes mathematical modeling including the processes of data collection, representation, interpretation, prediction, and simulation. The modeling perspective permits students to experience mathematics as a means of making sense of data and problems that arise in diverse contexts within and across cultures.

- **Access and Challenge** The curriculum is designed to make mathematics accessible to more students while at the same time challenging the most able students. Differences in student performance and interest can be accommodated by the depth and level of abstraction to which core topics are pursued, by the nature and degree of difficulty of applications, and by providing opportunities for student choice of homework tasks and projects.

- **Technology** Numeric, graphic, and symbolic manipulation capabilities such as those found in *CPMP-Tools*® and on many graphing calculators are assumed and appropriately used throughout the curriculum. *CPMP-Tools* is a suite of software tools that provide powerful aids to learning mathematics and solving mathematical problems. (See page xvii for further details.) This use of technology permits the curriculum and instruction to emphasize multiple linked representations (verbal, numerical, graphical, and symbolic) and to focus on goals in which mathematical thinking and problem solving are central.

- **Active Learning** Instructional materials promote active learning and teaching centered around collaborative investigations of problem situations followed by teacher-led whole-class summarizing activities that lead to analysis, abstraction, and further application of underlying mathematical ideas and principles. Students are actively engaged in exploring, conjecturing, verifying, generalizing, applying, proving, evaluating, and communicating mathematical ideas.

- **Multi-dimensional Assessment** Comprehensive assessment of student understanding and progress through both curriculum-embedded formative assessment opportunities and summative assessment tasks support instruction and enable monitoring and evaluation of each student's performance in terms of mathematical processes, content, and dispositions.

Integrated Mathematics

Core-Plus Mathematics is an international-like curriculum. It replaces the traditional Algebra-Geometry-Advanced Algebra/Trigonometry-Precalculus sequence of high school mathematics courses with a sequence of courses that features a coherent and connected development of important mathematics drawn from four strands.

The *Algebra and Functions* strand develops student ability to recognize, represent, and solve problems involving relations among quantitative variables. Central to the development is the use of functions as mathematical models. The key algebraic models in the curriculum are linear, exponential, power, polynomial, logarithmic, rational, and trigonometric functions. Modeling with systems of equations, both linear and nonlinear, is developed. Symbolic reasoning and manipulation are integral to this work.

The primary goal of the *Geometry and Trigonometry* strand is to develop visual thinking and ability to construct, reason with, interpret, and apply mathematical models of patterns in visual and physical contexts. The focus is on describing patterns in shape, size, and location; representing patterns with drawings, coordinates, or vectors; predicting changes and invariants in shapes under transformations; and organizing geometric facts and relationships through deductive reasoning.

The primary role of the *Statistics and Probability* strand is to develop student ability to analyze data intelligently, to recognize and measure variation, and to understand the patterns that underlie probabilistic situations. The ultimate goal is for students to understand how inferences can be made about a population by looking at a sample from that population. Graphical methods of data analysis, simulations, sampling, and experience with the collection and interpretation of real data are featured.

The *Discrete Mathematics* strand develops student ability to solve problems using recursion, matrices, vertex-edge graphs, and systematic counting methods (combinatorics). Key themes are discrete mathematical modeling, optimization, and algorithmic problem-solving.

Each of these strands of mathematics is developed within focused units connected by fundamental ideas such as functions, matrices, geometric transformations and symmetry, data analysis, and mathematical modeling. The strands also are connected across units by CCSS's mathematical practices and by mathematical habits of mind. These important mathematical practices include disposition toward, and proficiency in:

- making sense of problems and persevering in solving them

- reasoning both quantitatively and algebraically

- constructing sound arguments and critiquing the reasoning of others

- using mathematics to model problems in everyday life, society, and in careers

- selecting and using appropriate tools, especially technological tools (graphing calculator, spreadsheet, computer algebra system, statistical packages, and dynamic geometry software)

- communicating precisely and expressing calculations with an appropriate precision

- searching for and making use of patterns or structure in mathematical situations

- identifying structure in repeated calculations, algebraic manipulations, and reasoning patterns

Additionally, mathematical habits of mind such as visual thinking, recursive thinking, searching for and explaining patterns, making and checking conjectures, reasoning with multiple representations, inventing mathematics, and providing convincing justifications are integral to each strand.

Important mathematical ideas are frequently revisited through this attention to connections within and across strands, enabling students to develop a robust and connected understanding of, and proficiency with, mathematics.

Active Learning and Teaching

The manner in which students encounter mathematical ideas can contribute significantly to the quality of their learning and the depth of their understanding. *Core-Plus Mathematics* units are designed around multi-day lessons centered on big ideas. Each lesson includes 2–5 mathematical investigations that engage students in a four-phase cycle of classroom activities, described in the following paragraph—*Launch, Explore, Share and Summarize*, and *Check Your Understanding*. This cycle is designed to engage students in investigating and making sense of problem situations, in constructing important mathematical concepts and methods, in generalizing and proving mathematical relationships, and in communicating, both orally and in writing, their thinking and the results of their efforts. Most classroom activities are designed to be completed by students working collaboratively in groups of two to four students.

The Launch phase of a lesson promotes a teacher-led class discussion of a problem situation and of related questions to think about, setting the context for the student work to follow and providing important information about students' prior knowledge. In the second or Explore phase, students investigate more focused problems and questions related to the launch situation. This investigative work is followed by a teacher-led class discussion in which students summarize mathematical ideas developed in their groups, providing an opportunity to construct a shared understanding of important concepts, methods, and justifications. Finally, students are given tasks to complete on their own, to check their understanding of the concepts and methods.

Each lesson also includes homework tasks to engage students in applying, connecting, reflecting on mathematical practices, extending, and reviewing their mathematical understanding. These *On Your Own* tasks are central to the learning goals of each lesson and are intended primarily for individual work outside of class. Selection of tasks should be based on student performance and the availability of time and technology access. Students can exercise some choice of tasks to pursue, and at times they should be given the opportunity to pose their own problems and questions to investigate.

Formative and Summative Assessment

Assessing what students know and are able to do is an integral part of *Core-Plus Mathematics*. There are opportunities for formative assessment in each phase of the instructional cycle. Initially, as students pursue the investigations that comprise the curriculum, the teacher is able to informally assess through observation and interaction student understanding of mathematical processes and content and their disposition toward mathematics. At the end of each investigation, a class discussion to Summarize the Mathematics provides an opportunity for the teacher to assess levels of understanding that various groups of students and individuals have reached as they share, explain, and discuss their findings. Finally, the Check Your Understanding tasks and the tasks in the On Your Own sets provide further opportunities for formative assessment of the level of understanding of each individual student. Quizzes, in-class tests, take-home assessment tasks, and extended projects are included in the teacher resource materials for summative assessments.

UNIT **1** Reasoning and Proof

Reasoning and Proof develops student understanding of formal reasoning in geometric, algebraic, and statistical contexts and of basic principles that underlie those reasoning strategies.

Topics include inductive and deductive reasoning strategies; principles of logical reasoning—Affirming the Hypothesis and Chaining Implications; properties of line reflections; relation among angles formed by two intersecting lines or by two parallel lines and a transversal; rules for transforming algebraic expressions and equations; design of experiments; use of data from a randomized experiment to compare two treatments, sampling distribution constructed using simulation, randomization test, and statistical significance; inference from sample surveys, experiments, and observational studies and how randomization relates to each.

Lesson 1 Reasoning Strategies

Lesson 2 Geometric Reasoning and Proof

Lesson 3 Algebraic Reasoning and Proof

Lesson 4 Statistical Reasoning

Lesson 5 Looking Back

UNIT **2** Inequalities and Linear Programming

Inequalities and Linear Programming develops student ability to reason both algebraically and graphically to solve inequalities in one and two variables, introduces systems of inequalities in two variables, and develops a strategy for optimizing a linear function in two variables within a system of linear constraints on those variables.

Topics include inequalities in one and two variables, number line graphs, interval notation, systems of linear inequalities, and linear programming.

Lesson 1 Inequalities in One Variable

Lesson 2 Inequalities in Two Variables

Lesson 3 Looking Back

UNIT **3** Similarity and Congruence

Similarity and Congruence extends student understanding of similarity and congruence and their ability to use those relations to solve problems and to prove geometric assertions with and without the use of coordinates.

Topics include connections between Law of Cosines, Law of Sines, and sufficient conditions for similarity and congruence of triangles; connections between transformations and sufficient conditions for congruence and similarity of triangles; centers of triangles, applications of similarity and congruence in real-world contexts; necessary and sufficient conditions for parallelograms, sufficient conditions for congruence of parallelograms, and midpoint connector theorems.

Lesson 1 Reasoning about Similar Figures

Lesson 2 Reasoning about Congruent Figures

Lesson 3 Looking Back

UNIT DESCRIPTIONS AND TOPICS

UNIT 4 — Samples and Variation

Samples and Variation develops student ability to use the normal distribution as a model of variation, introduces students to the binomial distribution and its use in making inferences about population parameters based on a random sample, and introduces students to the probability and statistical inference used in industry for statistical process control.

Topics include normal distribution, standardized scores and estimating population percentages, binomial distributions (shape, expected value, standard deviation), normal approximation to a binomial distribution, odds, statistical process control, and the Central Limit Theorem.

UNIT 5 — Polynomial and Rational Functions

Polynomial and Rational Functions extends student ability to represent and draw inferences about polynomial and rational functions using symbolic expressions and manipulations.

Topics include definition and properties of polynomials, operations on polynomials; completing the square, proof of the quadratic formula, solving quadratic equations (including complex number solutions), vertex form of quadratic functions; definition and properties of rational functions, operations on rational expressions.

UNIT 6 — Circles and Circular Functions

Circles and Circular Functions develops student understanding of properties of special lines, segments, angles, and arcs in circles and the ability to use those properties to solve problems; develops student understanding of circular functions and the ability to use those functions to model periodic change; and extends student ability to reason deductively in geometric settings.

Topics include properties of chords, tangent lines, and central and inscribed angles and their intercepted arcs; linear and angular velocity; radian measure of angles; and circular functions as models of periodic change.

UNIT **7** Recursion and Iteration

Recursion and Iteration extends student ability to model, analyze, and solve problems in situations involving sequential and recursive change.

Topics include iteration and recursion as tools to model and solve problems about sequential change in real-world settings, including compound interest and population growth; arithmetic, geometric, and other sequences together with their connections to linear, exponential, and polynomial functions; arithmetic and geometric series; finite differences; linear and nonlinear recurrence relations; and function iteration, including graphical iteration and fixed points.

UNIT **8** Inverse Functions

Inverse Functions develops student understanding of inverses of functions with a focus on logarithmic functions and their use in modeling and analyzing problem situations and data patterns.

Topics include inverses of functions; logarithmic functions and their relation to exponential functions, properties of logarithms, equation solving with logarithms; and inverse trigonometric functions and their applications to solving trigonometric equations.

TABLE OF CONTENTS

TABLE OF CONTENTS

Have you ever wondered ...

- How engineers design roller coaster tracks so that they provide exciting but safe rides for the passengers?

- How predictions about population change or climate change are made?

- How civil engineers locate service centers so that they are equally-distant from the areas being served?

- How medical researchers decide whether using a new treatment is better than doing nothing?

- How large companies organize the many variables in their operation to give efficient service at a price that yields maximum profit?

- How FM radio broadcasts differ from AM broadcasts?

- How much money you would need to begin saving now, with compound interest, to buy a car or pay for college later?

- How some manufacturers are able to produce better automobiles than other manufacturers?

The mathematics you will learn in *Core-Plus Mathematics* Course 3 will help you answer questions like these.

Because real-world situations and problems like those above often involve data, shape, quantity, change, or chance, you will study concepts and

methods from several interwoven strands of mathematics. In particular, you will develop an understanding of broadly useful ideas from algebra and functions, geometry and trigonometry, statistics and probability, and discrete mathematics. In the process, you will also see and use many connections among these strands.

In this course, you will learn important mathematics as you investigate and solve interesting problems. You will develop the ability to reason and communicate about mathematics as you are actively engaged in understanding and applying mathematics. You will often be learning mathematics in the same way that many people work in their jobs—by working in teams and using technology to solve problems.

In the 21st century, anyone who faces the challenge of learning mathematics or using mathematics to solve problems can draw on the resources of powerful information technology tools. Calculators and computers can help with calculations, drawing, and data analysis in mathematical explorations and solving mathematical problems.

Graphing calculators and computer software tools will be useful in your work on many of the investigations in *Core-Plus Mathematics*. Set as one of your goals to learn how to make strategic decisions about the choice and use of technological tools in solving particular problems.

The curriculum materials include computer software called *CPMP-Tools* that will be of great help in learning and applying the mathematical topics of each CPMP course. You can access the software at www.wmich.edu/cpmp/CPMP-Tools/.

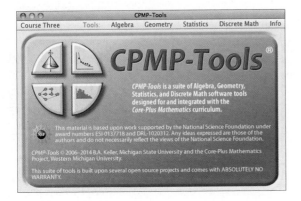

The software toolkit includes four families of programs:

- Algebra—The software for work on algebra problems includes a spreadsheet and a computer algebra system (CAS) that produces tables and graphs of functions, manipulates algebraic expressions, and solves equations and inequalities.

- Geometry—The software for work on geometry problems includes an interactive drawing program for constructing, measuring, and manipulating geometric figures and a set of custom apps for exploring properties of figures and geometric models of mechanical devices.

- Statistics—The software for work on data analysis and probability problems provides tools for graphic display and analysis of data, simulation of probabilistic situations, and mathematical modeling of quantitative relationships.

- Discrete Mathematics—The software provides tools for constructing, manipulating, and analyzing discrete mathematical models.

In addition to the general purpose tools provided for work on tasks in each strand of the curriculum, *CPMP-Tools* includes files of most data sets essential for work on problems in each *Core-Plus Mathematics* course. When the opportunity to use computer tools for work on a particular investigation seems appropriate, select the *CPMP-Tools* menu corresponding to the content of, and your planned approach to, the problem. Then select the submenu items corresponding to the appropriate mathematical operations and required data set(s).

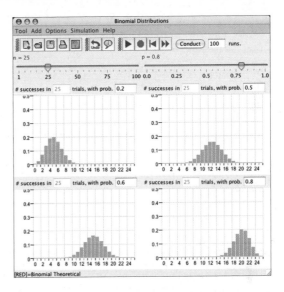

In Course 3, you are going to learn a lot of useful mathematics. It will make sense to you and you can use it to make sense of your world. You are going to learn a lot about working collaboratively on problems and communicating with others as well. You are also going to learn how to use technological tools strategically and effectively. Finally, you will have plenty of opportunities to be creative and inventive. Enjoy!

Reasoning and Proof

Life presents many opportunities to use logical reasoning in solving problems and drawing conclusions from information. Whether it is developing a winning strategy in a favorite game, figuring out how to build or repair something, or learning how to drive defensively, the ability to ask yourself "What will happen if … ?" is fundamental. Quite often, it is the ability of an athlete, a detective, or a lawyer to use reasoning that makes that person stand out from the crowd.

In this unit, you will examine more carefully the reasoning strategies you have used in your prior mathematics study. You will learn some basic principles of logical reasoning that underlie those strategies and develop skill in applying that understanding to mathematical questions in geometry, algebra, and statistics. The key ideas will be developed through work on problems in four lessons.

LESSONS

1 Reasoning Strategies

Analyze and use deductive and inductive reasoning strategies in everyday situations and in mathematical contexts.

2 Geometric Reasoning and Proof

Use inductive reasoning to discover and deductive reasoning to prove relations among angles formed by two intersecting lines or by two parallel lines and a transversal.

3 Algebraic Reasoning and Proof

Use symbolic notation to represent numerical patterns and relationships and use rules for transforming algebraic expressions and equations to prove those facts.

4 Statistical Reasoning

Know the characteristics of a well-designed experiment and use statistical reasoning to decide whether one treatment causes a better result than a second treatment.

Reasoning Strategies

In Courses 1 and 2 of *Core-Plus Mathematics*, you frequently used *inductive reasoning* to discover general patterns or principles based on evidence from experiments or several cases. You also used *deductive reasoning* to justify statements or conclusions based on accepted facts and definitions. In this unit, you will expand your ability to re ason carefully in geometric, algebraic, and statistical contexts.

Careful reasoning is important not only in mathematics. It is a key to success as a consumer, in careers, and even in recreational activities. Setting up a play in basketball or volleyball, winning a friendly game of CLUE® or MONOPOLY®, or solving a crossword puzzle involves careful strategic reasoning.

Consider the game of Sudoku (pronounced sue-doe-koo), a number game similar to a crossword puzzle, that has recently become popular around the world. Sudoku is a Japanese word meaning "single number." The goal of the game is to fill in the empty squares with the digits 1 to 9 so that each digit appears exactly once in every row, column, and outlined 3 × 3 block.

8				6				1
4	5	7	8					9
			2	5	9			4
2	3	9		8				
6	8						9	3
	7			9		1	6	8
7			9	2	8			
9			4		5	7	3	
1				7	3			2

Adapted from: www.knightfeatures.com

Think about strategies you would use to solve the Sudoku puzzle shown at the bottom of the previous page.

a How would you decide where to begin?

b Which square would you fill in first? Which one would you fill in next? Explain your reasoning.

c Describe a strategy (or a combination of strategies) you would use to fill in the remaining squares.

d When the game is completed, what will be true about the sums of the row entries, the sums of the column entries, and the sums of the 3 × 3 block entries? Explain.

In this lesson, you will learn how to examine arguments in terms of reasoning strategies, assumptions, and logical soundness. You will also learn how to use *if-then* reasoning patterns in deductive arguments or proofs.

INVESTIGATION 1

Reasoned Arguments

Careful reasoning, whether it is concerned with mathematics, science, history, or daily affairs, is important if you want to have confidence in the conclusions reached. A **valid** *argument* shows that a conclusion follows logically from accepted definitions and assumptions or previously established facts. If the assumptions are true, you can be confident that the argument is *sound* and the conclusion is true.

As you analyze the situations and arguments in this investigation, look for answers to this question:

How can you determine whether a conclusion follows logically
from information and facts you know are correct
or on which everyone would agree?

1 **Reasoning about Crime Scenes** In popular television shows like Criminal Minds and CSI that involve crime investigations, the detectives use careful reasoning to identify suspects, motives, and evidence that can be used to solve cases.

Consider the following plot from a crime case.

At 7:00 P.M., Mrs. Wilson's maid served her tea in the library. The maid noticed that Mrs. Wilson seemed upset and a little depressed. At 8:45 P.M., the maid knocked on the library door and got no answer. The door was locked from the inside. The maid called Inspector Sharpe and a professor friend. When the door was forced open, Mrs. Wilson was found dead. The maid burst into tears, crying, "I feel so bad that we haven't been getting along lately!" Nearby was a half-empty teacup, a tiny unstoppered vial, and a typewritten note that said, "Blessed are the poor for they shall be happy." The window was open. When the two men went out-side to inspect the grounds, Charles, the wealthy widow's sole heir, arrived. He was told his aunt was poisoned and said, "How terrible! Poisoned? Who did it? Why was the door locked? Had my aunt been threatened?" He explained he had been working late at the office and was stopping by on his way home.

Source: Adapted from Ripley, Austin. (1976). *Minute Mysteries.* New York: Harper & Row Publishers. pages 22–23.

a. Who are the possible suspects in this case?

b. For each suspect, identify the evidence that exists that could be used to charge them with the crime.

c. Write a convincing argument to charge the prime suspect with the crime.

d. Compare your prime suspect and argument with those of others. Resolve any differences.

2 **Reasoning about Games** Games based on strategy—such as Tic-Tac-Toe, Checkers, and Chess—have been played for thousands of years. The following two-person game can be played on any regular polygon. For this problem, assume the game is played on a regular nonagon. To play, place a penny on each vertex of the polygon. Take turns removing one penny or two pennies from adjacent vertices. The player who picks up the last coin(s) is the winner.

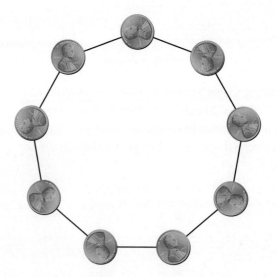

a. Working with a partner, play the nonagon game a few times. Make mental notes of strategies you used.

b. Tianna, Jairo, Nicole, and Connor each thought they found a strategy that would guarantee that they could always win the game if they played *second*. Analyze their reasoning. In each case, decide if the proposed strategy will always result in a win for the second player. Explain your reasoning in each case.

Tianna: *Each time the first player removes one or two pennies, I'll remove only one penny. That way, there will be more pennies left to choose from and at least one for me at the end.*

Jairo: *Each time the first player removes one penny, I'll remove one penny. If the first player removes two pennies, then I'll remove two pennies. Then the remaining number of pennies will be an odd number, so there will be at least one penny for me at the end.*

Nicole: *Each time the first player removes one penny, I'll remove two pennies; if the first player removes two pennies, I'll remove one penny as shown by the triangles in the diagram below. Since there are 9 pennies, I can always win on the third round.*

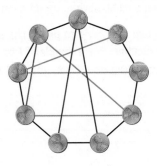

Connor: *If the first player removes one penny, I'll visualize the line of symmetry containing the "empty" vertex and remove the two adjacent pennies on opposite sides of the symmetry line. If the first player removes two pennies on the first move, then I'll visualize the line of symmetry between the two empty vertices and remove the penny on the symmetry line.*

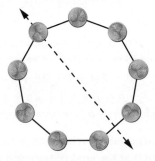

This strategy will always leave three pennies on each side of the symmetry line after we each make our first move. After that, I'll match each play the first player makes by choosing the mirror image. So, I'll be able to remove the final coin(s).

3 **Reasoning about Numbers** You may have noticed that when you add two odd numbers, for example $3 + 7$, the sum always seems to be an even number. Alex, Maria, Nesrin, and Teresa were asked to write an argument to justify the following claim.

If a and b are odd numbers, then a + b
(the sum) is an even number.

Carefully study each student's argument.

Alex: I entered odd numbers in list L₁ of my calculator and different odd numbers in list L₂. I then calculated L₁ + L₂. Scanning the calculator screen, you can see that in every case the sum is an even number.

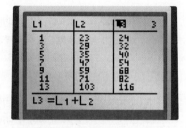

Maria: Odd numbers end in 1, 3, 5, 7, or 9. When you add any two of these, the answer will end in 0, 2, 4, 6, or 8. So, the sum of two odd numbers must always be an even number.

Nesrin: I can use counters to prove the sum of any two odd numbers is an even number. For example, if I take the numbers 5 and 11 and organize the counters as shown, you can see the pattern.

You can see that when you put the sets together (add the numbers), the two extra counters will form a pair and the answer is always an even number.

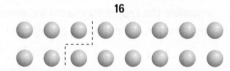

Teresa: I know that a and b are odd numbers.
By definition, $a = 2m + 1$ and $b = 2n + 1$, where m and n are integers.
So, $a + b = 2m + 1 + 2n + 1$.
Then $a + b = 2m + 1 + 2n + 1 = 2m + 2n + 2 = 2(m + n + 1)$.
Therefore, $a + b$ is an even number since the sum is a multiple of 2.

a. Of the four arguments, which *one* is closest to the argument you would give to prove that the sum of two odd numbers is an even number?

b. For each of the arguments, answer the following questions.

   i. Does the argument have any errors in it?

   ii. Does the argument show the statement is *always true* or does the argument *only* show the statement is true for *some* numbers?

   iii. Does the argument show *why* the statement is true?

   iv. Does the argument provide an easy way to convince someone in your class who is uncertain of the claim?

c. Select one of the arguments you think is correct. How, if at all, would you modify the argument to justify that the sum of *any* two odd numbers that are square numbers (like 9 and 25) is an even number? Explain your reasoning.

4 **Reasoning about Areas** In the Course 1 *Patterns in Shape* unit, you saw that by assuming the formula $A = bh$ for the area of a rectangle with a base of length b and height h, you could derive a formula for the area of a parallelogram. You also saw that if you knew the formula for the area of a parallelogram, you could derive a formula for the area of a triangle. A standard formula for calculating the area of a **trapezoid**—a quadrilateral with two opposite sides parallel—is given by:

$$A = \frac{1}{2}(b_1 + b_2)h$$

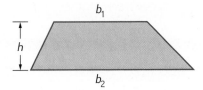

where b_1, b_2, and h represent the lengths of the two bases and the height of the trapezoid.

Study each of the following five arguments offered by students as justification of this formula for the area of a trapezoid.

Angela: I can split the trapezoid into two triangles by drawing a diagonal. One triangle has area $\frac{1}{2}b_1h$. The other has area $\frac{1}{2}b_2h$. So, the area of the trapezoid is $\frac{1}{2}b_1h + \frac{1}{2}b_2h$ or $\frac{1}{2}(b_1 + b_2)h$.

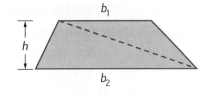

Dylan: In a parallelogram, opposite sides are the same length. Any side can be used as the base. In the trapezoid shown,

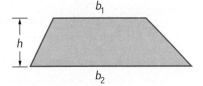

b_1h will underestimate the area.
b_2h will overestimate the area.

To find the correct area, you average the two estimates.
$$\frac{b_1h + b_2h}{2} = \frac{1}{2}(b_1h + b_2h)$$
$$= \frac{1}{2}(b_1 + b_2)h$$

Hsui: If I rotate the trapezoid 180° about the midpoint M of one side, the trapezoid and its image form a parallelogram.

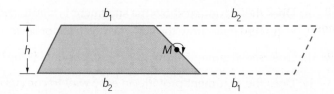

The length of the base of the parallelogram is $b_2 + b_1$ and the height is h. The area of the parallelogram is $(b_2 + b_1)h$. The area of the trapezoid is $\frac{1}{2}$ of this area, or $\frac{1}{2}(b_1 + b_2)h$.

Barbara:

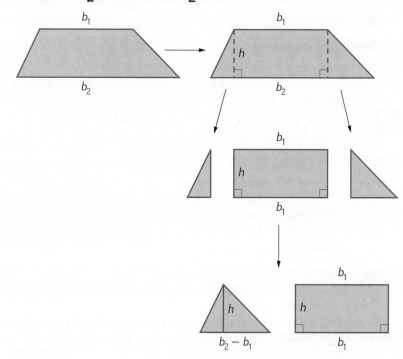

So, the area of the trapezoid is $\frac{1}{2}(b_2 - b_1)h + b_1h$, which equals $\frac{1}{2}(b_1 + b_2)h$.

Jorge: The area of the trapezoid is $\frac{1}{2}(b_1 + b_2)h$ because you can cut up the shape and find the areas of the individual pieces.

a. Which of the five arguments is closest to the argument you would have provided to justify the formula for the area of a trapezoid?

b. Which of these arguments show correct reasoning and which do not? Compare your responses with those of others and resolve any differences.

c. Select one of the arguments you think provides a correct proof of the area formula. Describe the features of the argument that you thought were good. What, if anything, would you add to that argument to make it easier to understand?

SUMMARIZE THE MATHEMATICS

In this investigation, you examined reasoning strategies and arguments in mathematical and nonmathematical contexts.

a Look back at the mathematical statements that the students were attempting to prove in Problems 3 and 4. In each case, answer the following questions.

    i. What information was given?

    ii. What conclusion was to be established?

    iii. How was the given information used by Teresa to reason logically to the conclusion in Problem 3? By Angela to reason to the conclusion in Problem 4?

b How can you tell whether an argument provides a correct proof of a claim?

Be prepared to share your ideas with the rest of the class.

 CHECK YOUR UNDERSTANDING

Analyze each student attempt to prove the following statement.

The sum of two even numbers is an even number.

Art: *I tried many different pairs of even numbers. It was impossible to find a counterexample to the claim that the sum of two even numbers is an even number. So, the claim must be true.*

Sherita: *Even numbers are numbers that can be divided by 2. When you add numbers with a common factor of 2, the answer will have a common factor of 2.*

Bill: *If a and b are any two numbers, then 2a and 2b are any two even numbers.*

$$2a + 2b = 2(a + b)$$

Katrina: *Even numbers can be represented by rectangular arrays of counters with two rows.*

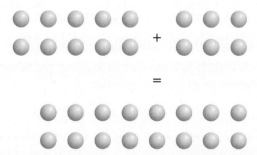

The sum of any two even numbers will be a rectangular array of counters with 2 rows. So, the sum is even.

a. Which argument is closest to the argument you would give to prove that the sum of two even numbers is always an even number?

b. Which arguments are *not* correct proofs of the claim? Explain your reasoning.

c. Modify one of the correct proofs to justify the following statement.

The sum of an even number and an odd number is always an odd number.

INVESTIGATION 2

If-Then Statements

Statements of the form "*If … , then …*" occur frequently in everyday life and in mathematics. For example, consider the two statements: "*If* it is raining on game day, *then* the game will be rescheduled for next Tuesday," and "*If* $x > 5$, *then* $2x > 10$." Other mathematical statements, such as definitions, can be interpreted in *if-then* form. For example, consider the definition of a trapezoid from Investigation 1.

A trapezoid is a quadrilateral with two opposite sides parallel.

This definition means that   *If* a quadrilateral is a trapezoid, *then* two opposite sides are parallel,

and the *converse*   *If* two opposite sides of a quadrilateral are parallel, *then* the quadrilateral is a trapezoid.

If-then statements are frequently used in deductive arguments because they imply that if some condition (called the **hypothesis**) is satisfied, some other condition (called the **conclusion**) follows. As you work on the following problems, look for answers to these questions:

> *How can you use if-then statements in deductive reasoning?*
>
> *How is deductive reasoning with if-then statements different from inductive reasoning from patterns?*

1 **Reasoning with If-Then Statements** You may recall that the numbers 2, 3, 7, and 11 are called prime numbers. By definition, a **prime number** is an integer greater than 1 that has exactly two factors, 1 and itself.

a. Write two if-then statements that together mean the same thing as the above definition of a prime number.

b. Is 23 a prime number? Explain your reasoning.

c. Which of the two if-then statements in Part a was used in your reasoning in Part b?

The reasoning you used in Problem 1 Part b is based on a fundamental principle of logic called **modus ponens** (Latin: mode that affirms) or **Affirming the Hypothesis**.

If you have a known fact   an (if-then) statement that is always true,
and you also know   the "if" part is true in a particular case,
you can conclude   the "then" part is true in that case.

2 Decide what can be concluded, if anything, from each of the following sets of statements. Be prepared to explain how reaching your conclusion involved Affirming the Hypothesis.

a. *Known fact:*   *If* a person has a Michigan driver's license, *then* the person is 16 years of age or older.

     Given: Andy has a Michigan driver's license.

     Conclusion: ?

b. *Known fact:*   *If* a person in Michigan has a driver's license, *then* the person is 16 years of age or older.

     Given: Janet is 18 years old.

     Conclusion: ?

c. *Known fact:*   *If* points A, B, and C are collinear and point B is between points A and C, *then* $AB + BC = AC$.

     Given: Points R, S, and T are collinear and point R is between points S and T.

     Conclusion: ?

d. *Known fact:*   *If* two sides of a triangle are the same length, *then* the triangle is isosceles.

     Given: $\triangle ABC$ has sides of length 2 cm, 5 cm, 5 cm.

     Conclusion: ?

e. *Known fact:*   *If* $f(x) = ax^2 + bx + c$ is a quadratic function with $a < 0$, *then* $f(x)$ has a maximum value.

     Given: $g(x) = -8x^2 + 5x - 2$

     Conclusion: ?

f. *Known fact:*   *If* a data set with mean $\bar{x}$ and standard deviation s is transformed by adding a constant c to each value, *then* the mean of the transformed data set is $\bar{x} + c$ and the standard deviation is s.

     Given: The Oak Park hockey team has a mean height of 5 feet 9 inches and a standard deviation of $2\frac{1}{2}$ inches. Wearing ice skates adds approximately $1\frac{3}{4}$ inches to the height of a skater.

     Conclusion: ?

g. *Known fact:*   *If* S is a size transformation with center at the origin and magnitude k and A is the area of a figure, *then* $k^2 \cdot A$ is the area of the image of the figure under S.

     Given: $\triangle P'Q'R'$ is the image of $\triangle PQR$ under a size transformation with center at the origin and magnitude 3. $\triangle PQR$ has area 32 cm².

     Conclusion: ?

3 If-then statements can be represented symbolically as $p \Rightarrow q$ (read "if p, then q" or "p implies q") where p represents the hypothesis and q represents the conclusion. The arrow signals that you move from the hypothesis p to the conclusion q. The reasoning pattern you used in Problem 2 can be represented as follows.

Affirming the Hypothesis

	Words	Symbolic Form
Known fact:	"If p, then q" is always true,	$p \Rightarrow q$
Given:	**and** p is true in a particular case,	p
Conclusion:	**then** q is true in that case.	q

In the symbolic form, everything above the horizontal line is assumed to be correct or true. What is written below the line follows logically from the accepted information.

a. In Problem 2 Part c, identify p and q in the general statement $p \Rightarrow q$. Identify the specific case of p. Of q.

b. In Problem 2 Part d, identify p and q in the general statement $p \Rightarrow q$. Identify the specific case of p. Of q.

The "known facts" used in Problem 2 Parts c–g are definitions or principles and relationships you discovered and, as a class, agreed upon in previous mathematics courses. Your discoveries were probably based on studying several particular cases or conducting experiments and then searching for patterns.

4 Select one of the statements given as a known fact in Problem 2 Parts e–g. Discuss with classmates how you could explore specific cases that might lead to a discovery of a pattern suggesting the given statement. Be prepared to explain your proposed exploration to the class.

Reasoning from patterns based on analysis of specific cases as you described in Problem 4 is called **inductive reasoning**. This type of reasoning is a valuable tool in making discoveries in mathematics, science, and everyday life. However, inductive reasoning must be used with caution.

5 The famous mathematician Leonard Euler (1707–1783) worked on a wide range of problems including questions of traversability of networks as you may have studied in Course 1. Like others of his time, he was interested in finding a formula to create prime numbers. An early attempt was:

If n is a positive integer, then $n^2 - n + 41$ is a prime number.

a. Test this conjecture by examining some specific cases. Choose several positive values for n and see if the expression gives a prime number. Share the work with others.

b. Based on your calculations, does the conjecture seem correct? Can you conclude for sure that it is always true? Explain your reasoning.

c. Test $n^2 - n + 41$ when $n = 41$. Is the result a prime number? Why or why not?

d. To prove an if-then statement is *not* true, you only have to find one *counterexample*. What is a counterexample to the statement, "if n is a positive integer, then $n^2 - n + 41$ is a prime number"?

e. How could you find a second counterexample?

Discovering and Proving If-Then Statements Now it is your turn to do some mathematical research. In Course 2, you investigated coordinate representations of rigid transformations—translations, rotations, line reflections, and glide reflections. Your reasoning with those transformations involved coordinate methods such as use of the distance and slope formulas. In Problems 6–9, you will re-examine line reflections, but in a plane without coordinates using *synthetic methods*. Make note of your use of inductive reasoning.

6 In Escher's *Magic Mirror*, figures and their reflected images reside in the same plane. The relationship between a figure and its reflected image agrees with the following definition of a *line reflection*.

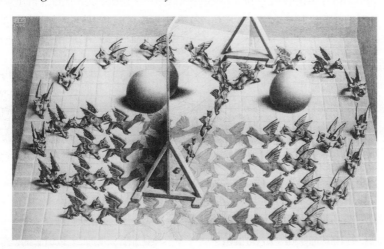

A **reflection across line ℓ** is a transformation that maps each point P of the plane onto an image point P' as follows:

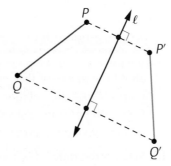

- If point P is not on ℓ, then ℓ is the perpendicular bisector of $\overline{PP'}$.

- If point P is on ℓ, then $P' = P$. That is, P is its own image.

Line Reflection Principle
To reflect a set of points, you reflect each point in the set.

a. The reflection image of a point A across a line ℓ is point B. Explain as precisely as you can why the reflection image of point B across ℓ is point A.

b. Use the line reflection tool of interactive synthetic geometry software (or a tinted Plexiglass mirror) to produce reflection images of points. Use the tool for three different lines of reflection. For each line, find the image of several points—some on one side of the line, some on the other side, and some points on the line. Check if the relation between the preimage point-image point pairs agree with the definition of a line reflection. Compare your findings with your classmates.

c. Look back again at Escher's *Magic Mirror*. Notice how as the winged dog closest to the top of the viewer's side of the mirror moves away from the mirror, so does the mirror-image dog on the other side. On the viewer's side, is the outside path the dogs are walking *clockwise* or *counterclockwise* in direction? What about the path of their reflection images?

d. In *Magic Mirror*, Escher strikingly captures the idea that reflection in a mirror *reverses orientation*. Use interactive geometry software to draw a line, a triangle *ABC*, and its reflection across the line. Reading the vertices of the preimage and of the image in alphabetical order, how does a line reflection affect orientation? Check this observation using a different line and a different type of polygon and its reflection image across that line. Write a statement summarizing your observations beginning:
Line reflections _____.

7 Now conduct each of the following experiments using interactive geometry software (or a tinted Plexiglas mirror).

a. Draw a line of reflection called ℓ and a second line different from the first called *m*. Mark and label five points on line *m*. Find and label the reflection images of the five points across line ℓ.

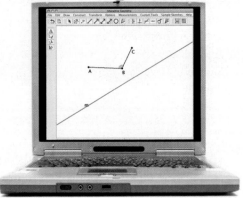

  i. What appears to be true about the five image points? Compare your finding with that of others and agree on a written *if-then* summary statement.

  ii. If a point on line *m* is between two other points, what do you observe about the image of that point in relation to the images of the other two points? Does this relationship hold for other triples of points on line *m*? Compare your finding with that of others and agree on a written summary statement in *if-then* form.

  iii. What is the reflection image of line *m* across line ℓ? Draw another line and find its reflection image across line ℓ. Does your result agree with your previous observation?

b. On a clear screen, draw a reflection line called ℓ. For each case below, compare *AB* and *A'B'*, where the reflection image of point *A* across ℓ is *A'* and the reflection image of point *B* across ℓ is *B'*.

  i. Draw points *A* and *B* so that they are on the same side of ℓ.

  ii. Draw points *A* and *B* so that they are on opposite sides of ℓ.

iii. Draw points *A* and *B* so that exactly one of the points is on ℓ.

iv. Draw points *A* and *B* so that both points are on ℓ.

Compare your findings with those of your classmates and agree on a written summary statement in *if-then* form.

c. On a clear screen, draw a reflection line *m* and an angle *ABC* as at the right. Find the reflection image of ∠*ABC*.

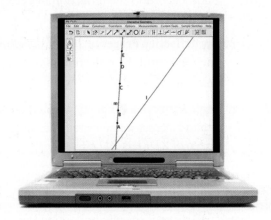

i. Why would you expect the reflected image to be an angle?

ii. Compare the measures of the angle and its reflected image.

iii. Check if your finding in part ii holds for other angles and their reflection images across different lines. Compare your findings with those of your classmates and agree on a written *if-then* summary statement.

Inductive reasoning may lead to if-then statements that are plausible, or seem true. However, as you saw in Problem 5, a statement may not be true for all cases. **Deductive reasoning** involves reasoning *from* facts, definitions, and accepted or assumed properties *to* conclusions using principles of logic. Using correct deductive reasoning, the conclusions reached are certain, not just plausible.

Your work in Problems 6 and 7 suggests the reasonableness of the following statements which will be assumed.

Assumptions about Line Reflections

1. Line reflections preserve collinearity of points.
2. Line reflections preserve betweenness of points.
3. Line reflections preserve distance between pairs of points.
4. Line reflections preserve angles and angle measure.
5. Line reflections reverse orientation.

You will use these assumptions in the remainder of this unit and in future units.

8 Give a reason for each statement in the following *chain of reasoning* that justifies this property of line reflections.

Line Reflection Property If *A′* is the reflection image of point *A* across a line ℓ and *B′* is the reflection image of a point *B* across line ℓ, then $\overline{AB}$ is the image of $\overline{A'B'}$.

Reasoning Chain:

1. The reflection image of $\overleftrightarrow{AB}$ is a line.

2. The reflection image of $\overleftrightarrow{AB}$ is found by reflecting each of the points of $\overleftrightarrow{AB}$. So, points *A′* and *B′* are in the image.

3. Since only one line contains both points *A′* and *B′*, $\overleftrightarrow{A'B'}$ is the image of $\overleftrightarrow{AB}$.

9 Dividing the work up among your classmates, prepare reasoning chains that provide arguments justifying each of the following properties. Report your arguments and critique the arguments of others.

    a. Segment Reflection Property If the reflection image of a point P across a line ℓ is P' and the reflection image of point Q is Q', then the reflection image of $\overline{PQ}$ is $\overline{P'Q'}$.

    b. Ray Reflection Property If the reflection image of point X across a line ℓ is X' and the reflection image of point Y is Y', then the reflection image of $\overrightarrow{XY}$ is $\overrightarrow{X'Y'}$.

    c. Angle Reflection Property If the reflection images of noncollinear points D, E, and F across a line ℓ are D', E', and F', respectively, then the reflection image of $\angle DEF$ is $\angle D'E'F'$.

SUMMARIZE THE MATHEMATICS

Inductive reasoning and deductive reasoning are each important; they are complementary aspects of mathematical reasoning. Inductive reasoning often leads to conjectures of new relationships or properties that can be proven using deductive reasoning. Consider this conjecture.

        The sum of any two consecutive odd numbers is divisible by 4.

a How could you arrive at this conjecture by using inductive reasoning?

b Write this conjecture in if-then form.

    i. What is the hypothesis of your statement?

    ii. What is the conclusion?

c How could you use deductive reasoning to prove this conjecture?

Be prepared to share your ideas and reasoning strategies with the class.

 CHECK YOUR UNDERSTANDING

Make a conjecture about what happens when you choose any four consecutive whole numbers, add the middle two, and then subtract the smallest of the four from that sum.

32, 33, 34, 35

a. Describe the procedure you used to create your conjecture.

b. Write your conjecture in if-then form.

c. If n represents the smallest of four consecutive whole numbers, how would you represent each of the next three numbers?

d. Use your representations in Part c to write an argument that proves your conjecture is always true.

APPLICATIONS

These tasks provide opportunities for you to use and strengthen your understanding of the ideas you have learned in the lesson.

1 Carefully study the following plot from a detective story.

> Detectives were called to the IT division of the Polaris Medical complex where it was discovered that someone had tampered with a sensitive USB flash drive. Complex security chief Krawitz, said she always kept IT a restricted area. However Riley, Vice President of IT, said that the flash drive was fine when used last Wednesday Oct. 30, but noticed it included unauthorized data when used last Friday Nov. 1. Senior Engineer Madziar discovered the problem and immediately reported it to Riley. She confirmed and reported it to Diaz, the executive Vice President of Operations, who in turn contacted the detectives.
>
> Upon further investigation, the detectives learned that the door to the room housing the PCs, flash drives, and password-protected disks was unforced. Thus it must have been accessed by Riley, Madziar, Diaz, or one of the other two senior engineers, Greenleaf and Nelson. However Greenleaf was new to the company and for now could only access the room with verification by Krawitz who was out with the flu for the past week. The company was able to verify use of the drive on the 30th and on November 1, the date Madziar discovered the problem. The drive was not used on the PC between those times. Using the flash drive on any personal computer required a decryption code which was changed on the 28th and automatically issued to senior engineers (standard practice). Krawitz is cleared to issue the code to other higher-ups but conducted no business while out sick.
>
> Prior to their arrival at Polaris Medical, the detectives were given a tip that an identify-theft scheme was in place which involved hacking Polaris Medical's patient records. With this information and the evidence they had discovered at the crime scene, the detectives were certain they would identify the culprit and crack the case.

a. Identify the prime suspect in this case. Explain the evidence you could use to charge that person and write a convincing argument to do so.

b. Write a series of arguments that could be used to exonerate or clear the other possible suspects of blame.

c. In what ways is the reasoning used to identify the prime suspect similar to the reasoning used to solve the Sudoku puzzle on page 2?

2 The following two-person game is played on a rectangular board, like a large index card. To play, players take turns placing pennies on the board. The coins may touch but cannot overlap or extend beyond the game board. A player cannot move the position of an already placed coin. The player who plays the last coin is the winner. The diagrams below show the first play of each of two players.

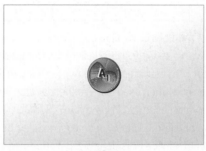

<table>
<tr><td>**1st Player**</td><td>**2nd Player**</td></tr>
</table>

a. Play the game a few times with a partner, noting how the shape of the game board influences your play.

b. Use the symmetry of the game board to devise a strategy that will always result in a win for the *1st player* when he or she places the first coin in position A_1 as above.

c. On a copy of the second diagram at the right above, show the next play of a 1st player who is using your winning strategy. Then show the next two possible plays of each player.

d. Write a description of your winning strategy for the 1st player. Provide an argument for why that strategy will guarantee a win for the 1st player.

e. What other game board shapes can you make so that the strategy will work?

3 **Consecutive numbers** are adjacent integers on a number line, such as 5 and 6. Nathan, Trina, Kasib, and Ivana were trying to prove the following statement.

The sum of any two consecutive numbers is always an odd number.

Study each of the arguments below.

Nathan: If the first number is even, then the second number must be odd. This combination will always add up to an odd number.

Kasib: Two consecutive numbers are of the form n and $n + 1$. $n + (n + 1) = 2n + 1$ which is, by definition, the form of an odd number.

Trina: No matter what two consecutive numbers you take, their sum is always odd as shown below.

$$5 + 6 = 11$$
$$22 + 23 = 45$$
$$140 + 141 = 281$$

Ivana: *For any two consecutive numbers, one will be even and the other will be odd. In Part c of the Check Your Understanding task on page 9, I gave an argument justifying that the sum of an even number and an odd number is always an odd number. So, the sum of any two consecutive numbers is always an odd number.*

a. Which proof is the closest to the argument you would give to prove that the sum of any two consecutive numbers is always an odd number?

b. Which arguments are *not* correct proofs of the statement? Explain your reasoning.

c. Give a "visual proof" of the statement using arrays of counters.

d. How would you prove or disprove the assertion, "The sum of three consecutive numbers is always an odd number"?

4 Examine each of the arguments below. Assuming the if-then statement is true, state whether the argument is correct or incorrect. Give a reason for your answers.

a. If the price of gas rises, then demand for gas falls. Demand for gas has risen. Therefore, the price of gas has risen.

b. If the price of gas rises, then demand for gas falls. The price of gas has risen. Therefore, the demand for gas has fallen.

c. If the price of gas rises, then demand for gas falls. The price of gas has fallen. Therefore, the demand for gas has fallen.

5 Suppose it is true that "all members of the senior class are at least 5 feet 2 inches tall." What, if anything, can you conclude with certainty about each of the following students?

a. Darlene, who is a member of the senior class

b. Trevor, who is 5 feet 10 inches tall

c. Anessa, who is 5 feet tall

d. Ashley, who is not a member of the senior class

6 Suppose it is true that "all sophomores at Calvin High School enroll in physical education."

a. Write this statement in if-then form. What is the hypothesis? The conclusion?

b. If Tadi is a sophomore at Calvin, what can you conclude?

c. If Rosa is enrolled in a physical education class at Calvin, what can you conclude? Explain your reasoning.

7 In 1742, number theorist Christian Goldbach (1690–1764) wrote a letter to mathematician Leonard Euler in which he proposed a conjecture that people are still trying to prove or disprove. Goldbach's Conjecture states:

Every even number greater than or equal to 4 can be expressed as the sum of two prime numbers.

a. Verify Goldbach's Conjecture is true for 12. For 28.

b. Write Goldbach's Conjecture in if-then form.

c. Write the converse of Goldbach's Conjecture. Prove that the converse is not true.

8 Examine the following if-then statements about properties of numbers.

i. If a, b, and c are consecutive positive numbers, then $a + b + c$ is divisible by 3.

ii. If a, b, and c are consecutive positive numbers, then $a + b + c$ is divisible by 6.

iii. If x is a real number, then $-x < x$.

iv. If x is a nonzero real number, then $x > \frac{1}{x}$.

v. If x is the degree measure of the smallest angle of a triangle, then $\cos x > 0$.

a. Use inductive reasoning to help you decide which statements might be correct and which are incorrect. For each statement that is incorrect, give a counterexample.

b. For each correct statement, use deductive reasoning to write a proof that could convince a skeptic that it is true.

9 Tonja made the following conjecture about consecutive whole numbers.

For any four consecutive whole numbers, the product of the middle two numbers is always two more than the product of the first and last numbers.

a. Test Tonja's conjecture for a set of four consecutive whole numbers.

b. Find a counterexample or give a deductive proof of Tonja's conjecture.

10 Let ℓ be a line of reflection and let points R', S', and T' be the reflection images of noncollinear points R, S, and T, respectively.

a. Write an argument justifying that the reflection image of $\triangle RST$ is $\triangle R'S'T'$.

b. How do $\triangle RST$ and $\triangle R'S'T'$ appear to be related?

These tasks will help you connect the ideas of mathematical reasoning in this lesson with other mathematical topics and contexts that you know.

11 *Factorial* notation is a compact way of writing the product of consecutive positive whole numbers. For example, $5! = 5 \times 4 \times 3 \times 2 \times 1$. $5!$ is read "5 factorial." In general, $n! = n \times (n - 1) \times (n - 2) \times \cdots \times 2 \times 1$.

  a. Calculate $3!$.

  b. The names of three candidates for the same office are to be listed on a ballot. How many different orderings of the names are possible? Compare your answer to that found in Part a.

  c. Provide an argument that $(n + 1)! = (n + 1) \times n!$.

  d. Kenneth Ruthven of Cambridge University proposed the following conjectures about $100!$ to a class in Great Britain. For each conjecture, write an argument that proves the conjecture or explain why it is not true.

    i. $100!$ is an even number.   **ii.** $100!$ is divisible by 101.

    iii. $100!$ is larger than $101! - 100!$.   **iv.** $100!$ is larger than $50! \times 10^{50}$.

12 Problem 3 of Investigation 2 (page 12) illustrated a symbolic model for reasoning with an if-then statement. You can also represent if-then statements geometrically using *Venn diagrams*.

  a. Examine these if-then statements and the corresponding Venn diagrams.

$$p \Rightarrow q$$
If a creature is a butterfly,
then it is an insect.

$$p \Rightarrow q$$
If a quadrilateral is a square,
then it is a parallelogram.

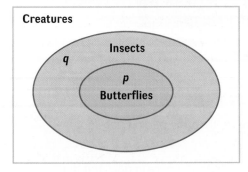
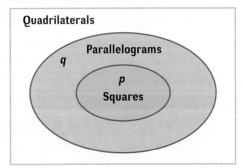

    i. If you know that a monarch is a butterfly, what can you conclude?

    ii. If you know that quadrilateral $WXYZ$ is a square, what can you conclude?

  b. Refer to Applications Task 6 (page 19). Represent the if-then statement you wrote in Part a with a Venn diagram. How can you use the Venn diagram to reason about Tadi's situation in Part b?

c. A common error in deductive reasoning is to assume that whenever an if-then statement $p \Rightarrow q$ is true, the *converse* statement $q \Rightarrow p$ is also true.

    i. For each statement in Part a, write the converse and decide whether or not the converse statement is true. If not, give a counterexample.

    ii. How do the Venn diagrams in Part a show that the converse of a statement is not always true? How is this analysis related to your reasoning about the case of Rosa in Applications Task 6 Part c (page 19)?

13 Suppose a statement $p \Rightarrow q$ and its converse $q \Rightarrow p$ are both true. Then the statement $p \Leftrightarrow q$ (read "*p* if and only if *q*") is true. In Course 1, you proved that both the Pythagorean Theorem and its converse were true. This fact can be stated in if-and-only-if form. *In $\triangle ABC$, $\angle C$ is a right angle if and only if $a^2 + b^2 = c^2$.*

a. Write the definition of a trapezoid from page 7 in if-and-only-if form.

b. Write the definition of a prime number from page 10 in if-and-only-if form.

c. Consider this statement about real numbers. If $a = b$, then $a + c = b + c$.

    i. Write the converse of this statement.

    ii. Is the converse always true? Explain your reasoning.

    iii. Write an if-and-only-if statement summarizing this property of equality.

14 Consider the true statement, "If a person lives in Chicago, then the person lives in Illinois," and the corresponding Venn diagram below.

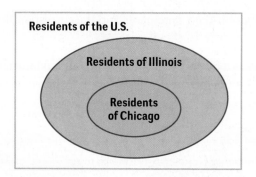

a. Which of the following statements are always true?

    i. If a person does not live in Chicago, then the person does not live in Illinois.

    ii. If a person does not live in Illinois, then the person does not live in Chicago.

b. How are your answers to Part a illustrated by the Venn diagram?

c. The if-then statement in Part ai is called the *inverse* of the original statement. In symbols, the **inverse** of $p \Rightarrow q$ is *not* $p \Rightarrow$ *not* q. Use a Venn diagram to explain why the inverse of a true if-then statement may not always be true.

d. The if-then statement in Part aii is called the *contrapositive* of the original statement. In symbols, the contrapositive of $p \Rightarrow q$ is *not* $p \Rightarrow$ *not* q. (An implication and its **contrapositive** are *logically equivalent statements*.) Use a Venn diagram to explain why the contrapositive of a true if-then statement is always true.

REFLECTIONS

These tasks provide opportunities for you to re-examine your thinking about ideas in the lesson and your use of mathematical practices and habits of mind that are useful throughout mathematics.

15 Look back at Connor's strategy to guarantee that the second player can always win the nonagon game. (Investigation 1, page 5)

a. Will his strategy work if the game is played on the vertices of a regular pentagon? Explain your reasoning.

b. Will his strategy work if the game is played on the vertices of a regular octagon? Explain.

c. Describe as precisely as you can all regular polygons for which Connor's strategy will work.

16 Look back to page 6 at the arguments that Nesrin and Teresa provided to justify that the sum of two odd numbers is always an even number.

a. How can Nesrin's counter model help you to better understand Teresa's argument?

b. How could Nesrin's argument be revised to make it more general?

17 If-then statements are sometimes called *conditional statements*. Why does that term make sense?

18 In this age of the Internet, advertising has become big business. Advertisers often use if-then statements to sell their products and services. The straightforward ad:

> *Use your money wisely—shop at FlorMart superstore.*

is worded to suggest the implication:

> *If you shop at FlorMart, then you use your money wisely.*

Often, with some added help from the advertiser, the statement is interpreted by consumers:

> *If you do not shop at FlorMart, then you do not use your money wisely.*

a. Why might the wording of the second implication have a stronger psychological effect upon most shoppers than the first implication?

b. Are the two statements logically the same? Explain.

19 If you think about it, inductive reasoning is a common form of reasoning in the world around you. Give an example of how inductive reasoning might be used by the following people.

a. An automobile driver   **b.** A consumer

c. A medical researcher   **d.** A teacher

20 Explain how inductive and deductive reasoning differ. In doing mathematics, how does one form of reasoning support the other?

21 Look back at the Properties of Line Reflections in Problems 8 and 9 of Investigation 2 on page 16. Illustrate how each property is a useful tool in sketching reflection images.

22 In this lesson, you engaged in important *mathematical habits of mind* such as:

- visual thinking,

- searching for and explaining patterns,

- making and checking conjectures,

- reasoning with different representations of the same idea, and

- providing sound arguments or proofs.

Carefully look back at your work in this lesson and describe a specific example where you found each of these habits of mind useful.

EXTENSIONS

These tasks provide opportunities for you to explore further or more deeply the ideas you studied in this lesson.

23 Look back at the nonagon game in Investigation 1 (page 4). Another student, Sofia, claimed she found a strategy using symmetry that guaranteed that the *first player* could always win the game. What is wrong with her argument below?

As the first player, I'll remove the one penny at the top. Then in my mind, I divide the remaining 8 pennies by the line of symmetry determined by the removed penny. Now, whatever the other player does, I'll do the symmetric move. So, there is always a move for me to make. Therefore, I can never lose by having no coins to remove.

24 In his book, *Proofs without Words*, mathematician Roger Nelsen offers the following two visual "proofs." Although not proofs in the strictest sense, the diagrams he provides help you see why each particular mathematical statement is true.

a. How does the diagram to the left help you see that for any positive integer n, the sum of the integers from 1 to n is $\frac{1}{2}n(n + 1)$?

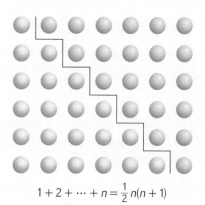

$$1 + 2 + \cdots + n = \frac{1}{2}n(n + 1)$$

b. How does the diagram at the right help you see that the "infinite sum" of fractions of the form $\left(\frac{1}{2}\right)^n$, $n \geq 1$ is 1?

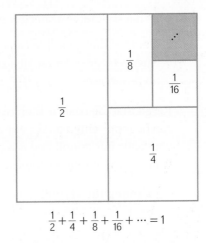

$$\frac{1}{2} + \frac{1}{4} + \frac{1}{8} + \frac{1}{16} + \cdots = 1$$

25 Many people, such as auto technicians, make a living out of repairing things. Often, repairs can be done at home if you have the right tools and can reason deductively. The first step in making a repair is to identify the problem. Examine the troubleshooting chart below for diagnosing problems that often occur with small engines, such as the one on a lawn mower.

a. What are the first things you should check if the engine runs but the mower does not?

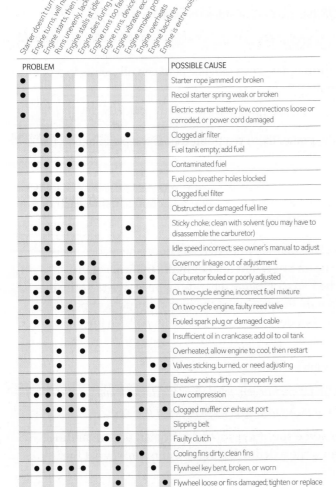

Starter doesn't turn engine	Engine turns, will not start	Engine starts, then stalls	Runs unevenly, lacks power	Engine stalls at idle speed	Engine dies during use	Engine runs too fast	Engine runs, device doesn't	Engine vibrates excessively	Engine smokes profusely	Engine overheats	Engine backfires	Engine is extra-noisy	POSSIBLE CAUSE
●													Starter rope jammed or broken
●													Recoil starter spring weak or broken
●													Electric starter battery low, connections loose or corroded, or power cord damaged
	●	●	●	●			●						Clogged air filter
	●	●		●									Fuel tank empty; add fuel
	●	●	●	●	●								Contaminated fuel
	●	●		●									Fuel cap breather holes blocked
	●	●	●	●									Clogged fuel filter
	●	●		●									Obstructed or damaged fuel line
	●	●	●	●			●						Sticky choke; clean with solvent (you may have to disassemble the carburetor)
	●		●										Idle speed incorrect; see owner's manual to adjust
			●		●	●							Governor linkage out of adjustment
	●	●	●	●	●	●		●	●	●			Carburetor fouled or poorly adjusted
	●	●	●	●				●	●				On two-cycle engine, incorrect fuel mixture
	●		●	●						●			On two-cycle engine, faulty reed valve
	●	●	●	●	●								Fouled spark plug or damaged cable
				●				●		●			Insufficient oil in crankcase; add oil to oil tank
				●									Overheated; allow engine to cool, then restart
		●		●					●	●			Valves sticking, burned, or need adjusting
	●	●	●					●	●				Breaker points dirty or improperly set
	●	●	●	●			●						Low compression
		●	●	●	●			●		●			Clogged muffler or exhaust port
					●								Slipping belt
					●	●							Faulty clutch
						●							Cooling fins dirty; clean fins
	●	●	●	●	●		●		●				Flywheel key bent, broken, or worn
					●			●					Flywheel loose or fins damaged; tighten or replace
●	●					●							Crankshaft bent; have engine replaced

Four-cycle engine with horizontal crankshaft

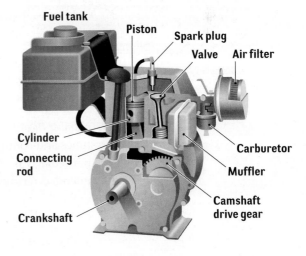

Fuel tank · Piston · Spark plug · Valve · Air filter · Cylinder · Connecting rod · Crankshaft · Carburetor · Muffler · Camshaft drive gear

b. What should you check if the engine runs too fast? If it backfires?

c. Some of the "possible causes" listed in the troubleshooting chart often suggest a next course of action. Explain how if-then reasoning is used in these cases.

26 Recall that the basic if-then reasoning pattern, Affirming the Hypothesis, can be represented as shown on the left below. In symbolic form, everything above the horizontal line is assumed to be correct or true. What is written below the line follows logically from the accepted information.

Affirming the Hypothesis **Denying the Conclusion**

$$\begin{array}{c} p \Rightarrow q \\ p \\ \hline q \end{array} \qquad \begin{array}{c} not\ q \Rightarrow not\ p \\ not\ q \\ \hline not\ p \end{array} \qquad \begin{array}{c} p \Rightarrow q \\ not\ q \\ \hline not\ p \end{array}$$

a. Explain why the second reasoning pattern above is valid.

b. Use the result of Connections Task 14 Part d to explain why Denying the Conclusion, shown above on the right, is a valid reasoning pattern.

c. Use Denying the Conclusion to decide what can be concluded from the following two statements.

Known fact: If a triangle is an isosceles triangle, then it has two sides the same length.

Given: $\triangle PQR$ has no pair of sides the same length.

Conclusion: ?

27 Formally, a **transformation** is a correspondence between all points in a plane and themselves such that:

- each point in the plane has a unique image point in the plane, and

- for each point in the plane, there is a point that is the preimage of the given point.

A line reflection was defined as a transformation with certain characteristics (see page 15). Verify that the definition of a line reflection satisfies each of the two conditions for a transformation.

28 *Number theory* is a branch of mathematics that has flourished since ancient times and continues to be an important field of mathematical activity, particularly in the applied area of *coding* (encrypting messages so only the intended recipients can read them). One of the first definitions appearing in the theory of numbers is a definition for *factor* or *divisor*. An integer b is a factor or divisor of an integer a provided there is an integer c such that $a = bc$.

a. One of the first theorems in number theory follows.

If a, b, and c are integers where a is a factor of b and a is a factor of c, then a is a factor of b + c.

Test this theorem for some specific cases to develop an understanding for what it says. Then write a deductive argument to prove that the theorem is always true.

b. Form a new if-then statement as follows. Use the hypothesis of the theorem in Part a, and replace the conclusion with "a is a factor of $bm + cn$ for all integers m and n."

  i. Do you think this new if-then statement is always true? Explain your reasoning.

  ii. If you think the statement is true, write a proof of it. If not, give a counterexample.

c. Prove or disprove this claim.

  If a is a factor of b and b is a factor of c, then a is a factor of c.

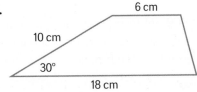

REVIEW

These tasks provide opportunities for you to review previously learned mathematics and to refine your skills in using that mathematics.

29 Find the area of each trapezoid.

a.

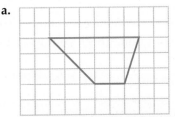

b.

6 cm

10 cm

30°

18 cm

30 Evaluate each expression for the values $x = \frac{5}{6}$, $y = -3$, and $z = 6$.

a. xz^2

b. x^{-1}

c. $|y - z| + y^2$

d. $x + \frac{4}{y}$

31 Determine if the pairs of lines are parallel, perpendicular, or neither.

a. The line containing $(8, 2)$ and $(-4, 6)$ and the line containing $(-2, 3)$ and $(1, 4)$

b. The lines with equations $y = 6x + 5$ and $y = -6x - 5$

c. The lines with equations $4x + 3y = 10$ and $3x - 4y = 10$

32 Recall that the converse of an if-then statement reverses the order of the two parts of the statement. Consider this statement.

  If a polygon is a regular polygon, then all its sides are the same length.

a. Is this a true statement?

b. Write the converse of this statement.

c. Is the converse a true statement?

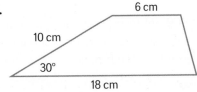

33 Write each of these expressions in equivalent expanded form.

a. $5(2x + 7)$ b. $(x + 9)(2x + 3)$ c. $(x - 4)(x + 3)$

d. $(x + 4)(x - 4)$ e. $(2x + 3)^2$ f. $(2x - 3)^2$

34 Using a compass and straightedge (no measuring), construct each of the following.

a. A segment congruent to $\overline{AB}$

b. A triangle congruent to $\triangle PQR$

c. An angle congruent to $\angle FGH$

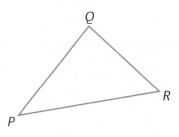
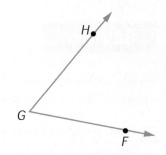

35 Without using your calculator, match each graph with the correct function rule. The scales on all axes are in increments of 1 unit. Check your work by graphing the functions using your calculator.

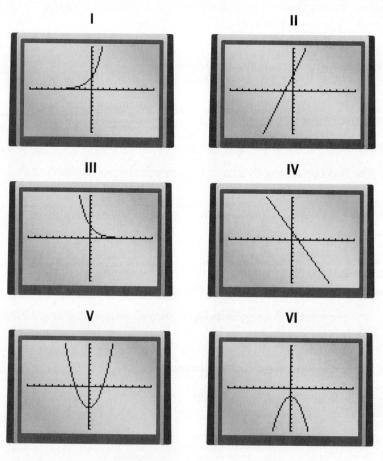

a. $y = 3x + 4$ b. $y = 3(2^x)$ c. $y = -x^2 - 2$

d. $y = -2x + 2$ e. $y = 3(0.5^x)$ f. $y = x^2 - 5$

2

Geometric Reasoning and Proof

In your previous studies, you used informal geometric reasoning to help explain the design of structures and the functioning of mechanical systems. You also explored how if-then reasoning could be used to derive geometric properties. For example, you saw that if you know the formula for the area of a rectangle, you can then derive formulas for the area of a square and of a right triangle. Knowing how to calculate the areas of these shapes enabled you to complete a proof of the Pythagorean Theorem.

The interplay among shapes, their properties, and their function is a recurring theme in geometry. The design of the lift-bed truck shown above is based on two of the most common figures in a plane, lines and angles.

Analyze the lift-bed truck mechanism shown on the previous page.

a How do you think the mechanism works?

b As the truck bed is raised or lowered, what elements change?

c As the truck bed is raised or lowered, what segment lengths and what angle measures remain unchanged?

d As the truck bed is raised or lowered, will it always remain parallel to the flat-bed frame of the truck? Explain your reasoning.

In this lesson, you will explore how from a few basic assumptions you can prove important properties of angles formed by intersecting lines and by two parallel lines intersected by a third line.

INVESTIGATION 1

Reasoning about Intersecting Lines and Angles

Skill in reasoning, like skill in sculpting, playing a musical instrument, or playing a sport, comes from practicing that skill and reflecting on the process. In this lesson, you will sharpen your reasoning skills in geometric settings. As you progress through the lesson, pay particular attention to the assumptions you make to support your reasoning, as well as to the validity of your reasoning.

When two lines intersect at a single point, special pairs of angles are formed. For example, a pair of adjacent angles formed by two intersecting lines like ∠AEC and ∠CEB, shown at the right, are called a **linear pair** of angles. Pairs of angles like ∠AEC and ∠BED are called **vertical angles**.

As you work on the problems in this investigation, make notes of answers to these questions:

How are linear pairs of angles related?

How are vertical angles related and why is that the case?

1 In the diagram at the right, lines *k* and *n* intersect at the point shown, forming angles numbered 1, 2, 3, 4.

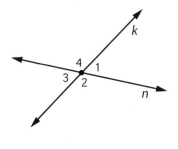

a. If m∠1 = 72°, what can you say about m∠2? About m∠3? About m∠4? What assumptions are you using to obtain your answers?

b. If m∠2 = 130°, what can you say about m∠1? About m∠3? About m∠4?

c. In general, what relationships between pairs of angles do you think are true? Make a list of them.

d. Will the general relationships you listed for Part c hold for any pair of intersecting lines? Test your conjectures using specific examples.

e. Write an if-then statement about linear pairs of angles that you think is *always* correct. You may want to begin as follows. If two angles are a linear pair, then … .

f. Write an if-then statement about vertical angles that you think is *always* correct. You may want to begin as follows. If two lines intersect, then … .

In the remainder of this lesson, you will continue to use inductive reasoning to discover possible relations among lines and angles, but you will also use deductive reasoning to prove your conjectures are always true. To reason deductively, you must first have some basic facts from which to reason. In mathematics, statements of basic facts that are accepted as true without proof are called **postulates** (or *axioms*). These assumed facts will be helpful in supporting your reasoning in the remainder of this unit and in future units. Begin by assuming the following postulate concerning linear pairs of angles.

Linear Pair Postulate If two angles are a linear pair, then the sum of their measures is 180°.

2 Study the attempt at the right by one group of students at Washington High School to prove the conjecture they made in Part f of Problem 1. Based on the labeling of the diagram, they set out to prove the following.

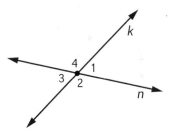

> **If lines *n* and *k* intersect at the point shown, then m∠1 = m∠3.**

They reasoned as follows.

(1) Since lines *n* and *k* intersect, ∠1 and ∠2 are a linear pair. So, m∠1 + m∠2 = 180°.

(2) Since lines *n* and *k* intersect, ∠2 and ∠3 are a linear pair. So, m∠2 + m∠3 = 180°.

(3) If m∠1 + m∠2 = 180° and m∠2 + m∠3 = 180°, then m∠1 + m∠2 = m∠2 + m∠3.

(4) If m∠1 + m∠2 = m∠2 + m∠3, then m∠1 = m∠3.

a. Explain why each of the statements in the students' reasoning is or is not correct.

b. Now write an argument to show the following: If lines *n* and *k* intersect at the point shown, then m∠2 = m∠4. Give reasons justifying each of your statements.

In mathematics, a statement that has been proved using deductive reasoning from definitions, accepted facts, and relations is called a **theorem**. The statement proved in Problem 2 is sometimes referred to as the **Vertical Angles Theorem**, *vertical angles have equal measure*.

In Course 1, you proved the Pythagorean Theorem. (Not all theorems are given names.) After a theorem has been proved, it may be used to prove other conjectures. As your geometric work in Course 3 progresses, you will want to know which theorems have been proved. Thus, you should prepare a geometry toolkit by listing assumptions, such as the Linear Pair Postulate, and proven theorems. Add each new theorem to your toolkit as it is proved.

3 Recall that two intersecting lines (line segments or rays) are **perpendicular** (⊥) if and only if they form a right angle.

a. Rewrite this definition as two if-then statements.

b. Claim: *Two perpendicular lines form four right angles.* Is this claim true or false? Explain your reasoning.

c. Study the following strategy that Juanita used to prove the claim in Part b.

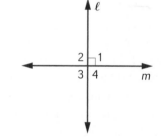

- First, she drew and labeled the diagram at the right.

- Next she developed a *plan for proof* based on her diagram.

   I know that if $\ell \perp m$, they form a right angle, say ∠1. A right angle has measure 90°. Use the fact that ∠1 and ∠3 are vertical angles to show m∠3 = 90°. Use the fact that ∠1 and ∠2 are a linear pair to show m∠2 = 90°. Then use the fact that ∠2 and ∠4 are vertical angles to show m∠4 = 90°.

- She then wrote her proof in a *two-column statement-reason form*.

Statements		Reasons
1. $\ell \perp m$	1.	Given
2. ℓ and m form a right angle. Call it ∠1.	2.	Definition of perpendicular lines
3. m∠1 = 90°	3.	Definition of right angle
4. ∠1 and ∠3 are vertical angles.	4.	Definition of vertical angles
5. m∠3 = m∠1 = 90°	5.	Vertical Angles Theorem
6. ∠1 and ∠2 are a linear pair.	6.	Definition of linear pair
7. m∠1 + m∠2 = 180°	7. _____
8. m∠2 = 180° − m∠1 = 90°	8. _____
9. ∠2 and ∠4 are vertical angles.	9. _____
10. m∠2 = m∠4	10. _____
11. m∠4 = 90°	11. _____
12. ∠1, ∠2, ∠3, and ∠4 are right angles.	12. _____

i. How does the diagram that Juanita drew show the information given in the claim?

ii. Why might it be helpful to develop a plan for a proof before starting to write the proof?

iii. Check the correctness of Juanita's reasoning and supply reasons for each of statements 7–12.

iv. Describe a *plan for proof* of the above claim that does *not* involve use of the Vertical Angles Theorem.

4 The design of buildings often involves perpendicular lines. In preparing a plan for a building like that below, an architect needs to draw lines perpendicular to given lines.

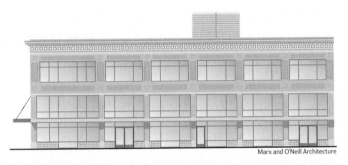

Marx and O'Neill Architecture

a. Draw a line ℓ on a sheet of paper. Describe how you would draw a line perpendicular to line ℓ through a point P in each case below.

   i. P is a point on line ℓ.

   ii. P is a point not on line ℓ.

b. Study the diagrams below which show a method for constructing a line perpendicular to a given line ℓ through a given point P on line ℓ. In the second step, the same compass opening is used to create both arcs.

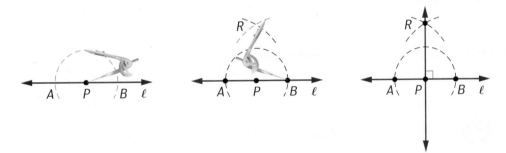

   i. On a separate sheet of paper, draw a line ℓ and mark a point P on ℓ. Use a compass and straightedge to construct a line perpendicular to line ℓ at point P.

   ii. Write an argument justifying that $\overleftrightarrow{PR} \perp \ell$ in the diagram above on the right. Start by showing that $\triangle APR \cong \triangle BPR$.

c. The following diagrams show a method for constructing a line perpendicular to a given line ℓ through a point P *not* on line ℓ. In the second step, the same compass opening is used to create both arcs.

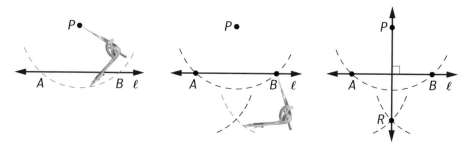

  i. On a separate sheet of paper, draw a line ℓ and a point P not on ℓ. Construct a line perpendicular to line ℓ through point P.

  ii. Write an argument justifying that $\overleftrightarrow{PR} \perp \ell$ in the diagram above on the right.

SUMMARIZE THE MATHEMATICS

In this investigation, you used deductive reasoning to establish relationships between pairs of angles formed by two intersecting lines. In the diagram at the right, suppose the lines intersect so that m∠DBA = m∠CBD.

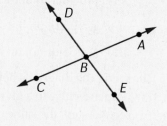

  a What can you conclude about these two angles? Prepare an argument to prove your conjecture.

  b What can you conclude about the other angles in the diagram? Write a proof of your conclusion.

  c What mathematical facts did you use to help prove your statements in Parts a and b? Were these facts definitions, postulates, or theorems?

  d Describe the relationship between $\overleftrightarrow{AC}$ and $\overleftrightarrow{DE}$.

Be prepared to share your conjectures and explain your proofs.

 CHECK YOUR UNDERSTANDING

In the diagram at the right, $\overline{AD}$ and $\overline{BE}$ intersect at point C and m∠ECD = m∠D.

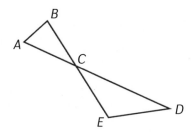

a. Is m∠ACB = m∠D? If so, prove it. If not, give a counterexample.

b. Is m∠A = m∠D? If so, prove it. If not, give a counterexample.

Reasoning about Parallel Lines and Angles

When a line intersects another line, four angles are formed. Some of the pairs of angles have equal measures, and some pairs are **supplementary angles**—they have measures that add to 180°. When a line intersects *two* lines, many more relationships are possible. Perhaps the most interesting case is when a line intersects two parallel lines, as with the various pairs of support beams on the faces of the John Hancock Center in Chicago, shown below.

Lines in a plane that do not intersect are called **parallel lines**. In the diagram below, line m is parallel to line n (written $m \parallel n$). Line t, which intersects the two lines, is called a **transversal**.

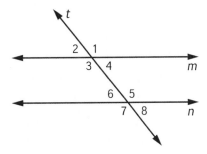

As you work on the problems of this investigation, look for answers to the following questions:

> *If two parallel lines are intersected by a transversal, what relations exist among the measures of the angles formed?*

> *What relations among the angles formed when two lines are cut by a transversal allow you to conclude that the lines are parallel?*

1 In the above diagram, the angles at each point of intersection are numbered so that they can be easily identified.

a. What pairs of angles, if any, appear to be equal in measure?

b. What angle pairs appear to be supplementary? (Supplementary angles need not be a linear pair.)

c. Draw another pair of parallel lines and a transversal with a slope different from the one above. Number the angles as in the figure above.

    i. Do the same pairs of numbered angles appear equal in measure?

    ii. Do the same pairs of numbered angles appear to be supplementary?

Angles that are in the same relative position with respect to each parallel line and the transversal are called **corresponding angles**. In the diagram on the previous page, angles 1 and 5 are corresponding angles; similarly, angles 3 and 7 are corresponding angles.

2 Examine the diagram you drew for Part c of Problem 1.

   a. Name two pairs of corresponding angles, other than angles 1 and 5 or angles 3 and 7. Were those corresponding angles among the pairs of angles that you thought had equal measure?

   b. Suppose $m\angle 1 = 123°$. Find the measures of as many other angles as you can in your diagram.

3 Descriptive names are also given to other pairs of angles formed by a transversal and two parallel lines. In the diagram below, $m \parallel n$ and t is a transversal intersecting m and n.

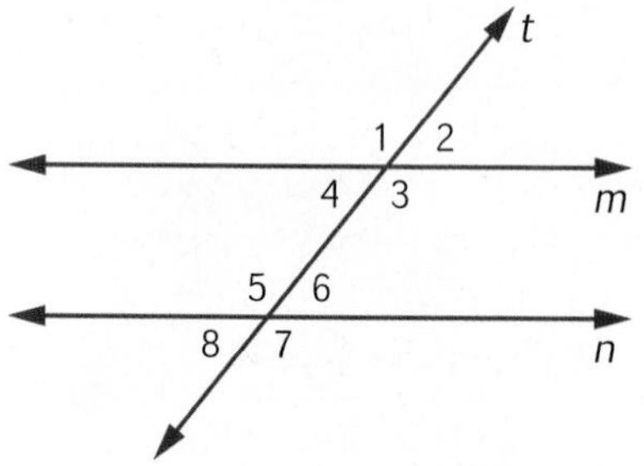

   a. For each pair of angles named below, describe how the pair can be identified in a diagram. Then give one more example of such a pair.

      i. Interior angles on the same side of the transversal: $\angle 4$ and $\angle 5$

      ii. Exterior angles on the same side of the transversal: $\angle 2$ and $\angle 7$

      iii. Alternate interior angles: $\angle 4$ and $\angle 6$

      iv. Alternate exterior angles: $\angle 1$ and $\angle 7$

   b. Identify a relationship that seems to exist for each type of angle pair named in Part a. Write your observations in if-then form, beginning each statement as follows. If two parallel lines are cut by a transversal, then … .

For Problems 4 and 5, assume the following statement as a known fact.

Corresponding Angles Assumption If two parallel lines are cut by a transversal, then corresponding angles have equal measure.

4 In completing Problem 3 Part b, one group of students at Brookwood High School made the following claim.

     If two parallel lines are cut by a transversal, then interior angles on the same side of the transversal are supplementary.

   a. Describe a plan for how you would prove this claim using the diagram in Problem 3 and the Corresponding Angles Assumption. Compare your plan for proof with that of others. Correct any errors in reasoning.

   b. The start of a proof given by the group of Brookwood students is on the next page.

      i. Supply a reason for each statement.

      ii. Continue the two-column statement-reason proof to show that $\angle 3$ and $\angle 6$ are supplementary.

Given: $\ell \parallel m$; t is a transversal
cutting ℓ and m

Prove: $\angle 4$ and $\angle 5$ are
supplementary.
$\angle 3$ and $\angle 6$ are
supplementary.

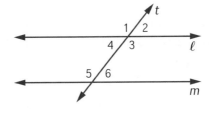

Statements	Reasons
1. $\ell \parallel m$; t is a transversal cutting ℓ and m	1.
2. $m\angle 4 + m\angle 1 = 180°$	2.
3. $m\angle 1 = m\angle 5$	3.
4. $m\angle 4 + m\angle 5 = 180°$	4.
5. $\angle 4$ and $\angle 5$ are supplementary.	5.
$\vdots$	$\vdots$

5 Describe plans for how you would prove that each of your three remaining conjectures in Part b of Problem 3 is correct. Share the task with others. Then discuss each other's plans for proof. Correct any errors in reasoning.

Using the Corresponding Angles Assumption, you can conclude that if two parallel lines are cut by a transversal, then certain relations among pairs of angles will always be true. In the next problem, you will consider the converse situation.

What relations among the angles formed when two lines are cut by a transversal allow you to conclude that the lines are parallel?

6 Conduct the following experiment.

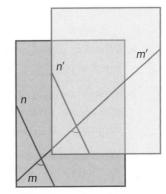

a. Draw and label two intersecting lines on a sheet of paper and mark an angle as shown.

b. Trace the lines and the angle marking on a second sheet of paper. Label the corresponding lines m' and n'. Slide the top copy so that line m' is a continuation of line m.

c. Why are the two marked angles congruent?

d. How is the pair of marked angles related to the two lines n and n' and transversal m? How do lines n and n' appear to be related? Check if those relationships hold when you slide the top copy to other positions, keeping line m' as a continuation of line m.

e. Write a conjecture in if-then form that generalizes the observations you made in the experiment.

In order to reason deductively about parallel lines and figures formed by parallel lines, you need to begin with some information about the conditions under which two lines are parallel. The conjecture you made in Problem 6 Part e could be stated this way.

If two lines are cut by a transversal so that corresponding angles have equal measure, then the lines are parallel.

This statement is the *converse* of the Corresponding Angles Assumption. There, you assumed that if two parallel lines are cut by a transversal, then corresponding angles have equal measure.

For the remainder of this unit and in future units, you can assume that both the Corresponding Angles Assumption and its converse are true. These two statements are combined as a single if-and-only-if statement called the *Parallel Lines Postulate*.

Parallel Lines Postulate In a plane, two lines cut by a transversal are parallel if and only if corresponding angles have equal measure.

7 It is reasonable to ask if there are other relations between two angles formed by a line intersecting two other lines that would allow you to conclude that the two lines are parallel. Consider the diagram below.

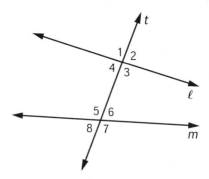

a. What condition on a pair of alternate interior angles would guarantee that line ℓ is parallel to line *m*? Write your conjecture in if-then form.

b. What condition on a pair of interior angles on the same side of the transversal *t* would guarantee that line ℓ is parallel to line *m*? Write your conjecture in if-then form.

c. What condition on a pair of exterior angles would guarantee that line ℓ is parallel to line *m*? Write your conjecture in if-then form.

d. Working with a classmate, write a proof for one of the statements in Parts a–c.

e. Be prepared to share and discuss the reasoning in your proof with the entire class. Correct any reasoning errors found.

8 In designing buildings such as the John Hancock Center, architects need methods for constructing parallel lines. In Investigation 1 (page 34), you examined how to use a compass and straightedge to construct a line perpendicular to a given line through a point not on the line.

a. On a copy of the diagram below, show how you could use the constructions of perpendiculars (pages 33–34) to construct a line *m* through point *P* that is parallel to *ℓ*.

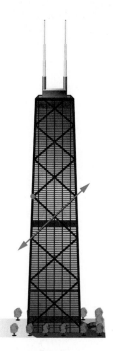

b. Write an argument that justifies that *m* || *ℓ*.

SUMMARIZE THE MATHEMATICS

In this investigation, you reasoned both inductively and deductively about angles formed by parallel lines and a transversal.

a What statements did you accept to be true without proof?

b What theorems and their converses were you able to prove about parallel lines and the angles they form with a transversal?

c Restate each theorem and its converse in Part b as a single if-and-only-if statement similar to the statement of the Parallel Lines Postulate.

Be prepared to compare your responses with those of others.

 CHECK YOUR UNDERSTANDING

Each of the following statements expresses a relationship between perpendicular and parallel lines. For each statement, draw and label a diagram. Then write an argument proving that the statement is true.

a. In a plane, if two lines are perpendicular to the same line, then they are parallel.

b. If a transversal is perpendicular to one of two parallel lines, then it is perpendicular to the other.

APPLICATIONS

1. In the diagram at the right, the lines $\overleftrightarrow{AD}$, $\overleftrightarrow{EC}$, and $\overleftrightarrow{FG}$ intersect at point B. $\overleftrightarrow{AD} \perp \overleftrightarrow{EC}$.

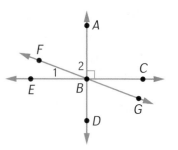

   a. What is m∠DBC?

   b. Suppose m∠1 = 27°. Find the measure of each angle.

      i. ∠2

      ii. ∠FBD

      iii. ∠EBG

   c. Suppose m∠2 = 51°. Find m∠ABG and m∠1.

   d. How would your answers to Part c change if m∠2 = p, where $0° < p < 90°$?

2. Computer-aided design (CAD) programs and interactive geometry software include tools for constructing perpendicular lines and parallel lines.

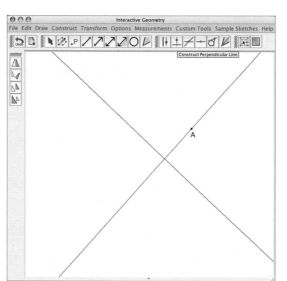

   a. Use one of those tools to draw a line. Then construct a line through a point A perpendicular to the drawn line in each case below.

      i. A is a point on the line.

      ii. A is a point *not* on the line.

   b. How do you think your software determines the perpendicular line in each case?

   c. Use one of those tools to construct a line parallel to a drawn line through a point not on the line. How do you think the software determines the parallel line?

3 Use the diagram below with separate assumptions for Part a and Part b.

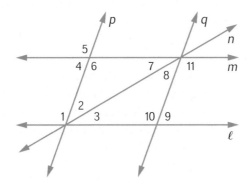

a. Assume $\ell \parallel m$, $p \parallel q$, m∠2 = 40°, and m∠3 = 35°.

   i. Find m∠8.   **ii.** Find m∠10.

   iii. Find m∠4.   **iv.** Find m∠7.

b. Do not assume any of the given lines are parallel. For each of the given conditions, which lines, if any, can you conclude are parallel?

   i. m∠2 = m∠8   **ii.** m∠6 = m∠1

   iii. m∠1 = m∠10   **iv.** m∠2 = m∠7

4 The photo below shows a carpenter's bevel, which is used to draw parallel lines.

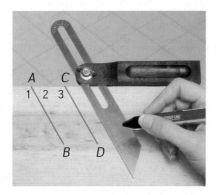

a. What part of the bevel guarantees that $\overleftrightarrow{AB} \parallel \overleftrightarrow{CD}$?

b. How are ∠2 and ∠3 related? How do you know?

5 In the diagram below, $\overleftrightarrow{AD}$ and $\overleftrightarrow{HE}$ are cut by transversal $\overleftrightarrow{GC}$. ∠1 and ∠2 are supplementary.

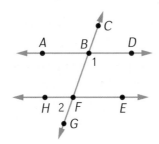

Can you conclude $\overleftrightarrow{AD} \parallel \overleftrightarrow{HE}$? If so, prove it. If not, find a counterexample.

6 The sparkle of a diamond results from light being reflected from facet to facet of the jewel and then directly to the eye of the observer. When a light ray strikes a smooth surface, such as a facet of a jewel, the angle at which the ray strikes the surface is congruent to the angle at which the ray leaves the surface. A diamond can be cut in such a way that a light ray entering the top will be parallel to the same ray as it exits the top.

Examine the cross sections of the two diamonds shown below. The diamond at the left has been cut too deeply. The entering and exiting light rays are not parallel. The diamond on the right appears to be cut correctly. The entering and exiting light rays are parallel.

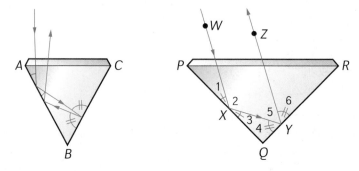

a. Use deductive reasoning to determine the measure of ∠Q at the base of the diamond to ensure that the entering and exiting light rays will be parallel.

b. Explain how you could use inductive reasoning to conjecture what the measure of ∠Q should be to ensure that the entering and exiting light rays will be parallel.

c. What are the advantages of deductive reasoning in this case?

d. As you have seen before, when modeling a situation it often helps to make simplifying assumptions. Consult with a science teacher about reflection and refraction of light and the modeling of the diamond cut. What simplifying assumptions were made in this situation?

7 Figure *PQRS* is a trapezoid with $\overline{PQ} \parallel \overline{SR}$.

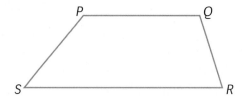

a. What can you conclude about ∠S and ∠P? About ∠Q and ∠R?

b. Write an if-then statement summarizing one of your observations in Part a. Prove your statement is correct.

c. Suppose m∠S = m∠R. What can you conclude about ∠P and ∠Q? Write an argument proving your claim.

8 Re-examine the mechanism of the lift-bed truck shown here. Assume the frame of the truck is parallel to the ground, and E is the midpoint of $\overline{AB}$ and $\overline{DC}$. You may want to make a simple model by attaching two linkage strips of the same length at their midpoints.

a. How is $\angle AEC$ related to $\angle BED$? Why?

b. How is $\triangle AEC$ related to $\triangle BED$? Why?

c. How is $\angle BAC$ related to $\angle ABD$? Why?

d. As a hydraulic ram pushes point B in the direction of point D, what happens to the relations in Parts a–c? Explain your reasoning.

e. Explain as precisely as you can why the truck bed $\overline{AC}$ remains parallel to the truck frame $\overline{BD}$ as the hydraulic ram is extended or contracted.

CONNECTIONS

9 Draw a line m and mark points A and B on opposite sides of line m. Use a compass and straightedge to construct the reflection images of points A and B across m.

10 When two lines intersect, two pairs of vertical angles are formed.

a. How many pairs of vertical angles are formed when three lines (in a plane) intersect at the same point?

b. How many pairs of vertical angles are formed when four lines (in a plane) intersect at the same point?

c. Suppose a fifth line was added to a diagram for Part b and the line intersected the other lines at the same point. How many *additional* pairs of vertical angles are formed? What is the total number of pairs of vertical angles formed by the five lines?

d. Make a conjecture about the number of pairs of vertical angles formed by n lines (in a plane) that intersect at the same point. Assume $n \geq 2$.

11 In Investigation 2, you saw that if you *assume* the Parallel Lines Postulate, then it is possible to prove other conditions on angles that result when parallel lines are cut by a transversal. It is also possible to prove other conditions on angles that guarantee two lines are parallel.

If the Parallel Lines Postulate is rewritten, replacing each occurrence of the phrase "corresponding angles" with "alternate interior angles," the new statement is often called the Alternate Interior Angles Theorem. You proved the two parts of this theorem in Problem 5 on page 37 and in Problem 7 on page 38 using the Parallel Lines Postulate.

Suppose next year's math class *assumes* the Alternate Interior Angles Theorem as its Parallel Lines Postulate.

a. Could the class then *prove* the two parts of the Parallel Lines Postulate? Explain your reasoning.

b. Could the class also prove the relationships between parallelism of lines and angles on the same side of the transversal? Explain.

12 In the diagram below, $\ell \parallel m$ and $m \parallel n$. Line p intersects each of these lines.

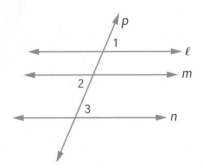

a. How are the measures of angles 1 and 2 related? Explain your reasoning.

b. How are the measures of angles 2 and 3 related? Explain your reasoning.

c. Using your deductions in Parts a and b, prove that lines ℓ and n are parallel.

d. Write an if-then statement that summarizes the theorem you have proved.

13 In the coordinate plane diagram below, lines ℓ, m, and n are parallel and contain the points shown.

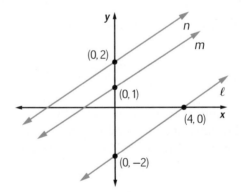

a. Write an equation for each line.

b. On a copy of this diagram, draw a line perpendicular to each line through its y-intercept.

c. How do the three lines you drew in Part b appear to be related to each other?

d. Prove your conjecture in Part c using postulates and/or theorems you proved in Investigation 2.

e. Prove your conjecture in Part c using the equations of the lines and coordinate methods.

14 In previous courses, you used the well-known fact that the sum of the measures of the angles of a triangle is 180°. That property was developed through experimentation and inductive reasoning based on several cases. One possible experiment is illustrated below. But how can you be sure this is true for *all* triangles?

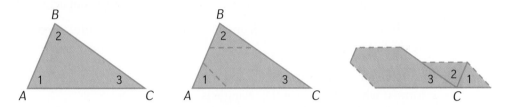

There is an important connection between the Triangle Angle Sum Property and the Alternate Interior Angles Theorem (Connections Task 11) that you proved using the Parallel Lines Postulate. To establish that connection, an additional assumption is needed.

Angle Addition Postulate If P is a point in the interior of $\angle ABC$, then $m\angle ABP + m\angle PBC = m\angle ABC$.

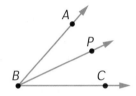

The diagram below shows a triangle ABC and a line k drawn parallel to $\overleftrightarrow{AC}$ through point B.

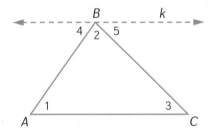

a. What construction did you carry out earlier that shows that line k can be drawn parallel to $\overleftrightarrow{AC}$ through point B?

b. How are $\angle 1$ and $\angle 4$ related? How are $\angle 3$ and $\angle 5$ related?

c. Write a deductive proof that $m\angle 1 + m\angle 2 + m\angle 3 = 180°$.

15 In Connections Task 14, you proved the **Triangle Angle Sum Theorem**: the sum of the measures of the angles of a triangle is 180°. For each of the following statements, decide if it is true. If true, explain why. Otherwise, give a counterexample. (Recall that an acute angle is an angle whose measure is less than 90°.)

   a. If the measures of two angles of one triangle are equal to the measures of two angles of another triangle, then the measures of the third angles are equal.

   b. The sum of the measures of the acute angles of a right triangle is 90°.

   c. If two angles of a triangle are acute, then the third angle is not acute.

   d. If a triangle is equiangular (all angles have the same measure), then each angle has measure 60°.

16 Recall that an *exterior angle* of a triangle is formed when one side of the triangle is extended as shown at the right. $\angle A$ and $\angle B$ are called *remote interior angles* with respect to the exterior angle, $\angle ACD$.

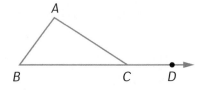

   a. How is m$\angle ACD$ related to m$\angle A$ and m$\angle B$?

   b. Write an argument to support your claim.

   c. Using the term "remote interior angles," write a statement of the theorem you have proved. This theorem is often called the **Exterior Angle Theorem for a Triangle**.

REFLECTIONS

17 In Investigation 1, you explored how to construct a line perpendicular to a given line through a point on the line and through a point not on the line. In Investigation 2, you discovered a method to construct a line parallel to a given line through a point not on the line. Think about these ideas algebraically in the context of a coordinate plane. Consider the line ℓ with equation $2x - y = 11$.

   a. Is $P(8, 5)$ on line ℓ? Write an equation of a line m perpendicular to line ℓ through point P.

   b. Write an equation of a line perpendicular to line ℓ through point $Q(8, 0)$.

   c. Compare your methods for answering Parts a and b.

   d. Write an equation of a line parallel to line ℓ through the point $Q(8, 0)$.

18 In this lesson, you proved mathematical statements in geometric settings. How do you decide where to start in constructing a proof? How do you know when the proof is completed?

19 Parallel lines were defined to be lines in a plane that do not intersect.

    a. Rewrite the definition of parallel lines as an if-and-only-if statement.

    b. Is it possible for two lines in three-dimensional space neither to be parallel nor to intersect? Illustrate your reasoning.

20 Determining if edges or lines are parallel is very important in design and construction.

    a. Describe at least three methods you could use to test if a pair of lines are parallel.

    b. Libby claimed that you could determine if two lines are parallel by measuring the perpendicular distance between the lines at two places. Describe how you would perform this test. Provide an argument that justifies Libby's claim.

21 On the John Hancock Center building, are angles of the type marked $\angle ABC$ and $\angle BAC$ congruent or not?

    a. Prove your claim.

    b. On what assumptions does your proof rely?

22 How are the terms *postulate* and *theorem* as used in this lesson related to the terms *assumptions* and *properties* as used in Lesson 1, Investigation 2 (pages 14–15)?

EXTENSIONS

23 What is the sum of the measures of $\angle A$, $\angle B$, $\angle C$, $\angle D$, and $\angle E$? What geometric assumptions and theorems did you use in answering this question?

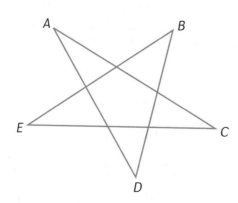

24 If you look at a map of flight paths of an airline, you will see that the flight paths are not straight lines. Since the Earth is a *sphere*, the shortest path between cities, staying on or slightly above the Earth's surface, follows a **great circle**—a circle on the surface of a sphere formed by a plane passing through the center of the sphere. For example, the equator is a great circle.

Think about how the geometry of a sphere (called *spherical geometry*) differs from the Euclidean geometry of a plane. In spherical geometry, all "lines" (shortest paths on the surface) are great circles.

Consider the equator as line ℓ and the North Pole as point P. In spherical geometry:

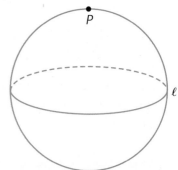

a. How many "lines" can be drawn through point P perpendicular to ℓ?

b. How many "lines" can be drawn through point P parallel to ℓ?

c. In spherical geometry, "segments" are parts of great circles. Draw a diagram of a "triangle" PAB for which $m\angle A + m\angle B + m\angle P > 180°$.

d. Draw a diagram to show that the Pythagorean Theorem is not true in spherical geometry.

e. Explain how a "triangle" in spherical geometry can have three right angles.

25 In the diagram below, $\ell \parallel m$. Prove that $m\angle BCD = m\angle 1 + m\angle 2$.
(*Hint*: Introduce a new line in the diagram as was done in the proof of the Triangle Angle Sum Theorem (page 45). Such a line is called an **auxiliary line**. Its addition to the diagram must be justified by a postulate or previously proved theorem.)

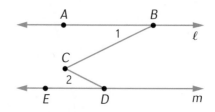

26 Drawing an altitude in a triangle or parallelogram to calculate its area rests on the following theorem.

> *Through a point not on a given line, there is exactly one line perpendicular to the given line.*

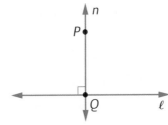

A proof of this theorem has two parts. First, establish there is one line *n* through *P* perpendicular to ℓ. Then establish there is no more than one line perpendicular to ℓ. You established the first part in Investigation 1, Problem 4 Part c.

a. Study the following plan for proof of the second part. Give reasons that will support each proposed step.

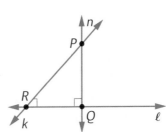

Plan for Proof:
Use indirect reasoning. Suppose there is a second line *k* through *P* perpendicular to ℓ. Show that this will lead to a contradiction. Conclude there is only one line *n* through *P* perpendicular to ℓ.

b. How does this theorem compare with the corresponding situation in spherical geometry (Extensions Task 24 Part a)?

27 In the process of proving the Triangle Angle Sum Theorem (Connections Task 14), you used the previously established fact: Through a point *B* not on a given line $\overleftrightarrow{AC}$, there is a line *k* through the point parallel to the given line.

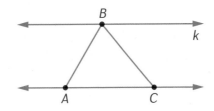

a. Do you think there is more than one line through point *B* parallel to $\overleftrightarrow{AC}$? Compare your answer with that for Extensions Task 24 Part b.

b. To show there is no more than one parallel line in Part a, you can use *indirect reasoning* in a **proof by contradiction**.

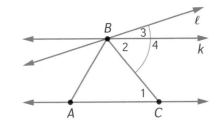

   i. *Assume* there is a second line, call it ℓ, through point *B* such that ℓ ∥ $\overleftrightarrow{AC}$.

   - Why is m∠4 = m∠1?

   - Why is m∠4 = m∠2 + m∠3?

   - Why is m∠1 = m∠2?

   - Why is m∠4 = m∠1 + m∠3?

   ii. Identify a *contradiction* in part i.

   iii. That contradiction shows the assumption in part i is false. So, what must be true?

REVIEW

28 When trying to solve real-world problems, it is often necessary to write symbolic expressions or rules that indicate relationships described in words. Write a symbolic rule that matches each description. Indicate what your letter symbols represent.

a. There are 3 more cats than dogs in the veterinarian's waiting room.

b. There are 25 students for each teacher on the field trip.

c. There are 3 fewer calculators than there are students in the classroom.

d. There are twice as many sophomores as there are juniors on the organizing committee.

29 The density of an object can be found by dividing the mass of the object by the volume of the object. Using symbols, this can be written $d = \frac{m}{V}$.

a. Rock 1 and Rock 2 have the same volume. The density of Rock 1 is greater than the density of Rock 2. How do their masses compare?

b. Two objects have the same mass but the volume of Object 1 is greater than the volume of Object 2. Compare the densities of the two objects.

c. Rewrite the formula in the form $m = \dots$.

d. Rewrite the formula in the form $V = \dots$.

30 Use properties of triangles and right triangle trigonometric ratios to find an approximate value of z in each triangle.

a.

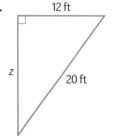

12 ft
z
20 ft

b.

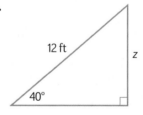

12 ft
z
40°

c.

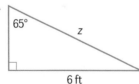

65°
z
6 ft

d.

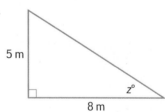

5 m
$z°$
8 m

e.

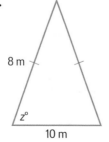

8 m
$z°$
10 m

31 Write each of these expressions in equivalent factored form.

a. $3x + 15$ b. $5x^2 + 15x$ c. $x^3 - x^2$

d. $4x^2 - 9$ e. $x^2 + 6x + 9$ f. $6x^2 - 18x - 60$

g. $-2x^2 - 12x - 18$ h. $x^4 + 2x^2 - 3$

32 There are 16 marbles in a bag. There are red and blue marbles. If a marble is randomly selected from the bag, the probability of selecting a red one is $\frac{1}{4}$.

a. How many blue marbles are in the bag?

b. Suppose that you reach into the bag, draw one marble, note the color, and return it to the bag. You then draw a second marble from the bag. What is the probability that the two marbles you draw are the same color? Explain your reasoning or show your work.

c. How many yellow marbles would you need to add to the bag to make the probability of choosing a red or a yellow marble $\frac{1}{2}$? Explain your reasoning.

33 Write each of these expressions in different equivalent form.

a. $(a^2)(a^5)$ b. $\dfrac{a^5}{a^3}$ c. $(a^2)^5$

d. a^{-2} e. $(2a^2)^3$ f. $a^5(a) - a^5$

34 Refer to the coordinate diagram shown at the right.

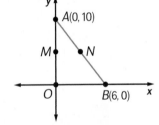

a. Find the coordinates of the midpoint M of $\overline{AO}$. The midpoint N of $\overline{AB}$.

b. Use coordinates to explain why $\overline{MN} \parallel \overline{OB}$.

c. How does the length of $\overline{MN}$ compare to the length of $\overline{OB}$?

d. How would your answers for Parts a–c change if point A has coordinates $(0, a)$ and point B has coordinates $(b, 0)$.

35 Solve each equation.

a. $3(2x - 5) = 6 - 2x$ b. $5(3^x) = 45$

c. $(x + 3)(x - 2) = 6$ d. $(x^2 + 4)(x^2 - 9) = 0$

36 Determine if the relationship in each table is linear, exponential, or quadratic. Then write a rule that matches each table. Explain your reasoning or show your work.

a.

x	2	3	4	5	6
y	6	11	18	27	38

b.

x	4	6	8	10	12
y	2	4.5	7	9.5	12

c.

x	1	2	3	4	5
y	200	100	50	25	12.5

Algebraic Reasoning and Proof

Algebraic calculations and reasoning can be used to solve many kinds of practical problems. But they can also be used to perform some amazing magic tricks with numbers. For example, do these numerical operations.

> Pick an integer between 0 and 20.
>
> Add 5 to your number.
>
> Multiply the result by 6.
>
> Divide that result by 3.
>
> Subtract 9 from that result.

If each class member reports the final number that results from those operations, your teacher will be able to tell the starting number of each student.

If you believe that your teacher performs this number magic by memorizing all the possible starting and ending number combinations, you can increase the range of starting numbers to 0–50 and then 0–100. Your teacher will still be able to find each start number.

Data in the table at the right show some combinations of starting and ending numbers.

Start Number	End Number
4	9
11	23
17	35
33	67
45	91

Oote Boe Photography/Alamy

In this lesson, you will explore ways to use algebraic reasoning to discover and prove number patterns and relationships like the one used in the number trick. Then you will see how the same reasoning methods can be used to prove other important mathematical principles.

INVESTIGATION 1

Reasoning with Algebraic Expressions

In earlier algebra units of *Core-Plus Mathematics*, you developed a toolkit of techniques for writing algebraic expressions in useful equivalent forms. As you work on the problems in this investigation, look for answers to this question:

> *How can strategies for manipulating algebraic expressions into equivalent forms be used to explain interesting number patterns?*

Algebra and Number Magic Algebraic reasoning skills can be used to create and explain many different number tricks.

1 There is a simple way to discover start numbers in the Think About This Situation number trick: *Subtract one from the end number and divide that result by two.* Explaining why that decoding strategy works and proving that it will *always* work requires some algebraic reasoning.

a. Use the letter n to represent the start number and build an algebraic expression that shows the steps for calculating the result that will be reported to the teacher.

    i. Adding 5 to your number is expressed by … .

    ii. Multiplying that result by 6 is expressed by … .

    iii. Dividing that result by 3 is expressed by … .

    iv. Subtracting 9 from that result is expressed by … .

b. Use what you know about algebra to write the final expression from Part a in simplest form. What relationship between starting and ending numbers is shown by the result?

c. How can the relationship between starting and ending numbers be used to find starting numbers when only ending numbers are known?

d. Suppose that the third step of the number trick, "Divide that result by 3," is replaced by "Divide that result by 2." How would that change in the procedure affect the decoding strategy?

2 Consider this different number trick.

> **Pick a number.**
>
> **Double it.**
>
> **Add 3 to the result.**
>
> **Multiply that result by 5.**
>
> **Subtract 7 from that result.**

a. Explore the way that this procedure transforms start numbers into final results. Find a decoding strategy that could be used to find the start number when only the final result is known.

b. Use algebraic reasoning to explain why your decoding strategy will always work.

3 Now that you have analyzed some number tricks with algebraic reasoning, you can adapt those arguments to design your own trick.

a. Create a similar number trick. Test it with other students to see if it works as intended.

b. Develop an algebraic argument to prove the trick that you created will work in every case.

Explaining Number Patterns Number patterns have fascinated amateur and professional mathematicians for thousands of years. It is easy to find interesting patterns, but usually more challenging to explain why the patterns work.

4 Consider the sequence of square numbers that begins 0, 1, 4, 9, 16, … .

a. Complete the following table to show how that sequence continues.

n	0	1	2	3	4	5	6	7	8	…	n
S_n	0	1	4	9	16					…	

b. Calculate the differences between consecutive terms in the sequence of square numbers (for example, $S_4 - S_3 = 16 - 9 = 7$). Then describe the pattern that develops in the sequence of differences.

c. If n and $n + 1$ represent consecutive whole numbers, the difference between the squares of those numbers can be given by $S_{n+1} - S_n = (n + 1)^2 - n^2$. Use your ability to simplify algebraic expressions to prove that the pattern of differences you described in Part b is certain to continue as the sequence of square numbers is extended.

d. How would you describe the pattern of differences between successive square numbers as a proposition in the form *If ... , then ...* ?

5 The pattern that you analyzed in Problem 4 can also be justified with a visual proof.

a. The diagrams shown here have shaded and unshaded regions. Express the area of each unshaded region as a difference of two square numbers.

Diagram I   **Diagram II**

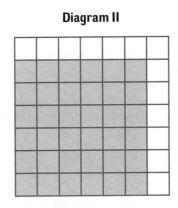

b. Explain how you could use the next diagram to justify the formula for the difference of any two consecutive square numbers that you found in work on Problem 4.

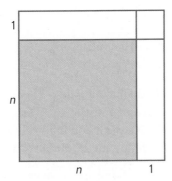

6 From prior work with linear functions, you know that the values of those functions change at a constant rate as values of the independent variable increase.

a. Consider the linear function $f(x) = 3x - 5$. Find these differences in values of $f(x)$.

   i. $f(7) - f(2)$ **ii.** $f(4) - f(-1)$

   iii. $f(-4) - f(-9)$ **iv.** $f(x + 5) - f(x)$

b. Consider the linear function $g(x) = -0.5x + 2$. Find these differences in values of $g(x)$.

   i. $g(8) - g(5)$ **ii.** $g(2) - g(-1)$

   iii. $g(-2) - g(-5)$ **iv.** $g(x + 3) - g(x)$

c. Use algebraic reasoning to complete and prove the property of linear functions illustrated in the patterns of Parts a and b.

For any linear function $h(x) = mx + b$, $h(x + k) - h(x) = \ldots$.

d. Explain how your reasoning in Parts a, b, and c shows that linear functions change by equal differences over equal intervals.

Proving Properties of Numbers and Functions In the Course 2 unit, *Nonlinear Functions and Equations*, you learned how to use *logarithms* to describe the intensity of sound and earthquakes and the pH of liquids. Historically, however, logarithms were first used as an aid to scientific calculation involving large numbers. That application of logarithms depends on a very useful property of the function $y = \log x$.

7 Recall that $y = \log x$ means $10^y = x$. For example, $3 = \log 1{,}000$ because $10^3 = 1{,}000$. Use reasoning about powers of 10 or the **log** function on your calculator to complete the following table that compares $\log a$, $\log b$, and $\log ab$ for a sample of positive numbers a and b.

a	b	ab	log a	log b	log ab
100	1,000				
100	0.001				
0.1	0.001				
75	100				
75	0.01				
75	120				
50	20				

What pattern do you see that relates $\log a$, $\log b$, and $\log ab$? Describe the pattern in words and algebraically with symbols.

8 Rows in the table of Problem 7 illustrate a property of logarithms that connects multiplication and addition in a very useful way: For any a and $b > 0$, $\log ab = \log a + \log b$. Use what you know about logarithms and exponents to justify each step in the following algebraic proof of that property.

(1) Since $a > 0$ and $b > 0$, there are numbers x and y so that
     $x = \log a$ and $y = \log b$.
(2) So, $a = 10^x$ and $b = 10^y$.
(3) $\log ab = \log (10^x 10^y)$
(4) $\qquad = \log 10^{x+y}$
(5) $\qquad = x + y$
(6) $\qquad = \log a + \log b$
(7) Therefore, $\log ab = \log a + \log b$.

This property of the logarithm function enabled mathematicians and scientists to use a table giving the logarithms of many numbers to convert any multiplication problem into an easier addition problem. This strategy was particularly useful for calculations involving very large numbers, such as those that occur in sciences, because it is much simpler to add logarithms of two large numbers than to find the product of the numbers.

9 The speed of light travels at about 186,000 miles per second, and there are 31,536,000 seconds in one year. To find the length in miles of a *light year*, you need to multiply 186,000 × 31,536,000.

a. To calculate, apply the standard multiplication algorithm.

$$31,536,000$$
$$\times\ 186,000$$

b. Use the facts log 186,000 ≈ 5.26951 and log 31,536,000 ≈ 7.49881 to find log (186,000 × 31,536,000). Then use the definition $y = \log x$ when $10^y = x$ to find 186,000 × 31,536,000.

c. Which procedure for finding the product of large numbers involves the simplest arithmetic?

SUMMARIZE THE MATHEMATICS

In this investigation, you explored ways that algebraic reasoning explains interesting number patterns.

a What were the key steps in explaining the number tricks discovered in Problems 1 and 2?

b What overall strategy and algebraic properties were used to prove the generality of number patterns discovered in Problem 4?

c What overall strategy and algebraic properties were used to prove that log ab = log a + log b for all positive values of a and b?

Be prepared to explain your ideas to the class.

 CHECK YOUR UNDERSTANDING

Use your understanding of equivalent expressions and algebraic reasoning to complete these tasks.

a. Prove that, whatever the starting number, the following calculations always lead to a result that is one greater than the starting number.

> **Think of a number.**
>
> **Double that number and add 7.**
>
> **Multiply that result by 5 and subtract 25.**
>
> **Divide that result by 10.**

b. Consider the sequence $S_n = (n-1)^2 + 1$.

    i. Write the first seven terms of this sequence.

    ii. Calculate the differences of consecutive terms in that sequence and describe the pattern that develops.

    iii. Use algebraic reasoning to prove that the pattern you described in part ii is certain to continue as the number sequence is extended.

INVESTIGATION 2

Reasoning with Algebraic Equations

Many important mathematical facts are expressed in the form of equations or inequalities relating two or more variables. For example, the coordinates (x, y) of points on a line will always be related by an equation in the form $ax + by = c$. In this investigation, look for answers to this question:

> *How can strategies for reasoning with algebraic equations*
> *be used to explain and prove important principles*
> *in algebra, geometry, and trigonometry?*

1 Mathematical work often requires solving linear equations, so it might be convenient to write a short calculator or computer program that will perform the necessary steps quickly when given only the equation coefficients and constants.

a. What formula shows how to use the values of a, b, and c to find the solution for any linear equation in the form $ax + b = c$ with $a \neq 0$? Show the algebraic reasoning used to derive the formula. Check your work by asking a computer algebra system (CAS) to **solve(a*x+b=c,x)** and reconcile any apparent differences in results.

b. What formula shows how to use the values of a, b, c, and d to find the solution for any linear equation in the form $ax + b = cx + d$ with $a \neq c$? Show the algebraic reasoning used to derive the formula. Check your work by asking a CAS to **solve(a*x+b=c*x+d,x)** and reconcile any apparent differences in results.

c. What statements would express the formulas of Parts a and b in if-then form?

2 When a problem requires analysis of a linear equation in two variables, it often helps to draw a graph of the equation. Information about x- and y-intercepts and slope can be used to make a quick sketch. For example, the diagram at the right shows the slope and intercepts of the graph for solutions of the equation $3x + 6y = 12$.

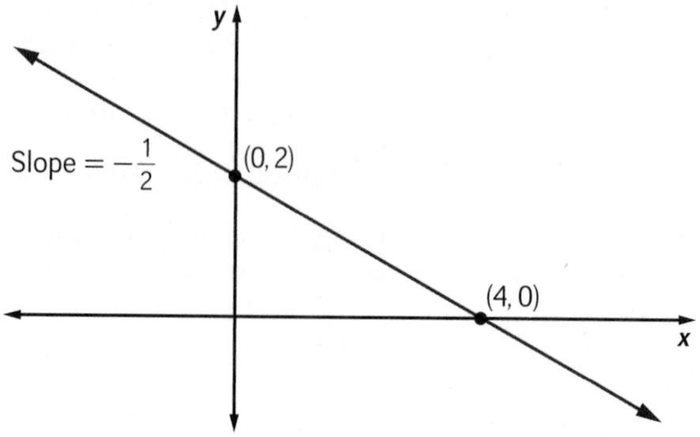

Slope $= -\dfrac{1}{2}$

$(0, 2)$

$(4, 0)$

Find the x- and y-intercepts and the slopes for lines with equations given in Parts a–d. Use that information to *sketch* graphs of the equations, labeling the x- and y-intercepts with their coordinates.

a. $-6x + 4y = 12$

b. $6x + 3y = 15$

c. $5x - 3y = -5$

d. $4y + 5x = -8$

3 To find formulas that give intercepts and slope of any line with equation in the form $ax + by = c$, you might begin by reasoning as follows.

(1) The y-intercept point has coordinates in the form $(0, y)$ for some value of y.

(2) If $a(0) + by = c$, then $y = \dots$.

a. Complete the reasoning to find a formula for calculating coordinates of the y-intercept. Describe any constraints on a and b.

b. Use reasoning like that in Part a to find a formula for calculating coordinates of the x-intercept of any line with equation in the form $ax + by = c$. Describe any constraints on a and b.

c. Use reasoning about equations to find a formula for calculating the slope of any line with equation in the form $ax + by = c$ with $b \neq 0$.

d. Write CAS solve commands that could be used to check your work in Parts a, b, and c.

4 You probably recall the Pythagorean Theorem that relates lengths of the legs and the hypotenuse in any right triangle. It says, "If a and b are lengths of the legs and c the length of the hypotenuse of a right triangle, then $a^2 + b^2 = c^2$."

There are over 300 different proofs of the Pythagorean Theorem. In the Course 1 unit, *Patterns in Shape*, you completed a proof of the Pythagorean Theorem that was based on a comparison of areas. The diagram at the right shows how to start on a proof that makes use of geometric and algebraic reasoning in important ways. It shows a large square divided into four triangles and a smaller quadrilateral inside that square.

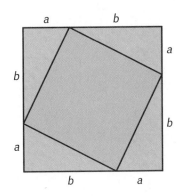

a. Use only the fact that the whole colored figure is a square and the division of its edges into lengths a and b to prove that the four triangles are congruent right triangles.

b. How can you use the fact that the outside figures are congruent right triangles to prove that the interior quadrilateral is a square?

c. Write two different algebraic expressions that show how to find the area of the large square.

- In the first, use the fact that the sides of the large square are of length $(a + b)$.

- In the second, label the edges of the inner square with length c. Use the fact that the large square is made up of four triangles, each with base a and height b, and an inner square with sides of length c.

d. Equate the two expressions from Part c and apply algebraic reasoning to that equation in order to show that $a^2 + b^2 = c^2$.

5 The kind of algebraic reasoning used to prove the Pythagorean Theorem can be adapted to prove the very useful Law of Cosines, introduced in the Course 2 *Trigonometric Methods* unit.

In any triangle *ABC* with sides of length *a*, *b*, and *c* opposite $\angle A$, $\angle B$, and $\angle C$, respectively, $c^2 = a^2 + b^2 - 2ab \cos C$.

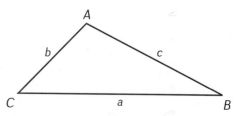

Analyze the following argument to identify and explain ways that algebraic principles (and a small amount of geometry and trigonometry) can be used to prove the special case pictured. Here $\angle C$ is an acute angle and $\overline{AD}$ is an altitude of the triangle with point *D* between points *C* and *B*.

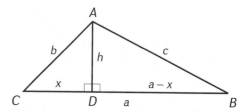

Step 1. In $\triangle ADC$: $\frac{x}{b} = \cos C$, so $x = b \cos C$.

Why are these relationships true?

Step 2. In $\triangle ABD$: $h^2 = c^2 - (a - x)^2$
$$= c^2 - (a^2 - 2ax + x^2)$$
$$= c^2 - a^2 + 2ax - x^2$$
Why are these equations true?

Step 3. In $\triangle ADC$: $h^2 = b^2 - x^2$

Why is this equation true?

Step 4. From Steps 2 and 3, we can conclude that
$$c^2 - a^2 + 2ax - x^2 = b^2 - x^2.$$
Why is this equation true?

Step 5. From the equation in Step 4, we can conclude that
$$c^2 = a^2 + b^2 - 2ax.$$
What rules of algebra justify this conclusion?

Step 6. Finally, combining Steps 1 and 5, we conclude that
$$c^2 = a^2 + b^2 - 2ab \cos C.$$
What property justifies this conclusion?

You can check that the same formula results in the cases where point *D* coincides with or is to the right of point *B* and that, as long as $\angle C$ is acute, point *D* cannot lie to the left of point *C*. A complete proof of the Law of Cosines requires considering cases when $\angle C$ is a right angle or an obtuse angle. See Extensions Task 33.

In this investigation, you explored ways that reasoning with equations can be used to prove important general principles in algebra, geometry, and trigonometry.

a What general properties of numbers, operations, and equations did you use to discover and prove a formula for solving equations in the form $ax + b = c$ with $a \neq 0$? In the form $ax + b = cx + d$ with $a \neq c$?

b What general properties of numbers, operations, and equations did you use to discover and prove formulas for slope and intercepts of graphs for linear equations in the form $ax + by = c$?

c What algebraic principles were used to justify steps in the proof of the Pythagorean Theorem?

d What is the main idea behind the proof of the Law of Cosines?

Be prepared to explain your ideas to the class.

 CHECK YOUR UNDERSTANDING

Use your understanding of equations and algebraic reasoning to complete these tasks.

a. Consider linear equations that occur in the form $a(x - b) = c$.

   i. Solve the particular example $3(x - 7) = 12$.

   ii. Write a formula that will give solutions to any equation in the form $a(x - b) = c$ with $a \neq 0$, in terms of a, b, and c. Show the algebraic reasoning used to derive your formula. Check your work by using a computer algebra system to **solve(a*(x−b)=c,x)**.

b. Many problems in geometry require finding the length of one leg in a right triangle when given information about the lengths of the other leg and the hypotenuse.

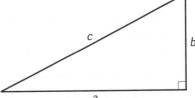

   i. If the hypotenuse of a right triangle is 20 inches long and one leg is 15 inches long, what is the length of the other leg? Show the reasoning that leads to your answer.

   ii. Starting with the algebraic statement of the Pythagorean Theorem $a^2 + b^2 = c^2$, use algebraic reasoning to derive a formula that shows how to calculate the length a when lengths b and c are known.

APPLICATIONS

1 Whenever three consecutive integers are added together, the sum is always equal to three times the middle number. For example, $7 + 8 + 9 = 3(8)$ and $-4 + -3 + -2 = 3(-3)$.

a. If n represents the smallest number in a sequence of three consecutive integers, what expressions represent:

   i. the other two numbers?

   ii. the sum of all three numbers?

b. Simplify the expression for the sum in Part a to prove that the pattern relating the sum and the middle number occurs whenever three consecutive integers are added.

c. What pattern will relate the sum of any five consecutive integers to one of the integer addends? Use algebraic reasoning to prove your idea.

2 By definition, any odd number can be expressed in the form $2m + 1$ for some integer m. Any even number can be expressed in the form $2m$ for some integer m.

a. Express 5, 17, and 231 in the form $2m + 1$.

b. Express 6, 18, and 94 in the form $2m$.

c. Prove that the product of an odd number and an even number is always an even number. Begin by writing the odd number as $2m + 1$ and the even number as $2n$.

d. Prove that the product of any two odd numbers is always an odd number. Use different letters in representing the two numbers symbolically.

e. Why are different letters, m and n, used to represent integers in Parts c and d?

3 The two-digit number 53 can be written in *expanded* form as $5(10) + 3$. In general, a two-digit number with tens digit a and ones digit b can be written as $a(10) + b$. Use this fact to prove the number patterns described below.

a. Pick any two-digit number and reverse the order of the digits. Then add the new number to the original number. The result will always be a multiple of 11.

b. Pick any two-digit number and reverse the order of the digits. Then find the difference between the new number and the original number. The result will always be a multiple of 9.

4 Consider the sequence 1, 3, 9, 27, 81, 243, 729, ... generated by powers of 3.

   a. Let T_0 represent the initial term of the sequence. So, $T_0 = 1$. Write expressions for term n, T_n, and for term $n + 1$, T_{n+1}, of the sequence.

   b. Use the results from Part a to prove that the differences between successive terms of the sequence can be calculated using the expression $2(3^n)$.

5 The beginnings of two number sequences are given here.

   Sequence I   1, 4, 16, 64, 256, 1,024, 4,096, ...
   Sequence II 1, 5, 25, 125, 625, 3,125, 15,625, ...

   For each sequence:

   a. Identify the terms t_0, t_4, and t_5. Then find an algebraic rule that shows how to calculate term n in the sequence.

   b. Study the pattern of differences between successive terms and find a formula that gives the nth difference.

   c. Prove that your formula is correct.

6 If the logarithm of the product of two positive numbers is the sum of the logarithms of the individual numbers, you might expect that the logarithm of the quotient of the two numbers is the difference of their logarithms. Use reasoning about powers of 10 or the **log** function of your calculator to check the property $\log\left(\frac{a}{b}\right) = \log a - \log b$ for the sample of positive numbers a and b in this table.

a	b	$\frac{a}{b}$	$\log a$	$\log b$	$\log\left(\frac{a}{b}\right)$
1,000	10				
0.01	100				
0.001	0.1				
75	100				
75	0.01				
75	3				

7 Adapt the reasoning in Problem 8 of Investigation 1 to prove that for any positive numbers a and b, $\log\left(\frac{a}{b}\right) = \log a - \log b$. (*Hint*: Recall that $\left(\frac{1}{10}\right)^y = 10^{-y}$.)

8 Consider linear equations that occur in the form $a(x + b) = c$ with $a \neq 0$.

   a. Solve the particular example $7(x + 3) = 49$.

   b. Write a formula that will give solutions for any equation in the form $a(x + b) = c$ with $a \neq 0$, in terms of a, b, and c. Show the algebraic reasoning used to derive the formula.

9 In Lesson 2, you used geometric reasoning to prove that if two lines in a plane are perpendicular to the same line, then they are parallel to each other. Prove that statement using algebraic reasoning. Begin by assuming ℓ, p, and m are lines with $\ell \perp p$, $m \perp p$, and the slope of ℓ is a where $a \neq 0$.

10 If a line has slope m and passes through the point (p, q), it is sometimes convenient to write the equation for that line in the form $(y - q) = m(x - p)$.

    a. Write an equation in this form for the line that has slope 2 and passes through the point $(4, 7)$.

        • Check that coordinates of the points $(4, 7)$ and $(5, 9)$ satisfy your equation.

        • Use the two points to check that the slope of the graph will be 2.

    b. Explain how you know that the point (p, q) is always on the graph of $(y - q) = m(x - p)$.

    c. Use algebraic reasoning to write the equation from Part a in $y = mx + b$ form.

    d. Show how any equation in the form $(y - q) = m(x - p)$ can be written in equivalent $y = mx + b$ form.

    e. Look back at your work in Part d. How is b related to p and q?

11 You have seen in the problems of this lesson that algebraic reasoning usually involves writing symbolic expressions or equations in equivalent forms using rules for symbol manipulation. The most useful symbol manipulations are based on the list of algebraic properties of equality and operations shown in the following charts.

Algebraic Properties

Addition

For any numbers a, b, and c:

    Commutative Property of Addition: $a + b = b + a$

    Associative Property of Addition: $(a + b) + c = a + (b + c)$

    Additive identity element (0): $a + 0 = a$

    Additive Inverse Property: There is a number $-a$ such that $a + (-a) = 0$.

Multiplication

For any numbers a, b, and c:

    Commutative Property of Multiplication: $ab = ba$

    Associative Property of Multiplication: $(ab)c = a(bc)$

    Multiplicative identity element (1): $a1 = a$

    Multiplicative Inverse Property: For each $a \neq 0$, there is a number a^{-1} such that $a^{-1}a = 1$.

Distributive Property of Multiplication over Addition

    For any numbers a, b, and c: $a(b + c) = ab + ac$

Properties of Equality

For any numbers a, b, and c:

    Transitive Property of Equality: If $a = b$ and $b = c$, then $a = c$.

    Addition Property of Equality: If $a = b$, then $a + c = b + c$.

    Subtraction Property of Equality: If $a = b$, then $a - c = b - c$.

    Multiplication Property of Equality: If $a = b$, then $ac = bc$.

    Division Property of Equality: If $a = b$, then $a \div c = b \div c$ (whenever $c \neq 0$).

State the properties that justify steps in the following derivations of equivalent expressions. Some steps may involve more than one property. Arithmetic and substitution may also be used to justify steps.

a. For any x, $3x + 5x = (3 + 5)x$ (1)

$\qquad\qquad\qquad = 8x$ (2)

b. For any x, $(3x + 5) + 7x = (5 + 3x) + 7x$ (1)

$\qquad\qquad\qquad = 5 + (3x + 7x)$ (2)

$\qquad\qquad\qquad = 5 + 10x$ (3)

c. If $7x + 5 = 5x + 14$, then $(7x + 5) + (-5) = (5x + 14) + (-5)$ (1)

$\qquad\qquad 7x + (5 + (-5)) = 5x + (14 + (-5))$ (2)

$\qquad\qquad\qquad 7x + 0 = 5x + 9$ (3)

$\qquad\qquad\qquad 7x = 5x + 9$ (4)

$\qquad\qquad -5x + 7x = -5x + (5x + 9)$ (5)

$\qquad\qquad\qquad 2x = (-5x + 5x) + 9$ (6)

$\qquad\qquad\qquad 2x = 0x + 9$ (7)

$\qquad\qquad\qquad 2x = 9$ (8)

$\qquad\qquad \left(\frac{1}{2}\right)(2)x = \left(\frac{1}{2}\right)9$ (9)

$\qquad\qquad\qquad 1x = 4.5$ (10)

$\qquad\qquad\qquad x = 4.5$ (11)

12 In your work on Task 11 Part c, you probably used a property of multiplication that seemed so obvious it did not need justification: For any n, $0n = 0$. While not stated as one of the basic number system properties, this result can be proven from the listed properties. Provide explanations for each step in the following proof.

$0 + 0 = 0$ (1)

$(0 + 0)n = 0n$ (2)

$0n + 0n = 0n$ (3)

$0n + 0n = 0n + 0$ (4)

$0n = 0$ (5)

13 In your previous coursework, you used the **Zero Product Property**:

$$\text{If } ab = 0, \text{ then } a = 0 \text{ or } b = 0.$$

a. Show how this property is used to solve $(x + 6)(2x - 8) = 0$.

b. Use properties listed in Tasks 11 and 12 to justify each step in the following proof of the Zero Product Property.

Suppose that $ab = 0$. If $a = 0$, the result is clearly true.
Suppose $a \neq 0$.

(1) Then there is a number a^{-1} so that $a^{-1}a = 1$.

(2) $a^{-1}(ab) = a^{-1}0$

(3) $(a^{-1}a)b = 0$

(4) $1b = 0$

(5) $b = 0$

c. What is the logic of the argument given in Part b to prove the claim that "if $ab = 0$, then $a = 0$ or $b = 0$"?

14 An **isosceles trapezoid** is shown in the coordinate diagram below. It has a pair of nonparallel opposite sides the same length. Use ideas from geometry and trigonometry and algebraic reasoning to complete the following tasks.

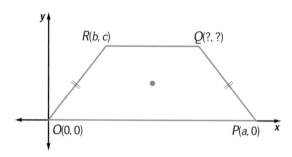

a. Draw and label a similar diagram on a coordinate grid. Determine the coordinates of point Q.

b. Prove that the *base angles*, $\angle O$ and $\angle P$, have the same measure.

c. Draw diagonals $\overline{OQ}$ and $\overline{PR}$. How do the diagonals appear to be related? Prove your conjecture.

d. Find the coordinates of the midpoint M of $\overline{OR}$ and the midpoint N of $\overline{PQ}$. Draw $\overline{MN}$. How does $\overline{MN}$ appear to be related to $\overline{OP}$? Prove your conjecture.

e. Write three general statements that summarize what you have proven about isosceles trapezoids.

15 Because proportional reasoning is required to solve many practical, scientific, and mathematical problems, it is important to be able to solve proportions in specific and general cases.

a. Solve these proportions for x.

    i. $\dfrac{3}{x} = \dfrac{9}{15}$ **ii.** $\dfrac{x}{12} = \dfrac{10}{30}$ **iii.** $\dfrac{4}{5} = \dfrac{10}{x}$

b. What algebraic reasoning justifies the fact that for any a, b, and c ($a \neq 0$, $b \neq 0$, and $x \neq 0$) if $\dfrac{a}{b} = \dfrac{c}{x}$, then $x = \dfrac{bc}{a}$?

CONNECTIONS

16 Suppose that in one school term, you have five mathematics quiz scores x_1, x_2, x_3, x_4, and x_5.

a. What algebraic expression shows how to calculate the mean of those scores?

b. Use algebraic reasoning to show how the mean of your quiz scores is affected if each score is multiplied by some common factor p.

c. Use algebraic reasoning to show how the mean of your quiz scores is affected if a constant k is added to each score.

17 Suppose that $\bar{x}$ is the mean of five mathematics quiz scores $x_1, x_2, x_3, x_4,$ and x_5.

  a. Describe in words how to calculate the standard deviation $s = \sqrt{\dfrac{\Sigma(x - \bar{x})^2}{n}}$ for this set of scores.

  b. Use algebraic reasoning to show how the standard deviation is affected if each score is multiplied by some common factor p.

  c. Use algebraic reasoning to show how the standard deviation is affected if a constant k is added to each score.

18 The kind of algebraic reasoning you used to explain and create number tricks can also be used to explain some useful procedures for rewriting algebraic expressions in convenient equivalent forms. For example, you might impress friends or parents with your mental arithmetic skills by backing up the following claim.

> "Pick two numbers that are equidistant from 100, like 112 and 88, and I'll find their product almost immediately without using a calculator."

Do you know how to do this kind of calculation in your head quickly? You can use the fact that for any two numbers r and s, the product $(r + s)(r - s) = r^2 - s^2$.

  a. Explain why each step in the following algebraic proof is justified.

$$
\begin{aligned}
(r + s)(r - s) &= (r + s)r - (r + s)s & (1) \\
&= r^2 + sr - rs - s^2 & (2) \\
&= r^2 + rs - rs - s^2 & (3) \\
&= r^2 - s^2 & (4)
\end{aligned}
$$

  b. How could the above algebraic relationship be used to calculate 105×95 or 112×88?

  c. How could you quickly calculate products like $990 \times 1{,}010$ or $996 \times 1{,}004$?

  d. How could you use the relationship in Part a to help factor $4p^2 - 9q^2$?

19 The expanded form of any expression like $(x + p)^2$ is $x^2 + 2px + p^2$.

  a. Check this relationship by calculating the value of each expression when:

   i. $x = 3$ and $p = 7$

   ii. $x = 3$ and $p = -7$

   iii. $x = -4$ and $p = -3$

  b. Adapt the reasoning in Task 18 to develop an algebraic proof of expanding perfect squares, $(x + p)^2 = x^2 + 2px + p^2$.

  c. How could the algebraic relationship in Part b be used to quickly compute 35^2?

  d. How could you use the relationship in Part b to help expand $(3x + 2y)^2$?

20 Recall that for any two points on a number line, the distance between them is the absolute value of the difference of their coordinates. For example, on the number line below, the distance between points A and B is given by $|5 - 2| = |2 - 5| = 3$, and the distance between points B and C is given by $|5 - (-4)| = |(-4) - 5| = 9$.

a. Find the numbers represented by the following absolute value expressions.

i. $|12 - 3|$

ii. $|3 - 12|$

iii. $|7 - (-4)|$

iv. $|(-4) - 7|$

v. $|(-7) - (-3)|$

vi. $|(-3) - (-7)|$

b. What can you conclude about the relationship between numbers p and q in each case below?

i. $|p - q| = p - q$

ii. $|p - q| = q - p$

iii. $|p - q| = 0$

c. Use your answers to Part b to prove that the following calculations can be used to compare any two distinct numbers p and q.

i. The larger of numbers p and q will always be given by $\dfrac{p + q + |p - q|}{2}$.

ii. The smaller of numbers p and q will always be given by $\dfrac{p + q - |p - q|}{2}$.

21 Algebraic reasoning can be used to prove one of the most useful relationships in trigonometry called the **Pythagorean Identity**. Justify each step in the following proof that for any angle θ, $\sin^2 \theta + \cos^2 \theta = 1$. The notation $\sin^2 \theta$ means $(\sin \theta)^2$.

Proof: Suppose θ is an angle in standard position in the coordinate plane, and $P(a, b)$ is a point on the terminal side of θ. Let c be the distance from the origin to point P.

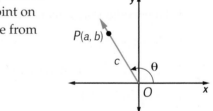

Then:

$$\sin \theta = \frac{b}{c} \text{ and } \cos \theta = \frac{a}{c} \qquad (1)$$

$$\sin^2 \theta = \frac{b^2}{c^2} \text{ and } \cos^2 \theta = \frac{a^2}{c^2} \qquad (2)$$

$$\sin^2 \theta + \cos^2 \theta = \frac{b^2}{c^2} + \frac{a^2}{c^2} \qquad (3)$$

$$\sin^2 \theta + \cos^2 \theta = \frac{b^2 + a^2}{c^2} \qquad (4)$$

$$\sin^2 \theta + \cos^2 \theta = \frac{c^2}{c^2} \qquad (5)$$

$$\sin^2 \theta + \cos^2 \theta = 1 \qquad (6)$$

REFLECTIONS

22 Students in one Colorado school came up with a simple rule of thumb for deciding on the kinds of algebraic manipulations that would transform given expressions into forms that they could be sure were equivalent to the original. They reasoned, "You can do anything to algebraic expressions that you can do to expressions that involve only specific numbers." Does this seem like a safe guideline for symbol manipulation?

23 A mathematician once said that there are three stages in proving the truth of some mathematical proposition. First, you have to convince yourself. Second, you have to convince your friends. Third, you have to convince your enemies. The kinds of evidence that you find convincing for yourself might be different than the evidence that will convince someone else.

   a. What kinds of evidence do you find to be most persuasive: a collection of specific examples, a visual image of the pattern, or an algebraic argument that links given conditions to conclusions by a sequence of equivalent equations or expressions?

   b. Describe the value of each of these types of persuasive arguments.

      i. several specific illustrative examples

     ii. a visual image

    iii. formal algebraic reasoning

24 In mathematics, it is often helpful to solve problems and discover general principles by studying results from repeated calculations with specific numerical values.

   a. Where did you use that practice in the problems of this lesson?

   b. In what ways is the strategy helpful to your thinking?

25 The problems of this lesson called for reasoning and proof based on mathematical principles without the aid of cues from familiar problem contexts.

   a. In what ways is that kind of abstract mathematical reasoning more difficult or easier than work on questions that focus on familiar real-life contexts?

   b. What is the value of proving results with mathematical reasoning that does not rely on cues from a familiar situation?

26 The problems of this lesson called for reasoning and proof based on mathematical principles without the aid of tools like spreadsheets, technology-produced graphs, or computer algebra systems.

   a. What is gained by such reasoning without aid of mathematical tools?

   b. How could exploration with mathematical tools play a role in discovery and proof of mathematical principles?

EXTENSIONS

27 Recall that a **rational number** is a real number that can be expressed in the form $\frac{m}{n}$, where m and n are integers, $n \neq 0$. An **irrational number** is a real number that is *not* rational. Let $\frac{m}{n}$ and $\frac{p}{n}$ be any two rational numbers and s an irrational number. Decide if each of the following claims is true or false. If false, give a counterexample. If true, provide an argument justifying the claim.

a. The sum (product) of two rational numbers is a rational number.

b. The sum (product) of a rational number and an irrational number is an irrational number. For multiplication, assume the rational number is nonzero.

28 The following table shows days and dates for January 2010. Several sets of four related dates have been enclosed in boxes.

Sun	Mon	Tue	Wed	Thu	Fri	Sat
					1	2
3	4	5	6	7	8	9
10	11	12	13	14	15	16
17	18	19	20	21	22	23
24	25	26	27	28	29	30
31						

a. In each box, identify the pairs of date numbers that are at opposite corners. Find the products of those pairs, then find the differences of the products.

b. Describe the pattern relating all results of the calculations prescribed in Part a and check some more 2 × 2 boxes in the calendar page to test your pattern conjecture.

c. If the number in the upper-left corner of a box is x, how can the other three numbers in the box be expressed in terms of x?

d. Use the expressions in Part c and algebraic reasoning to prove the pattern that you described in Part b will hold for any such 2 × 2 box of dates on any calendar page arranged in rows like that above.

29 Explore the output of the following algorithm.

> Start with the first three digits of your phone number (not the area code).
>
> Multiply that entry by 80 and add 1.
>
> Multiply that result by 250.
>
> Double the number given by the last four digits of your phone number and add that number to your result from the previous step.
>
> Subtract 250 from that result.
>
> Divide that result by 2.

a. What does the algorithm seem to do with your phone number?

b. Use x to represent the number defined by the first three digits of a phone number and y to represent the number defined by the last four digits of a phone number and write an expression that shows how the algorithm operates.

c. Simplify the expression in Part b to explain the connection between the entry and result of the algorithm.

30 A number trick found on www.flashpsychic.com directs visitors to follow these steps.

> Pick any two-digit number.
> Add together both digits.
> Subtract that total from your original number.

Then the program asks you to find your result in a table that pairs each two-digit number with a special symbol. The program proceeds to tell you the symbol corresponding to your ending number.

a. Visit the Web site and see if the program works as advertised.

b. Use algebraic reasoning to explain how the number trick works.

31 The largest planet in our solar system, Jupiter, has volume of about 333,000,000,000,000 cubic miles, while our Earth has volume of about 268,000,000,000 cubic miles.

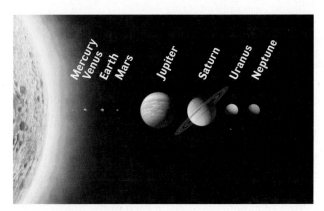

a. Explain why log 333,000,000,000,000 = 14 + log 3.33. Use a calculator or software **log** function to approximate this value.

b. Explain why log 268,000,000,000 = 11 + log 2.68. Use a calculator or software **log** function to approximate this value.

c. Use the results of Parts a and b to approximate log (333,000,000,000,000 ÷ 268,000,000,000).

d. Use the result of Part c and the fact that $y = \log x$ when $10^y = x$ to find an approximate value of the quotient 333,000,000,000,000 ÷ 268,000,000,000. Then explain what the result tells about the ratio of the volume of Jupiter to the volume of Earth.

e. Explain why 333,000,000,000,000 ÷ 268,000,000,000 = 333,000 ÷ 268. Compare this result to your answer in Part d.

f. The average radius of Jupiter is about 43,000 miles and that of Earth is about 4,000 miles. What do those figures imply about the ratio of the volumes of the two planets, and how does your answer compare to that in Part d?

32 Explain how the diagram below gives a visual "proof without words" of the fact that the product of two odd numbers is always an odd number.

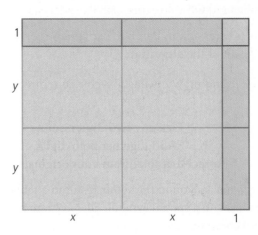

33 In Investigation 2, you completed a proof of the Law of Cosines for the case of an acute angle in a triangle.

   a. Why is the Law of Cosines true for the case where $\angle C$ is a right angle?

   b. Modify the diagram and its labeling on page 60 for the case where $\angle C$ is an obtuse angle. How would you modify the argument in the text for this case?

REVIEW

34 Use algebraic reasoning to solve the following equations for x.

   a. $12x - 9 = -7x + 6$ **b.** $3x^2 + 7 = 55$

   c. $3x^2 + 7x = 0$ **d.** $x^2 + 5x - 14 = 0$

   e. $4(x - 5) = 5 + 7x$ **f.** $x^4 - 3x^2 - 10 = 0$

35 Use properties of exponents to write each of the following expressions in equivalent simplest form, using only positive exponents.

   a. $p^3 p^7$ **b.** $(p^3)^2$

   c. $(p^2 q^3)(p^{-1} q^2)$ **d.** $\dfrac{p^4 q^3}{q^2 p^2}$

   e. $\left(\dfrac{p^2 q}{pq^3}\right)^4$ **f.** $\left(\dfrac{p^{\frac{4}{3}} q^2}{p^2 q^{\frac{1}{3}}}\right)^3$

36 A is the point with coordinates $(6, 3)$, B is the point $(-4, 8)$, and O is the origin $(0, 0)$.

   a. Plot these points in the coordinate plane. Draw $\overleftrightarrow{OA}$ and $\overleftrightarrow{OB}$. How do these two lines appear to be related?

   b. Justify your conjecture in Part a in two different ways.

      i. using slopes

      ii. using the distance formula

37 In the diagram at the right, $\angle C$ is a right angle. Use information given to find the requested trigonometric function values. Express your answers as fractions in simplest form.

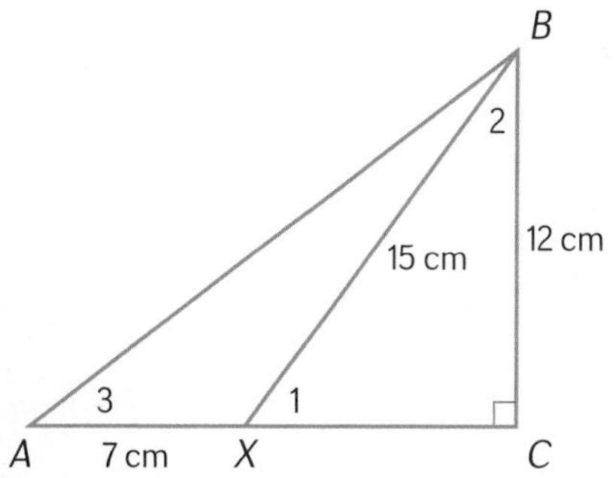

a. $\sin \angle 1$ **b.** $\cos \angle 1$

c. $\cos \angle 2$ **d.** $\sin \angle 2$

e. $\sin \angle 3$ **f.** $\cos \angle 3$

g. $\tan \angle 1$ **h.** $\tan \angle 2$

i. $\tan \angle 3$

38 Suppose that in the diagram of Task 37, $m\angle 1$ is 53° and $m\angle 3$ is 37°. Find the measures of these angles:

a. $\angle 2$

b. $\angle AXB$

c. $\angle XBA$

39 Find algebraic rules for functions with these properties.

a. A linear function $f(x)$ for which $f(0) = -5$ and $f(4) = 7$

b. A linear function $g(x)$ for which $g(3) = -5$ and $g(-5) = -1$

c. A quadratic function $h(x)$ for which $h(0) = 4$, $h(4) = 0$, and $h(1) = 0$

d. An exponential function $j(x)$ for which $j(0) = 5$ and $j(1) = 15$

40 Keisha has created a game in which the player rolls two dice and must roll a sum of 3, 4, or 5 in order to move a gameboard piece into play.

a. What is the probability that a person will be able to move a piece into play on the first roll?

b. While testing the game, Keisha rolled the dice until she got a sum of 3, 4, or 5 and kept track of the number of rolls she needed. The results of her 100 trials are shown in the histogram at the right. Using the data in the histogram, calculate the mean number of rolls needed to get a sum of 3, 4, or 5.

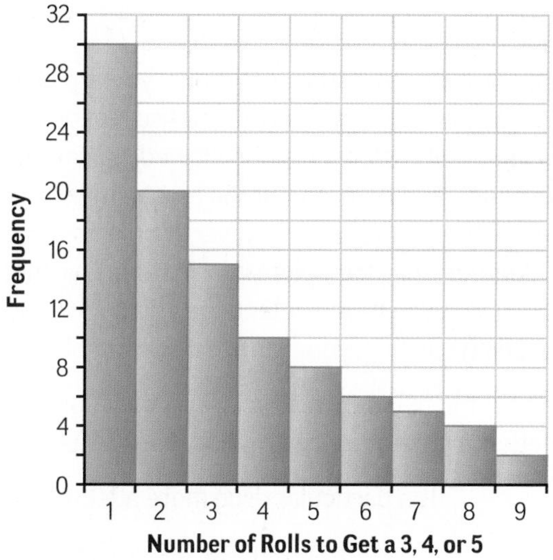

c. Using Keisha's data, estimate the probability that a player would have to roll more than five times before getting a 3, 4, or 5.

d. Using probability formulas and rules, rather than Keisha's data, find the probability that a player would first get a 3, 4, or 5 on one of the first three rolls of the dice.

Statistical Reasoning

In the first three lessons of this unit, you explored how mathematical statements can be proved using algebraic and geometric reasoning. The reasoning was deductive—it involved showing that the conclusion followed logically from definitions, accepted postulates or properties, and theorems. If the reasoning was correct, you could be certain about the truth of the conclusion. However, problems in life that involve data also involve variability, randomness, or incomplete information. This means that you can never be completely certain that you have come to the correct conclusion. In such situations, statistical reasoning can help you come to a reasonable conclusion, backed by convincing evidence.

Advances in fields such as science, education, and medicine depend on determining whether a change in treatment causes better results.

Is a particular drug more effective against acne than doing nothing?

Does a sprain heal faster if the part is exercised or if it is rested?

Does listening to Mozart make people smarter?

You can get convincing evidence about such questions only by carrying out properly designed experiments.

With your class, perform an experiment to decide whether students can, on average, stack more pennies with their dominant hand (the hand with which they write) or with their nondominant hand. First, agree on the rules. Can you touch a penny again after you have stacked it? Can your elbow rest on the table? Then, randomly divide your class into two groups of about equal size. The students in one group will stack pennies using only their dominant hand. The students in the other group will use their nondominant hand. Each student will count the number of pennies he or she stacks before a penny falls.

THINK ABOUT THIS SITUATION

Think about the design of this penny-stacking experiment and how you would interpret the results.

a Why is it important to agree on rules (a *protocol*) for how you must stack the pennies?

b Why is it important to divide your class into the two groups at random?

c Complete this experiment and then organize your data using plots and summary statistics. (Save the data, as you will need them in Investigation 2.)

d What can you conclude? Have you *proved* that, for students your age, one hand tends to be better than the other in stacking pennies? Why or why not?

In this lesson, you will learn how to design a good experiment and how to use statistical reasoning to decide whether one treatment (using the dominant hand) causes a better result (number of pennies stacked) than another treatment (using the nondominant hand). You will also explore reasoning from a sample to a population as in the case of predicting the results of an election from polling a sample of voters.

INVESTIGATION 1

Design of Experiments

Statistical reasoning involves these steps.

- Formulate a question that can be answered with data.

- Collect data.

- Display and summarize the data.

- Interpret the results and make generalizations.

In earlier statistics units of *Core-Plus Mathematics*, you concentrated mostly on the third and fourth steps, plotting data that had already been collected, computing summary statistics, and interpreting the results. In this investigation, you will focus on the second step. As you work on the problems in this investigation, look for answers to this question:

How can you design an experiment that provides convincing evidence that one treatment causes a different response than another treatment?

1. A common science experiment attempts to determine if mung bean seeds that are given a gentle zap in a microwave oven are more likely to sprout than mung bean seeds that are not given a zap.

Mung beans that were zapped in a microwave oven

a. In such an experiment, what are the **treatments** (conditions you want to compare)? What is the **response variable** (outcome you are measuring)?

b. For his experiment, Carlos zapped 10 mung bean seeds and 8 sprouted. Explain why Carlos should not conclude that mung bean seeds zapped in a microwave are more likely to sprout than if they had not been zapped.

c. For her experiment, Mia took 20 mung bean seeds, picked out 10 that looked healthy and zapped them. Of the 10 that were zapped, 8 sprouted. Of the 10 that were not zapped, 3 sprouted. Explain why Mia should not conclude that mung bean seeds zapped in a microwave are more likely to sprout than if they had not been zapped.

d. For her experiment, Julia took 4 mung bean seeds, selected 2 at random to be zapped, and zapped those 2. Both seeds that were zapped sprouted. The 2 seeds that were not zapped did not sprout. Explain why Julia should not conclude that mung bean seeds zapped in a microwave are more likely to sprout than if they had not been zapped.

e. Design an experiment to determine if mung bean seeds are more likely to sprout if they are zapped in a microwave.

2 In a typical experiment, two or more treatments are randomly assigned to an available group of people (or animals, plants, or objects) called **subjects**. The purpose of an experiment is to establish cause and effect. Does one treatment cause a different response than the other treatment? A well-designed experiment must have three characteristics.

- *Random assignment*: Treatments are assigned randomly to the subjects.

- *Sufficient number of subjects*: Subjects will vary in their responses, even when they are treated alike. If there are not enough subjects, this variability within each treatment may obscure any difference between the effects of the treatments. Deciding how many subjects are sufficient is one of the more difficult tasks that statisticians do.

- *Comparison group or control group*: Either the group that gets the treatment is compared to a group that gets no treatment (a **control group**) or two groups that get different treatments are compared.

a. Which characteristic(s) of a well-designed experiment was (were) missing in Problem 1 in the mung bean seed study of:

   i. Carlos?   **ii.** Mia?   **iii.** Julia?

b. Which characteristics of a well-designed experiment, if any, were missing from your penny-stacking experiment?

c. What can go wrong if treatments are not assigned randomly to the subjects?

3 In 1954, a huge medical experiment was carried out to test whether a newly developed vaccine by Jonas Salk was effective in preventing polio. Over 400,000 children participated in the portion of the study described here. Children were randomly assigned to one of two treatments. One group received a *placebo* (an injection that looked—and felt!—like a regular immunization but contained only salt water). The other group received an injection of the Salk vaccine.

(**Source:** Paul Meier, "The Biggest Public Health Experiment Ever," in *Statistics: A Guide to the Unknown*, 3rd ed. Edited by Judith Tanur, *et al.* Pacific Grove, CA: Wadsworth and Brooks/ Cole Advanced Books and Software, 1989, pp. 3–14 or at www.stat.luc.edu/StatisticsfortheSciences/ MeierPolio.htm.)

Jonas Salk examining vials of polio vaccine

a. What are the treatments in the Salk experiment? What is the response variable?

b. Did the test of the Salk vaccine have the three characteristics of a well-designed experiment?

4 Many difficulties in testing the Salk vaccine had been anticipated. Which of the three characteristics of a well-designed experiment helped overcome each difficulty described below? Explain.

a. The incidence of polio was very low, even without immunization.

b. The vaccine was not expected to be 100% effective.

c. One possible approach would have been to immunize all children in the study and compare the incidence of polio to that of children the same age the previous year. However, the incidence of polio varied widely from year to year.

d. One possible experimental design would have been to let parents decide whether their child was vaccinated and compare the rates of polio of the vaccinated and unvaccinated children. In the United States, polio was primarily a disease of children from middle- and upper-income families and so those children's parents were especially anxious to get them vaccinated.

5 Many studies have shown that people tend to do better when they are given special attention or when they believe they are getting competent medical care. This is called the **placebo effect**. Even people with post-surgical pain report less discomfort if they are given a pill that is actually a placebo (a pill containing no medicine) but which they believe contains a painkiller.

One way to control for the placebo effect is to make the experiment **single blind**, the person receiving the treatment does not know which treatment he or she is getting. That is, subjects in both treatment groups appear to be treated exactly the same way.

In a **double-blind** experiment, neither the subject nor the person who evaluates how well the treatment works knows which treatment the subject received.

a. The Salk experiment was double blind. One reason this was necessary was because the diagnosis of polio is not clear-cut. Cases that cause paralysis are obvious, but they are the exception. Sometimes polio looks like a bad cold, and so professional judgment is needed. How might a doctor's knowledge of whether or not a child had been immunized affect his or her diagnosis? How might this lead to the wrong conclusion about how well the vaccine works?

b. Could you make the penny-stacking experiment in the Think About This Situation single blind? Double blind? Explain.

6 A **lurking variable** helps to explain the association between the treatments and the response but is not the explanation that the study was designed to test. Treatments are assigned randomly to subjects to equalize the effects of possible lurking variables among the treatment groups as much as possible. Analyze each of the following reports of studies with particular attention to possible lurking variables.

a. Researchers from the Minnesota Antibiotic Resistance Collaborative reported an attempt to deal with the problem that bacteria are becoming resistant to antibiotics. One reason for increasing resistance is that some people want antibiotics when they have a cold, even though cold viruses do not respond to antibiotics.

   Five medical clinics distributed colorful kits containing Tylenol®, decongestant, cough syrup, lozenges, powdered chicken soup, and a tea bag to patients with cold symptoms. At five other medical clinics, patients with similar symptoms were not given these kits. Patients with colds who visited clinics that made the kits available were less likely to fill prescriptions for antibiotics in the next three days than patients with colds who visited clinics where the kits were not available. This study involved nearly 11,000 patients. (**Source:** sciencedaily.com/releases/ 2004/02/040229231510.htm; *The Los Angeles Times,* March 8, 2004.)

   i. What are the treatments in the study? What is the response variable?

   ii. Why is this not a well-designed experiment? How could you improve it?

   iii. What lurking variable might account for the difference in response?

b. Researchers supplied 238 New York City households with hand-washing soaps, laundry detergents, and kitchen cleansers. Half of the households, selected at random, were given antibacterial products, and the other half received products that were identically packaged but without the antibacterial ingredient.

   The participants were asked weekly about any disease in the household. The researchers found no difference in frequency of infectious disease symptoms over one year. (**Source:** Elaine L. Larson *et al.* "Effect of Antibacterial Home Cleaning and Handwashing Products on Infectious Disease Symptoms," *Annals of Internal Medicine,* Vol. 140, No. 5, March 2, 2004.)

   i. Does this study have the three characteristics of a well-designed experiment?

   ii. This study was double blind. Explain how that must have been done.

   iii. Suppose that instead of assigning the treatments at random to the households, the researchers simply compare the frequency of infectious disease symptoms over a year in households that use antibacterial products and those that do not. Describe lurking variables that might invalidate the conclusion of the study.

c. An article on washingtonpost.com entitled "In AP-vs.-IB Debate, A Win for the Students" reports on a study by the National Center for Educational Accountability. The study found that "even students who fail AP examinations in high school are twice as likely to graduate from college in five years as students who never try AP." This study followed 78,079 students in Texas.

    i. What are the treatments? What is the response variable?

    ii. Do you think that the conclusion came from a well-designed experiment?

    iii. What lurking variables could account for the difference in responses for the two groups?

    iv. Can you design an experiment to establish that taking an AP course, even if you fail the exam, means you are more likely to graduate from college in five years?

SUMMARIZE THE MATHEMATICS

In this investigation, you examined characteristics of well-designed experiments.

a What are the three characteristics of a well-designed experiment? Why is each necessary?

b Why is it desirable for an experiment to be single or double blind?

c What is the placebo effect? How can you account for it when designing an experiment?

Be prepared to share your ideas with the class.

 CHECK YOUR UNDERSTANDING

To find out if students do better on exams where the easier problems come first, a teacher wrote two versions of an exam. One had the more difficult problems first. The second contained the same problems only with the easier ones first. He gave the first version to his first-period class and the second version to his second-period class. He will compare the scores on the first version with the scores on the second version.

a. What are the treatments? What is the response variable?

b. Does this study have the three characteristics of a well-designed experiment?

c. Is this study single blind? Is it double blind?

d. Name at least one lurking variable for this study.

e. Describe how the teacher could improve the design of his study.

By Chance or from Cause?

In the previous investigation, you learned the importance of randomly assigning treatments to the subjects in an experiment. With randomization, you trust that any initial differences among the subjects get spread out fairly evenly between the two treatment groups. Consequently, you feel justified in concluding that any large difference in response between the two groups is due to the effect of the treatments. In this investigation, you will learn a technique for making *inferences* (drawing valid conclusions) from an experiment.

As you work on the problems in this investigation, look for answers to this question:

> *How can you decide whether the difference in the mean responses*
> *for the two groups in your experiment happened just by chance*
> *or was caused by the treatments?*

1 To illustrate what happens when treatments are equally effective, your class will perform a simple but well-designed experiment to determine whether a calculator helps students perform better, on average, on a test about factoring numbers into primes.

- Each student should write his or her name on a small card.

- Divide your class at random into two groups of about equal size by having one person shuffle the name cards and deal them alternatively into two piles. One group will get the "calculator" treatment and the other group will get the "pencil-and-paper" treatment.

- Using the treatment assigned to you, answer the questions on the test provided by your teacher. If you do not know an answer for sure, make the best guess you can so that you have an answer for all eight questions.

- Grade your test from the answers provided by your teacher. Report the number of questions that you got correct as your response.

a. Was a calculator any help on this test? Should the treatment you received make any difference in your response?

b. Make a list of the responses for each treatment. Then, for each treatment, prepare a plot of the number correct and calculate appropriate summary statistics.

c. To the nearest hundredth, compute the difference: *mean of calculator group − mean of pencil-and-paper group*. Is there a difference in the mean response from the two treatment groups?

d. Why did you get a difference that is not zero even though which treatment you received should not matter?

2 Now, explore what the size of the difference in the mean responses in Problem 1 Part c would have been if the randomization had placed different students in the two treatment groups, but each student's response was exactly the same.

a. On each name card, record the number of correct responses for that student.

    i. Divide your class at random once again by shuffling the name cards and dealing them into two piles representing the calculator treatment and the pencil-and-paper treatment. To the nearest hundredth, what is the difference: *mean of calculator group − mean of pencil-and-paper group*?

    ii. Construct an approximate distribution of the possible differences of means by repeating the process above. Do this, sharing the work, until you have generated a total of at least 50 differences.

b. Where is your distribution centered? Why does this make sense?

c. On your distribution, locate the difference in the means from the actual experiment (Problem 1 Part c). Would this difference be unusually extreme if the treatments made no difference in the responses? In other words, is the probability small (say, less than 5%) of obtaining a difference this large or larger in absolute value by chance alone?

d. If so, what can you conclude? If not, what can you conclude?

3 Researchers at the Smell & Taste Foundation were interested in the following question.

Can pleasant aromas improve ability to complete a task?

They randomly assigned volunteers to wear an unscented mask or to wear a floral-scented mask. The subjects then completed two pencil-and-paper mazes. The time to complete the two mazes was recorded. Data were recorded separately for smokers and nonsmokers as smoking affects the sense of smell. Results for the 13 nonsmokers are given in the tables below.

Unscented Mask (in seconds)	Scented Mask (in seconds)
38.4	38.0
72.5	35.0
82.8	60.1
50.4	44.3
32.8	47.9
40.9	46.2
56.3	

Source: Hirsch, A. R., and Johnston, L. H. "Odors and Learning," Smell & Taste Treatment and Research Foundation, Chicago. DASL library, lib.stat.cmu.edu/DASL/Datafiles/Scents.html

a. Does this study have the three characteristics of a well-designed experiment?

b. Why did some subjects have to wear an unscented mask?

c. Compute summary statistics and make plots to help you decide whether one type of mask results in shorter times to complete the mazes.

d. From your summaries, you found that people who wore the unscented mask took an average of 8.19 seconds longer to complete the mazes than people who wore the scented mask. Do you think that the scented mask causes a shorter mean time to complete the mazes or do you think that the difference of 8.19 seconds is no more extreme than you would expect just by chance?

4 Assume that the type of mask in Problem 3 makes absolutely no difference in how long it takes a person to complete the mazes.

a. Why would you expect there to be a nonzero difference anyway in the mean times for the two treatments?

b. How long should it take the people who wore the unscented masks to complete the mazes if they had worn scented masks instead?

c. Now suppose that you are the researcher beginning this experiment and need to pick 7 of the 13 nonsmokers to wear the unscented mask.

   • Write, on identical slips of paper, the 13 maze-completing times given in the tables on page 82.

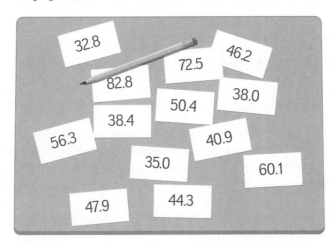

   • Mix them up well. Draw 7 of them to represent the 7 people who will wear the unscented mask. Compute their mean time. Compute the mean time of the remaining 6 people.

   • Subtract: *unscented mean − scented mean*.

   Compare your result with that of others. Why does your finding make sense?

d. A faster way to construct an approximate distribution from many randomizations is to use statistical software to randomly select groups. The "Randomization Distribution" feature of *CPMP-Tools* created this approximate distribution of possible differences

(*unscented mean − scented mean*).

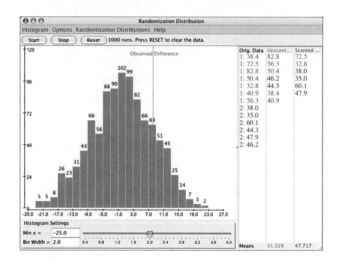

The display shows 1,000 runs of randomly assigning the 13 times required to complete the mazes to either the unscented mask treatment (7 times) or the scented mask treatment (6 times). Out of these 1,000 differences, about how many times is the difference at least as extreme as 8.19 seconds? If the scent made absolutely no difference, estimate the probability that you would get a difference as extreme as 8.19 seconds just by random chance. This probability is called *p*-**value**.

e. Which of the following is the best conclusion?

- The difference of 8.19 from the actual experiment is extreme, so you should abandon your supposition that the scent made no difference.

- It is quite plausible that the scent does not affect the time to complete the maze. In other words, a difference of 8.19 seconds would not be unusual if the scent made no difference and you randomly divide the subjects into two groups.

5 Look back at the results of your penny-stacking experiment in the Think About This Situation on page 75.

a. Compute the difference for your class: *mean number of pennies stacked by those using their dominant hand − mean for those using their nondominant hand*. Suppose that the hand people use makes absolutely no difference in how many pennies they can stack. Why, then, would there almost always be a nonzero difference in every class that does this experiment?

b. If which hand you used made absolutely no difference, how many pennies would you have been able to stack if you had used your other hand?

c. Use the "Randomization Distribution" feature of *CPMP-Tools* to create an approximate distribution of the possible differences (*dominant hand mean − nondominant hand mean*). Run at least 500 random assignments, continuing until the shape of the distribution stabilizes.

d. How many times did you get a simulated difference as extreme or more extreme than you got in your experiment? If which hand used makes no difference, estimate the probability that you would get a difference as extreme or more extreme than you got in your experiment just by random chance.

e. What should you conclude?

The reasoning that you followed in Problems 1–5 to decide whether the results of an experiment provide convincing evidence that different treatments cause a different mean response is called a **randomization test**. The steps below summarize this reasoning.

Step 1. Assume that which treatment each subject gets makes absolutely no difference in his or her response. In other words, assume the subjects in the experiment would give the same response no matter which treatment they receive. In the next two steps, you will see if this assumption is plausible.

Step 2. Simulate the experiment.

- Write the name of each subject along with his or her response on a card.

- Randomly divide the cards into two treatment groups.

- Compute the mean for each treatment group.

- Find the difference of these means.

- Repeat this many times until you can see the shape of the distribution of differences.

Step 3. Locate the difference from the actual experiment and its opposite on the distribution you generated in Step 2. Estimate the *p*-value. This is the probability of getting, by chance alone, a difference in means as large or larger, in absolute value, than the difference from the actual experiment.

Step 4. If the *p*-value is less than 5%, conclude that the results are **statistically significant**. That is, you have statistically significant evidence that the treatments caused the difference in the mean response. If the *p*-value is 5% or greater, conclude that your original assumption was plausible. The difference can be reasonably attributed solely to the particular random assignment of treatments to subjects.

6 Chrysanthemums with long stems are likely to have smaller flowers than chrysanthemums with shorter stems. An experiment was conducted at the University of Florida to compare growth inhibitors designed to reduce the length of the stems, and so increase the size of the flowers. Growth inhibitor A was given to 10 randomly selected plants. Growth inhibitor B was given to the remaining 10 plants. The plants were grown under nearly identical conditions, except for the growth inhibitor used. The table below gives the amount of growth during the subsequent 10 weeks.

Growth by Plants Given A (in cm)	Growth by Plants Given B (in cm)
46	51
41.5	55
45	57
44	57.5
41.5	53
50	45.5
45	53
43	54.5
44	55.5
30.5	45.5

Source: Ann E. Watkins, Richard L. Scheaffer, and George W. Cobb, *Statistics in Action*, 2nd Ed. Key Curriculum Press, 2008, p. 802.

a. Does this experiment have the three characteristics of a well-designed experiment?

b. Examine the following summary statistics and plots. Which growth inhibitor treatment appears to be better?

Treatment	Mean	Standard Deviation	Number of Plants
A	43.05	5.04	10
B	52.75	4.28	10

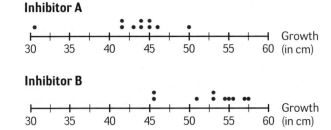

c. Describe how to use a randomization test to decide whether, on average, one growth inhibitor works better than the other.

d. Use the "Randomization Distribution" feature of *CPMP-Tools* to create an approximate distribution of possible differences (*growth inhibitor A mean − growth inhibitor B mean*). Run at least 500 random assignments.

e. If the type of growth inhibitor makes no difference, what is your estimate of the probability of getting a difference at least as extreme as the difference from the actual experiment?

f. What is your conclusion? Is the difference statistically significant?

SUMMARIZE THE MATHEMATICS

In this investigation, you explored the randomization test. This test is one method of determining whether a difference between two treatment groups can be reasonably attributed to the random assignment of treatments to subjects or whether you should believe that the treatments caused the difference.

a Explain why this statement is true: Even if the response for each subject would be the same no matter which treatment he or she receives, there is almost always a nonzero difference in the means of the actual responses from the two treatments.

b What does it mean if the results of an experiment are called "statistically significant"?

c Explain the reasoning behind the steps of a randomization test.

d Explain how this statement applies to the reasoning in this unit: Statistical reasoning is different from mathematical proof because in statistics, you can never say you are certain.

Be prepared to share your responses and thinking with the class.

CHECK YOUR UNDERSTANDING

Forty-nine volunteer college students were randomly assigned to two treatments. Twenty-five students were told that they would view a video of a teacher who other students thought was "charismatic": lively, stimulating, and encouraging. The remaining twenty-four students were told that the instructor they would view was thought to be "punitive": not helpful, not interested in students, and a hard grader. Then all students watched the same twenty-minute lecture given by the same instructor. Following the lecture, subjects rated the lecturer. The students' summary ratings are given below. Higher ratings are better.

Charismatic: $1\frac{2}{3}$, 3, $1\frac{2}{3}$, $2\frac{1}{3}$, 4, $2\frac{1}{3}$, 2, $2\frac{2}{3}$, $2\frac{2}{3}$, $2\frac{1}{3}$, $3\frac{1}{3}$, $2\frac{1}{3}$,

$2\frac{1}{3}$, $2\frac{2}{3}$, 3, $2\frac{2}{3}$, 3, 2, $2\frac{1}{3}$, $2\frac{2}{3}$, 3, $3\frac{1}{3}$, 3, $2\frac{2}{3}$, $2\frac{1}{3}$

Punitive: $2\frac{2}{3}$, 2, 2, $1\frac{1}{3}$, $1\frac{2}{3}$, $2\frac{1}{3}$, $2\frac{2}{3}$, 2, $1\frac{2}{3}$, $1\frac{1}{3}$, $2\frac{1}{3}$, 2,

$2\frac{1}{3}$, $2\frac{1}{3}$, $2\frac{1}{3}$, $2\frac{1}{3}$, $1\frac{2}{3}$, $3\frac{2}{3}$, $2\frac{1}{3}$, $2\frac{2}{3}$, 2, $2\frac{1}{3}$, $2\frac{1}{3}$, $3\frac{1}{3}$

Summary statistics and plots are on page 88.

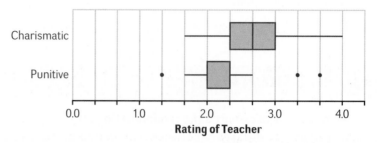

Treatment	Mean	Standard Deviation	Number of Students
Charismatic	2.61	0.53	25
Punitive	2.24	0.54	24

Source: www.ruf.rice.edu/%7Elane/case_studies/instructor_reputation/index.html
(Their source: Towler, Annette., & Dipboye, R. L. The effect of instructor reputation and
need for cognition on student behavior—poster presented at American Psychological
Society conference, May 1998.)

a. What are the two treatments?

b. From the box plots and summary statistics, does it look like the two treatments cause different mean responses? Explain.

c. Describe how to perform one run for a randomization test to decide whether the two different treatments result in different mean ratings.

d. The randomization distribution below shows *mean charismatic − mean punitive* for 100 runs. Use the histogram to estimate the probability of getting a difference in means at least as extreme as that from the actual experiment if the treatment makes no difference.

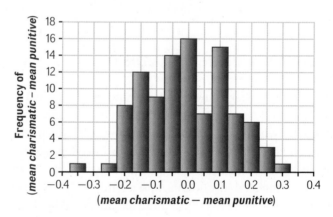

e. What is your conclusion? Is the difference in the two means statistically significant?

Statistical Studies

Experiments are one of the three major types of statistical studies. The other two are *sample surveys* and *observational studies*. As you work on the problems in this investigation, look for answers to these questions:

What are the differences between sample surveys, experiments, and observational studies?

What kind of conclusions can be made from each?

The three main types of statistical studies are described below.

- *sample survey or poll*: You observe a random sample in order to estimate a characteristic of the larger population from which the sample was taken. Getting a **random sample of size** n is equivalent to writing the name of every member of the population on a card, mixing the cards well, and drawing n cards.

- *experiment*: You randomly assign two (or more) treatments to the available subjects in order to see which treatment is the most effective.

- *observational study*: The conditions you want to compare come already built into the subjects that are observed. Typically, no randomization is involved.

1 Suppose you want to investigate the effects of exercise on the blood pressure of students in your school. You have thought about three different study designs. Classify each design as a sample survey, an experiment, or an observational study.

Study 1: You ask for volunteers from the students in your school and get 30 students willing to participate in your study. You randomly divide them into two groups of 15 students. You ask one group not to exercise at all for the next week, and you ask the other group to do at least 30 minutes of exercise each day. At the end of the week, you find that everyone complied with your instructions. You then take each student's blood pressure. You find that the mean blood pressure of the students who exercised is lower than the mean blood pressure of the students who did not exercise.

Study 2: You get a list of all students in your school and use a random digit table to select 30 of them for your study. You take these students' blood pressure and then have them fill out a questionnaire about how much exercise they get. You divide them into those who exercise a lot and those who exercise less. You find that the mean blood pressure of the students who exercise more is lower than the mean blood pressure of the students who exercise less.

Study 3: You discover that the nurse in the health office at your school has taken the blood pressure of 157 students who have visited the health office over the past year for a variety of reasons. In some cases, they felt sick; and in other cases, they had to turn in routine paperwork. You get the names of these students and have them fill out a questionnaire about how much exercise they get. You find that the mean blood pressure of the students who exercise more is lower than the mean blood pressure of the students who exercise less.

2 In each study in Problem 1, there was an association between amount of exercise and blood pressure. Assume that in each case the difference in mean blood pressure was statistically significant. Answer the following questions for each study in Problem 1.

a. Is it reasonable to conclude that it was the exercise that caused the lower blood pressure? Explain your thinking.

b. Can you generalize the results of this study to all of the students in your school? Explain your thinking.

c. Exactly what can you conclude from this study?

3 Refer to Problem 6 Part c on page 80 about students who take AP examinations in high school.

a. What type of study is this?

b. State the conclusion that can be drawn.

4 You want to predict the winner of the election for mayor of your town. You do this by first creating a list of all household phone numbers in your town. You then phone households using random digit dialing, calling back if no one answers. A voter is selected at random from each household and interviewed about whether he or she intends to vote and for whom.

a. What type of study is this?

b. Explain why the households cannot be selected from your town's phone book.

c. Are all voters in your town equally likely to be in the sample? Explain.

SUMMARIZE THE MATHEMATICS

In this investigation, you examined the three main types of statistical studies.

a What is a random sample?

b How is randomization used in a sample survey? In an experiment? In an observational study?

c What kind of conclusion can you draw from a sample survey? From an experiment? From an observational study?

Be prepared to share your responses with the class.

✔ CHECK YOUR UNDERSTANDING

The British Doctors Study was one of the earliest studies to establish a link between smoking and lung cancer. In 1951, all male doctors in the United Kingdom were contacted. About two-thirds, or 34,439 doctors, agreed to participate. Eventually, researchers found that the doctors who smoked were more likely to get lung cancer than doctors who did not smoke. The difference was statistically significant. (**Source:** Richard Doll, *et al.* "Mortality in Relation to Smoking: 50 Years' Observations on Male British Doctors," *British Medical Journal,* Vol. 22, June 2004, www.bmj.com/content/328/7455/1519)

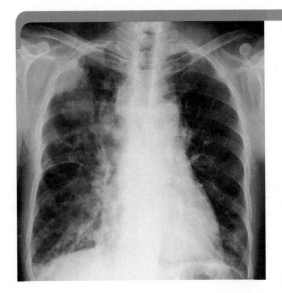

a. What type of study is this?

b. Can you conclude from this study that smoking causes lung cancer? Explain your thinking.

c. Can you generalize the results of this study to some larger population? Explain your thinking.

d. Describe exactly what you can conclude from this study.

1 Suppose that the manufacturer of a cough medicine wants to conduct a randomized, double-blind experiment to determine if adding a new ingredient results in a reduction in the mean number of coughs per hour. Fifty adult volunteers with persistent coughs are available. Describe how the manufacturer should conduct this experiment.

2 There can be many factors that contribute to automobile accidents and the nature of driver injuries.

The table below gives the overall driver death rate for various weights of cars. There is a strong association between the weight of a car and driver death rate. However, there are many advantages to lighter cars. They tend to be easier to park, to get better gas mileage, to pollute less, to have cheaper insurance, and to be less expensive to buy. So, it is worth carefully considering whether they are really less safe.

Weight of Car (in lbs)	Overall Driver Death Rate (per million registered vehicle years)
2,500 or less	71
2,501–3,000	70
3,001–3,500	51
3,501–4,000	47
4,001–4,500	41

Source: *Dying in a Crash*, Insurance Institute for Highway Safety, Status Report special issue: June 9, 2011.

a. Describe the association between the weight of a car and driver death rate.

b. Suppose that 4,000,000 of a certain model of car are registered, and there were 467 driver deaths over a three-year period. What would be the overall driver death rate per million registered vehicle years for that particular model of car?

c. Name a possible lurking variable. Describe how it could account for the association between the weight of a car and driver death rate.

d. How could you design an experiment to provide convincing evidence that lighter cars cause a larger overall driver death rate than heavier cars? Would this be ethical?

3 Psychrotrophic bacteria cause meat to spoil. Six beef steaks were randomly assigned to be packaged using commercial plastic wrap or to be vacuum packaged. The following table gives the logarithm (log) of the number of psychrotrophic bacteria per square centimeter on the meat after nine days of storage at controlled temperature.

Commercial Plastic Wrap log (count/cm²)	Vacuum Packaged log (count/cm²)
7.66	5.26
6.98	5.44
7.80	5.80

Source: Robert O. Kuehl, *Statistical Principles of Research Design and Analysis*, Duxbury Press, Belmont, CA, 1994, p. 31. Original source: B. Nichols, *Comparison of Grain-Fed and Grass-Fed Beef for Quality Changes When Packaged in Various Gas Atmospheres and Vacuum*, M.S. thesis, Department of Animal Science, University of Arizona, 1980.

a. How many bacteria per square centimeter were on the steak with a log of 7.66?

b. Does this study have the three characteristics of a well-designed experiment? What else would you like to know about how it was conducted?

c. What is the difference *mean (log) response for commercial plastic wrap − mean (log) response for vacuum packaged*?

d. Describe how to conduct a randomization test to decide whether the different packaging causes different mean numbers of bacteria. Perform 10 runs and add them to a copy of the randomization distribution below, which shows the results of 90 runs.

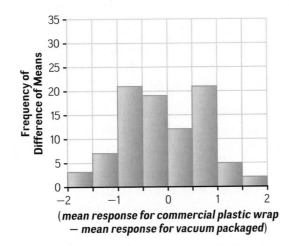

e. Use the randomization distribution to estimate the probability that random assignment alone will give you a difference that is at least as extreme as that from the actual experiment. Do you have statistically significant evidence from this experiment that one type of packaging is better than the other?

4 For a science project, Brian wanted to determine whether eleventh-graders did better when they took a math test in silence or when Mozart was being played. Twenty-six students were randomly divided into the two treatment groups. Part of Brian's results are in the table below.

Mozart (percentage correct)	Silence (percentage correct)
65	44
80	70
72	68
68	58
38	58
58	47
45	54
42	44
58	61
81	61
40	9
41	52
27	30
Mean = 55	**Mean ≈ 50.46**

a. What is the difference *mean response for Mozart group − mean response for silence group*?

b. Describe how to conduct a randomization test to decide whether the different treatments cause different performance on the math test.

c. Use the "Randomization Distribution" feature of your data analysis software to create an approximate distribution of the possible differences (*mean response for Mozart group − mean response for silence group*). Run at least 500 random assignments.

d. Use the randomization distribution to estimate the probability that you could get a difference, just by the random assignment, that is at least as extreme as that from the real experiment.

e. Brian concluded that students who listen to Mozart during a test tend to do better. Do you agree with this conclusion or do you think the difference in means can reasonably be attributed to the random assignment alone?

5 You conduct a test to see if an inexperienced person gets a bigger geyser if he or she drops Mentos® candy into a liter bottle of cola by hand or by using a paper funnel. You find ten friends who have never done this demonstration and are willing to participate. You write "by hand" on five slips of paper and "funnel" on five slips of paper. Each person draws a slip and is shown how to use that method. Then, each person drops Mentos® into his or her bottle of cola using the method assigned. The maximum height of each geyser is measured.

You find that the difference between the mean heights of the geysers produced by your friends who used their hand and your friends who used a paper funnel is not statistically significant.

a. What type of study is this?

b. What are the treatments? What are the subjects?

c. Can you generalize the results of this study to some larger population? Explain your thinking.

d. Describe exactly what you can conclude from this study.

6 Researchers wanted to determine whether social class is related to smoking behavior. They conducted telephone interviews with 1,308 Massachusetts adolescents aged 12 to 17, selected by dialing at random. They found a statistically significant association between whether the adolescent smoked or not and the household income. Adolescents from households with less income were more likely to smoke, and this was true across all ages, for both sexes, for all races, and for all amounts of disposable income the adolescent had. (**Source:** Elpidoforos S. Soteriades and Joseph R. DiFranza. "Parent's Socioeconomic Status, Adolescents' Disposable Income, and Adolescents' Smoking Status in Massachusetts," *Journal of Public Health*, Vol. 93, July 2003, pp. 1155–1160, www.pubmedcentral.nih.gov/articlerender. fcgi?artid=1447926)

a. What type of study is this?

b. Can you conclude from this study that smoking is caused by an adolescent's social class? Can you think of a lurking variable that might be responsible for both?

c. Can you generalize the results of this study to some larger population? Explain your thinking.

d. Describe exactly what you can conclude from this study.

CONNECTIONS

7 In mathematics, if you have one counterexample, it disproves a conjecture. For example, in Lesson 1, you disproved the conjecture that all numbers of the form $n^2 - n + 41$ are prime, where n is a whole number 0, 1, 2, 3, … . To disprove it, all you had to do was find one value of n for which this conjecture is not true.

However, anecdotal evidence, looking at just one counterexample, is the *worst* kind of statistical reasoning. For example, you should not conclude that smoking does not cause lung cancer just because you know someone who smoked all his life and died at age 98 in a traffic accident. Even though not everyone who smokes gets lung cancer, explain in what sense it is valid to say that smoking causes lung cancer.

8 Look back at Applications Task 3. Because the sample sizes are so small, you can list all possible randomizations.

a. List all 20 of the possible selections of three of the six steaks to get the commercial plastic wrap and three to get vacuum packaging. Assuming that the type of packaging did not affect the response, compute the 20 possible differences *mean response for commercial plastic wrap* − *mean response for vacuum packaged*.

b. What is the probability that you get a difference just through a random assignment of steaks to packaging that is as extreme as that from the real experiment?

c. Do you have statistically significant evidence from this experiment that one type of packaging is better than the other? Explain.

d. Is your conclusion consistent with that from the randomization test in Applications Task 3?

9 The reasoning of a randomization test in statistics is similar to a special form of indirect reasoning in mathematics called *proof by contradiction*. (See Extensions Tasks 26 and 27 on page 49.) To prove that a statement is not true using *proof by contradiction*, you follow these steps.

- Assume the statement is true.

- Show that this assumption leads to a contradiction.

- Conclude that the statement is not true.

It is a theorem (proven fact) that if a number p is a prime, then when you divide $2^p - 2$ by p, the remainder is 0.

a. Show that this theorem holds for the primes 5, 7, and 11.

b. Use this theorem and the three steps above to prove by contradiction that 21 is not a prime. Begin by assuming that 21 is a prime.

c. Explain how using the reasoning of a randomization test to conclude that two treatments cause different mean responses is similar to proof by contradiction.

10 In the Salk experiment, 82 of the 200,745 children who received the Salk vaccine were diagnosed with polio. Of the 201,229 children who received the placebo, 162 were diagnosed with polio.

  a. What proportion of children who received the Salk vaccine were diagnosed with polio? What proportion who received the placebo were diagnosed with polio?

  b. Do you think the result of this experiment is statistically significant or do you think the difference in the proportions is about the size that you would expect just by chance?

  c. The number of children in the Salk experiment may seem excessive, but it was necessary. In the 1950s, before the Salk vaccine, the rate of polio in the U.S. was about 50 per 100,000 children. Suppose the experiment had "only" 4,000 children in the placebo injection group and 4,000 children in the Salk vaccine group. Also, suppose the vaccine is 50% effective; that is, it eliminates half of the cases of polio.

    i. What is the expected number of children in the placebo group who would get polio?

    ii. What is the expected number of children in the Salk vaccine group who would get polio?

    iii. Does the difference now appear to be statistically significant?

  d. As another part of the Salk experiment, in some schools, parents of second-grade children decided whether the child would be vaccinated or not. Among the 123,605 second-grade children whose parents did not give permission for them to receive the Salk vaccine, there were 66 cases of polio. In other schools, children were selected at random to receive the vaccine or the placebo from among those children whose parents gave permission for them to be in the experiment. In those schools, there were 162 cases of polio among the 201,229 children who received the placebo. How do you explain this result? (You may want to reread Problem 4 on page 78.)

REFLECTIONS

11 You may have seen a child fall and skin his knee, and his mother picks him up and says she will "kiss it and make it well." On what is the mother depending?

12 Mathematical arguments frequently use sentences beginning with the phrase *it follows that* … . They also often involve sentences connected with words like *because, therefore, so,* and *consequently*. Statistical arguments often involve phrases like *it is reasonable to conclude that* … , *there is statistically significant evidence that* … , and *the data strongly suggest that* … . What is it about mathematical reasoning and statistical reasoning that explains the difference in choice of words and phrases?

13 How are the experiments as described in this lesson similar to and different from experiments you have conducted in your previous mathematical studies?

14 Why is it the case that an experiment can never *prove* without any doubt that two treatments cause different responses?

EXTENSIONS

15 Joseph Lister (1827–1912), surgeon at the Glasgow Royal Infirmary, was one of the first to believe in the theory of Louis Pasteur (1822–1895) that germs cause infection. In an early medical experiment, Lister disinfected the operating room with carbolic acid before 40 operations. He did not disinfect the operating room before another 35 operations. Of the 40 operations in which carbolic acid was used, 34 patients lived. Of the 35 operations in which carbolic acid was not used, 19 patients lived.

a. Why did Lister need to have one group of patients for whom he did not disinfect the operating room?

b. What is the difference in the proportion who lived if carbolic acid was used and the proportion who lived if carbolic acid was not used?

c. Does Lister's study provide convincing evidence that disinfecting operating rooms with carbolic acid results in fewer deaths than not disinfecting? Explain your thinking. Is there anything else you would like to know about how Lister conducted his experiment before you decide?

d. To begin a randomization test, you can let 0 represent a response of "died" and 1 represent a response of "lived." Describe how to finish the randomization test.

e. The randomization distribution below shows the results of 200 runs of the randomization. It records *proportion who survived when carbolic acid used — proportion who survived when carbolic acid not used*. Are the results of Lister's experiment statistically significant? Explain.

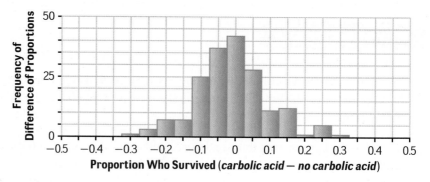

16 In a classic psychology experiment, a group of 17 female college students were told that they would be subjected to some painful electric shocks. A group of 13 female college students were told they would be subjected to some painless electric shocks. The subjects were given the choice of waiting with others or alone. (In fact, no one received any shocks.) Of the 17 students who were told they would get painful shocks, 12 chose to wait with others. Of the 13 students told they would get painless shocks, 4 chose to wait with others. (**Source:** Stanley Schachter. *The Psychology of Affiliation*, Stanford, CA: Stanford University Press, 1959, pp. 44–45.)

a. Describe how to use a randomization test to see if the difference in the proportions of students who choose to wait together is statistically significant. You can let 0 represent a response of waiting alone and 1 represent waiting with others.

b. Conduct 5 runs for your test and add them to a copy of the randomization distribution below, which shows 195 runs.

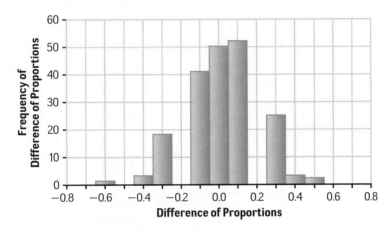

c. Is the difference in the proportions who choose to wait together statistically significant?

d. Why are there no differences between 0.15 and 0.25?

17 A market research experiment was designed to determine how much a subtle color change in white tennis shoes mattered to people who wore them. Twenty people who volunteered to try a new brand of tennis shoe were randomly assigned to get one of two colors of the same shoe. They wore the shoes for a month and then were told they could buy the pair of tennis shoes at a greatly reduced price or return them. Of the 10 people who got color A, 9 decided to buy them. Of the 10 people who got color B, 4 decided to buy them.

a. What kind of a study was this?

b. Describe the logic of a randomization test to determine if the difference is statistically significant and how to do such a test. You can let 0 represent a response of not buying the tennis shoes and 1 represent a response of buying the shoes.

c. Use the "Randomization Distribution" feature of your data analysis software to create an approximate distribution of the differences *mean number deciding to buy color A − mean number deciding to buy color B*. You will need to enter the data in a data sheet. Run at least 1,000 random assignments. What is your conclusion?

REVIEW

18 Draw the graph of each equation.

  a. $3x + 4y = 12$

  b. $x - 5y = 7$

  c. $x + y = 0$

  d. $x = 5$

  e. $y = 3$

  f. $y = |x|$

19 Triangle ABC has vertex matrix $\begin{bmatrix} 0 & 5 & 10 \\ 2 & 6 & 2 \end{bmatrix}$.

  a. Is $\triangle ABC$ an isosceles triangle? Explain your reasoning.

  b. The transformation $(x, y) \rightarrow (-x, y)$ is applied to $\triangle ABC$. Find the vertex matrix for the image $\triangle A'B'C'$ and describe this transformation.

  c. Give a vertex matrix for a triangle that has an area 9 times the area of $\triangle ABC$.

20 The Venn diagram below indicates the number of seniors who have taken Drama and Public Speaking. There are 178 seniors.

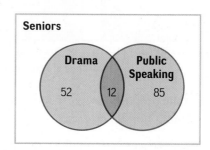

  a. What percentage of the seniors have taken both Drama and Public Speaking?

  b. How many seniors have taken at least one of these courses?

  c. How many seniors have not taken Drama and have not taken Public Speaking? Where should this number be placed in the Venn diagram?

21 Solve each equation.

  a. $x^2 + 10x + 22 = 6$

  b. $2x - 3 = \frac{1}{x}$

  c. $2x^2 - x + 3 = 6$

  d. $-\frac{1}{4}x^2 + \frac{1}{2}x + 2 = 0$

  e. $4x^3 - 12x^2 + 8x = 0$

22 Solve each inequality and graph the solution on a number line.

a. $3(x + 5) \geq 3$

b. $7x + 3 < 11x - 5$

c. $4x + 2(3 - 5x) > -12$

d. $10 - 2x \leq 6(x - 3) + 10$

 23 Achmed kept track of the number of miles n he drove for several consecutive weeks and the amount of gas g in gallons he purchased each week. He found the linear regression equation relating these two variables to be $g = 0.035n + 0.4$.

a. Is the correlation between these two variables positive or negative? Explain your reasoning.

b. Explain the meaning of the slope of the regression equation in terms of these two variables.

c. During one week, Achmed drove 228 miles and bought 7.36 gallons of gas. Find the residual for that week.

 24 Without using a calculator or computer graphing tool, sketch graphs of these quadratic functions. Then, for each, use the graph to determine the number of solutions for the related quadratic equation $f(x) = 0$.

a. $f(x) = x^2 - 4$

b. $f(x) = x^2$

c. $f(x) = x^2 + 4$

d. $f(x) = -x^2 - 5$

e. $f(x) = -x^2$

f. $f(x) = -x^2 + 4$

 25 The height h in feet of a thrown basketball is a function of the time t in seconds since it was released. Suppose that Samuel shoots a free throw for which the height of the ball can be approximated by the function $h(t) = 6 + 40t - 16t^2$.

a. Find the value of $h(2)$ and explain what it tells you about the path of the basketball.

b. How high above the floor was the basketball when it was released?

c. For what times t was the basketball 27 feet above the floor?

d. When was the basketball at its highest point and how high was it?

Looking Back

In this unit, you examined and practiced reasoning principles and strategies that mathematicians and statisticians use to prove or justify their claims. In particular, you learned how to use definitions and assumed geometric facts to prove important properties of perpendicular and parallel lines and line reflections. You learned how to use operations on algebraic expressions to prove patterns in number sequences, prove properties of real numbers, write equivalent expressions, and solve equations. You also learned how statisticians design experiments and use reasoning like the randomization test to determine whether one treatment causes a different outcome than another treatment.

The following tasks will help you to review and apply some of the key ideas involved in geometric, algebraic, and statistical reasoning.

1. **Reasoning about Shapes** In your earlier coursework, you saw that there were at least two different, but equivalent, ways to describe a parallelogram. One of those descriptions was: A **parallelogram** is a quadrilateral with two pairs of opposite sides parallel.

   a. Write two if-then statements that together mean the same thing as that description.

   b. Suppose you are told that in the case of quadrilateral $PQRS$, $\overline{PQ} \parallel \overline{SR}$ and $\overline{QR} \parallel \overline{PS}$. What can you conclude? Which of the two if-then statements in Part a was used in your reasoning?

   c. A definition of trapezoid was given in Lesson 1. A **trapezoid** is a quadrilateral with a pair of opposite sides parallel.

      i. Why is every parallelogram a trapezoid but not every trapezoid a parallelogram?

      ii. Use a Venn diagram to illustrate the relationship of parallelograms, trapezoids, and quadrilaterals.

2 In a parallelogram *ABCD*, angles that share a common side, like ∠*A* and ∠*B*, are called *consecutive angles*. Angles that do not share a common side, like ∠*A* and ∠*C*, are called *opposite angles*.

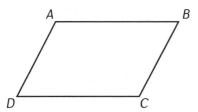

a. Using the description of a parallelogram in Task 1 and properties of parallel lines, prove that in ▱*ABCD*, ∠*A* and ∠*B* are supplementary. Could you use a similar argument to prove other pairs of consecutive angles are supplementary? Explain your reasoning.

b. Use what you proved in Part a to help you prove that in ▱*ABCD*, m∠*A* = m∠*C*. Does the other pair of opposite angles have equal measure? Explain your reasoning.

c. You have seen that there is often more than one correct way to prove a statement. Use the numbered angles in the diagram below to provide a different proof that m∠*A* = m∠*C*.

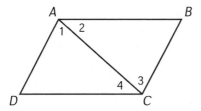

d. Write two statements summarizing what you proved in Parts a and b.

e. Give counterexamples to show the statements in Part d are not necessarily true for trapezoids.

3 Let ▱*WXYZ* be a square and *m* a line in the same plane.

a. Make a sketch of the given information.

b. Find and sketch the reflection image of ▱*WXYZ* across line *m*. Label the images of the vertices *W'*, *X'*, *Y'*, and *Z'*, respectively.

c. Write an argument proving that the reflection image of ▱*WXYZ* is a square.

4 **Reasoning about Patterns in Sequences** In Lesson 3, you discovered and proved an interesting pattern in the sequence of square numbers 1, 4, 9, 16, 25, 36, ... You found that the differences of successive terms in that sequence create the sequence of odd numbers 3, 5, 7, 9, 11, ... The *n*th number in that sequence of differences is given by the expression $2n + 1$.

a. Look again at the sequence of square numbers and study the pattern formed by calculating the differences that begins $9 - 1 = 8$, $16 - 4 = 12$, $25 - 9 = 16$, ...

   i. What are the next several terms in this sequence of differences?

ii. What expression in the form "*an* + *b*" shows how to calculate the *n*th term in the sequence of differences?

iii. Use algebraic reasoning to prove that the expression $(n + 2)^2 - n^2$ is equivalent to the simpler expression you proposed in part ii.

b. Now consider the sequence of differences that begins $16 - 1 = 15$, $25 - 4 = 21$, $36 - 9 = 27$, $49 - 16 = 33$, ...

  i. What are the next several terms in this sequence?

  ii. What simple expression shows how to calculate the *n*th term in this sequence of differences?

  iii. Use algebraic reasoning to prove that the expression proposed in part ii will give the *n*th term in the sequence of differences.

c. **i.** Try to generalize the patterns you examined in Parts a and b. What simple expression will give the *n*th term in the sequence formed by calculating the differences of numbers that are *k* steps apart in the sequence of square numbers?

  ii. Use algebraic reasoning to show that your idea is correct.

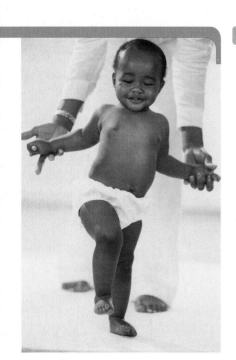

5 Reasoning about Data An experiment was designed to see whether a program of special stepping and foot-placing exercises for 12 minutes each day could speed up the process of babies learning to walk. As part of this study, 12 baby boys were randomly assigned to the special exercise group or to the "exercise control" group. For the control group, parents were told to make sure their infant sons exercised at least 12 minutes per day. But they were not given any special exercises to use, and they were not given any other instructions about exercise.

The researchers recorded the age, in months, when each baby first walked without help.

Age that Baby Boys Learned to Walk	
Special Exercise (in months)	**Exercise Control** (in months)
9	11
9.5	10
9.75	10
10	11.75
13	10.5
9.5	15

Source: Phillip R. Zelazo, Nancy Ann Zelazo, and Sarah Kolb. "Walking in the Newborn," Science 176, 1972, pp. 314–315.

a. What are the treatments? What is the response?

b. Does this study have the three characteristics of a well-designed experiment? Could it have been double blind? Is the placebo effect a possible issue here?

c. What is the difference: *mean walking age for exercise control group − mean walking age for special exercise group*?

d. Suppose that the treatment makes no difference in the age that baby boys learn to walk. Would you expect there to be a nonzero difference in *mean walking age for exercise control − mean walking age for special exercise* for almost any two groups of six baby boys? Explain.

e. Describe how to conduct a randomization test to determine if the different treatments cause a difference in the mean age that baby boys learn to walk. Perform 10 runs and add them to the randomization distribution below, which shows the results of 990 runs.

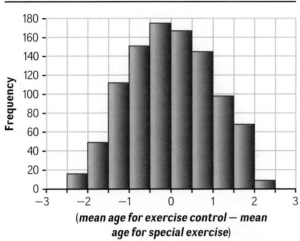

(*mean age for exercise control − mean age for special exercise*)

f. What is your estimate of the probability that just by the random assignment of treatments to subjects, you get a difference that is as least as extreme as that from the real experiment? What is your conclusion about whether you have statistically significant evidence from this experiment that the special exercises are better than just reminding parents to be sure their baby boys exercise at least 12 minutes per day?

6 **Reflecting on Mathematical Practices** In this unit, you have engaged in a number of important mathematical practices that characterize the work of mathematicians and others in mathematics-related fields. Carefully look back at your work in this unit and describe specific examples of where you engaged in the mathematical practices below.

a. Made sense of a problem situation and persevered in solving it

b. Reasoned abstractly and quantitatively, making sense of quantities and their relationships in problem situations

c. Constructed a sound mathematical argument and critiqued the reasoning of others

d. Created a mathematical model for a real-world problem

e. Used appropriate tools such as computer software and calculators to explore and deepen your understanding of ideas and solve problems

f. Attended to precision when communicating with classmates and your teacher or when writing solutions to problems

g. Looked for and made use of mathematical patterns and structure in reasoning

h. Looked for and expressed regularity in repeated calculations or in reasoning patterns.

SUMMARIZE THE MATHEMATICS

In this unit, you learned some basic reasoning principles and strategies that are useful in justifying claims involving concepts of geometry, algebra, and statistics.

a How is *deductive* reasoning different from *inductive* reasoning? Why are both types of reasoning important in mathematics?

b When trying to prove an if-then statement, with what facts do you begin? What do you try to deduce?

c What strategies and geometric properties can you use to prove that two lines are perpendicular? That two lines are parallel? Draw and label sketches to illustrate how those properties are used.

d How would you go about proving a statement like "The sum of the measures of the angles of a trapezoid is 360°"?

e What does it mean to say that line reflections preserve betweenness of points? Distance? Angle measure?

f What overall strategy and rules of algebra can you use to prove that an equation like $2ab + c^2 = (a + b)^2$ implies that $c^2 = a^2 + b^2$?

g If you want to see whether two algebraic expressions like $(n + 2)^2 - n^2$ and $4n + 4$ are equivalent, you could begin by comparing tables and graphs of the functions $y = (n + 2)^2 - n^2$ and $y = 4n + 4$. How would an algebraic proof give different evidence that the expressions are equivalent?

h What are the mathematical conventions about *order of operations* when numerical patterns and relationships are represented with symbolic expressions and equations?

i What are the differences between sample surveys, experiments, and observational studies?

j How do you use a randomization test to determine if the result of an experiment to compare two treatments is statistically significant?

k How is statistical reasoning similar to algebraic and geometric reasoning and how is it different?

Be prepared to share your responses and reasoning with the class.

 CHECK YOUR UNDERSTANDING

Write, in outline form, a summary of the important mathematical concepts and methods developed in this unit. Organize your summary so that it can be used as a quick reference in future units and courses.

UNIT 2

Inequalities and Linear Programming

For some people, athletes and astronauts in particular, selection of a good diet is a carefully planned scientific process. Each person wants maximum performance for minimum cost. The search for an optimum solution is usually constrained by available resources and outcome requirements.

The mathematics needed to solve these and other similar optimization problems involves work with *inequalities* and a technique called *linear programming*. The essential understandings and skills required for this work are developed in two lessons of this unit.

LESSONS

1 Inequalities in One Variable

Use numeric and graphic estimation methods and algebraic reasoning to solve problems that involve linear and quadratic inequalities in one variable.

2 Inequalities with Two Variables

Use graphic and algebraic methods to determine solution sets for systems of linear inequalities in two variables. Recognize problems in which the goal is to find optimum values of a linear objective function, subject to linear constraints on the independent variables. Represent both objective and constraints in graphic and algebraic form, and use linear programming techniques to solve the optimization problems.

Inequalities in One Variable

In previous courses, you learned how to solve a variety of problems by representing and reasoning about them with algebraic equations and inequalities. For example, suppose that plans for a fundraising raffle show that profit P will depend on ticket price x according to the function $P(x) = -2{,}500 + 5{,}000x - 750x^2$. A graph of profit as a function of ticket price is shown here.

Raffle Fundraiser Profit

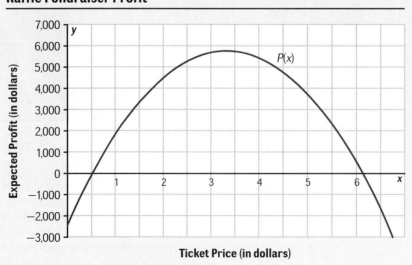

In this lesson, you will learn how to use graphical reasoning and algebraic methods to solve inequalities in one and two variables. You will also learn how to represent the solutions symbolically and graphically and how to interpret them in the contexts of the questions that they help to answer.

INVESTIGATION 1

Getting the Picture

You learned in earlier work with inequalities that solutions can be found by first solving related equations. For example, in the raffle fundraiser situation, the solutions of the equation

$$-2{,}500 + 5{,}000x - 750x^2 = 2{,}500$$

are approximately $1.23 and $5.44. The reasonableness of these solutions can be seen by scanning the graph of the profit function and the constant function $y = 2{,}500$.

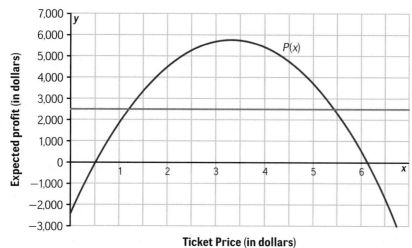

The solutions of the inequality $-2{,}500 + 5{,}000x - 750x^2 \geq 2{,}500$ are all values of x between \$1.23 and \$5.44. Those solutions to the inequality can be represented using symbols or a *number line graph*.

$$1.23 \leq x \leq 5.44$$

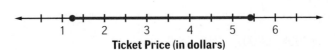

Ticket Price (in dollars)

Similarly, the solutions of the inequality $-2{,}500 + 5{,}000x - 750x^2 < 2{,}500$ are all values of x that are either less than \$1.23 or greater than \$5.44. Those solutions can also be represented using symbols or a number line graph.

$$x < 1.23 \text{ or } x > 5.44$$

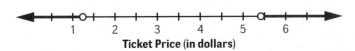

Ticket Price (in dollars)

As you work on problems of this investigation, look for answers to these questions:

How can you solve inequalities in one variable?

How can you record the solutions in symbolic and graphic form?

1 The next graph shows the height of the main support cable on a suspension bridge. The function defining the curve is $h(x) = 0.04x^2 - 3.5x + 100$, where x is horizontal distance (in feet) from the left end of the bridge and $h(x)$ is the height (in feet) of the cable above the bridge surface.

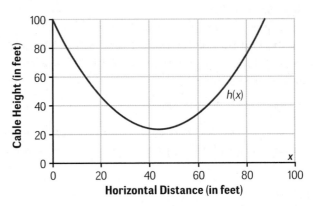

For the questions in Parts a–d:

- Write an algebraic calculation, equation, or inequality whose solution will provide an answer to the question.

- Then use the graph above to estimate the solution and calculator- or computer-generated tables and graphs of $h(x)$ to sharpen the accuracy to the nearest tenth.

- Express your answer with a symbolic expression and (where appropriate) a number line graph.

  a. Where is the bridge cable less than 40 feet above the bridge surface?

  b. Where is the bridge cable at least 60 feet above the bridge surface?

c. How far is the cable above the bridge surface at a point 45 feet from the left end?

d. Where is the cable 80 feet above the bridge surface?

2 The graph below shows the height of a bungee jumper's head above the ground at various times during her ride on the elastic bungee cord. Suppose that $h(t)$ gives height in feet as a function of time in seconds.

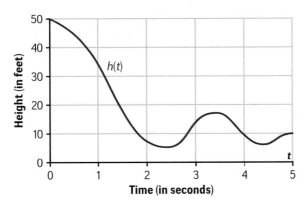

For each Part a–d:

- Write a question about the bungee jump that can be answered by the indicated mathematical operation.

- Use the graph to estimate the answer.

- Express your answer (where appropriate) with a number line graph.

a. Evaluate $h(2)$.　　　　　　**b.** Solve $h(t) = 10$.

c. Solve $h(t) \geq 10$.　　　　　**d.** Solve $h(t) < 10$.

SUMMARIZE THE MATHEMATICS

In this investigation, you developed strategies for solving problems by estimating solutions for equations and inequalities in one variable. You also used symbols and number line graphs to record the solutions.

a Describe strategies for solving inequalities in the form $f(x) \leq c$ and $f(x) \geq c$ when given a graph of the function $f(x)$.

b If the solution of an inequality is described by $2 \leq x$ and $x < 5$, what will a number line graph of that solution look like?

c What inequality statement(s) describe the numbers represented in this number line graph?

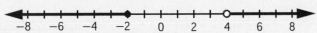

Be prepared to share your ideas and reasoning with the class.

✔️ CHECK YOUR UNDERSTANDING

Use information from this graph of the function $f(x)$ to answer the questions that follow.

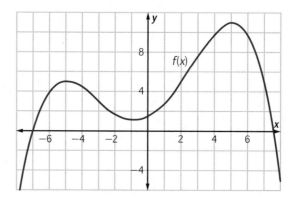

a. Estimate the values of x for which $f(x) = 5$.

b. Describe the values of x for which $f(x) \geq 5$ using the following.

    i. words   **ii.** symbols   **iii.** a number line graph

c. Explain how your answers to Part b would change if you were asked to consider the values of x for which $f(x) > 5$?

INVESTIGATION 2

Quadratic Inequalities

Inequalities that involve functions with familiar rules can often be solved by algebraic reasoning. As you work on the problems of this investigation, look for answers to these questions:

> *What are the solution possibilities for quadratic inequalities?*
>
> *How can solution strategies for quadratic equations*
> *be applied to solution of inequalities?*

1 Consider the inequality $t^2 - t - 6 \leq 0$.

a. Which of these diagrams is most like what you would expect for a graph of the function $g(t) = t^2 - t - 6$? How can you decide without using a graphing tool?

Graph I

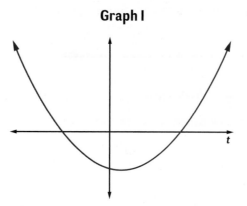

Graph II

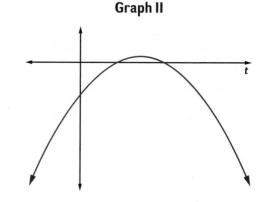

b. The expression $t^2 - t - 6$ can be written in equivalent factored form as $(t - 3)(t + 2)$. How can this fact be used to solve the equation $t^2 - t - 6 = 0$? What do those solutions tell about the graph of $g(t)$?

c. Use your answers from Parts a and b to solve the inequality $t^2 - t - 6 \leq 0$. Describe the solution using symbols and a number line graph.

d. Use similar reasoning to solve the inequality $t^2 - t - 6 > 0$ and record the solutions using symbols and a number line graph.

2 Your answers to the questions in Problem 1 show how two key ideas reveal solutions to any quadratic inequality in the form $ax^2 + bx + c \leq 0$ or $ax^2 + bx + c > 0$.

a. How does sketching the graph of $f(x) = ax^2 + bx + c$ help in solving a quadratic inequality in the form above?

b. How does solving the equation $ax^2 + bx + c = 0$ help in solving a quadratic inequality like those shown?

To use these strategies effectively, you need to recall the cues that tell shape and location of a quadratic function graph and the techniques for solving quadratic equations. Problems 3–5 provide review of the ideas required by those tasks.

3 For each of these quadratic functions, use the coefficients of the terms to help determine the shape and location of the graph.

- whether the graph has a maximum or minimum point

- the location of the line of symmetry

- where the graph will cross the y-axis

a. $f(x) = x^2 - 4x - 5$ **b.** $g(x) = -x^2 + 2x + 8$

c. $h(x) = x^2 + 6x + 9$ **d.** $j(x) = 2x^2 + 2x + 1$

4 What are the possible numbers of x-intercepts for the graph of a quadratic function? Sketch graphs to illustrate each possibility.

5 Solutions for quadratic equations can be estimated by scanning tables or graphs of the related quadratic functions. Solutions can also be found exactly by factoring the quadratic expression or by use of the quadratic formula. Practice using those exact solution strategies to solve these quadratic equations. Be prepared to explain your choice of strategy.

a. $x^2 + 5x + 4 = 0$ **b.** $x^2 + 2x - 8 = 0$

c. $-x^2 + 6x - 9 = 0$ **d.** $2x^2 + 2x + 1 = 0$

e. $x^2 - 6x + 11 = 2$ **f.** $x^2 + 1 = -3x$

6 Combine algebraic and graphic reasoning to solve the following inequalities. For each inequality:

- sketch a graph showing the pattern of change you expect for the function involved.

- use algebraic reasoning to locate key intersection points.

- combine what you learn from your sketch and algebraic reasoning to solve the inequality.

- record the solution using symbols and number line graphs.

a. $x^2 + 2x > 0$ **b.** $x^2 + 2x < 0$

c. $n^2 + 2n - 24 \leq 0$ **d.** $n^2 + 2n - 24 > 0$

e. $-s^2 + 4s - 6 < 0$ **f.** $-s^2 + 4s - 6 > 0$

g. $8r - r^2 \geq 15$ **h.** $3x^2 - 5x > 8$

i. $z^2 - 6z + 7 < 2$ **j.** $-p^2 + 10p - 7 \leq 14$

SUMMARIZE THE MATHEMATICS

In this investigation, you developed strategies for solving quadratic inequalities.

a   Describe strategies for solving inequalities in the form $ax^2 + bx + c \leq d$ and $ax^2 + bx + c \geq d$ by algebraic and graphic reasoning.

b   What are possible solutions for quadratic inequalities, and how can each solution possibility be expressed in algebraic and number line graph form?

Be prepared to share your ideas and reasoning with the class.

 CHECK YOUR UNDERSTANDING

For each inequality: (1) sketch a graph of the function involved in the inequality, (2) use algebraic reasoning to locate x-intercepts of the graph, (3) combine what you learn from your sketch and algebraic reasoning to solve the inequality, and (4) record the solution using symbols and a number line graph.

a. $k^2 - 3k - 4 > 0$

b. $-b^2 + 8b - 10 \geq 0$

Complex Inequalities

Many problems require comparison of two functions to find when values of one are greater or less than values of the other. As you work on the problems of this investigation, look for answers to this question:

How can the reasoning developed to deal with inequalities involving a single function be adapted to find solutions for more complex cases?

1 The diagram below shows the graphs of functions $h(x)$ and $k(x)$. Assume that all points of intersection are shown and that the functions have no breaks in their graphs.

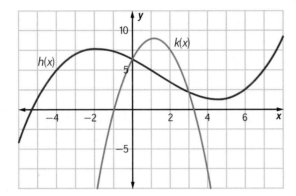

a. What are the approximate values of x for which $h(x) = k(x)$?

b. What are the values of x for which $h(x) \leq k(x)$? Express your answer using symbols and a number line graph.

c. What are the values of x for which $h(x) > k(x)$? Express your answer using symbols and a number line graph.

2 Harriet Tubman Elementary School needs to hire staff for a new after-school program. The program has a budget of $1,000 per week to pay staff salaries.

The function $h(p) = \dfrac{1,000}{p}$ shows how the number of staff that can be hired depends on the weekly pay per staff member p. Research suggested that the number of job applicants depends on the weekly pay offered according to the function $a(p) = -5 + 0.1p$.

a. Without using a graphing calculator, sketch a graph showing how $h(p)$ and $a(p)$ depend on p.

b. Solve the equation $\dfrac{1,000}{p} = -5 + 0.1p$ algebraically. Then explain what the solution tells about the staffing situation for the after-school program at Harriet Tubman Elementary.

c. The program director wants to ensure that she will be able to choose her staff from a large enough pool of applicants.

   i. Write an inequality that can be solved to determine the weekly pay per staff member for which the number of applicants will be more than the number of staff that can be hired.

   ii. Use your responses to Parts a and b to solve the inequality.

3 For each inequality given below:

   • sketch a graph showing how you expect the two component functions to be related.

   • use algebraic reasoning, a computer algebra system (CAS), or estimation using function tables or graphs to locate the points of intersection of the graphs.

   • apply what you learn about the relationship of the functions to solve the inequality.

   • record the solution using symbols and number line graphs.

   a. $q^2 + 3q - 6 \leq q + 2$ **b.** $c^2 - 4c - 5 \geq 2c + 2$

   c. $v^2 - v + 3 > 2v - 1$ **d.** $2m + 3 > 4 - m^2$

   e. $7 - x < \dfrac{10}{x}$ **f.** $\sqrt{d} > 2d - 1$

   g. $|x - 3| > 7$ **h.** $|2x + 5| \leq 3$

Interval Notation The solution of an inequality in one variable is generally composed of one or more intervals on a number line. Mathematicians use *interval notation* as a kind of shorthand to describe those solutions. For example, consider your response to Problem 1 Part b.

$$h(x) \leq k(x) \text{ when } 0 \leq x \leq 3.$$

Using interval notation,

$$0 \leq x \leq 3 \text{ can be written as } [0, 3].$$

The square brackets [,] indicate that the endpoints 0 and 3 are included in the solution set as well as all points between those end points. On the other hand, in Part c of Problem 1, you found that:

$$h(x) > k(x) \text{ when } x < 0 \text{ or } x > 3.$$

Using interval notation,

$$x < 0 \text{ or } x > 3 \text{ can be written as } (-\infty, 0) \cup (3, +\infty).$$

Here, the round brackets (,) indicate that the end points are not included in the solution set. The symbol ∞ represents infinity. Round brackets are always used with $-\infty$ and $+\infty$. The symbol $\cup$ indicates the *union* of the two intervals, that is, the numbers that are in one interval *or* the other. If an inequality has no real number solutions, the solution set is the *empty set*, denoted by the symbol Ø.

4 Here are descriptions of solutions for several inequalities. Describe each solution using interval notation.

a. $x < -2$ or $x > 0$

b.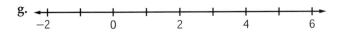

c. $0 \le x < 1$

d.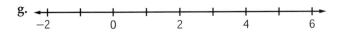

e. $x \le -1$ or $x \ge 7$

f. The inequality is true for all values of x.

g.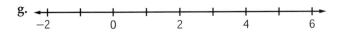

SUMMARIZE THE MATHEMATICS

In this investigation, you developed a strategy for solving complex inequalities in one variable. You used symbols, number line graphs, and interval notation to record the solution sets.

a Describe a general strategy for solving an inequality $f(x) < g(x)$.

b Suppose that $f(x) \ge g(x)$ for values of x from a up to and including b, where b is greater than a. Represent the solution using symbols, a number line graph, and interval notation.

c Suppose that $h(t) < j(t)$ for values of t less than c or greater than d where d is greater than c. Represent the solution using symbols, a number line graph, and interval notation.

Be prepared to share your ideas and reasoning with the class.

 CHECK YOUR UNDERSTANDING

For each inequality: (1) sketch a graph of the functions involved in the inequality, (2) use algebraic reasoning to locate intersection points of the graphs, (3) combine what you learn from your sketch and algebraic reasoning to solve the inequality, and (4) record the solution using symbols, a number line graph, and interval notation.

a. $2 - w \le w^2 - 2w$

b. $\dfrac{6}{x} < x + 5$

c. $|x - 2| \le 6$

APPLICATIONS

1. The graph below shows the path of a football kick, with height above the ground a function of horizontal distance traveled, both measured in yards. The function defining the path is $h(x) = -0.02x^2 + 1.3x + 1$.

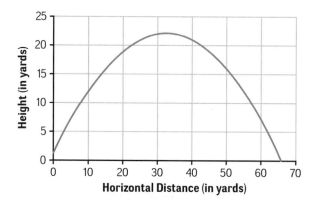

For each of the questions in Parts a–d:

- Write an algebraic calculation, equation, or inequality whose solution will provide an answer to the question.

- Use the graph above to estimate the solution and calculator-generated tables and graphs of $h(x)$ to sharpen the accuracy to the nearest tenth.

- Express your answer with a symbolic expression and a number line graph.

a. When is the kicked ball 20 yards above the field?

b. When is the ball less than 10 yards above the playing field?

c. How far is the ball above the field when it has traveled horizontally 40 yards?

d. When is the ball at least 15 yards above the field?

2. The next graph shows the depth of water alongside a ship pier in a tidal ocean harbor during one 24-hour period. Suppose that $d(t)$ gives depth in feet as a function of time in hours.

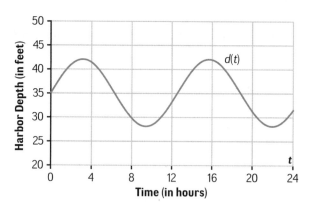

For each Part a–d:

- Write a question about the water depth that can be answered by the indicated mathematical operation.

- Use the graph to estimate the answer.

- Express your answer (where appropriate) with a number line graph.

a. Evaluate $h(4)$.

b. Solve $h(t) = 40$.

c. Solve $h(t) \geq 40$.

d. Solve $h(t) < 30$.

3 Use this graph of a function $S(n)$ to estimate the values of n that satisfy each inequality below. Describe those values using words, symbols, and number line graphs.

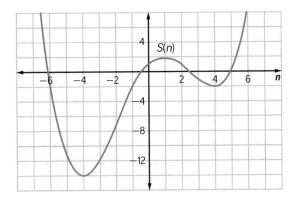

a. $S(n) \geq -2$

b. $S(n) < -2$

c. $S(n) > -14$

d. $S(n) \leq -14$

4 Describe the solutions of these inequalities using symbols and number line graphs.

a. $7t - t^2 < 0$

b. $a^2 + 4a \geq 0$

c. $h^2 + 2h - 3 \leq 0$

d. $x^2 + 0.5x - 3 > 0$

e. $-d^2 - 12d - 20 > 0$

f. $3 + 2r - r^2 \geq 0$

5 Describe the solutions of these inequalities using symbols and number line graphs.

a. $7t - t^2 < 10$

b. $a^2 + 4a \geq 12$

c. $5h^2 + 14h < 3$

d. $-x^2 + 6x - 8 < 1$

e. $-d^2 - 11d - 20 > 4$

f. $3 + 2r - r^2 \geq 8$

6 Describe the solutions of these inequalities using symbols and a number line graph.

a. $k + 10 \geq \dfrac{24}{k}$

b. $w - 1 < \dfrac{20}{w}$

c. $x + 6 < \dfrac{7}{x}$

d. $u + 1 \leq \dfrac{12}{u}$

7 Shown below are the graphs of two functions $P(n)$ and $Q(n)$. Assume that the graph shows all points of intersection of $P(n)$ and $Q(n)$.

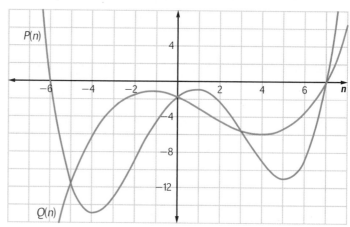

a. Describe the values of n for which $P(n) < Q(n)$ using symbols and a number line graph.

b. Describe the values of n for which $P(n) > Q(n)$ using symbols and a number line graph.

8 The Riverdale Adventure Club has planned a skydiving lesson and first jump for new members. They plan to make and sell videos of each jump.

Income from sale of videos of the jump and expenses for producing the videos will depend on price p charged according to these rules.

$$I(p) = p(118 - p)$$
$$E(p) = 6{,}400 - 50p$$

a. Sketch graphs showing how you expect the functions $I(p)$ and $E(p)$ to be related.

b. Find coordinates of the intersection points algebraically.

c. Describe the solution of the inequality $I(p) \geq E(p)$ using symbols and a number line graph.

d. Explain what your work suggests about how much the club should charge for videos of the jump.

9 For each of the following inequalities:

- sketch graphs showing how you expect the two functions in the inequality to be related.

- use algebraic reasoning, a CAS, or estimation using tables and graphs of the functions to locate the points of intersection of the graphs.

- use what you learn about the relationship of the functions to solve the inequality.

- record the solution using symbols and a number line graph.

a. $2n^2 + 3n < 8n + 3$ b. $9 + 6x - x^2 \leq x - 5$

c. $d^2 - 5d + 10 \geq 1 - 2d$ d. $j^2 + 7j + 20 < -j^2 - 5j$

e. $5 - u^2 \geq u^2 - 3$ f. $\sqrt{b} > 6 - b$

g. $|a + 5| < 10$ h. $|2x - 1| \geq 3$

10 Graph these intervals on number lines. Write inequalities that express the same information.

a. $[-2, 5)$

b. $(-\infty, 0) \cup [4, \infty)$

c. $[3, 7]$

d. $(-4, -1) \cup (1, \infty)$

11 Below are descriptions of the solutions for six inequalities. Describe each solution using interval notation.

a. $k \leq -3$ or $k > -1$

b. All numbers between negative 1 and positive 3.5

c.

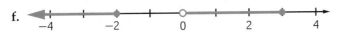

d. $2 < g < 6$

e. All numbers less than 4 or greater than 7

f.

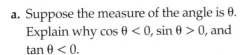

CONNECTIONS

12 The diagram at the right shows an angle in standard position whose terminal side lies in Quadrant II.

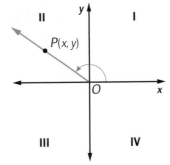

a. Suppose the measure of the angle is θ. Explain why $\cos \theta < 0$, $\sin \theta > 0$, and $\tan \theta < 0$.

b. For each of the following conditions, state the quadrant in which the terminal side of the angle in standard position must lie.

  i. $\cos \theta > 0$, $\sin \theta > 0$

  ii. $\cos \theta > 0$, $\sin \theta < 0$

  iii. $\cos \theta < 0$, $\tan \theta > 0$

  iv. $\sin \theta < 0$, $\tan \theta < 0$

13 A firework contains a time-delay fuse that burns as the firework soars upward. The length of this fuse must be made so that the firework does not explode too close to the ground.

The height (in feet) of a firework t seconds after it is launched is given by $h(t) = 3 + 180t - 16t^2$.

a. The firework is to explode at a height of at least 450 feet. Write an inequality whose solution gives the possible explosion times for the firework.

b. Draw a graph of the height of the firework over time, indicating the times when the firework is at a height of at least 450 feet.

14 In solving quadratic equations by factoring, you used the fact that if $ab = 0$, then $a = 0$ or $b = 0$. Consider whether similar reasoning can be used to solve a quadratic inequality.

a. Complete a table like this.

if ...	and ... ,	then ... .
$a > 0$	$b > 0$	$ab\,?\,0$
$a > 0$	$b < 0$	$ab\,?\,0$
$a < 0$	$b > 0$	$ab\,?\,0$
$a < 0$	$b < 0$	$ab\,?\,0$

b. Based on your work in Part a, if $ab > 0$, then what can be said about a and b? What if $ab < 0$?

c. For what values of v is the inequality $(v + 3)(v - 4) > 0$ true?

d. How might you expand this kind of reasoning to determine the values of w for which the inequality $(w + 5)(w + 1)(w - 2) < 0$ is true?

15 The graphs below show the way two different variables depend on x. One function is given by $f(x) = \dfrac{50}{x^2}$. The other function is given by $g(x) = x^2 - 5x$.

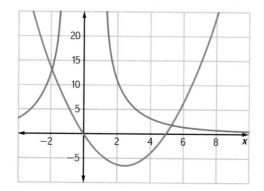

a. Make a copy of these graphs. Identify and label the graphs of $f(x)$ and $g(x)$.

b. What connections between function rules and their graphs allow you to match $f(x)$ and $g(x)$ to their graphs in this case?

c. Color the x-axis of your copy of the graphs to highlight the points corresponding to solutions of the following equation and inequalities. If possible, use the colors suggested.

   i. In blue: $\dfrac{50}{x^2} = x^2 - 5x$

   ii. In red: $\dfrac{50}{x^2} > x^2 - 5x$

   iii. In green: $\dfrac{50}{x^2} < x^2 - 5x$

16 In the *Reasoning and Proof* unit, you completed an argument to prove one case of the Law of Cosines.

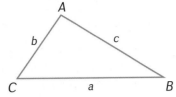

In $\triangle ABC$, $c^2 = a^2 + b^2 - 2ab \cos C$.

Completing the tasks below will help you to see a connection between the Law of Cosines and the Triangle Inequality: $a + b > c$. Recall that $-1 < \cos C < 1$.

a. Explain why it must be the case that $a^2 + b^2 - 2ab \cos C < a^2 + b^2 - 2ab(-1)$.

b. Explain why $c^2 < (a + b)^2$.

c. Explain why it follows that $a + b > c$.

d. What form of the Law of Cosines would you use to prove this form of the Triangle Inequality: $a + c > b$?

17 You learned in Investigation 3 that the symbol $\cup$ indicates the *union* of two sets. The symbol $\cap$ indicates the *intersection* of two sets.

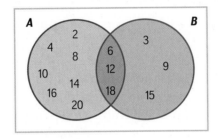

a. Consider the Venn diagram above.

    i. How would you describe the numbers in circle A? In circle B?

    ii. How would you describe the numbers in $A \cup B$?

    iii. Why are the numbers 6, 12, and 18 in $A \cap B$?

b. Consider the integers from -10 to 10. Create Venn diagrams that satisfy the conditions below for C and D. Then describe $C \cup D$ and $C \cap D$ using inequalities.

    i. C consists of integers greater than -5. D consists of integers less than 7.

    ii. C consists of integers less than -3. D consists of integers greater than 4.

    iii. C consists of integers less than 2. D consists of integers less than -1.

REFLECTIONS

18 Explain why the values of x indicated by $-3 < x < 7$ are *not* the same as the values indicated by $-3 < x$ or $x < 7$.

19 If $f(x) > c$ only when $a < x < b$, then what are the values of x for which $f(x) \leq c$?

20 In this lesson, the notation (a, b) has been used in two different ways.

- Graph the point $(-3, 2)$ on a coordinate plane.

- Graph the interval $(-3, 2)$ on a number line.

How can you tell, in any particular problem, whether (a, b) refers to a point in the plane or an interval on a number line?

EXTENSIONS

21 In your work on problems of this lesson, you focused on inequalities involving linear, quadratic, and inverse variation functions. But inequalities involving other powers and exponential functions are important as well. For each inequality given below:

- sketch a graph of the functions involved.

- use what you know about the functions involved to solve the inequality exactly or by accurate approximation.

- record the solution using symbols and a number line graph.

a. $\sqrt{x} > x \ (x \geq 0)$ **b.** $x^3 < x^2$

c. $2^x < x^2$ **d.** $0.5^x < 0.5x$

22 In the *Trigonometric Methods* unit of Course 2, you studied the patterns of values for the sine, cosine, and tangent functions. In particular, for an angle θ in standard position with point $P(x, y)$ on the terminal side, if $r = \sqrt{x^2 + y^2}$, then $\sin \theta = \frac{y}{r}$, $\cos \theta = \frac{x}{r}$, and $\tan \theta = \frac{y}{x}$, if $x \neq 0$. For each inequality given below:

- use what you know about the functions involved to solve the inequality exactly or by accurate approximation for $0 \leq \theta \leq 180°$.

- record the solution using symbols and a number line graph.

a. $\cos \theta > 0$ **b.** $\cos \theta > \sin \theta$

c. $\tan \theta > 1$ **d.** $\sin \theta < 0.5$

23 For each *compound inequality* below, indicate whether the solution is the union or the intersection of the solutions to the individual inequalities. Then graph the solution of the compound inequality on a number line.

a. $v > -0.5$ and $v \leq 3$ **b.** $x < -2$ or $x \geq 7$

c. $3d + 2 \leq 5$ or $6 - d < 7$ **d.** $2p - 1 < 3$ and $p + 6 \geq 10$

24 Can the solution to an equation ever be an inequality? Can the solution to an inequality ever be an equation? Use sketches of graphs to illustrate your response.

REVIEW

25 For each of these quadratic functions, use the coefficients of the terms to help determine the shape and location of the graph: (1) whether the graph has a maximum or minimum point, (2) the location of the line of symmetry, and (3) where the graph will cross the y-axis.

a. $f(x) = x^2 + 4x - 5$

b. $g(x) = -x^2 - 2x + 8$

c. $h(x) = x^2 - 10x + 25$

d. $j(x) = -0.5x^2 + 2x + 4$

26 Solve each of these quadratic equations in two ways, by factoring and by use of the quadratic formula. Check the solutions you find by substitution in the original equation or by use of a CAS **solve** command.

a. $x^2 - 6x + 5 = 0$

b. $x^2 + 4x = 0$

c. $2x^2 - x - 6 = 0$

d. $-x^2 - 5x - 4 = 0$

e. $-x^2 + 7x + 6 = 2x$

f. $x^4 - 19x^2 = -3x^2$

27 Solve each of the following for x.

a. $3x - 12 < 24$

b. $-2x + 19 > 5x - 2$

c. $10 - 6x < 2x + 90$

d. $x - 3 = \dfrac{18}{x}$

28 Without using your calculator, sketch a graph of each function below. Then check your sketches using a calculator or computer graphing tool.

a. $f(x) = \dfrac{1}{x}$

b. $g(x) = x^2$

c. $f(t) = 3(2^t)$

d. $f(p) = -p^2$

e. $h(n) = -\dfrac{1}{n^2}$

f. $g(t) = \sqrt{t}$

29 Without using the graphing feature of your calculator or computer, find the coordinates of the intersection point for each pair of lines.

a. $x = 5$
   $3x + 2y = 24$

b. $3x + 2y = 24$
   $y = -0.5x + 6$

c. $y = 8$
   $x = 5$

d. $x + y = 200$
   $2x + y = 280$

30 You may recall that you can use the Law of Cosines and the Law of Sines to find side lengths and angle measures in any triangle. Below is one form of each of those useful laws for $\triangle ABC$.

Law of Cosines

$c^2 = a^2 + b^2 - 2ab \cos C$

Law of Sines

$\dfrac{a}{\sin A} = \dfrac{b}{\sin B} = \dfrac{c}{\sin C}$

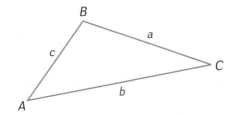

For each triangle below, decide if you should use the Law of Cosines or the Law of Sines to find the indicated side length or angle measure. Then use your choice to find the indicated side length or angle measure.

a. Find *AB*.

b. Find m∠*B*.

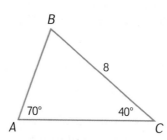

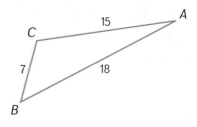

31 For the linear equations below:

- find coordinates of the *y*-intercept of the graph.

- find coordinates of the *x*-intercept of the graph.

- sketch the graph of all solutions.

a. $3x + 2y = 6$

b. $8y - 4x = 24$

32 It seems reasonable that the length of a person's stride is related to the length of his or her legs. Manish collected these two measurements for 20 different people.

a. Would you expect the correlation for the data to be positive or negative? Explain your reasoning.

b. Using his data, Manish created the scatterplot below. Describe the direction and strength of the relationship.

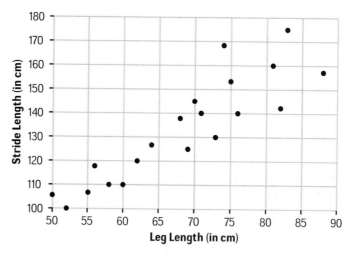

c. The linear regression line for these data is $y = 1.78x + 12$. Explain the meaning of the slope of this line in terms of the context.

LESSON 2

Inequalities with Two Variables

Important decisions in business often involve many variables and relations among those variables. The key to making good decisions is finding a way to organize and compare options.

For example, suppose that the manager of an electronics company must plan for and supervise production of two video game systems, the basic CP•I and the advanced CP•II. Assume that demand for both game systems is high, so the company will be able to sell whatever is produced. To plan the work schedule, the manager has to consider the following conditions.

- Assembly of each CP•I model takes 0.6 hours of technician time, and assembly of each CP•II model takes 0.3 hours of technician time. The plant can apply at most 240 hours of technician time to assembly work each day.

- Testing for each CP•I model takes 0.2 hours, and testing of each CP•II model takes 0.4 hours. The plant can apply at most 160 hours of technician time each day for testing.

- Packaging time is the same for each model. The packaging department of the plant can handle at most 500 game systems per day.

- The company makes a profit of $50 on each CP•I model and $75 on each CP•II model.

The production planning challenge is to maximize profit while operating under the constraints of limited technician time and packaging time.

Yann Layma/The Image Bank/Getty Images

Many problems like those facing the electronics plant manager are solved by a mathematical strategy called *linear programming*. In this lesson, you will learn how to use this important problem-solving technique and the mathematical ideas and skills on which it depends.

INVESTIGATION **1**

Solving Inequalities

Many problems that arise in making plans for a business involve functions with two or more independent variables. For example, income for the Old Dominion Music Festival is given by the function $I = 8a + 12g$, where a is the number of admission tickets sold in advance and g is the number of tickets sold at the gate.

If expenses for operating the two-day festival total $2,400, then solutions of $8a + 12g = 2,400$ give (a, g) combinations for which festival income will equal operating expenses. But festival organizers are probably interested in earning *more than* $2,400 from ticket sales. They want solutions to the inequality $8a + 12g > 2,400$.

As you work on the problems of this investigation, look for answers to these questions:

> *How can one find and graph the solutions of a linear inequality in two variables?*
>
> *How can one find the solutions of a system of inequalities in two variables?*

Eric Nathan/Alamy

1 The next diagram gives a first quadrant graph of the line representing solutions for the equation $8a + 12g = 2{,}400$, the combinations of tickets sold in advance and at the gate that will give festival income of exactly \$2,400. Use that graph as a starting point in solving the inequality $8a + 12g > 2{,}400$.

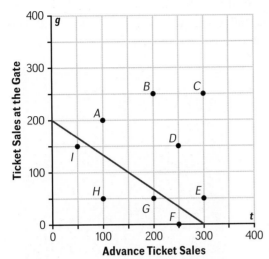

a. For each point on the diagram, give the (a, g) coordinates and the festival income from that combination of tickets sold in advance and tickets sold at the gate.

b. Based on your work in Part a, describe the graphical representation of inequalities.

  i. $8a + 12g > 2{,}400$

  ii. $8a + 12g < 2{,}400$

c. Solve the equation $8a + 12g = 2{,}400$ for g in terms of a.

d. Use what you know about manipulation of inequality statements to express each inequality in Part b in an equivalent form $g < \ldots$ or $g > \ldots$. Then explain how those equivalent inequalities help to describe the solution graphs for the original inequalities.

2 A local bank is giving away souvenir Frisbees and sun visors at the Old Dominion Music Festival. The Frisbees will cost the bank \$4 each, and the visors will cost the bank \$2.50 each. The promotional cost for the bank depends on the number of Frisbees F and the number of visors V given away at the festival. The bank has budgeted \$1,000 for the purchase of these items.

a. Write an inequality that represents the question, "How many souvenir Frisbees and visors can the bank give away for a total cost of no more than \$1,000?"

b. Draw a graph that shows the region of the first quadrant in the coordinate plane containing all points (F, V) that satisfy the inequality in Part a.

Graphing Conventions In Problems 1 and 2, the only meaningful solutions of inequalities were in the first quadrant. Negative values of a, g, F, and V make no sense in those situations. However, the ideas and graphing techniques you developed in work on those problems can be applied to situations without the constraint of only positive values.

For example, the following graph shows solutions with both positive and negative values for $x + 2y < -2$ and $x + 2y > -2$.

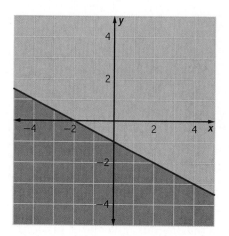

3 Compare the inequalities and the indicated graph regions.

a. Which region corresponds to solutions of $x + 2y < -2$? How did you decide?

b. How would you graph the solutions to a linear inequality like $x - y > 3$?

Just as with inequalities in one variable, the solution region of an inequality in two variables is generally bounded by the solution of a corresponding equation. Sometimes the boundary is *included* in the solution of the inequality, indicated on a graph by a *solid* boundary. A *dashed* boundary on a graph indicates that points on the boundary are *excluded* from the solution.

4 Draw graphs that show all points with x- and y-coordinates between -10 and 10 and satisfying these linear inequalities in two variables. Use either solid or dashed boundary lines as appropriate.

a. $3x - 2y < 12$ **b.** $2x + y \geq 4$

c. $8x - 5y > 20$ **d.** $4x + 3y \leq 15$

5 Suppose that the organizers of the Old Dominion Music Festival can sell no more than 1,000 admission tickets due to space constraints.

a. Write an inequality whose solutions are the (a, g) pairs for which the number of tickets sold will be no more than 1,000.

b. Draw a graph that uses appropriate boundary lines to show the region of the first quadrant in the coordinate plane that contains all points (a, g) satisfying both the inequality $8a + 12g > 2,400$ (from Problem 1) and the inequality from Part a of this problem.

6 Suppose that the bank wants to give away at least 300 promotional items at the Old Dominion Music Festival.

  a. Write an inequality whose solutions are the (F, V) combinations for which the total number of promotional items will be 300 or more.

  b. Draw a graph that shows the region of the first quadrant in the coordinate plane containing all points (F, V) that satisfy both the inequality $4F + 2.5V \leq 1{,}000$ (from Problem 2) and the inequality from Part a of this problem.

You have solved systems of equations in previous work. Recall that the goal in solving a system of equations is to find values of the variables that satisfy all equations in the system. Similarly, the goal in solving a *system of inequalities* is to find values of the variables that satisfy all inequalities in the system. As with systems of equations, systems of inequalities can be solved by graphing the solution of each inequality in the system and finding the points of intersection as you did in Problems 5 and 6.

7 Draw graphs that show the solutions of these systems of inequalities.

  a. $\begin{cases} 2x - 3y > -12 \\ x + y \geq -2 \end{cases}$ **b.** $\begin{cases} 3x - 4y > 18 \\ 5x + 2y \leq 15 \end{cases}$

  c. $\begin{cases} y > 4 - x^2 \\ 2x - y > -3 \end{cases}$ **d.** $\begin{cases} y > x^2 - 4x - 5 \\ 2x - y \geq -2 \end{cases}$

SUMMARIZE THE MATHEMATICS

In this investigation, you developed strategies for graphing the solution of a linear inequality in two variables and for graphing the solution of a system of inequalities in two variables.

  a Describe a general strategy for graphing the solution of a linear inequality in two variables.

  b How do you show whether the points on the graph of the corresponding linear equation are included as solutions?

  c Describe the goal in solving a system of inequalities and a general strategy for finding the solutions.

Be prepared to share your ideas and reasoning with the class.

✔️ CHECK YOUR UNDERSTANDING

Jess has made a commitment to exercise in order to lose weight and improve overall fitness. Jess would like to burn *at least* 2,000 calories per week through exercise but wants to schedule *at most* five hours of exercise each week.

Jess will do both walking and bike riding for exercise. Brisk walking burns about 345 calories per hour, and biking at a moderate pace burns about 690 calories per hour.

a. Write a system of linear inequalities that describes Jess' exercise goals.

b. Identify at least three (*number of hours walking, number of hours biking*) combinations that satisfy both of Jess' conditions.

c. Graph the set of all points that satisfy the system from Part a.

INVESTIGATION 2

Linear Programming—A Graphic Approach

Linear programming problems, like that faced by managers of the video game factory described at the start of this lesson, involve finding an optimum choice among many options. As you work on the problems in this investigation, look for an answer to this question:

> *How can coordinate graphs be used to display and analyze the options in linear programming decision problems?*

Production Planning The production-scheduling problem at the electronics plant requires the plant manager to find a combination of CP•I and CP•II models that will give greatest profit. But there are **constraints** or limits on the choice. Each day, the plant has capacity for:

- *at most* 240 hours of assembly labor with each CP•I requiring 0.6 hours and each CP•II requiring 0.3 hours.

- *at most* 160 hours of testing labor with each CP•I requiring 0.2 hours and each CP•II requiring 0.4 hours.

- packing *at most* 500 video game systems with each model requiring the same time.

The company makes profit of $50 on each CP•I model and $75 on each CP•II model.

1 One way to search for the production plan that will maximize profit is to make some guesses and test the profit prospects for those guesses. Here are three possible production plans for CP•I and CP•II video game models.

> **Plan 1:** Make 100 of model CP•I and 200 of model CP•II.
> **Plan 2:** Make 200 of model CP•I and 100 of model CP•II.
> **Plan 3:** Make 400 of model CP•I and 100 of model CP•II.

a. Check each production plan by answering the following questions.

i. Will the required assembly time be within the limit of 240 hours per day?

ii. Will the required testing time be within the limit of 160 hours per day?

iii. Will the number of systems produced fall within the packing limit of 500 units per day?

iv. If the constraints are satisfied, what profit will be earned?

b. Design and check a plan of your own that you think will produce greater profit than Plans 1, 2, and 3 while satisfying the assembly, testing, and packing constraints.

As you compared the possible production plans in Problem 1, you checked several different constraints and evaluated the profit prospects for each combination of CP•I and CP•II video game systems. However, you checked only a few of many possible production possibilities.

It would be nice to have a systematic way of organizing the search for a maximum profit plan. One strategy for solving **linear programming** problems begins with graphing the options. If x represents the number of CP•I game systems produced and y represents the number of CP•II game systems produced, then the scheduling goal or *objective* is to find a pair (x, y) that meets the constraints and gives maximum profit.

Using a grid like the one below, you can search for the combination of CP•I and CP•II video game system numbers that will give maximum profit. The point (100, 200) represents production of 100 CP•I and 200 CP•II game systems. These numbers satisfy the constraints and give a profit of

$$\$50(100) + \$75(200) = \$20,000.$$

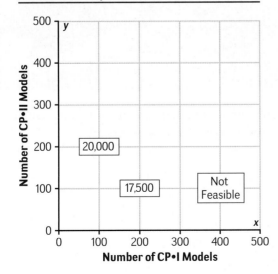

Video Game System Profits (in dollars)

2 Each **lattice point** on the grid (where horizontal and vertical grid lines intersect) represents a possible combination of CP•I and CP•II video game system numbers. Points with coordinates satisfying all the constraints are called **feasible points**.

a. What do the labels on the points with coordinates (200, 100) and (400, 100) tell about the production planning options?

b. Collaborate with others to check the remaining lattice points on the given graph to see which are feasible points and which are not. For each feasible point, find the profit for that production plan and record it on a copy of the graph. Label each nonfeasible point "NF."

c. Based on the completed graph:

   i. describe the region where coordinates of lattice points satisfy all three constraints.

   ii. pick the combination of CP•I and CP•II video game systems that you think the factory should produce in order to maximize daily profit. Be prepared to explain why you believe your answer is correct.

Balancing Astronaut Diets Problems like the challenge of planning video game system production to maximize profit occur in many quite different situations. For example, think about the variables, constraints, and objectives in choosing the foods you eat. Usually, you choose things that taste good. But it is also important to consider the cost of the food and the ways that it satisfies your body's dietary needs.

3 For some people, athletes and astronauts in particular, selection of a good diet is a carefully planned scientific process. Each person wants a high-performance diet at minimal cost. In the case of astronauts, the goal might be minimal total food weight onboard the spacecraft.

Consider the following simplified version of the problem facing NASA flight planners who must provide food for astronauts.

Suppose there are two kinds of food to be carried on a space shuttle trip, special food bars and cartons of a special drink.

- Each food bar provides 5 grams of fat, 40 grams of carbohydrate, and 8 grams of protein.

- Each drink carton provides 6 grams of fat, 25 grams of carbohydrate, and 15 grams of protein.

- Minimum daily requirements for each astronaut are *at least* 61 grams of fat, *at least* 350 grams of carbohydrate, and *at least* 103 grams of protein.

- Each food bar weighs 270 grams, and each drink carton weighs 300 grams.

The goal is to find a combination of food bars and drinks that will fulfill daily requirements of fat, carbohydrate, and protein with minimum total weight onboard the spacecraft.

This probably seems like a complicated problem. But you can get a good start toward a solution by doing some systematic testing of options.

a. For each of these numbers of food bars and drink cartons, check to see if they provide at least the daily minimums of fat, carbohydrate, and protein. Then find the total weight of each feasible combination.

  i. 4 food bars and 10 cartons of drink

  ii. 10 food bars and 4 cartons of drink

  iii. 4 food bars and 4 cartons of drink

  iv. 10 food bars and 10 cartons of drink

b. Record your findings on a copy of the following graph. The case of 4 food bars and 10 drink cartons has been plotted already.

Food and Drink Weights (in grams)

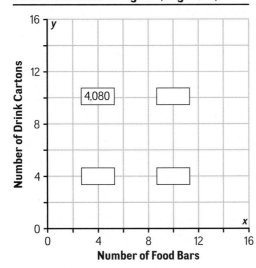

c. Collaborate with others to test the remaining lattice points on the grid to get a picture of the **feasible region**, the points with coordinates that meet all constraints. For each feasible point, find the total weight of the food and drink. Plot it on a copy of the grid.

4 Now analyze the pattern of *feasible points* and *objective values* (weights) for possible astronaut diets.

a. Describe the shape of the feasible region.

b. Study the pattern of weights for points in the feasible region. Decide on a combination of food bars and drink cartons that you think will meet the diet constraints with minimum weight. Be prepared to explain why you believe your answer is correct.

SUMMARIZE THE MATHEMATICS

Problems that can be solved by linear programming have several common features: variables, constraints, and an objective.

a What are the variables, constraints, and objective in the video game system production problem?

b What are the variables, constraints, and objective in the astronaut diet problem?

c What are *feasible points* and the feasible region in a linear programming problem?

d What do coordinates of the feasible points and the "not feasible" points tell you in the video game system production problem? In the astronaut diet problem?

Be prepared to explain your ideas to the class.

 CHECK YOUR UNDERSTANDING

Suppose that a new store plans to lease selling space in a mall. The leased space would be divided into two sections, one for books and the other for music and videos.

- The store owners can lease up to 10,000 square feet of floor space.

- Furnishings for the two kinds of selling space cost $5 per square foot for books and $7 per square foot for music and videos. The store has a budget of at most $60,000 to spend for furnishings.

- Each square foot of book-selling space will generate an average of $75 per month in sales. Each square foot of music- and video-selling space will generate an average of $95 per month in sales.

The store owners have to decide how to allocate space to the two kinds of items, books or music/video, to maximize monthly sales.

a. Identify the variables, constraints, and objective in this situation.

b. Find three feasible points and one point that is not feasible.

c. Evaluate the predicted total monthly sales at the three feasible points.

d. Plot the three feasible points and the nonfeasible point on an appropriate grid and label each feasible point with the corresponding value of predicted total monthly sales.

e. Which of the three sample feasible points comes closest to the problem objective?

INVESTIGATION 3

Linear Programming—Algebraic Methods

As you worked on the production-planning problem for the video game factory and the astronaut diet problem, you probably thought, "There's got to be an easier way than testing all those possible combinations." Computers have been programmed to help explore the options. In fact, *CPMP-Tools* software offers such a program. However, in order to use such tools, it is essential to express the problem in algebraic form.

As you complete the problems in this investigation, look for answers to these questions:

> *How can constraints and objectives of linear programming problems be expressed in symbolic form?*

> *How can algebraic and graphical methods be combined to help solve the problems?*

Video Game System Production Revisited Think again about the objective and the constraints in the production-planning problem. Each CP•I video game system earns $50 profit, and each CP•II system earns $75 profit. But daily production is constrained by the times required for assembly, testing, and packaging.

1 If x represents the number of CP•I models and y represents the number of CP•II models produced in a day, what algebraic rule shows how to calculate total profit P for the day? This rule is called the **objective function** for the linear programming problem because it shows how the goal of the problem is a function of, or depends on, the independent variables x and y.

2 Recall that assembly time required for each CP•I model is 0.6 hours, and assembly time required for each CP•II model is 0.3 hours. Assembly capacity is limited by the constraint that the factory can use at most 240 hours of technician time per day.

a. Explain how the linear inequality $0.6x + 0.3y \leq 240$ represents the assembly time constraint.

b. The graph below shows points meeting the assembly time constraint. *Points that are **not** feasible have been shaded out of the picture.*

Video Game System Assembly Time (in hours)

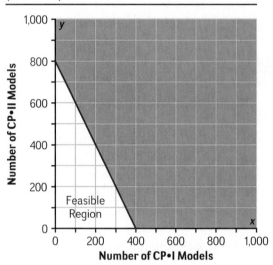

i. Why does the graph only show points in the first quadrant?

ii. Which feasible point(s) do you think will lead to greatest daily profit for the company? Test your ideas by evaluating the objective function at a variety of feasible points.

3 Next, recall that testing of each CP•I model requires 0.2 hours of technician time and testing of each CP•II model requires 0.4 hours. The factory can apply at most 160 hours of technician time to testing each day.

a. Write an inequality that expresses the testing time constraint. Be prepared to explain how you know that your algebraic representation of the constraint is correct.

b. Graph the solutions of that inequality, shading *out* the region of non feasible points.

c. Which feasible point(s) do you think will lead to greatest daily profit for the factory? Test your ideas by evaluating the objective function at a variety of feasible points.

4 Finally, recall that the factory can package and ship at most 500 video game systems each day.

    a. Write an inequality that expresses the packaging/shipping constraint. Be prepared to explain how you know that your algebraic expression of the constraint is correct.

    b. Graph the solutions of that inequality, shading *out* the region of nonfeasible points.

    c. Which feasible point(s) do you think will lead to greatest daily profit for the factory? Test your ideas.

5 In work on Problems 2–4, you developed some ideas about how to maximize profit while satisfying each problem constraint separately. The actual production-planning problem requires maximizing profit while satisfying all three constraints. The feasible region for this problem is shown as the unshaded region in the diagram below.

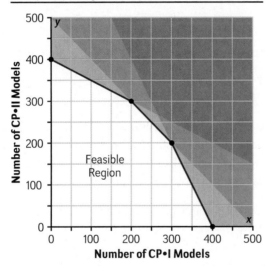

Video Game System Production Feasibility

    a. On a copy of the graph, use results from Part a of Problems 2, 3, and 4 to label each segment of the feasible region boundary with its corresponding linear equation.

    b. Use the graph to estimate coordinates of points at the corners of the feasible region. Then explain how you could check the accuracy of your estimates and how you could calculate the coordinates by algebraic methods.

c. Now think about the objective function $P = 50x + 75y$.

    i. How does the value of P change as the numbers x and y increase within the feasible region?

    ii. Where in the feasible region would you expect a combination of production numbers (x, y) to give maximum profit?

    iii. List coordinates of some likely maximum profit points. Evaluate the profit corresponding to each point.

Video Game System Profit		
Number of CP•I Models	Number of CP•II Models	Profit (in dollars)

    iv. Of these production options, which gives the maximum profit?

d. Compare your choice of maximum profit production plan to those of others in your class and resolve differences.

Astronaut Diet Planning Revisited Think again about the objective and constraints in the astronaut diet-planning problem. Each food bar weighs 270 grams, and each drink carton weighs 300 grams. But the diet plan is constrained by the minimum daily requirements of fat, carbohydrate, and protein.

6 If f represents the number of food bars and d represents the number of drink cartons in a daily diet, what algebraic rule shows how to calculate total weight W of that food? (This rule is the objective function for the diet linear programming problem.)

7 Recall that each food bar provides 5 grams of fat, 40 grams of carbohydrate, and 8 grams of protein. Each drink carton provides 6 grams of fat, 25 grams of carbohydrate, and 15 grams of protein. Write inequalities that express the constraints of providing the daily astronaut diet with:

a. at least 61 grams of fat.

b. at least 350 grams of carbohydrate.

c. at least 103 grams of protein.

NASA

8 The next graph shows the feasible region for the astronaut diet problem. The feasible region for this problem is shown as the unshaded region in the graph.

Astronaut Food and Drink Feasibility

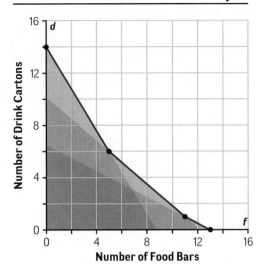

a. On a copy of the graph, label each segment of the feasible region boundary with its corresponding linear equation.

b. Use the graph to estimate coordinates of points at the corners of the feasible region. Then check the accuracy of your estimates by algebraic methods.

c. Now think about the objective function $W = 270f + 300d$.

   i. How does the value of W change as the numbers f and d decrease within the feasible region?

   ii. Where in the feasible region do you expect a pair of numbers (f, d) to give minimum weight?

   iii. List coordinates of some likely minimum weight points. Evaluate the weight corresponding to each point.

Astronaut Diet Options		
Number of Food Bars	Number of Drink Cartons	Weight (in grams)

   iv. Of these diet options, which gives minimum weight?

d. Compare your choice of minimum weight diet plan to those of others in your class and resolve differences.

In this investigation, you explored ways that reasoning with symbolic expressions, inequalities, and graphs helps to solve linear programming problems.

 a Why are the constraints in the video game system production and astronaut diet planning problems most accurately expressed as inequalities rather than equations?

b What shapes would you expect for the feasible regions in other linear programming problems where the goal is to maximize some objective function? In problems where the goal is to minimize some objective function?

c What seems to be the best way to locate feasible points that maximize (or minimize) the objective function in a linear programming problem?

Be prepared to explain your ideas to the class.

✓ CHECK YOUR UNDERSTANDING

Paisan's Pizza makes gourmet frozen pizzas to sell to supermarket chains. The company makes two deluxe pizzas, one vegetarian and the other with meat.

- Each vegetarian pizza requires 12 minutes of labor, and each meat pizza requires 6 minutes of labor. The plant has at most 3,600 minutes of labor available each day.

- The plant freezer can handle at most 500 pizzas per day.

- Vegetarian pizza is not quite as popular as meat pizza, so the company can sell at most 200 vegetarian pizzas each day.

- Sale of each vegetarian pizza earns Paisan's $3 profit. Each meat pizza earns $2 profit.

Jandke/Alamy

Paisan's Pizza would like to plan production to maximize daily profit.

a. Translate the objective and constraints in this situation into symbolic expressions and inequalities.

b. On a copy of a grid like the one below, graph the system of constraints to determine the feasible region for Paisan's linear programming problem. Label each segment of the feasible region boundary with the linear equation that determines it.

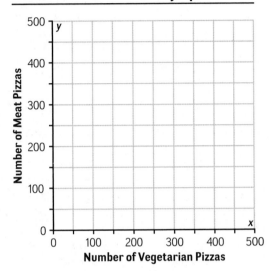

Pizza Production Feasibility Options

c. Evaluate the objective function at the point(s) that you believe will yield greatest daily profit for Paisan's Pizza. Compare that profit figure to the profit for several other feasible points that you think might be "next best."

APPLICATIONS

1 Match each inequality listed in Parts a–j with the region I–X that shows points (x, y) that satisfy the inequality. The scales on the coordinate axes are 1.

    a. $y \geq x$ **b.** $y \leq x$

    c. $2x + 3y \leq 3$ **d.** $2x + 3y \geq 3$

    e. $y \geq 0.5x - 2$ **f.** $y \leq 0.5x - 2$

    g. $x \leq 2$ **h.** $x \geq 2$

    i. $3x + 2y \geq 4$ **j.** $3x + 2y \leq 4$

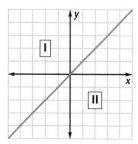 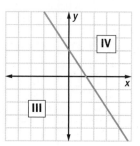

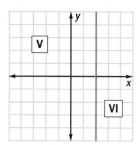 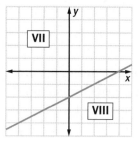

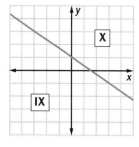

2 Graph the solution of $5x - 2y > 10$.

3 The symbol of Columbia, Maryland, is a people tree that stands by Lake Kittamaqundi. The tree is 14 feet tall with 66 gilded people as branches. As part of a renovation of downtown Columbia in 1992, residents purchased engraved brick pavers to pay for regilding of the people tree. The pavers were used to cover a new plaza around the tree.

Brick pavers engraved with one line of text sold for $25, and pavers engraved with two lines of text sold for $30. Each brick cost $18 to buy, engrave, and install. Regilding the people tree sculpture cost $11,000.

a. Write an inequality whose solution gives the combinations of *one-line pavers* n_1 and *two-line pavers* n_2 that would cover the cost of regilding the people tree sculpture.

b. Draw a graph that uses shading to show the region of the coordinate plane containing all points (n_1, n_2) that satisfy the inequality.

4 Refer back to Applications Task 3. Suppose that the Columbia people tree plaza has space for 1,200 engraved pavers.

a. Write a system of inequalities whose solution will give the (n_1, n_2) combinations for which space is available and the cost of regilding the people tree sculpture will be covered.

b. Draw a graph that shows the region of the first quadrant in the coordinate plane that contains all points (n_1, n_2) satisfying the system of inequalities from Part a.

5 Graph the solutions of the following systems of inequalities in a coordinate plane.

a. $\begin{cases} x - y < 7 \\ 3x + 2y > 9 \end{cases}$

b. $\begin{cases} y \geq x^2 + 5x \\ 3x + y \leq 15 \end{cases}$

6 The Bestform Ring Company makes class rings for high schools and colleges all over the country. The rings are made in production runs that each yield 100 rings. Each production run is a three-step process involving molding, engraving, and polishing.

The following chart gives information about the time required for the steps in each production run of high school or college rings and the time that machines and operators are available during one day.

Class Ring Production			
Stage in Ring Making	**Time to Make 100 High School Rings** (in hours)	**Time to Make 100 College Rings** (in hours)	**Machine and Operator Time Available Each Day** (in hours)
Molding	1.2	2	14
Engraving	0.6	3	15
Polishing	2.0	2	20

a. Test the feasibility of these possible daily production plans.

Plan 1: 2 production runs of high school rings and 3 production runs of college rings

Plan 2: 3 production runs of high school rings and 5 production runs of college rings

Plan 3: 7 production runs of high school rings and 1 production run of college rings

Plan 4: 4 production runs of high school rings and 4 production runs of college rings

Plan 5: 5 production runs of high school rings and 3 production runs of college rings

b. The company makes $500 profit on each production run of high school rings and $525 profit on each production run of college rings.

　i. Compute the daily profit from each feasible production plan in Part a.

　ii. See if you can find a feasible production plan that results in higher profit than any of those in Part a.

7 The Junior Class of Oakland Mills High School sells drinks at the Columbia Fair to raise funds for the Junior Prom. The juniors mix and sell two drinks, Carnival Juice and Lemon Punch.

- Each batch of Carnival Juice uses two liters of orange juice and two liters of lemonade.

- Each batch of Lemon Punch uses one liter of orange juice and three liters of lemonade.

- The students have 120 liters of orange juice and 200 liters of lemonade to use.

- The profit is $9 per batch on the Carnival Juice and $12 per batch on the Lemon Punch.

a. Test the feasibility of these plans for mixing Carnival Juice and Lemon Punch.

Plan 1: 40 batches of Carnival Juice and 20 batches of Lemon Punch
Plan 2: 30 batches of Carnival Juice and 30 batches of Lemon Punch
Plan 3: 40 batches of Carnival Juice and 50 batches of Lemon Punch
Plan 4: 50 batches of Carnival Juice and 20 batches of Lemon Punch

b. Find the profit that the Junior Class will earn from each feasible combination of batches of Carnival Juice and Lemon Punch in Part a.

c. See if you can you find a feasible combination of drink batches that results in higher profit than any of those in Part a.

8 When candidates for political office plan their campaigns, they have choices to make about spending for advertising. Suppose that advisers told one candidate that radio and television ads might reach different audiences in the following ways.

- Every dollar spent on radio advertising will assure that the candidate's message will reach 5 Democratic voters, 2 Republican voters, and 4 Independent voters.

- Every dollar spent on television advertising will assure that the candidate's message will reach 4 Democratic voters, 4 Republican voters, and 1 Independent voter.

The candidate's goals are to reach at least 20,000 Democratic voters, 12,000 Republican voters, and 8,000 Independent voters with a minimum total advertising expense.

a. Test the feasibility of these campaign spending plans for reaching voters.

Plan 1: $2,000 on radio advertising and $3,000 on television advertising
Plan 2: $3,000 on radio advertising and $2,000 on television advertising
Plan 3: $2,000 on radio advertising and $2,000 on television advertising
Plan 4: $4,000 on radio advertising and $3,000 on television advertising

b. i. Find the total advertising cost for each feasible campaign plan in Part a.

ii. See if you can find a feasible combination of advertising dollars that results in a lower total expense for the candidate than any of those in Part a.

9 Refer back to Applications Task 6. Use h to represent the number of production runs of high school rings and c to represent the number of production runs of college rings.

   a. Write inequalities or rules that represent:

      i. the constraint on time available for molding rings.

      ii. the constraint on time available for engraving rings.

      iii. the constraint on time available for polishing rings.

      iv. the objective function.

   b. Use the constraint inequalities you wrote in Part a to create a graph showing the feasible region of the ring production planning problem. Label each segment of the feasible region boundary with the equation of that line.

   c. Find the production plan that will maximize profit for the Bestform Ring Company. Be prepared to explain how you know that your answer is correct.

10 Refer back to Applications Task 7. Use C to represent the number of batches of Carnival Juice and L to represent the number of batches of Lemon Punch to be mixed.

   a. Write inequalities or expressions that represent:

      i. the constraint on amount of orange juice available.

      ii. the constraint on amount of lemonade available.

      iii. the objective function.

   b. Use the constraint inequalities you wrote in Part a to create a graph showing the feasible region of the juice sale problem. Label each segment of the feasible region boundary with the equation of that line.

   c. Find the plan that will maximize profit for the Junior Class from drink sales. Be prepared to explain how you know that your answer is correct.

11 Refer back to Applications Task 8. Use r to represent the number of dollars spent on radio advertising and t to represent the number of dollars spent on television advertising.

   a. Write symbolic expressions that represent:

      i. the constraint on number of Democratic voters to be reached.

      ii. the constraint on number of Republican voters to be reached.

      iii. the constraint on number of Independent voters to be reached.

      iv. the objective function.

   b. Use the constraint inequalities you wrote in Part a to create a graph showing the feasible region of the campaign message problem. Label each segment of the boundary with the equation of that line.

   c. Use the results of your work on Parts a and b to find the plan that will minimize total advertising expense. Be prepared to explain how you know that your answer is correct.

12 Sketch the feasible regions defined by the following inequalities. Use the given equations for profit P and cost C to find (x, y) combinations yielding maximum profit or minimum cost within the feasible regions.

a. $3y - 2x \geq 6$
    $0 \leq x \leq 4$
    $y \leq 5$

    $P = 5x + 3y$

b. $x \leq 10$
    $2x + y \geq 20$
    $y \leq 14$

    $C = 20x + 5y$

13 The director of the Backstage Dance Studio must plan for and operate many different classes, 7 days a week, at all hours of the day.

Each Saturday class fills up quickly. To plan the Saturday schedule, the director has to consider the following facts.

- It is difficult to find enough good teachers, so the studio can offer at most 8 tap classes and at most 5 jazz classes.

- Limited classroom space means the studio can offer at most 10 classes for the day.

- The studio makes profit of $150 from each tap class and $250 from each jazz class.

a. Write and graph the constraint inequalities.

b. The director wants to maximize profit. Write the objective function for this situation.

c. Find the schedule of classes that gives the maximum profit.

d. The director of Backstage Dance Studio really wants to promote interest in dance, so she also wants to maximize the number of children who can take classes. Each tap class can accommodate 10 students, and each jazz class can accommodate 15 students. Find the schedule that gives maximum student participation.

14 A city recreation department offers Saturday gymnastics classes for beginning and advanced students. Each beginner class enrolls 15 students, and each advanced class enrolls 10 students. Available teachers, space, and time lead to the following constraints.

- There can be at most 9 beginner classes and at most 6 advanced classes.

- The total number of classes can be at most 7.

- The number of beginner classes should be at most twice the number of advanced classes.

a. What are the variables in this situation?

b. Write algebraic inequalities giving the constraints on the variables.

c. The director wants as many children as possible to participate. Write the objective function for this situation.

d. Graph the constraints and outline the feasible region for the situation.

e. Find the combination of beginner and advanced classes that will give the most children a chance to participate.

f. Suppose the recreation department director sets new constraints for the schedule of gymnastics classes.

- The same limits exist for teachers, so there can be at most 9 beginner and 6 advanced classes.

- The program should serve at least 150 students with 15 in each beginner class and 10 in each advanced class.

The new goal is to minimize the cost of the program. Each beginner class costs $500 to operate, and each advanced class costs $300. What combination of beginner and advanced classes should be offered to achieve the objective?

CONNECTIONS

15 Identify the geometric shapes with interiors defined by these systems of inequalities.

a. $x \leq 1, y \leq 1, x \geq -1$, and $y \geq -1$

b. $-2x + y \leq 2, 2x + y \geq -2, 2x + y \leq 2$, and $-2x + y \geq -2$

c. $y \leq x, y \leq 1, y \geq x - 2$, and $y \geq 0$

d. $x + y \leq 1, x - y \leq 1, -x + y \geq 1$, and $x + y \geq 1$

16 Identify the geometric shapes with interiors defined by these systems of inequalities.

a. $x^2 + y^2 \leq 25, x \geq 0$, and $y \leq 0$

b. $x^2 + y^2 \leq 25$ and $y \geq x$

17 The linear programming problems in this lesson involve maximizing or minimizing a linear objective function in two variables. In previous work, you sought to find the maximum value of a quadratic function in one variable. For example, when Miguel Tejada hit a homerun in the 2005 major league baseball All-Star game, the function $h(t) = -16t^2 + 64t + 3$ might have been a reasonable model for the relationship between height of the ball (in feet) and time (in seconds) after it left his bat.

a. At what time was his hit at its maximum height and what was that height?

b. At what height did his bat hit the ball?

c. If the ball reached the outfield seats at a point 20 feet above the playing field, what was the approximate time it landed there?

18 Suppose that the following diagram shows the feasible region for a linear programming problem and that the objective function is $P = 2x + 3y$.

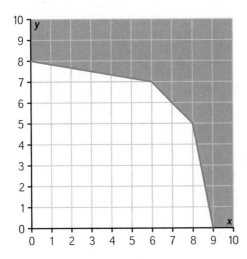

a. On a copy of the diagram, plot the line $2x + 3y = 6$. Explain why the objective function has the value 6 at each point on that line.

b. On the same diagram, plot the line $2x + 3y = 12$. What is the value of the objective function for each point on that line?

c. On the same diagram, plot the line $2x + 3y = 21$. What is the value of the objective function for each point on that line?

d. On the same diagram, plot the line $2x + 3y = 33$. What is the value of the objective function for each point on that line?

e. Explain how the pattern of results in Parts a–d suggests that the lines with equation $2x + 3y = k$ contain no feasible points if $k > 33$.

f. Explain how your work on Parts a–e illustrates the fact that the maximum objective function value in a linear programming problem like this will always occur at a "corner" of the feasible region.

19 The assembly and packaging of CP·I and CP·II video game systems is only one part of the business activity needed to develop a new product. The table below shows some other key tasks, their times to completion, and other tasks on which they depend. Use the information to find the *minimal time to completion* (earliest finish time) for the whole venture of designing and bringing to market products like the CP·I and CP·II video game systems.

Task	Time to Completion	Prerequisites
T1: Design of Game Concept	6 months	
T2: Market Research	3 months	T1
T3: Game Programming	3 months	T1
T4: Design of Production	2 months	T3
T5: Advance Advertising	4 months	T2
T6: Production and Testing of Initial Units	1 months	T4

REFLECTIONS

20 In what ways do you agree and/or disagree with this statement about inequalities and their solutions?

> *Inequalities give less precise answers to questions than equations, but they are more realistic models of practical problems.*

21 Solving linear programming problems includes finding the boundary of the feasible region. Describe at least three different ways to find the points where the boundary lines defining the region intersect.

22 How are the goals and constraints of linear programming problems similar to other optimization problems like finding a minimal spanning tree in a graph, analyzing a PERT chart, or finding selling prices that maximize profit from sale of a product?

23 Look back at your solutions of the linear programming problems in this lesson.

   a. What kind of values made sense for the variables in these problems?

   b. How would you proceed if you determine the optimal solution to a linear programming problem occurs at a point that has noninteger coordinates?

24 Why do you think linear programming has the word *linear* in its name? Do you think the process would work to find optimal solutions if any of the constraints or the objective function were not linear? Why or why not?

25 Realistic linear programming problems in business and engineering usually involve many variables and constraints. Why do you think that linear programming was not used in business and engineering until fairly recently?

EXTENSIONS

26 Regular cell phone use involves a combination of talk and standby time. CellStar claims that its cell phone will get up to 200 minutes of talk time or up to 5 days (7,200 minutes) of standby time for a single battery charge. Consider the possible combinations of talk time and standby time before the battery would need to be recharged.

   a. Write an inequality whose solution will give the (t, s) combinations that will not completely drain a battery that begins use fully charged.

   b. Draw a graph that uses appropriate boundary lines to show the region of the coordinate plane that contains all points with coordinates (t, s) that satisfy the inequality in Part a.

   c. Solve the inequality from Part a for s to show how many minutes of standby time remain after t minutes of talk time before the battery will need to be recharged.

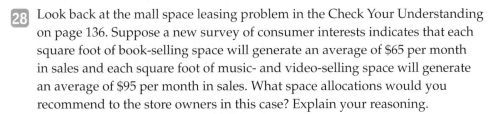

27 Graph the solution of $\frac{x}{y} > 1$.

28 Look back at the mall space leasing problem in the Check Your Understanding on page 136. Suppose a new survey of consumer interests indicates that each square foot of book-selling space will generate an average of $65 per month in sales and each square foot of music- and video-selling space will generate an average of $95 per month in sales. What space allocations would you recommend to the store owners in this case? Explain your reasoning.

29 Refer back to Extensions Task 26. Suppose that Jovan has a CellStar phone and is planning to take a bus from Durham, North Carolina, to San Antonio, Texas, to visit his aunt. The trip is scheduled to take 1 day, 8 hours, and 10 minutes. Jovan will most likely not be able to recharge the phone during the long bus trip and would like to make sure that the phone maintains a charge during the entire trip.

   a. Jovan decides that the solution to the following system will give the number of minutes t of talk time and s of standby time for which the cell phone will maintain a charge for the whole trip.

$$\begin{cases} \frac{t}{200} + \frac{s}{7{,}200} < 1 \\ t + s > 1{,}930 \end{cases}$$

   Do you agree with Jovan? Why or why not?

   b. Graph the solution to the system of inequalities from Part a.

   c. What number of talk-time minutes must Jovan stay below to ensure a charged cell phone for the entire trip?

30 Suppose the manufacturing company that supplies a chain of Packaging Plus stores receives a rush order for 290 boxes. It needs to fill the order in 8 hours or less.

   • The factory has a machine that can produce 30 boxes per hour and costs $15 per hour to operate.

   • The factory can also use two student workers from less urgent tasks. Together, those students can make 25 boxes per hour at a cost of $10 per hour.

   What combination of machine and student work times will meet the deadline with least total cost?

31 The Dutch Flower Bulb Company bags a variety of mixtures of bulbs. There are two customer favorites. The Moonbeam mixture contains 30 daffodils and 50 jonquils, and the Sunshine mixture contains 60 daffodils and 20 jonquils.

   The company imports 300,000 daffodils and 260,000 jonquils for sale each year. The profit for each bag of Moonbeam mixture is $2.30. The profit for each bag of Sunshine mixture is $2.50. The problem is deciding how many bags of each mixture the company should make in order to maximize profit without exceeding available supplies.

   Explore how to use the inequality graphing capability of a CAS like that in *CPMP-Tools* to help find the combination of Moonbeam and Sunshine bags that give maximum profit.

32 When architects design buildings, they have to consider many factors. Construction and operating costs, strength, ease of use, and style of design are only a few.

For example, when architects designing a large city office building began their design work, they had to deal with the following conditions.

• The front of the building had to use windows of traditional style to complement the surrounding historic buildings. There had to be at least 80 traditional windows on the front of the building. Those windows each had an area of 20 square feet and glass that was 0.25 inches thick.

• The back of the building was to use modern-style windows that had an area of 35 square feet and glass that was 0.5 inches thick. There had to be at least 60 of those windows.

• In order to provide as much natural lighting for the building as possible, the design had to use at least 150 windows.

a. One way to rate the possible designs is by how well they insulate the building from loss of heat in the winter. The heat loss R in BTU's (British Thermal Units) per hour through a glass window can be estimated by $R = \frac{5.8A}{t}$, where A stands for the area of the window in square feet and t stands for the thickness of the glass in inches.

   i. What are the heat flow rates of the traditional windows and the modern windows?

   ii. Without graphing the number of window constraints, find the combination of traditional and modern windows that will minimize heat flow from the building.

b. Minimizing construction cost is another consideration. The traditional windows cost $200 apiece, and the modern windows cost $250 each. Without graphing the number of window constraints, find the combination of traditional and modern windows that will minimize the total cost of the windows.

c. Now consider the constraints graphically.

   i. Write the constraint inequalities that match this situation and identify the feasible region.

   ii. Write the objective functions for minimizing heat flow and for minimizing construction costs. Find the combination of traditional and modern windows that will minimize heat flow and total cost of the windows.

   iii. Compare your responses to the minimization questions in Parts a and b to your region and objective functions. Adjust your solutions if necessary.

33 Graph the feasible regions for the following sets of inequalities in the first quadrant. A graphing calculator or CAS may be helpful in finding the "corner points" of the region.

a. $\begin{cases} y \leq 10 - 3x \\ y \leq x^2 - 4x + 5 \end{cases}$
      b. $\begin{cases} y \leq \dfrac{1}{x^2} \\ x + y \leq 4 \end{cases}$

c. $\begin{cases} y \geq 10(0.5^x) \\ y \geq 8 - 2x \end{cases}$

REVIEW

34 In the diagram below, $\ell \parallel m$ and p bisects $\angle ABC$. Find the measure of each numbered angle.

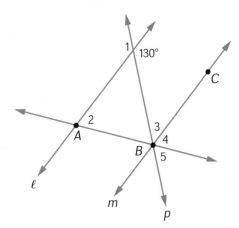

35 Rewrite each expression in equivalent factored form.

a. $x^2 - 16$
      b. $x^2 + 6x - 40$

c. $4x^2 + 12x + 9$
      d. $2x^2 + 7x + 3$

e. $18x^2 + 42x + 12$
      f. $160x^2 - 10$

36 Consider quadrilateral *ABCD* shown in the diagram below. The scale on both axes is 1.

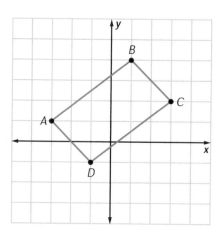

a. Explain why quadrilateral *ABCD* is *not* a rectangle.

b. Find the area of quadrilateral *ABCD*.

c. Quadrilateral *A′B′C′D′* is the image of quadrilateral *ABCD* under a size transformation of magnitude 2 centered at the origin. What is the coordinate matrix of quadrilateral *A′B′C′D′*?

d. How are the angle measures of quadrilateral *A′B′C′D′* related to the angle measures of quadrilateral *ABCD*?

e. Find the area of quadrilateral *A′B′C′D′*.

37 Cesar entered a summer reading challenge at his local library. A winner would be determined as the child who read the most pages during the summer. Each week, Cesar reported the minutes he had spent reading. The librarian told him his ranking and his percentile. After the fourth week, Cesar was ranked 52nd and was at the 94th percentile. How many people were entered in the summer reading contest?

38 Determine if each pair of algebraic expressions are equivalent. If they are equivalent, provide algebraic reasoning to prove your claim. If they are not equivalent, provide a counterexample.

a. $(x + b)^2$ and $x^2 + b^2$

b. $8x - 2(x - 4)^2$ and $32 - 2x^2$

c. $\dfrac{6x + 4}{4}$ and $1.5x + 1$

39 Rewrite each expression in equivalent form with the smallest possible integer under the radical sign.

a. $\sqrt{84}$

b. $\sqrt{160}$

c. $25\sqrt{18}$

d. $(3\sqrt{5})^2$

e. $\sqrt{\dfrac{125}{9}}$

f. $\sqrt{20a^2}$, $a \geq 0$

g. $\sqrt[3]{54}$

h. $\sqrt[3]{8a^3}$

40 For each triangle below, find the measures of the remaining side(s) and angle(s).

a.

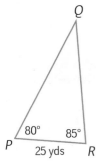

b.

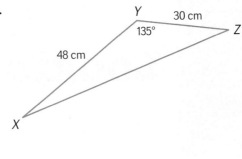

41 Consider the following two sets of data.

- the daily low temperature in International Falls, Minnesota, each day during the month of February

- the temperature of your math classroom at noon each day during the same month

a. Which set of data would you expect to have the greater mean? Explain your reasoning.

b. Which set of data would you expect to have the greater standard deviation? Explain your reasoning.

42 Which quadratic function below has the greatest maximum value?

   i. $f(x) = -(x - 3)^2 + 2$

ii.

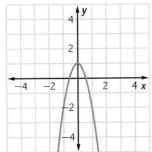

   iii. Quadratic function $h(x)$ whose graph contains the points $(0, -22)$, $(-5, 3)$, and $(-10, -22)$

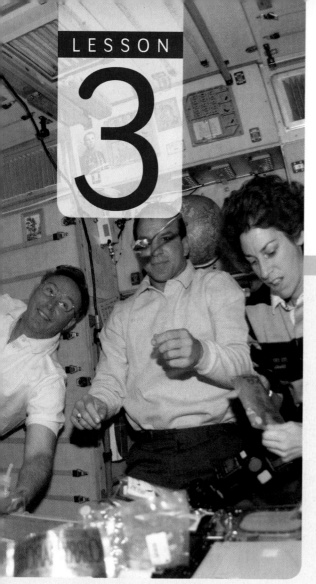

Looking Back

The lessons in this unit involved situations in which important relationships among variables could be expressed by inequalities in one and two variables. In Lesson 1, you combined numeric, graphic, and algebraic methods for representing quadratic functions and solving quadratic equations in order to solve inequalities in one variable. You learned how to use inequality symbols, number line graphs, and interval notation to describe the solutions of those inequalities. In Lesson 2, you used graphs to identify the solutions for linear inequalities and systems of inequalities in two variables. You applied that knowledge to develop graphic and algebraic strategies for solving linear programming problems, optimizing a linear objective function given several linear constraints on the variables.

The tasks in this final lesson will help you review and organize your knowledge about solving inequalities and linear programming problems.

1 For each inequality below:

- make a sketch to show how the functions defined by expressions on each side of the inequality sign are related.

- use algebraic reasoning to locate the key intercepts and points of intersection.

- combine what you learn from your sketch and algebraic reasoning to solve the inequality.

- describe each solution set using symbols, a number line graph, and interval notation.

a. $8 - x^2 \leq 6$ **b.** $x^2 + 4x + 4 < -3x - 8$

c. $7 + x > 2 + 3x - x^2$ **d.** $3 - x^2 \geq x^2 + 5$

e. $2\sqrt{x} > x - 3$ **f.** $|x - 4| \leq 9$

2 Write inequalities that express the same information as the interval notation in these situations.

a. $[3, 6]$ **b.** $(-1, 4)$

c. $(-7, 1.5]$ **d.** $(-\infty, -2) \cup [7, \infty)$

3 For each condition described below, write an inequality involving one linear function and one quadratic function that satisfies the condition. Then draw a graph that illustrates the relationship between the two functions.

   a. The inequality has no solutions.

   b. All real numbers are solutions of the inequality.

   c. Exactly one value satisfies the inequality.

4 Suppose that Lee wants to buy macadamia nuts and banana chips for an after-school snack. Both are available in the bulk foods section of a local grocery store. Macadamia nuts cost $8 per pound, and banana chips cost $5 per pound.

   a. If Lee has $4 to spend, write an inequality whose solution set gives the (*pounds of macadamia nuts m, pounds of banana chips b*) combinations that Lee can buy.

   b. Draw a graph that uses shading to show the region of the coordinate plane that contains all points (m, b) that satisfy the inequality.

5 The graphs representing the system of linear equations
$$\begin{cases} 3x + 5y = 12 \\ 2x - y = 10 \end{cases} \text{ are shown}$$
to the right. How could you replace each of the "=" symbols in the two equations with a "<" or ">" symbol so that the solution of the resulting system of inequalities would be:

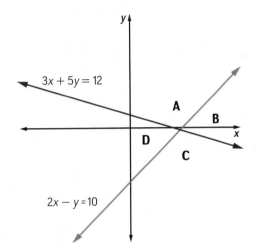

   • region A   • region B

   • region C   • region D

6 A small sporting goods manufacturer produces skateboards and in-line skates. Its dealers demand at least 30 skateboards per day and 20 pairs of in-line skates per day. The factory can make at most 60 skateboards and 40 pairs of in-line skates per day.

   a. Write inequalities expressing the given constraints on daily skateboard and in-line skate manufacturing.

   b. Graph the feasible region. Label each segment of the boundary of the region with the equation of that line.

   c. How many combinations of numbers of in-line skates and skateboards are possible?

   d. Suppose the total number of skateboards and pairs of in-line skates cannot exceed 90.

      i. What inequality expresses this constraint?

      ii. Show this new constraint on your graph.

e. Find coordinates of the corners for the new feasible region.

f. Suppose the profit on each skateboard is $12 and on each pair of in-line skates is $18. Write the profit function.

g. How many of each product should the company manufacture to get maximum profit?

SUMMARIZE THE MATHEMATICS

In this unit, you investigated problems that could be solved by writing and reasoning with inequalities in one or in two variables. In some cases, the problem could be represented with a single inequality. In other cases, as in linear programming problems, representation of the problem required a system of constraint inequalities and an objective function.

a Consider the two functions $f(x)$ and $g(x)$. Describe a general strategy for finding values of x for which each of the following is true.

i. $f(x) = g(x)$ **ii.** $f(x) < g(x)$ **iii.** $f(x) > g(x)$

b The inequality $x^2 - 5x - 24 > 0$ is true for values of x that are less than -3 or greater than 8. Describe these values of x using:

i. inequality symbols   **ii.** a graph on a number line   **iii.** interval notation

c In what ways is solving an inequality like $x^2 - 5x - 24 > 0$ different from finding the solution set for one like $y > x^2 - 5x - 24$? How is it the same?

d What steps are involved in graphing the solutions of an inequality like $3x + 5y \leq 10$?

e Describe the roles played by the following elements of linear programming problems.

i. constraints   **ii.** feasible region   **iii.** objective function

f In a linear programming problem, where in the feasible region will the objective function have its optimum (maximum or minimum) value? Why is that true?

Be prepared to explain your ideas and methods to the class.

 CHECK YOUR UNDERSTANDING

Write, in outline form, a summary of the important mathematical concepts and methods developed in this unit. Organize your summary so that it can be used as a quick reference in future units and courses.

3

Similarity and Congruence

Similarity and congruence are key related ideas in design, manufacturing, and repair of products, from cell phones and MP3 players to automobiles and space shuttles. When individual components that comprise these products are manufactured to design plans and specifications, the components are similar in shape to the design images and congruent to each other, whether manufactured at the same or different plants. Because individual components are congruent, they can be easily interchanged in assembly line production or in repair work.

In this unit, you will extend your understanding and skill in the use of similarity and congruence relations to solve problems involving shape and size. In each of the two lessons, you will have many opportunities to continue to develop skill and confidence in reasoning deductively using both synthetic and coordinate methods.

LESSONS

1 Reasoning about Similar Figures

Derive sufficient conditions for similarity of triangles using the Law of Cosines and the Law of Sines. Use similarity conditions to prove properties of triangles and use those conditions and properties to solve applied problems. Discover and formalize properties of size transformations (dilations).

2 Reasoning about Congruent Figures

Derive sufficient conditions for congruence of triangles as special cases of similarity. Use congruence conditions to prove properties of triangles, to establish sufficient conditions for quadrilaterals to be (special) parallelograms and for special parallelograms to be congruent, and to solve applied problems. Discover and formalize properties of rigid transformations. Revisit congruence and similarity using transformations.

Reasoning about Similar Figures

In Course 1 of *Core-Plus Mathematics*, you discovered combinations of side or angle measures that were sufficient to determine if two triangles were congruent, that is, whether they had the same shape and size. You also explored conditions under which congruent copies of triangles or other polygons could tile a plane.

The mathematically inspired print shown at the right was created by the Dutch artist M. C. Escher for a book about tilings of a plane. The print provides an interesting twist on use of a basic shape to cover a portion of a plane. It also represents an early attempt by the artist to depict infinity.

Escher's framework for the design of his print is shown below. Isosceles right triangle *ABC* is the starting point.

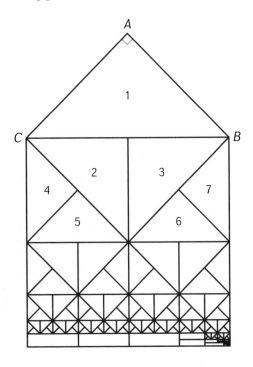

THINK ABOUT THIS SITUATION

Study the Escher print and the framework used to create it.

a How are the lizards in the Escher print alike? How are they different? How could you test your ideas?

b Starting with △*ABC*, the framework was created by next drawing right isosceles triangles 2 and 3. How do you think these two triangles were determined? How are triangles 2 and 3 related to each other? To triangle 1? Why?

c How do you think triangles 4 and 5 and triangles 6 and 7 were created? How are these pairs of triangles related? How are they related to triangle 1? Explain your reasoning.

d Describe a way to generate the remaining portion of the framework.

In the first lesson of this unit, you will explore similarity of polygons and use deductive reasoning to determine sufficient conditions for similarity of triangles. You will use those relationships to solve applied problems and prove important geometric theorems. Throughout, you will see the interconnectedness of geometry, trigonometry, and algebra.

In the second lesson, you will see one of the mathematical advantages of developing sufficient conditions for similarity prior to developing sufficient conditions for congruence of triangles.

When Are Two Polygons Similar?

Whenever the light source of a projector is positioned perpendicular to a screen, it enlarges images on the film into *similar* images on the screen. The enlargement factor from the original object to the projected object depends on the distance from the projector to the screen (and special lens features of the projector).

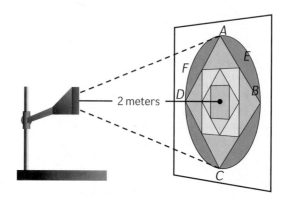

As you work on the problems of this investigation, look for answers to the following questions:

How can you test whether two polygons are similar?

How can you create a polygon similar to a given polygon?

1 Examine the test pattern shown on the screen.

a. How would you test if quadrilateral *ABCD* on the screen is similar to the original quadrilateral that is being projected?

b. How would you test if △*AEF* on the screen is similar to the original triangle that is being projected?

c. How would you move the projector if you wanted to make the test pattern larger? Smaller?

d. How could you determine how much larger projected △*AEF* is than the original triangle? Explain what measurements you would make and how you would use them.

e. Compare your answers with other students. Resolve any differences.

2 Cameras record images digitally and on film. When a camera lens is positioned perpendicular to the plane containing an object, the object and its recorded image are similar. This photograph was taken by a digital camera.

a. Describe how you could use information in the photograph to help determine the actual dimensions of the face of the cell phone.

b. What are those dimensions?

The following problem formalizes some of the ideas you used in Problems 1 and 2.

3 Two polygons with the same number of sides are **similar** provided their corresponding angles have the same measure and the ratios of lengths of corresponding sides is a constant.

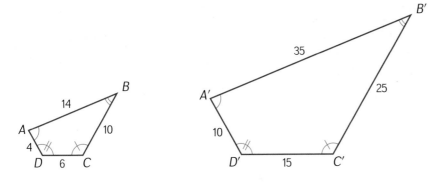

In the above diagram, *quadrilateral A'B'C'D' ~ quadrilateral ABCD*.
The symbol ~ means "is similar to."

$$m\angle A' = m\angle A, m\angle B' = m\angle B, m\angle C' = m\angle C, \text{ and } m\angle D' = m\angle D.$$

$\dfrac{A'B'}{AB} = \dfrac{35}{14} = \dfrac{5}{2}$ or equivalently $A'B' = \dfrac{5}{2}AB$.

$\dfrac{B'C'}{BC} = \dfrac{25}{10} = \dfrac{5}{2}$ or equivalently $B'C' = \dfrac{5}{2}BC$.

$\dfrac{C'D'}{CD} = \dfrac{15}{6} = \dfrac{5}{2}$ or equivalently $C'D' = \dfrac{5}{2}CD$.

$\dfrac{D'A'}{DA} = \dfrac{10}{4} = \dfrac{5}{2}$ or equivalently $D'A' = \dfrac{5}{2}DA$.

The constant $\dfrac{5}{2}$ is called the **scale factor** *from quadrilateral ABCD to quadrilateral A'B'C'D'*. It scales (multiplies) the length of each side of quadrilateral *ABCD* to produce the length of the corresponding side of quadrilateral *A'B'C'D'*.

a. What is the scale factor from quadrilateral *A'B'C'D'* to quadrilateral *ABCD*?

b. How, if at all, would you modify your similarity test in Problem 1 Part b using the above definition of similarity?

c. If two pentagons are similar, describe how to find the scale factor from the smaller pentagon to the larger pentagon. Then describe how to find the scale factor from the larger pentagon to the smaller pentagon.

d. Suppose △*PQR* ~ △*XYZ* and the scale factor from △*PQR* to △*XYZ* is $\dfrac{3}{4}$. Write as many mathematical statements as you can about pairs of corresponding angles and about pairs of corresponding sides. Compare your statements with other students. Resolve any differences.

Knowing that two triangles are similar allows you to conclude that the three pairs of corresponding angles are congruent and that the three pairs of corresponding sides are related by the same scale factor. Conversely, if you know that the three pairs of corresponding angles are congruent and the three pairs of corresponding sides are related by the same scale factor, you can conclude that the triangles are similar.

4 The diagram below is a portion of the framework for Escher's print that you examined in the Think About This Situation (page 163). Recall that $\triangle ABC$ is an isosceles right triangle. Assume $BC = 2$ units.

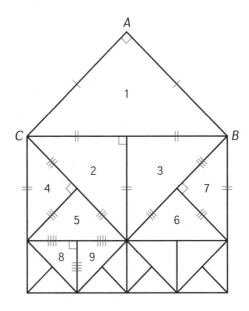

a. Compare the markings on the sides and angles of the triangles with your analysis of the framework. Explain why the markings are correct.

b. Determine if each statement below is correct. If so, explain why and give the scale factor from the first triangle to the second triangle. If the statement is not correct, explain why.

   i. $\triangle 1 \sim \triangle 3$ ii. $\triangle 2 \sim \triangle 6$

   iii. $\triangle 4 \sim \triangle 6$ iv. $\triangle 8 \sim \triangle 3$

   v. $\triangle 9 \sim \triangle 1$

5 Based on their work in Problem 4, several students at Black River High School made conjectures about families of polygons. Each student tried to outdo the previous student. For each claim, explain as precisely as you can why it is true or give a counterexample.

a. Monisha conjectured that *all isosceles right triangles are similar.*

b. Ahmed conjectured that *all equilateral triangles are similar.*

c. Loreen claimed that *all squares are similar.*

d. Jeff conjectured that *all rhombuses are similar.*

e. Amy claimed that *all regular hexagons are similar.*

6 The "Explore Similar Triangles" custom app is designed to help you construct similar triangles based on a chosen scale factor. As a class or in pairs, experiment with the software. Conduct at least three trials. Observe what happens to the angle and side measures of the triangle as you drag the red vertices.

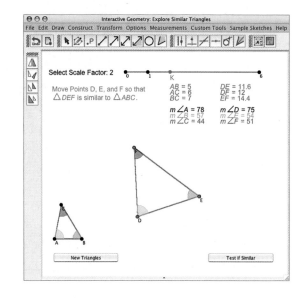

a. For each trial:

- choose a target scale factor and construct △DEF so that it is similar to △ABC.

- use the definition of similarity to justify that your triangles are similar.

b. Think about the process you used to create a triangle similar to △ABC.

i. When dragging △DEF to form a triangle similar to △ABC, how did you decide how to initially drag the vertices?

- Did you use the angle measurements?

- Did you use the length measurements?

ii. Once you knew that two pairs of corresponding sides were related by the same scale factor, did that guarantee that the third pair was automatically related by the same scale factor?

c. How do you think the software determines if your constructed triangle is similar to the original triangle?

d. How does the software visually demonstrate the similarity of two triangles?

SUMMARIZE THE MATHEMATICS

In this investigation, you explored similarity of polygons with a focus on testing for similarity of triangles.

a Explain why not all rectangles are similar.

b What is the fewest number of measurements needed to test if two rectangles are similar? Explain.

c Explain why any two regular *n*-gons are similar. How would you determine a scale factor relating the two *n*-gons?

d What needs to be verified before you can conclude that two triangles, △PQR and △UVW, are similar?

Be prepared to share your ideas and reasoning with the class.

Each triangle described in the table below is similar to △ABC. For each triangle, use this fact and the additional information given to:

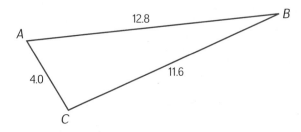

a. identify the correspondence between its vertices and those of △ABC.

b. determine the remaining table entries.

Triangle Angle Measures			Shortest Side Length	Longest Side Length	Third Side Length	Scale Factor from △ABC
m∠A = 64°	m∠B = 18°	m∠C = 98°	AC = 4.0	AB = 12.8	BC = 11.6	1
m∠D = ?	m∠E = 64°	m∠F = 18°				2
m∠G = ?	m∠H = ?	m∠I = ?		IG = 6.4	GH = 5.8	
m∠J = ?	m∠K = 18°	m∠L = 98°	JL = 14.0			

INVESTIGATION 2

Sufficient Conditions for Similarity of Triangles

In Investigation 1, you proposed a method by which computer software could test if two triangles were similar. In Lesson 2, you will examine more explicitly this transformation-based test. As you tested pairs of triangles, you compared corresponding angle measures and corresponding side lengths. You may have observed that if certain conditions on measures of some corresponding sides or angles are met, similar relationships must then hold for the remaining corresponding parts. In this investigation, you will explore minimal conditions that will ensure that two triangles are similar. You will find the previously proven Law of Sines and Law of Cosines helpful in your work. These two laws are reproduced here for easy reference.

Law of Sines

In any triangle ABC with sides of lengths a, b, and c opposite $\angle A$, $\angle B$, and $\angle C$, respectively, $\dfrac{\sin A}{a} = \dfrac{\sin B}{b} = \dfrac{\sin C}{c}$ or equivalently, $\dfrac{a}{\sin A} = \dfrac{b}{\sin B} = \dfrac{c}{\sin C}$.

Law of Cosines

In any triangle ABC with sides of lengths a, b, and c opposite $\angle A$, $\angle B$, and $\angle C$, respectively, $c^2 = a^2 + b^2 - 2ab \cos C$.

As you work on the problems of this investigation, look for answers to the following question:

What combinations of side or angle measures are sufficient to determine that two triangles are similar?

1 To begin, consider $\triangle ABC$ with b, c, and m$\angle A$ given as shown below.

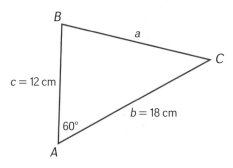

a. Calculate a, m$\angle B$, and m$\angle C$.

b. Are a, m$\angle B$, and m$\angle C$ *uniquely determined*? That is, is there exactly one value possible for each? Explain your reasoning.

2 Now consider the general case. Suppose you know values of b, c, and m$\angle A$.

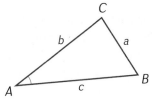

a. Can values for a, m$\angle B$, and m$\angle C$ always be found? Could any of a, m$\angle B$, or m$\angle C$ have two or more values when b, c, and m$\angle A$ are given?
Compare your answers and reasoning with others. Resolve any differences.

b. Summarize your work in an if-then statement that begins as follows:

In a triangle, if the lengths of two sides and the measure of the angle included between those sides are known, then … .

3 Next, examine $\triangle ABC$ and $\triangle XYZ$ shown below. In both cases, you have information given about two sides and an included angle. In $\triangle XYZ$, k is a constant. Test if this information is sufficient to conclude that $\triangle ABC \sim \triangle XYZ$ using Parts a–d as a guide.

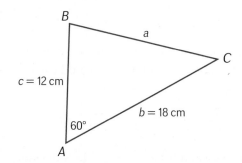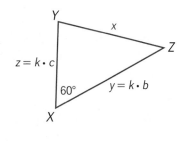

a. In $\triangle XYZ$, m$\angle X = $ m$\angle A = 60°$, $z = k \cdot c = 12k$ cm, and $y = k \cdot b = 18k$ cm, where k is a constant. Explain why x, m$\angle Y$, and m$\angle Z$ are uniquely determined.

b. Use the information given for △XYZ to write and then solve an equation to find length *x*. Based on your work in Problem 2, how is *x* related to *a*?

c. Find m∠Y. How are m∠B and m∠Y related?

d. Find m∠Z. How are m∠C and m∠Z related?

e. Considering all the information in Parts a–d, what can you conclude about △ABC and △XYZ?

4 Suppose you know that in △ABC and △DEF, m∠C = m∠F, $d = k \cdot a$, and $e = k \cdot b$.

a. Could you prove that △ABC ~ △DEF? If so, explain how. If not, explain why not.

b. Complete the following statement.

> *If an angle of one triangle has the same measure as an angle of a second triangle, and if the lengths of the corresponding sides including these angles are multiplied by the same scale factor k, then … .*

c. Compare your statement in Part b with those of your classmates. Resolve any differences.

The conclusion you reached in Part c of Problem 4 is called the **Side-Angle-Side (SAS) Similarity Theorem**. This theorem gives at least a partial answer to the question of finding minimal conditions that will guarantee two triangles are similar. In the next several problems, you will explore other sets of sufficient conditions.

5 Suppose you know that in △ABC and △XYZ, the lengths of corresponding sides are related by a scale factor *k* as in the diagram below.

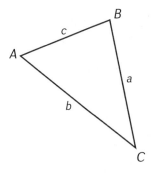
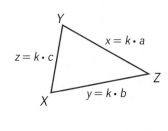

a. What additional relationship would you need to know in order to conclude that △ABC ~ △XYZ? What is a possible strategy you could use to establish the relationship?

b. Write a deductive argument proving the relationship you stated in Part a. What can you conclude?

c. The results of your work in Parts a and b establish a **Side-Side-Side (SSS) Similarity Theorem**. Write this SSS Similarity Theorem in if-then form.

6 In Problems 4 and 5, given two triangles, you needed to know three of their corresponding measures in order to determine that the triangles were similar. In this problem, you will examine a situation in which only two corresponding measures of two triangles are known.

Suppose you are given $\triangle ABC$ and $\triangle XYZ$ in which $m\angle X = m\angle A$ and $m\angle Y = m\angle B$.

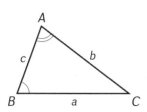
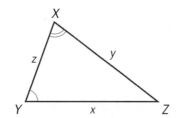

a. Do you think $\triangle ABC \sim \triangle XYZ$? Why or why not?

b. What additional relationship(s) would you need to know to conclude for sure that the triangles are similar?

c. Students at Edison High School believed the answer to Part a was "yes." Study their proof given below.

Try to provide reasons that support each of their statements. Correct any misstatements.

Proof:

We first looked at the ratio of the lengths a and x of a pair of corresponding sides.

If we let k represent the ratio $\frac{x}{a}$, then $x = ka$. (1)

For $\triangle ABC$, we know that $\frac{a}{\sin A} = \frac{b}{\sin B}$. So, $b = \frac{a \sin B}{\sin A}$. (2)

Similarly for $\triangle XYZ$, we know that $\frac{x}{\sin X} = \frac{y}{\sin Y}$. So, $y = \frac{x \sin Y}{\sin X}$. (3)

It follows that $y = \frac{ka \sin Y}{\sin X}$. (4)

Since $m\angle Y = m\angle B$ and $m\angle X = m\angle A$, it follows that $y = \frac{ka \sin B}{\sin A}$. (5)

Since $b = \frac{a \sin B}{\sin A}$, it follows that $y = kb$. (6)

Since $m\angle X = m\angle A$ and $m\angle Y = m\angle B$, then $m\angle Z = m\angle C$. (7)
We've shown $x = ka$, $y = kb$, and $m\angle Z = m\angle C$.

So, $\triangle ABC \sim \triangle XYZ$. (8)

d. Could the students have reasoned differently at the end of their argument? How?

e. Write an if-then statement of the theorem proved in this problem. What name would you give the theorem? Compare your answer with your classmates and resolve any differences.

In this investigation, you established three sets of conditions, each of which is sufficient to prove that two triangles are similar.

a State, and illustrate with diagrams, the sets of conditions on side lengths and/or angle measures of a pair of triangles that ensure the triangles are similar.

b For each set of conditions, identify the key assumptions and theorems used in deducing the result.

c How do the Law of Cosines and Law of Sines help in determining the sets of sufficient conditions for similarity of two triangles?

Be prepared to share your descriptions and reasoning with the class.

 CHECK YOUR UNDERSTANDING

The design of homes and office buildings often requires roof trusses that are of different sizes but have the same pitch. The trusses shown below are designed for roofs with solar hot-water collectors. For each Part a through e, suppose a pair of roof trusses have the given characteristics. Determine if $\triangle ABC \sim \triangle PQR$. If so, explain how you know that the triangles are similar and give the scale factor from $\triangle ABC$ to $\triangle PQR$. If not, give a reason for your conclusion.

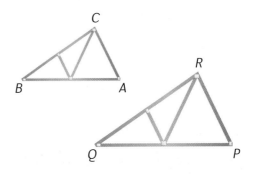

a. $m\angle A = 57°$, $m\angle B = 38°$, $m\angle P = 57°$, $m\angle R = 95°$

b. $AB = 12$, $BC = 15$, $m\angle B = 35°$, $PQ = 16$, $QR = 20$, $m\angle Q = 35°$

c. $AB = BC$, $m\angle B = m\angle Q$, $PQ = QR$

d. $AC = 4$, $BC = 16$, $BA = 18$, $PR = 10$, $QR = 40$, $QP = 48$

e. $AB = 12$, $m\angle A = 63°$, $m\angle C = 83°$, $m\angle P = 63°$, $m\angle Q = 34°$, $PQ = 12$

Reasoning with Similarity Conditions

Similarity and proportionality are important ideas in mathematics. They provide tools for designing mechanisms, calculating heights and distances, and proving mathematical relationships. For example, a *pantograph* is a parallelogram linkage used for copying drawings and maps to a different scale. The pantograph is fastened down with a pivot at *F*. A stylus is inserted at *D*. A pencil is inserted at *P*. As the stylus at *D* traces out a path on the map, the pencil at *P* draws a scaled copy of that path.

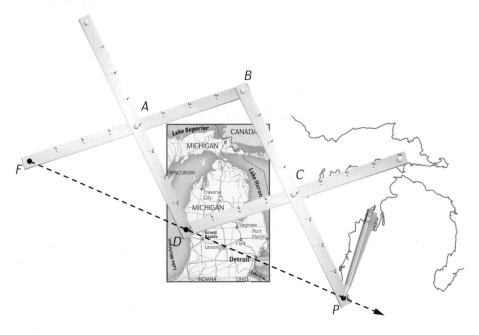

As you work on the problems of this investigation, look for answers to why a pantograph works as it does and the more general question:

What strategies are useful in solving problems using similar triangles?

1 Designing Pantographs Working with a partner, construct a model of a pantograph like the one shown below. Connect four linkage strips at points *A*, *B*, *C*, and *D* with paper fasteners.

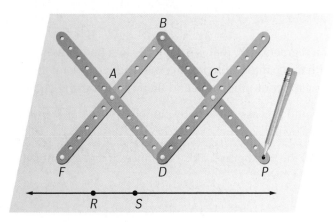

a. Mark points R and S on a sheet of paper about two inches apart. Draw $\overleftrightarrow{RS}$. Place the pantograph so that point F is on the line near the left end. Label your placement of point F for future reference. Hold point F *fixed*. Place point D somewhere else on the line. Where is point P in relation to the line?

As you move point D along the line:

   i. describe the path a pencil at point P follows.

   ii. what kind of quadrilateral is $ABCD$? Why?

   iii. how is the triangle determined by points F, B, and P related to the triangle determined by points F, A, and D? Explain your reasoning.

   iv. what can you conclude about the lengths FP and FD?

b. Now, holding point F fixed in the same place as before, find the image of point S by placing point D over point S and marking the location of point P. Label your mark as the image point S'. What is true about points F, S, and S'?

c. Keeping F fixed in the same position, find the image of point R in the same manner. Label it R'.

   i. What is true about the points F, R, and R'?

   ii. Compare the lengths RS and $R'S'$. How does the relationship compare with what you established in Part aiii?

d. Mark a point T on your paper so that $\triangle RST$ is a right triangle with right angle at R. Find the image of point T as you did for points R and S.

e. How is $\triangle R'S'T'$ related to $\triangle RST$? Explain your reasoning.

2 Examine the pantograph illustrated in the diagram below.

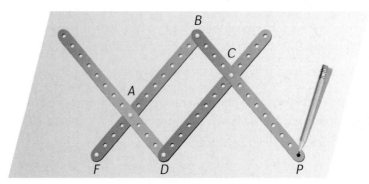

a. Draw a segment, $\overline{XY}$, that is 5 cm long. Imagine drawing the image of $\overline{XY}$ with this pantograph. Predict the length of the image $\overline{X'Y'}$. Explain the basis for your prediction and then check it using your pantograph.

b. How would you assemble your pantograph so that it multiplies distances by 4?

c. How will the pantograph affect distances if it is assembled so that $FB = k \cdot FA$? Why?

3 **Calculating Heights and Distances** In Course 2, you calculated the height of Chicago's *Bat Column* and other structures using trigonometry. Triangle similarity provides another method.

Suppose a mirror is placed on the ground as shown. You position yourself to see the top of the sculpture reflected in the mirror. An important property of physics states that in such a case, the *angle of incidence* is congruent to the *angle of reflection*.

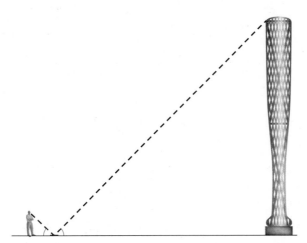

a. On a copy of the diagram above, sketch and label a pair of triangles which, if proven similar, could be used to calculate the height of the *Bat Column*. Write a similarity statement relating the triangles you identified.

b. Prove that the two identified triangles are similar.

c. Add to your diagram the following measurements.

  • The ground distance between you and the mirror image of the top of the column is 6 feet 5 inches.

  • The ground distance between the mirror image and the base of the column is 116 feet 8 inches.

  • Assume that the distance from the ground up to your eyes is 66 inches.

  About how tall is the column?

d. How could you use a trigonometric ratio and the ground distance from the mirror image to the column to calculate the height of the *Bat Column*? What measurement would you need? How could you obtain it?

4 As part of their annual October outing to study the changing colors of trees in northern Maine, several science club members from Poland Regional High School decided to test what they were learning in their math class by finding the width of the Penobscot River at a particular point *A* as shown on the next page.

Pacing from point *A*, they located points *D*, *E*, and *C* as shown in the diagram below.

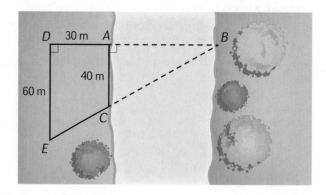

a. How do you think they used similarity to calculate the distance *AB*? Be as precise as possible in your answer.

b. What is your estimate of the width of the river at point *A*?

c. Another group of students repeated the measurement, again pacing from point *A* to locate points *D*, *E*, and *C*. In this case, *AD* = 18 m, *DE* = 26 m, and *AC* = 20 m. Would you expect that this second group got the same estimate for the width of the river as you did in Part b? Explain.

d. What trigonometric ratio could be used to calculate the width of the river? What measurements would you need and how could you obtain them?

5 **Discovering and Proving New Relationships** Study the diagram below of △*ABC*. $\overline{MN}$ connects the midpoints *M* and *N* of sides $\overline{AB}$ and $\overline{BC}$, respectively.

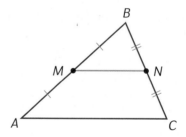

a. How does $\overline{MN}$ appear to be related to $\overline{AC}$?

b. How does the length of $\overline{MN}$ appear to be related to the length of $\overline{AC}$?

c. Use interactive geometry software or careful paper-and-pencil drawings to investigate if your observations in Parts a and b hold for triangles of different shapes. Compare your findings with those of your classmates.

d. Complete the following statement.

If a line segment joins the midpoints of two sides of a triangle, then it … .

e. Collaborating with others as needed, write a proof of the statement in Part d. That statement is often called the **Midpoint Connector Theorem for Triangles**.

6 Possible new theorems are often discovered by studying several cases and looking for patterns as you did in Problem 5. Another way of generating possible theorems is to modify the statement of a theorem you have already proven. For example, consider the following statement.

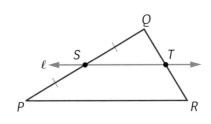

If a line intersects the midpoint of one side of a triangle and is parallel to a second side, then it intersects the third side at its midpoint.

Do you think this statement is always true? If so, prove it. If not, provide a counterexample.

Revisiting Size Transformations In Course 2, you studied size transformations with center at the origin, defined by coordinate rules of the form $(x, y) \rightarrow (kx, ky)$, where $k > 0$. Your work with pantographs suggests a way of defining size transformations without using coordinates and with *any* point in the plane as the center.

For any point C and real number $k > 0$, a **size transformation** (dilation) with **center** C and **magnitude** k is a function that maps each point of the plane onto an image point as follows:

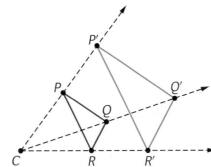

- Point C is its own image.

- For any point P other than C, the image of point P is the point P' on $\overrightarrow{CP}$ for which $CP' = k \cdot CP$.

Size Transformation Principle To size transform a set of points, you find the size-transformation image of each point in the set.

7 Now conduct each of the following experiments using interactive geometry software or paper, pencil, and a ruler.

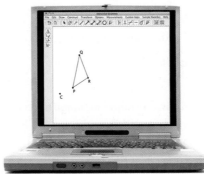

a. Draw a △PQR as shown at the right. Mark and label any two points on $\overline{PQ}$. Repeat for sides $\overline{QR}$ and $\overline{PR}$. Choose a point C in the *exterior* of △PQR as the center of a size transformation with magnitude 2.5. Find and label the image of △PQR and the six points on its sides.

   i. Does collinearity of points appear to be preserved by this size transformation? Does betweenness of points appear to be preserved? Compare your findings with those of others and agree on written *if-then* summary statements.

   ii. What kind of figures are the images of $\overline{PQ}$, $\overline{QR}$, and $\overline{PR}$? How are the lengths of the images related to the lengths of their preimages?

   iii. What kind of figures are the images of ∠PQR, ∠QRP, and ∠RPQ? How are the measures of images and preimages related?

b. Compare your findings in Parts ai–aiii with those of your classmates and resolve any differences.

c. Repeat Parts a and b for a size transformation with center *C* in the *interior* of △*PQR* and magnitude 0.5.

d. Based on your work in Parts a and c, how does the orientation of an image triangle compare to that of its preimage under a size transformation?

Your work in Problem 7 supports the reasonableness of the following assumed statements.

Assumptions about Size Transformations

1. Size transformations preserve:
   a. collinearity of points
   b. betweenness of points
   c. angles and angle measures
   d. orientation

2. Under a size transformation with center *C* and magnitude *k*, the distance between pairs of image points is *k* times the distance between their preimages.

These assumptions lead to useful properties of size transformations that aid in sketching images of figures.

8 Dividing the work up among your classmates, provide arguments justifying each of the following properties of size transformations with center *C* and magnitude *k*. Report your argument to the class and critique the arguments of others.

a. Size Transform Image of a Line Property If *A′* is the image of point *A* and *B′* is the image of point *B*, then $\overleftrightarrow{A'B'}$ is the image of $\overleftrightarrow{AB}$.

b. Size Transform Image of a Segment Property If the image of point *P* is *P′* and the image of point *Q* is *Q′*, then the image of $\overline{PQ}$ is $\overline{P'Q'}$.

c. Size Transform Image of a Ray Property If the image of point *X* is *X′* and the image of point *Y* is *Y′*, then the image of $\overrightarrow{XY}$ is $\overrightarrow{X'Y'}$.

d. Size Transform Image of an Angle Property If the images of noncollinear points *D*, *E*, and *F* are *D′*, *E′*, and *F′*, respectively, then the image of ∠*DEF* is ∠*D′E′F′*.

9 In the diagram below, a size transformation with enter *C* and magnitude *k* maps *A* onto *A′* and *B* onto *B′*. Prove that $\overleftrightarrow{A'B'} \parallel \overleftrightarrow{AB}$.

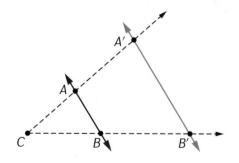

10 In the diagram below, △E'F'G' is the image of △EFG under a size transformation with center C and magnitude k.

a. How would you calculate the magnitude k?

b. How do △EFG and △E'F'G' appear to be related?

c. Using the definition of Similarity in Investigation 1, prove that under a size transformation with center C and magnitude k, a triangle and its image are similar.

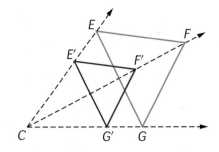

SUMMARIZE THE MATHEMATICS

In this investigation, you explored how conditions for similarity of triangles could be used to solve applied problems and to prove important mathematical relationships.

a How would you assemble a pantograph to make an enlargement of a shape using a scale factor of 8? Of r?

b How could you calculate the height of a structure like the *Bat Column* using shadows cast on a sunny day?

c Suppose in a diagram, △M'N'O' is the image of △MNO under a size transformation of magnitude $k > 0, k \neq 1$.

   i. How could you locate the center of the size transformation?

   ii. How could you determine the magnitude k?

   iii. How are the side lengths and angle measures of the two triangles related?

d What strategies are helpful in using similarity to solve problems in applied contexts? To prove mathematical statements?

Be prepared to explain your ideas and strategies to the class.

 CHECK YOUR UNDERSTANDING

Examine the diagram of a pantograph enlargement of △PQR.

a. Prove each statement.

   i. △FAR ~ △FBR'

   ii. △FRQ ~ △FR'Q'

   iii. △PQR ~ △P'Q'R'

b. How could you represent the scale factor of this enlargement?

c. Describe the center and magnitude of a size transformation that will map △P'Q'R' onto △PQR.

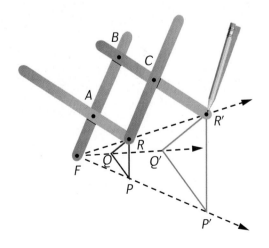

APPLICATIONS

1 A flashlight is directed perpendicularly at a vertical wall 24 cm away. A cardboard triangle with sides of lengths 3, 4, and 5 cm is positioned directly between the light and the wall, parallel to the wall so that its projected shadow image is similar to it.

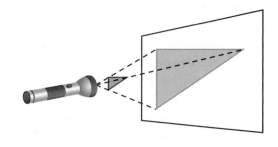

a. Suppose the shadow of the 4-cm side is 10 cm. Find the lengths of the shadows of the other two sides.

b. Suppose the shadow of the 5-cm side is 7.5 cm. Find the lengths of the other two sides of the shadow.

c. How far from the light source should you place the cardboard triangle so that the 4-cm side has a 12-cm shadow?

2 While pumping gas on his way to work, Josh witnessed a robber run out of the station, get into his car, and drive off. As an off-duty police detective, Josh knew that he needed to document the scene before the evidence was destroyed. Using his cell phone camera and whatever he could find in his pockets, Josh took the following four pictures.

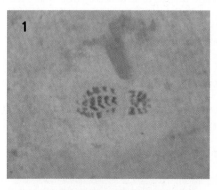

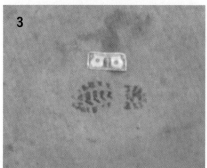

a. Consider the dimensions of the standard dollar bill in each of the photos. In each case below, which photo provides more useful or reliable information? Explain your reasoning.

   i. Photo 1 or Photo 2

   ii. Photo 3 or Photo 4

b. How could a technician use your selected photos to calculate the actual dimensions of the impression? What are those dimensions?

c. Would you need to consider two photos to determine the actual dimensions of the impression, or could you use just one? How might having two different photos strengthen the case?

3 Amber and Christina had different ideas about how to use grid paper to reduce triangles to half their original size. Each was given △ABC drawn on half-inch grid paper. Amber drew △A'B'C' as shown on quarter-inch grid paper. Christina drew △A"B"C" as shown on half-inch grid paper.

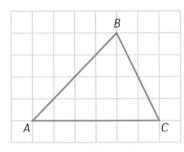
Original Triangle

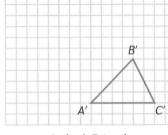
Amber's Triangle

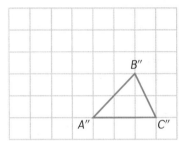
Christina's Triangle

a. Explain why Amber's triangle is similar to the original. What is the scale factor?

b. Explain why Christina's triangle is similar to the original and has a scale factor of $\frac{1}{2}$.

c. Are Christina's triangle and Amber's triangle similar to one another? If so, what is the scale factor?

4 Examine each pair of triangles. Using the information given, determine whether the triangles are similar or the information is not sufficient to draw a conclusion. Explain your reasoning.

a.
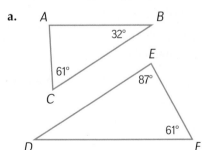

b.
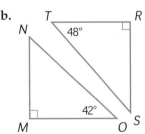

c.

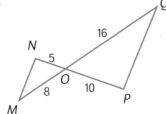

d.

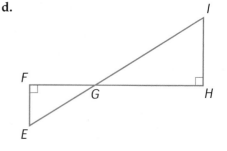

e.

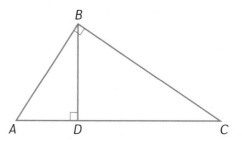

f.

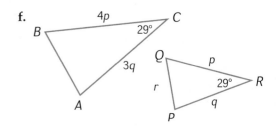

5 In the diagram below, $\triangle ABC$ is a right triangle. $\overline{BD}$ is the altitude drawn to the hypotenuse $\overline{AC}$.

a. Identify three pairs of similar triangles. Be sure that corresponding vertices are labeled in the same order.

b. Describe the strategies you would use to prove each pair of triangles similar.

6 In Investigation 2, you established three sets of conditions that are sufficient to prove that two triangles are similar: SAS, SSS, and AA. Here, you are asked to replace the word "triangle" in these statements with the word "parallelogram." Then decide if the new statement is true or not.

a. Rewrite the SAS Similarity Theorem for triangles as a statement about two parallelograms. Prove or disprove your statement.

b. Rewrite the SSS Similarity Theorem for triangles as a statement about two parallelograms. Prove or disprove your statement.

c. Rewrite the AA Similarity Theorem for triangles as a statement about two parallelograms. Prove or disprove your statement.

7 Wildlife photographers use sophisticated film cameras in their field work. Negatives from the film are then resized for publication in books and magazines. In this task, you will examine how a photographic enlarger uses the principles of size transformations. The light source is the center of the size transformation, a photographic negative or slide is the preimage, and the projection onto the photographic paper is the image. On one enlarger, the distance between the light source and film negative is fixed at 5 cm. The distance between the paper and the negative adjusts to distances between 5 cm and 30 cm.

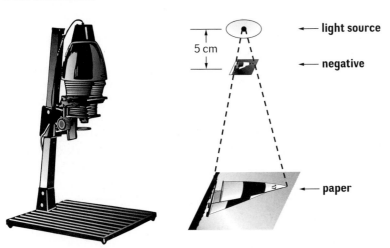

a. If the distance from the light source to the paper is 20 cm, what ratio gives the magnitude of the size transformation represented?

b. If the distance between the negative and the paper is 8 cm, what is the magnitude of the size transformation?

c. Deonna uses 35 mm film in her camera, so her negatives measure 35 mm by 23 mm.

  i. When the paper is 5 cm from the negative, her prints are 70 mm by 46 mm (7.0 cm by 4.6 cm). Use a diagram and your understanding of size transformations to explain why this is the case.

  ii. When she adjusts the distance between the paper and the negative to 10 cm, her prints are 105 mm by 69 mm (10.5 cm by 6.9 cm). Explain why this is the case.

  iii. What is the magnitude of the enlargement in each of the two settings above? How does this affect the distances involving the light source, negative, and photographic paper?

d. Recall that the distance between the paper and the negative can be adjusted to distances from 5 cm to 30 cm. What are the dimensions of the *smallest* print that Deonna can make rom a negative using this brand of enlarger? What are the dimensions of the *largest* print?

e. Suppose the paper is 20 cm from Deonna's negative. What size print is produced?

f. How far should the paper be from Deonna's negative to produce a print whose longest side is 12 cm?

8 Most ironing boards can be adjusted to different heights. One ironing board, with a design similar to the one shown here, has legs that are each 110 cm long and are hinged at a point that is 40 cm from the top end of each. Possible working heights are 90 cm, 85 cm, 80 cm, 75 cm, and 70 cm.

a. Make a sketch of the ironing board, labeling vertices of important triangles. Carefully explain why, for any of the working heights, the surface of the ironing board is parallel to the floor.

b. For a working height of 90 cm, determine the distance between the two points at which the legs connect to the ironing board. Find the measures of the angles of the top triangle.

9 In the diagram below, quadrilateral *PQRS* is a parallelogram, $\overline{SQ}$ is a diagonal, and $\overleftrightarrow{SQ} \parallel \overleftrightarrow{XY}$.

a. Prove that $\triangle XYR \sim \triangle SQR$.

b. Prove that $\triangle XYR \sim \triangle QSP$.

c. Identify the center and magnitude of a size transformation that maps $\triangle RXY$ onto $\triangle RSQ$.

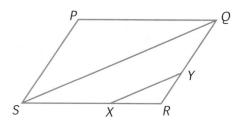

10 Using interactive geometry software (without coordinates) or paper, pencil, and a ruler, conduct the following experiment.

- Draw $\triangle ABC$.

- Mark a point *D* and then find the image of $\triangle ABC$ under a size transformation with center at *D* and scale factor 3. Label the image $\triangle A'B'C'$.

- Mark a second point *E* and then find the image of $\triangle A'B'C'$ under a size transformation with center *E* and scale factor $\frac{1}{2}$. Label the image $\triangle A''B''C''$.

a. How are $\triangle ABC$ and $\triangle A''B''C''$ related? Explain your reasoning.

b. Can you find a single size transformation that maps $\triangle ABC$ onto $\triangle A''B''C''$? If so, what is the center of the size transformation? What is the scale factor?

c. Repeat the experiment for a different $\triangle ABC$, different points *D* and *E*, and different scale factors. Include a case where the two centers are the same point. Answer Parts a and b for these new cases.

d. Repeat Part c. But this time, use scale factors 2 and $\frac{1}{2}$.

e. Conduct additional experiments as necessary in search of general patterns to complete this statement.

The composition of two size transformations with scale factors m and n is … .

CONNECTIONS

11 **Golden rectangles** have the special property that if you cut off a square from one side, the remaining rectangle is similar to the original rectangle. The ratio $\dfrac{\text{length of long side}}{\text{length of short side}}$ leads to the special proportion,

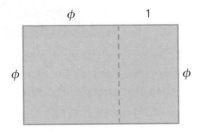

$$\frac{\phi}{1} = \frac{\phi + 1}{\phi}, \text{ or}$$

$$\phi = \frac{\phi + 1}{\phi}.$$

The French architect Le Corbusier incorporated the golden rectangle in the design of this villa.

The number ϕ (the Greek letter "phi") is called the **golden ratio**. Golden rectangles are often judged to be the most attractive rectangles from an artistic and architectural point of view.

a. Use algebraic reasoning to determine exact and approximate values of ϕ.

b. Determine which of the rectangles below are golden rectangles. Describe your test.

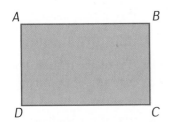

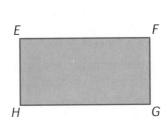

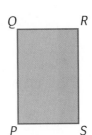

c. Is rectangle *ABCD* similar to rectangle *PQRS*? Explain your reasoning.

d. Prove or disprove that all golden rectangles are similar.

12 Suppose $\triangle ABC \sim \triangle XYZ$ with k as the scale factor from $\triangle ABC$ to $\triangle XYZ$.

a. Explain why $\dfrac{XY}{AB} = \dfrac{YZ}{BC} = \dfrac{XZ}{AC}$.

b. Explain why $\dfrac{AB}{XY} = \dfrac{BC}{YZ} = \dfrac{AC}{XZ}$.

c. Recall that a *proportion* is a statement of equality between ratios. Restate the definition of similar polygons (page 165) using the idea of proportion.

13 In similar triangles, the ratios of lengths of the corresponding sides are equal.

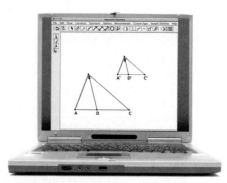

a. Use interactive geometry software to test each of the following claims made by a group of students at Smyrna High School. Which claims appear to be true statements?

Claim 1: In similar triangles, the ratio of the lengths of the bisectors of corresponding angles is the same as the ratio of the lengths of corresponding sides. For example, if $\triangle ABC \sim \triangle A'B'C'$, $\overline{BD}$ bisects $\angle ABC$, and $\overline{B'D'}$ bisects $\angle A'B'C'$, then $\frac{BD}{B'D'} = \frac{AC}{A'C'}$.

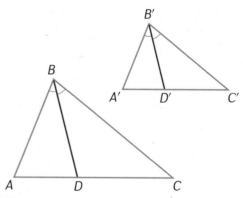

Claim 2: In similar triangles, the ratio of the lengths of the altitudes from corresponding vertices is the same as the ratio of the lengths of corresponding sides.

Claim 3: In similar triangles, the ratio of the lengths of the medians from corresponding vertices is the same as the ratio of the lengths of corresponding sides.

b. Write an argument proving one of the claims that you think is true. Be prepared to present and defend your argument to the class.

c. Restate the claim you proved in Part b using the language of proportions.

14 Look back at your work for Applications Task 5. Since $\triangle ABD \sim \triangle BCD$, it follows that $\frac{AD}{BD} = \frac{BD}{CD}$. Note that BD appears twice in the proportion. The length of $\overline{BD}$ is the *geometric mean* of the lengths of $\overline{AD}$ and $\overline{CD}$.

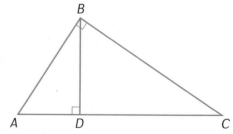

a. Using the language of geometric mean, state a theorem about the altitude to the hypotenuse of any right triangle.

b. The **geometric mean** of two positive numbers a and b is the positive number x such that $\frac{a}{x} = \frac{x}{b}$, or $x = \sqrt{ab}$. What is the geometric mean of 4 and 9? Of 7 and 12?

c. For any two positive numbers, describe the relation between the arithmetic mean and the geometric mean using $<, \le, =, >$, or $\ge$.

d. At the right is a circle with center T.

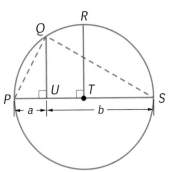

   i. Explain why one of the segments, $\overline{QU}$ and $\overline{RT}$, has length the geometric mean of a and b and the other has length the arithmetic mean of a and b.

   ii. How could you use the diagram to justify your answer to Part c?

e. Under what circumstances are the arithmetic mean and geometric mean of a and b equal?

15 In the Course 2 *Coordinate Methods* unit, you discovered that two nonvertical lines in a coordinate plane are perpendicular if and only if their slopes are opposite reciprocals. You now have the necessary tools to prove this discovery. In this task, you will prove that *if* the lines are perpendicular, *then* the slopes are opposite reciprocals. In Extensions Task 28, you will prove the converse.

In the diagram at the right, $\ell_1 \perp \ell_2$.

a. Why can line ℓ_3 be drawn through point A on ℓ_2 perpendicular to the x-axis?

b. What is the slope of ℓ_1? Of ℓ_2?

c. Prove that $\triangle AOB \sim \triangle OCB$.

d. Using Parts b and c, justify that the slopes of ℓ_1 and ℓ_2 are opposite reciprocals.

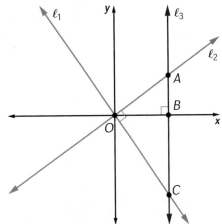

16 In Problem 5 of Investigation 3, you used similar triangles to prove this statement.

> *If a line segment joins the midpoints of two sides of a triangle, then it is parallel to the third side and one half its length.*

Your proof involved reasoning from a combination of assumed or established statements that were independent of a coordinate system. Such proofs are sometimes called *synthetic proofs*.

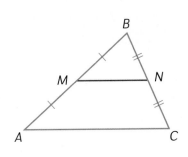
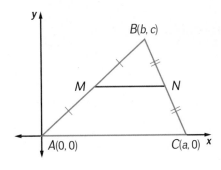

a. Use the labeled diagram on the bottom-right of the previous page to help you write a proof, using coordinates, that $\overleftrightarrow{MN} \parallel \overleftrightarrow{AC}$.

b. Use coordinate methods to prove $MN = \frac{1}{2}AC$.

c. Compare the proofs you constructed in Parts a and b with those you constructed in Problem 5 (page 176).

   • Which proof was easier for you to construct? Why?

   • Which proof would be easier for you to explain to someone else? Why?

d. Suppose M is $\frac{1}{3}$ the distance from point B to point A and N is $\frac{1}{3}$ the distance from point B to point C. How is $\overline{MN}$ related to $\overline{AC}$?

e. Write a proof of your conjecture in Part d. Did you use synthetic or coordinate methods? Why?

17 There are important connections between the perimeters of similar figures and between the areas of similar figures.

a. Suppose $\triangle P'Q'R'$ is the image of $\triangle PQR$ under a size transformation with center C and magnitude k.

   i. Write an argument to prove that
      perimeter $\triangle P'Q'R' = k(perimeter\ \triangle PQR)$.

   ii. Use a result from Connections Task 13 to help prove that
       area $\triangle P'Q'R' = k^2(area\ \triangle PQR)$.

b. Suppose one polygon is the image of another polygon under a size transformation with center C and magnitude k.

   i. How are their perimeters related? Explain your reasoning.

   ii. How are their areas related? Explain your reasoning.

REFLECTIONS

18 A quick test that engravers and photographers use to determine whether two rectangular shapes are similar is illustrated in the diagram below.

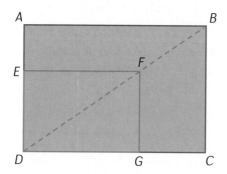 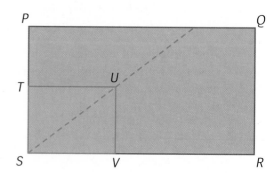

a. Explain why *rectangle ABCD ~ rectangle EFGD* but rectangle *PQRS* is not similar to rectangle *TUVS*.

b. Can this *diagonal test* be used to determine if two nonrectangular parallelograms are similar? Explain your reasoning.

19 A *Sierpinski triangle* is constructed through a sequence of steps illustrated by the figures below. At Stage $n = 0$, you construct an equilateral triangle whose sides are all of length 1 unit. In succeeding stages, you remove the "middle triangles," from each of the remaining colored triangles as shown in Stages $n = 1, 2,$ and 3. This process continues indefinitely.

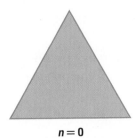

$n = 0$

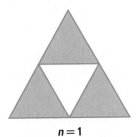

$n = 1$

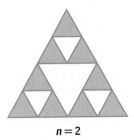

$n = 2$

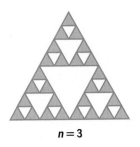

$n = 3$

The Sierpinski triangle is an example of a *fractal* in that small portions of the design are similar to the design as a whole.

a. How is the idea of "self-similarity" seen at Stage 3?

b. Using the Stage 3 triangle, attach, or imagine attaching, a congruent copy of that triangle to each side of the Stage 3 triangle. How is the larger triangle formed related to the stages of construction?

c. In what sense can the triangle at Stage 0 be considered similar to itself?

d. How many colored similar triangles are formed at Stage 1? At Stage 2?

e. Use a computer program like the interactive geometry "Sierpinski Triangle" sketch to explore cases where the initial triangle is not an equilateral triangle. How, if at all, would you change your answers for Parts a–d?

20 In this unit, you continued to engage in important ways of thinking mathematically, or *mathematical habits of mind*, such as:

- visual thinking,

- searching for and explaining patterns,

- making and checking conjectures,

- reasoning with different representations of the same idea, and

- providing sound arguments and proofs.

Carefully look back at the problems in the investigations of Lesson 1 and describe a specific example where you made use of each of the above habits of mind.

21 To find the size transformation image of each figure below, what is the minimum number of image points that you must find? Does it make any difference what points are used? Explain.

    a. a line   **b.** a line segment

    c. a ray   **d.** an angle

    e. a triangle   **f.** a polygon

22 In Problem 5 of Investigation 3, you wrote a synthetic proof of the Midpoint Connector Theorem for Triangles. In Connections Task 16, you prepared a coordinate proof of the theorem.

    a. Using the diagram in Problem 5 (page 176), describe how the theorem could be proved using a size transformation with center at vertex B of $\triangle ABC$.

    b. If you were asked to demonstrate a proof of this theorem later in the year, which proof method would you use? Why?

23 How could you use a size transformation to explain why a circle of radius 5 is similar to a circle of radius 12? Why any two circles are similar?

EXTENSIONS

24 Pietro's Pizza makes rectangular pizzas. The shop posted an inviting message on its sign, "Now! Our large is 20% bigger!" The original large pizza had dimensions 20 inches by 15 inches.

    a. Assume 20% bigger means that the dimensions of the new pizza are 120% of the original size.

        i. Why is the new pizza similar (in shape) to the original pizza?

        ii. What is the scale factor relating the original pizza to the new pizza?

        iii. What are the dimensions of the new pizza?

        iv. Calculate the ratio of the perimeter of the new pizza to that of the original. How is this ratio related to the scale factor?

        v. Calculate the ratio of the areas of the new pizza to the original pizza. How is this ratio related to the scale factor?

    b. Assume 20% bigger means that the area of the new pizza is 120% of the original size.

        i. What is the area of the new pizza?

        ii. If the new pizza is similar to the original, what are its dimensions?

        iii. What is the scale factor relating the original pizza to the new pizza?

    c. Which interpretation favors the consumer?

25 Study the following algorithm for constructing a golden rectangle. (See Connections Task 11.)

Step 1. Construct square $ABCD$.

Step 2. Locate the midpoint M of $\overline{AD}$.

Step 3. Draw $\overline{CM}$.

Step 4. Locate point E on $\overrightarrow{AD}$ so that $MC = ME$.

Step 5. Construct $\overrightarrow{EX} \perp \overline{AE}$.

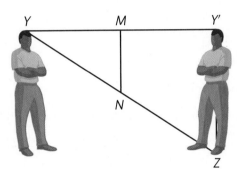

Step 6. Extend $\overline{BC}$ to intersect $\overline{EX}$ at point F.

Prove that rectangle $ABFE$ is a golden rectangle.

26 Using your knowledge of mirror reflections and the diagram below (in which $\overline{MN}$ represents the mirror), determine the wall mirror of minimum height that would permit you to see a full view of yourself. State any assumptions you made.

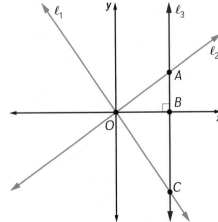

27 In Connections Task 14, you conjectured and then provided a geometric proof that if a and b are two positive numbers, their arithmetic mean is greater than or equal to their geometric mean. Use the facts that $(a + b)^2 \geq 0$ and $(a - b)^2 \geq 0$ along with algebraic reasoning to prove the **arithmetic-geometric mean inequality,** $\frac{a + b}{2} \geq \sqrt{ab}$.

28 In Connections Task 15, you proved that if two nonvertical lines are perpendicular, then their slopes are opposite reciprocals.

Using the diagram at the right, prove this converse statement.

If the slopes of two nonvertical lines are negative reciprocals, then the lines are perpendicular.

Assume the slope of ℓ_1 is m_1 and the slope of ℓ_2 is m_2, where $m_1 m_2 = -1$.

29 Using interactive geometry software, draw $\triangle ABC$ with altitudes $\overline{AD}$, $\overline{BE}$, and $\overline{CF}$.

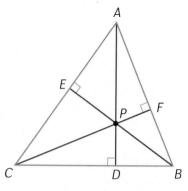

  a. Calculate the product $\left(\frac{AF}{FB}\right)\left(\frac{BD}{DC}\right)\left(\frac{CE}{EA}\right)$.

  b. Use the dragging capability of your software to explore if the relationship you found in Part a holds for other triangles.

  c. Make a conjecture based on your work in Parts a and b.

  d. Check your conjecture with that made by other students. Resolve any differences.

REVIEW

30 Answer these questions using only mental computation. Then check your work using pencil and paper or a calculator.

  a. What number is $\frac{2}{3}$ of 18?

  b. What number is $\frac{5}{4}$ of 12.8?

  c. 25 is $\frac{5}{4}$ of what number?

  d. 10 is $\frac{2}{5}$ of what number?

  e. If $x \neq 0$, which of the following ratios are equivalent to $\frac{1}{3}$?

$$\frac{1.2}{3.6} \qquad \frac{12}{4} \qquad \frac{2}{6} \qquad \frac{1}{3x} \qquad \frac{5x}{15x} \qquad \frac{x+6}{x+18} \qquad \frac{x+2}{3x+6}$$

31 For each statement, write the converse of the statement. Then decide if the converse is true or false. If the converse is false, provide a counterexample. If the converse is true, provide reasoning to support your conclusion.

  a. If $a > 0$, then $a^2 > 0$.

  b. If a is an even number and b is an even number, then ab is a multiple of 4.

  c. All quadrilaterals that are rectangles have four right angles.

  d. If the linear regression line for two variables has positive slope, then the correlation is positive.

32 Use algebraic reasoning to show that the expressions in each pair are equivalent.

  a. $(\sqrt{8k})^2$ and $8k$, for $k > 0$

  b. $(12k)^2 + (k\sqrt{5})^2$ and $149k^2$

  c. $\dfrac{(5a)^2 - (4a)^2}{2a^2}$ and $\dfrac{9}{2}$

  d. $\dfrac{(ab)^2 + 3b^2 - (4ab)^2}{2ab^2}$ and $\dfrac{-15a^2 + 3}{2a}$

  e. $\dfrac{5x^3 - 3x^2y + 10x^2}{x^2}$ and $5x - 3y + 10$

33 For each pair of triangles, identify a line reflection or a composition of line reflections that will map the preimage triangle onto the indicated image triangle. Where possible, specify a second composition that will also work.

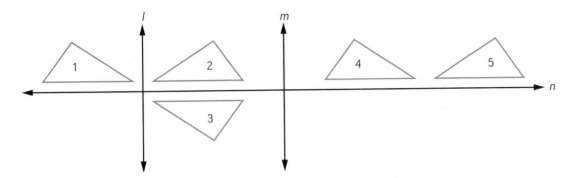

a. △1 onto △2   **b.** △2 onto △3   **c.** △1 onto △4

d. △1 onto △5   **e.** △3 onto △4   **f.** △4 onto △1

34 Solve each equation for the indicated variable.

a. $k = \frac{a}{b}$ for a

b. $\frac{a}{b} = \frac{c}{d}$ for d

c. $x^2 = 36k^2$ for x (assume $k > 0$)

d. $x^2 = y^2 + z^2 - 2yz \cos X$ for $\cos X$

35 Solve each proportion.

a. $\frac{t}{12} = \frac{6}{9}$

b. $\frac{m+4}{m} = \frac{12}{5}$

c. $\frac{y-3}{10} = \frac{2y+5}{3}$

36 Use the provided information to find the measure of each indicated angle.

a. If $m\angle A = 25°$ and $m\angle C = 120°$, find $m\angle B$.

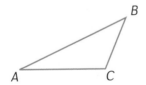

b. If $\ell \parallel m$, $a \parallel b$, and $m\angle ABD = 70°$, find $m\angle CEF$.

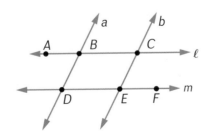

c. If m∠JLK = 123°, m∠LMN = 37°, and $\overline{JK} \parallel \overline{MN}$, find m∠$JKL$ and m∠MNL.

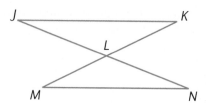

37 Rewrite each expression in $ax^2 + bx + c$ form.

   a. $(x - 5)(x + 5)$

   b. $3(2x + 1)(6 - x)$

   c. $x(8 - 3x) + (5x + 3)$

   d. $(10x - 6)^2$

38 Find the solution to each inequality. Display your solution two ways, using symbols and using a number line graph.

   a. $x^2 + 3x < x + 8$

   b. $\dfrac{20}{x} \geq 10$

   c. $2x + 3 > x^2 + 5$

39 $\triangle ABC \cong \triangle LMP$. Complete each statement about the relationships between the corresponding side lengths and corresponding angle measures of the two triangles.

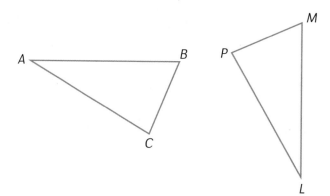

   a. $AB =$ _____   **b.** $MP =$ _____

   c. $CA =$ _____   **d.** m∠$B =$ _____

   e. m∠$A =$ _____   **f.** m∠$P =$ _____

LESSON
2

Reasoning about Congruent Figures

The Rock and Roll Hall of Fame is located in Cleveland, Ohio. The building complex was designed by I.M. Pei to "echo the energy of rock and roll." A glass pyramidal structure covers the interior exhibits. The front surface design of the "glass tent" is an isosceles triangle with a lattice framework like that shown here. The labeling of points helps to reveal some of the many similar and congruent triangles embedded in the framework.

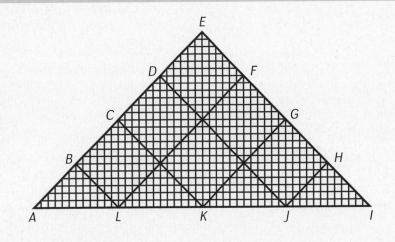

In this lesson, you will explore the relationship between sufficient conditions for congruence of triangles and sufficient conditions for similarity of triangles. You will then use congruence conditions for triangles to solve problems in applied and mathematical contexts. Finally, you will re-examine congruence and similarity of triangles from a transformation approach.

INVESTIGATION 1

Congruence of Triangles Revisited

Congruent triangles have the same shape and size. Your analysis of the Think About This Situation suggests the following equivalent definition of **congruent triangles**.

> *Two triangles are congruent if and only if they*
> *are similar with a scale factor of 1.*

In Course 1, you discovered sufficient conditions on side and angle measures to conclude that two triangles are congruent. You assumed those conditions were correct. As you work on the problems of this investigation, look for answers to the following questions:

> *How are sufficient conditions for congruence of triangles*
> *related to sufficient conditions for similarity of triangles?*

> *How is reasoning with congruence conditions for triangles similar to,*
> *and different from, reasoning with similarity conditions for triangles?*

1 Look back at the similarity theorems that you proved in Lesson 1.

    a. How would you modify the hypothesis of the SSS Similarity Theorem so that it can be used to show that two triangles are not only similar but also congruent? Write a statement of the corresponding **SSS Congruence Theorem**.

    b. Write a theorem for congruence of triangles corresponding to the SAS Similarity Theorem.

    c. What needs to be added to the AA Similarity Theorem to make it into a correct statement about conditions that ensure congruence of the triangles? There are two possible different additions. Find them both. Give names to those congruence conditions.

2 The architectural design used in the roof structure of this rest area building along Interstate 94 may remind you of portions of the design of the glass tent of the Rock and Roll Hall of Fame. Similarity, as well as congruence, is suggested visually in both structures.

    In the diagram at the right, △*ABC* is an isosceles triangle with $\overline{AB} \cong \overline{BC}$. Points *D*, *E*, and *F* are midpoints of $\overline{AB}$, $\overline{BC}$, and $\overline{AC}$, respectively.

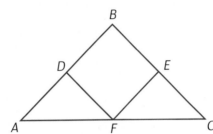

    a. Identify as many pairs of similar triangles as you can. Then carefully explain how you know they are similar.

    b. Identify as many pairs of congruent triangles as you can. Carefully explain how you know they are congruent.

3 For each set of conditions for △*ABC* and △*DEF* given below, draw and mark a copy of the diagram. Using only the given conditions, decide if △*ABC* ≅ △*DEF* or △*ABC* ~ △*DEF*. Explain your reasoning.

    a. ∠*B* ≅ ∠*E*, ∠*A* ≅ ∠*EDF*, and $\overline{AC} \cong \overline{FD}$

    b. $\overleftrightarrow{AB} \parallel \overleftrightarrow{DE}$ and $\overleftrightarrow{BC} \parallel \overleftrightarrow{EF}$

    c. $\overline{AD} \cong \overline{FC}$, $\overline{BC} \parallel \overline{EF}$, and ∠*B* ≅ ∠*E*

    d. $\overline{BC} \cong \overline{ED}$, $\overline{AB} \cong \overline{FE}$, and ∠*A* ≅ ∠*F*

    e. $\overline{AD} \cong \overline{FC}$, $\overline{BC} \parallel \overline{EF}$, and $\overline{AB} \cong \overline{DE}$

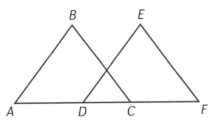

4 In the diagram below, *S* and *T* are the midpoints of $\overline{PQ}$ and $\overline{QR}$, respectively, and *ST* = *TU*.

    a. Describe a strategy that you would use to prove that *SQ* = *UR*.

    b. Describe a strategy that you would use to prove *ST* = $\frac{1}{2}$*PR*.

    c. What type of quadrilateral does *PSUR* appear to be? Describe a strategy that you would use to prove your conjecture.

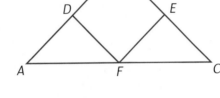

5 For each part below, draw and label two right triangles.

    a. What conditions on the hypotenuse and leg of one triangle would guarantee that it is similar to the other triangle? How would you prove your claim?

    b. What conditions on the hypotenuse and leg of one triangle would guarantee that it is congruent to the other triangle? How would you prove your claim?

    c. What conditions on the hypotenuse and an acute angle of one triangle would guarantee that it is similar to the other triangle? How would you prove your claim?

    d. What conditions on the hypotenuse and an acute angle of one triangle would guarantee that it is congruent to the other triangle? How would you prove your claim?

6 Plans for the location of a telecommunications tower that is to serve three northern suburbs of Milwaukee are shown below.

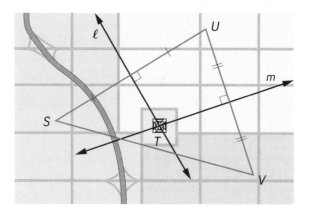

Design specifications indicate the tower should be located so that it is equidistant from the center S, U, and V of each of the suburbs. In the diagram, line ℓ is the *perpendicular bisector* of $\overline{SU}$. Line m is the perpendicular bisector of $\overline{UV}$. Lines ℓ and m intersect at point T.

    a. On a copy of the diagram, draw $\overline{TS}$ and $\overline{TU}$. Prove that $TS = TU$.

    b. Draw $\overline{TV}$ on your diagram. Prove that $TU = TV$.

    c. Explain why the tower should be located at point T.

7 Now try to generalize your finding in Problem 6.

    a. Is it the case that *any* point on the perpendicular bisector of a line segment is equidistant from the endpoints of the segment? Justify your answer in terms of the diagram at the right.

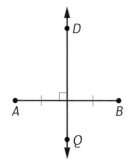

b. Is it the case that if a point is equidistant from the endpoints of a line segment, then it is on the perpendicular bisector of the segment? Justify your answer in terms of a modified diagram.

c. Summarize your work in Parts a and b by completing this statement.

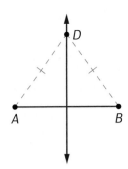

A point is on the perpendicular bisector of a segment if and only if … .

SUMMARIZE THE MATHEMATICS

In this investigation, you re-examined the definition of congruence of triangles and sets of minimal conditions sufficient to prove two triangles are congruent.

a Explain why congruence of triangles can be described as a special case of similarity.

b Construct a table that highlights the relationships between the sufficient conditions for similarity of triangles and the sufficient conditions for congruence of triangles.

c Knowing that two pairs of corresponding angles in two triangles are congruent ensures that the triangles are similar, but it does not ensure that they are congruent. Why?

d How were conditions for congruence of triangles used in your justification in Problem 7?

Be prepared to explain your ideas and reasoning to the class.

✔ CHECK YOUR UNDERSTANDING

Examine each of the following pairs of triangles and their markings showing congruence of corresponding angles and sides. In each case, decide whether the information given by the markings ensures that the triangles are congruent. If the triangles are congruent, write a congruence relation showing the correspondence between vertices. Cite an appropriate congruence theorem to support your conclusion.

a.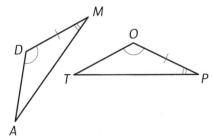

b.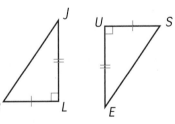

c.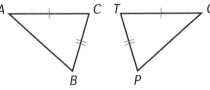

d.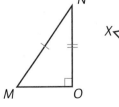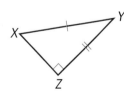

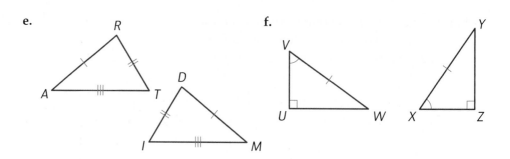

e.

f.

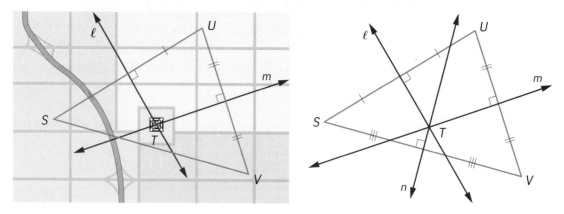

Congruence in Triangles

In Investigation 1, you located a telecommunications tower that was to be equidistant from the centers of three Milwaukee suburbs. The strategy involved considering the triangle formed by the centers of the suburbs, drawing perpendicular bisectors of two of its sides, and then reasoning with congruent triangles.

If the perpendicular bisector *n* of the third side of △*SUV* is drawn, it *appears* that the three lines are **concurrent**. That is, they intersect at a common point. Will this be the case for other triangles?

As you work on the problems in this investigation, look for answers to these questions:

> *Under what conditions will the perpendicular bisectors of the sides of a triangle be concurrent?*
>
> *Under what conditions will the bisectors of the angles of a triangle be concurrent?*
>
> *Under what conditions will the medians of a triangle be concurrent?*
>
> *What are special properties of these points of concurrency?*

1 **Focusing on Perpendicular Bisectors of Sides** Look more closely at your solution of the telecommunications tower location problem. Refer to the diagram on the center-right of the previous page.

a. If $\ell \perp \overline{SU}$ and $m \perp \overline{UV}$, then ℓ and m must intersect at a point T as shown. Why is it not possible that $\ell \parallel m$?

b. On a copy of the diagram, draw in a second color $\overline{TS}$, $\overline{TU}$, and $\overline{TV}$. Discuss with classmates how you used congruent triangles to show that $TU = TV$. That $TU = TS$.

c. From Part b, you were able to conclude that $TS = TU = TV$ and therefore a telecommunications tower located at point T would be equidistant from points S, U, and V. Why must point T be on line n, the perpendicular bisector of $\overline{SV}$?

d. Your work in Parts a–c proves that the perpendicular bisectors of the sides of $\triangle SUV$ are concurrent. $\triangle SUV$ is an acute triangle. Draw a diagram for the case where $\triangle SUV$ is a right triangle. Where $\triangle SUV$ is an obtuse triangle. Explain why the reasoning in Parts a–c applies to any triangle.

2 The point of concurrency of the perpendicular bisectors of the sides of a triangle is called the **circumcenter** of the triangle.

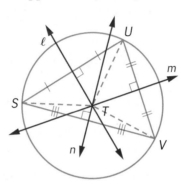

a. Why can a circle with center T and radius $\overline{TS}$ be *circumscribed* about $\triangle SUV$ as shown?

b. Explain with an illustration why a circle can be circumscribed about *any* triangle.

3 **Focusing on Angle Bisectors** In each of the four triangles below, the angle bisectors at each vertex have been constructed.

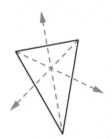  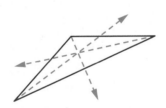 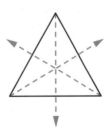

a. What appears to be true about the three angle bisectors in each case?

b. Use interactive geometry software or a compass and straightedge to test your ideas in the case of other triangles. Compare your findings with those of your classmates.

c. If you pick *any* point on the bisector of one of the angles, how does that point seem to be related to the sides of the angle? Compare your ideas with other students. Resolve any differences.

To find logical explanations for your observations in Problem 3, it is helpful to consider the angle bisectors of a triangle, one angle bisector at a time.

4 In the diagram at the right, $\overrightarrow{KX}$ is the bisector of $\angle JKL$ and P is a point on $\overrightarrow{KX}$. $\overline{PQ} \perp \overleftrightarrow{JK}$ and $\overline{PR} \perp \overleftrightarrow{KL}$. The **distance from a point to a line** is the length of the perpendicular segment from the point to the line.

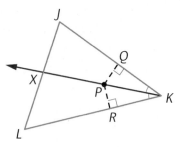

a. Prove that point P is equidistant from the sides of $\angle JKL$ by showing $PQ = PR$.

b. Is it also the case that if a point is equidistant from the sides of an angle, then it is on the bisector of the angle? Justify your answer in terms of a modified diagram.

c. Summarize your work in Parts a and b by completing this statement.

A point is on the bisector of an angle if and only if... .

5 The diagram in Problem 4 is reproduced at the right with $\overrightarrow{LY}$ the bisector of $\angle JLK$ added to the diagram.

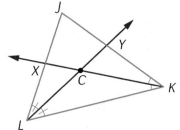

a. Why must angle bisectors $\overrightarrow{KX}$ and $\overrightarrow{LY}$ intersect at some point C?

b. How is point C related to the sides of $\angle JKL$? The sides of $\angle JLK$?

c. Why must point C be on the bisector of $\angle LJK$?

d. Your work in Parts a–c proves that the bisector of the angles of $\triangle JKL$ are concurrent. $\triangle JKL$ is an acute triangle. Draw a diagram for the case where $\triangle JKL$ is a right triangle. Where $\triangle JKL$ is an obtuse triangle. Explain why the reasoning in Parts a–c applies to *any* triangle.

6 The point of concurrency of the bisectors of the angles of a triangle is called the **incenter** of the triangle.

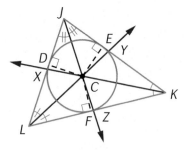

a. Why can a circle with center C and radius $\overline{CD}$ be inscribed in $\triangle JKL$ as shown?

b. Explain with an illustration why a circle can be inscribed in *any* triangle.

7 **Focusing on Medians** You have now been able to prove that the three perpendicular bisectors of the sides of a triangle are concurrent and so are the three angle bisectors. Now investigate to see if the three medians of a triangle are concurrent.

a. Use interactive geometry software or a compass and straightedge to draw a triangle. Then construct the three medians. What appears to be true?

b. Test your observations in the case of other triangles. Compare your findings with those of your classmates. Then write a statement summarizing your findings.

8 The point of concurrency of the medians of a triangle is called the **centroid** of the triangle.

- Cut a triangle from poster board or cardboard.

- Find and mark the centroid of your triangle.

- Verify that the centroid is the *center of gravity* of the triangle by balancing the triangle on the tip of a pencil or pen.

SUMMARIZE THE MATHEMATICS

In this investigation, you explored centers of triangles and their properties.

a Under what conditions will the three perpendicular bisectors of the sides of a triangle be concurrent? The three angle bisectors? The three medians?

b How can you locate the center of a circle inscribed in a triangle?

c How can you locate the center of a circle circumscribed about a triangle?

d How can you locate the center of gravity of a triangular shape of uniform density?

Be prepared to share your ideas and reasoning with the class.

 CHECK YOUR UNDERSTANDING

In the diagram, $\triangle ABC$ is an equilateral triangle. Point P is the circumcenter of $\triangle ABC$.

a. Use congruent triangles to prove that point P is also the incenter of $\triangle ABC$.

b. Prove that point P is also the centroid of $\triangle ABC$.

c. Suppose $AB = 12$ cm.

   i. Find the length of the radius of the circumscribed circle.

   ii. Find the length of the radius of the inscribed circle.

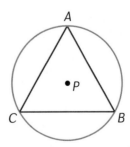

Congruence in Quadrilaterals

In Investigations 1 and 2, you saw that strategies for solving problems and proving claims using congruent triangles were similar to strategies you used in Lesson 1 when reasoning with similar triangles. You identified triangles that could be proven congruent and then used the congruence of some corresponding sides or angles to draw desired conclusions.

In this investigation, you will extend those reasoning strategies to solve more complex problems by drawing on your knowledge of relations among parallel lines. For example, when the midpoints of the sides of a quadrilateral are connected in order, a new quadrilateral is formed.

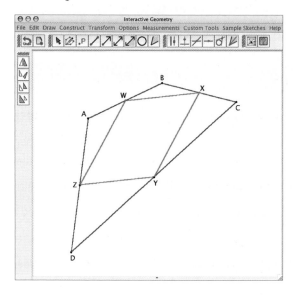

What kind of special quadrilateral is formed?
Will that always be the case?

As you work on the problems of this investigation, look for answers to those questions and the more general questions:

How can you use congruent triangles to establish properties of special quadrilaterals, and what are those properties?

1. Quadrilaterals come in a variety of specialized shapes. There are trapezoids, kites, parallelograms, rhombuses, rectangles, and squares, as well as quadrilaterals with no special additional characteristics.

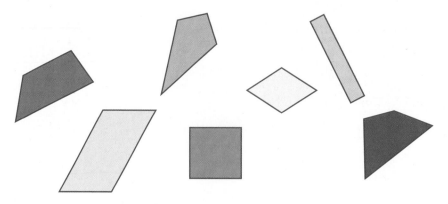

a. Complete a copy of the *quadrilateral* tree shown below by filling in each oval with the name of one of the six special quadrilaterals listed at the bottom of page 204. If the oval for a quadrilateral is connected to one or more ovals *above* it, then that quadrilateral must also have the properties of the quadrilateral(s) in the ovals above it to which it is connected.

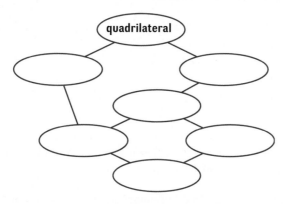

b. Suppose you know that a characteristic X (such as opposite sides congruent) is true for parallelograms. For which other quadrilaterals must characteristic X also be true? Explain your reasoning.

c. Suppose a characteristic Y is true for all kites. Which other quadrilaterals must also have characteristic Y?

2 In your work in Course 1 and Course 2, you showed that the following two statements are equivalent. That is, if you accepted one statement as the definition of a parallelogram, you could prove the other statement as a theorem.

> **I.** *A parallelogram is a quadrilateral that has two pair of opposite sides congruent.*

> **II.** *A parallelogram is a quadrilateral that has two pair of opposite sides parallel.*

Using the diagram at the right and your knowledge of congruent triangles and properties of parallel lines:

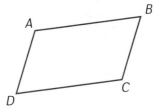

a. explain how you could prove Statement II using Statement I.

b. explain how you could prove Statement I using Statement II.

3 **Properties of Parallelograms** A class at Rockbridge High School was asked to investigate and make a list of common properties of *every* parallelogram. Students prepared the following list. Working in groups, examine each statement to see if you think it is correct. If you think it is correct, construct a proof. If you think it is incorrect, give a counterexample. Share the work. Be prepared to explain your conclusions and reasoning to the entire class.

If a quadrilateral is a parallelogram, then:

a. *each diagonal divides the parallelogram into two congruent triangles.*

b. *opposite angles are congruent.*

c. consecutive angles are supplementary.

d. the diagonals are congruent.

e. the diagonals bisect each other.

f. the diagonals are perpendicular.

4 How would your answers to Problem 3 change if "parallelogram" was replaced by:

a. "rectangle"?

b. "rhombus"?

c. "kite"?

5 How are your responses to Problem 4 reflected in the quadrilateral tree you completed for Problem 1 Part a?

6 **When Is a Quadrilateral a Parallelogram?** Here are additional conditions that may or may not ensure that a quadrilateral is a parallelogram. Investigate each conjecture by drawing shapes satisfying the conditions and checking to see if the quadrilateral is a parallelogram. Share the workload among members of your group. Compile a group list of conditions that seem to guarantee that a quadrilateral is a parallelogram. Be prepared to explain your conclusions and reasoning to the entire class.

a. If a diagonal of a quadrilateral divides it into two congruent triangles, then the quadrilateral is a parallelogram.

b. If a quadrilateral has two pairs of opposite angles congruent, then it is a parallelogram.

c. If a quadrilateral has one pair of opposite sides parallel and the other pair of opposite sides congruent, then it is a parallelogram.

d. If a quadrilateral has two distinct pairs of consecutive sides congruent, then it is a parallelogram.

e. If a quadrilateral has one pair of opposite sides congruent and parallel, then it is a parallelogram.

7 Now refer back to the interactive geometry software screen and questions posed at the beginning of this investigation. Compare the conjecture that you and your classmates made with the statement below.

If the midpoints of consecutive sides of any quadrilateral are connected in order, the resulting quadrilateral is a parallelogram.

a. Draw and label a diagram representing this statement. Then add a diagonal of the original quadrilateral to your diagram.

b. Write an argument to prove this **Midpoint Connector Theorem for Quadrilaterals**.

c. What was the key previously proven result on which your proof depended?

8 **Diagonal Connections** In your earlier work, you represented the diagonals of a quadrilateral with two linkage strips attached at a point. Complete the following table that relates conditions on diagonals to special quadrilaterals. Be prepared to explain how you can use congruent triangles to help justify your conclusion.

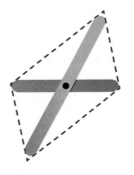

If the diagonals of a quadrilateral, ...	then the quadrilateral is a ....
a. bisect each other	_____
b. bisect each other and are equal in length	_____
c. are perpendicular bisectors of each other	_____
d. are the same length and are perpendicular bisectors of each other	_____
e. are such that one diagonal is the perpendicular bisector of the other	_____

SUMMARIZE THE MATHEMATICS

In this investigation, you explored how conditions for congruence of triangles could be used to prove properties of special quadrilaterals and sufficient conditions for a quadrilateral to be a parallelogram or a special parallelogram.

a Rectangles are a special type of parallelogram. List all the properties they have in common. Identify at least two properties of rectangles that are *not* properties of all parallelograms.

b A square is a special type of rhombus. What properties do they have in common? Identify at least two properties of squares that are *not* properties of all rhombuses.

c Summarize your work in Problems 3, 4, and 8 by writing if-and-only-if statements characterizing special quadrilaterals in terms of relationships involving their diagonals.

d What strategies are helpful in using congruence to establish properties of special quadrilaterals?

Be prepared to share your lists and reasoning with the class.

 **CHECK YOUR UNDERSTANDING**

For each set of conditions, explain why they do or do not ensure that quadrilateral *PQRS* is a parallelogram.

a. $\angle SPQ \cong \angle QRS$ and $\angle PQR \cong \angle PSR$

b. $\overline{PQ} \parallel \overline{RS}$ and $\overline{QR} \cong \overline{PS}$

c. $\overline{PQ} \cong \overline{RS}$ and $\overline{QR} \cong \overline{PS}$

d. $\overline{PQ} \cong \overline{QR}$ and $\overline{RS} \cong \overline{PS}$

e. *M* is the midpoint of $\overline{PR}$ and of $\overline{QS}$.

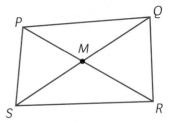

INVESTIGATION **4**

Congruence and Similarity: A Transformation Approach

In Course 2, you studied transformations of a plane defined by coordinate rules. You explored translations, rotations about the origin, and line reflections across the *x*- and *y*-axes, across lines parallel to the axes, and across lines with equations $y = x$ and $y = -x$. You also explored coordinate rules for size transformations. You used composition of these transformations to create interesting animations by repositioning and resizing figures on a computer screen.

In this investigation, you will explore properties and applications of these transformations without the use of coordinates and without restrictions on the positions of the lines of reflection, the centers of rotation, or the centers of size transformations.

As you work on the problems in this investigation, look for answers to these questions:

> *What are the connections between line reflections and both translations and rotations?*
>
> *How can you use rigid transformations to prove sufficient conditions for congruence of triangles?*
>
> *How can you use the composite of a size transformation and rigid transformations to prove sufficient conditions for similarity of triangles?*

The Reflection-Translation Connection Recall that a **translation** is a transformation that "slides" all points in the plane the same distance and same direction. That is, if points P' and Q' are the images of points P and Q under a translation, then $PP' = QQ'$ and $\overline{PP'} \parallel \overline{QQ'}$.

In Problems 1 and 2, you will investigate important connections between line reflections and translations.

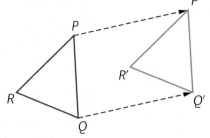

1 The diagram below shows the image of $\triangle ABC$ under a composition of two line reflections, first across line ℓ (image $\triangle A_1B_1C_1$) and then across line m, where $\ell \parallel m$ (image $\triangle A'B'C'$).

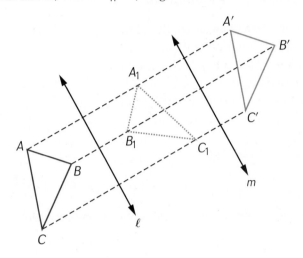

a. How does $\triangle A'B'C'$ appear to be related to $\triangle ABC$ by position?

b. What kind of special quadrilateral is $AA'C'C$? Justify your claim using the diagram below and what you learned about line reflections in Unit 1, *Reasoning and Proof*.

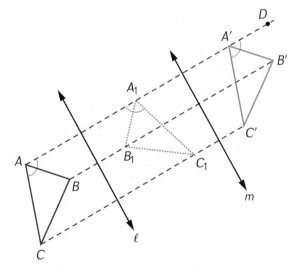

c. Explain why $AA' = CC'$ and $\overline{AA'} \parallel \overline{CC'}$.

d. Is it also the case that $BB' = CC'$ and $\overline{BB'} \parallel \overline{CC'}$? Justify your answer.

e. Suppose P is a point on ℓ and P' is the image of P under the composition of the two reflections. Draw a diagram. Explain why it is also the case that $PP' = AA'$ and $\overline{PP'} \parallel \overline{AA'}$.

2 Look back at your work in Problem 1 Parts c–e.

a. Explain why the composite of two reflections across parallel lines is a translation.

b. How would you describe the **magnitude** (distance points are translated) of the translation? The **direction**?

The Reflection-Rotation Connection Recall that a **rotation** is a transformation that "turns" all points in the plane about a fixed center point through a specified angle. That is, if points P' and Q' are the images of points P and Q under a counterclockwise rotation about point C, then $CP = CP'$, $CQ = CQ'$, $CR = CR'$, and $m\angle PCP' = m\angle QCQ' = m\angle RCR'$.

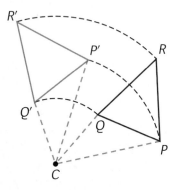

In Problems 3 and 4, you will investigate connections between line reflections and rotations.

3 The diagram below shows $\triangle A'B'C'$ the image of $\triangle ABC$ under a composition of two reflections across intersecting lines, first a reflection across ℓ and then a reflection across m.

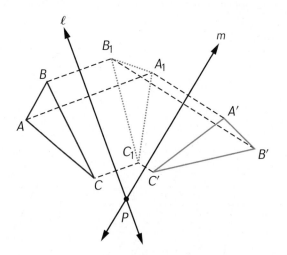

a. How does $\triangle A'B'C'$ appear to be related to $\triangle ABC$ by position?

b. A portion of the diagram at the bottom of the previous page is reproduced here.

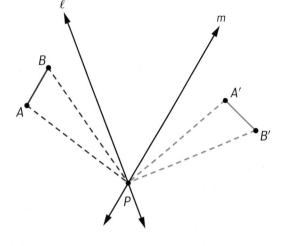

  i. Explain why $PA = PA'$ and $PB = PB'$.

  ii. Explain why $m\angle A'PB' = m\angle APB$.

  iii. Explain why $m\angle APA' = m\angle BPB'$.

c. Suppose X is a point on ℓ ($X \neq P$) and X' is the image of X under the composition of a reflection across line ℓ followed by a reflection across line m.

  i. Draw a diagram.

  ii. Explain why it is also the case that $m\angle XPX' = m\angle APA'$.

4 Look back at your work in Problem 3 Parts b and c.

a. Explain why the composite of reflections across two intersecting lines is a rotation about the point of intersection.

b. How would you describe the *directed angle of the rotation*?

5 Your work on the problems in this investigation suggests some important generalizations. Explain as precisely as you can or illustrate why each of the following statements is true.

a. The composite of two or more line reflections preserves distance. Preserves angle measure.

b. The image of a triangle under a line reflection or composite of two or more line reflections is a congruent triangle.

c. If, $\overline{AB} \cong \overline{CD}$, then $\overline{AB}$ can be mapped onto $\overline{CD}$ by the composition of *at most two* line reflections. (You might begin by showing you can always find a line ℓ so that the reflection image of point A across ℓ is point C. Locate the image of point B.)

The Rigid Transformations-Congruence Connection In Problem 5 Part b, you established that under a rigid transformation (line reflection, translation, rotation), a triangle and its image are congruent. In Problems 6 and 7, you will investigate how rigid transformations can be used to test if two given triangles are congruent. In the process, you will develop transformation approaches to proving the congruence conditions in Investigation 1 of this lesson.

6 Using the "Transformations and Congruence" custom app in *CPMP-Tools*, explore how rigid transformations might be used to test the congruence of two triangles satisfying each condition below. For pairs of triangles that are shown to be congruent, identify the rigid transformation(s) that mapped the first triangle onto the second.

a. SSS Congruence Condition

b. SAS Congruence Condition

c. ASA Congruence Condition

7 The "proofs without words" in Problem 6 may suggest to you strategies for proving each congruence condition using a deductive argument. Study the following proof plans prepared by different groups of students at Challenger Early College High School in North Carolina.

SSS Plan for Proof

To prove △*ABC* ≅ △*PQR*, we can begin by mapping $\overline{AB}$ onto $\overline{PQ}$ using the given fact that $\overline{AB} \cong \overline{PQ}$ and the result of Problem 5 Part c. Label the image of point *C*, *C'*. Then if we use a line reflection across $\overleftrightarrow{PQ}$, the image of point *P* is *P*, the image of point *Q* is *Q*, and the image of point *C'* is point *R*. So, △*ABC* is mapped onto △*PQR* by the composite transformation. Thus, the two triangles are congruent.

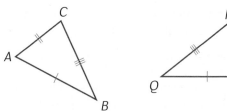

Given: $\overline{AB} \cong \overline{PQ}, \overline{AC} \cong \overline{PR}, \overline{BC} \cong \overline{QR}$

Prove: △*ABC* ≅ △*PQR*

SAS Plan for Proof

Here △*ABC* and △*XYZ* have the same orientation. To prove △*ABC* ≅ △*XYZ*, we can translate point *B* to point *Y*. Label the translation image of △*ABC*, △*A'YC'*. Now rotate △*A'YC'* counterclockwise about point *Y* through an angle with measure ∠*A'YX*. Then the rotation image of △*A'YC'* is △*XYZ*. So, the composite of the translation followed by the rotation maps △*ABC* onto △*XYZ* and the two triangles are congruent.

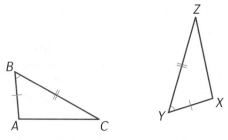

Given: $\overline{AB} \cong \overline{XY}, \angle B \cong \angle Y, \overline{BC} \cong \overline{YZ}$

Prove: △*ABC* ≅ △*XYZ*

ASA Plan for Proof

Begin by reflecting △ABC over ℓ, the perpendicular bisector of $\overline{AD}$. Label the image triangle △DB'C'. Then let m be the perpendicular bisector of $\overline{B'E}$. Show the image of △DB'C' is △DEF. Thus, the composite of the line reflections across ℓ and then m maps △ABC onto △DEF. So, the two triangles are congruent.

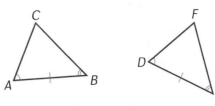

Given: $\angle A \cong \angle D$, $\overline{AB} \cong \overline{DE}$, $\angle B \cong \angle E$

Prove: △ABC ≅ △DEF

a. Dividing the work up among your classmates, use these Plans for Proof to write complete proofs of each congruence condition. Report your proofs and critique the proofs of others.

b. How would you modify your proof in Part a if the given orientation of the second triangle was reversed?

The Size Transformations-Similarity Connection In Lesson 1, you found that once the conditions for similarity of triangles were proved, it was very easy to quickly establish corresponding conditions for congruence of triangles. When using transformations, and the conditions for congruence of triangles are proven, you can easily establish the conditions for similarity of triangles. You will discover a general strategy as you complete Problems 8 and 9.

8 Use the "Transformations and Similarity" custom app to explore how transformations can be used to test the similarity of two triangles.

9 In Lesson 1, Investigation 3, Problem 10 Part c (page 179), you proved that under a size transformation, a triangle and its image are similar. In this problem, you will use a composite of a size transformation and a rigid transformation to prove the similarity conditions in Lesson 1, Investigation 2.

a. Based on your work with the "Transformations and Similarity" custom app, is △ABC ~ △PQR? If so, explain why using transformations.

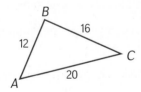 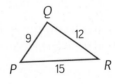

b. Using the given information and transformations, write an argument proving the **SSS Similarity Condition**.

Given: $\frac{XY}{AB} = \frac{YZ}{BC} = \frac{ZX}{CA} = k$

Prove: △ABC ~ △XYZ

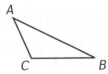 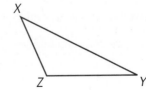

c. Using the given information and transformations, prove the **SAS Similarity Condition**.

> **Given:** $\dfrac{AB}{PQ} = \dfrac{AC}{PR} = k$ and $\angle A \cong \angle P$
>
> **Prove:** $\triangle ABC \sim \triangle PQR$

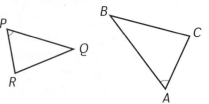

d. Using the given information and transformations, write a proof of the **AA Similarity Condition**.

> **Given:** $\angle E \cong \angle K$, $\angle F \cong \angle L$
>
> **Prove:** $\triangle DEF \sim \triangle JKL$

In the On Your Own set that follows, you will examine the power of rigid transformations and size transformations in helping to solve both applied and theoretical problems.

SUMMARIZE THE MATHEMATICS

In this investigation, you examined the connections between line reflections and both translations and rotations from a synthetic perspective and explored how conditions for congruence of triangles could be established using properties of those transformations. You also explored how conditions for similarity of triangles could be established using composites of size transformations and rigid transformations.

a Under what condition is the composite of two line reflections a translation? How can you determine the direction and magnitude of the translation?

b Under what condition is the composite of two line reflections a rotation? Where is the center of rotation? How can you determine the measure of the angle of rotation?

c Line reflections preserve segment lengths and angle measures and map a triangle onto a congruent triangle. Why must translations and rotations have these same properties?

d Explain as precisely as you can why the image of a triangle under the composite of a size transformation and a rigid transformation is a similar triangle.

e What are the key ideas in proving the SSS, SAS, and ASA congruence conditions using line reflections, translations, and/or rotations?

f What are the key ideas in proving the SSS, SAS, and AA similarity conditions using transformations?

Be prepared to share your ideas and reasoning with the class.

Using a copy of the diagram below, draw the image of $\triangle ABC$ under the composition of successive reflections across line ℓ, line m, and line n. Label the final image triangle $\triangle A'B'C'$.

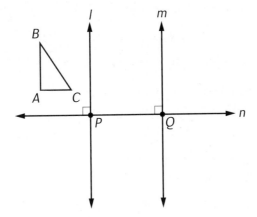

a. Explain why $\triangle A'B'C' \cong \triangle ABC$.

b. Compare the clockwise/counterclockwise orientation of the corresponding vertices of the two triangles. What explains this fact?

c. How would you describe this transformation?

d. Draw segments $\overline{AA'}$ and $\overline{BB'}$. What appears to be true about the midpoints of those segments? Is the same true for the midpoint of $\overline{C'C'}$?

e. How could you use Problem 6 on page 177 to prove that your observation in Part d is true for any point X and its image X' under this composite transformation? Start by drawing a new diagram showing lines ℓ, m, and n. Mark a point X and locate its image X'.

f. On a new copy of the diagram above, find the image of $\triangle ABC$ under the composite of a size transformation with center P and magnitude 2 followed by a size transformation with center Q and magnitude $\frac{1}{2}$. What type of transformation is this composite mapping? Why?

APPLICATIONS

1 Prove, if possible, that $\triangle ABD \cong \triangle CBD$ under each set of conditions below.

   a. $\overline{AB} \cong \overline{BC}$ and $\angle ABD \cong \angle CBD$.

   b. $\overline{BD}$ is the perpendicular bisector of $\overline{AC}$.

   c. $\overline{BD} \perp \overline{AC}$ and $m\angle A = m\angle C$.

   d. $AB = BC$ and D is the midpoint of $\overline{AC}$.

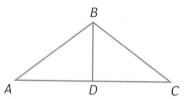

2 Suppose you are given the following information about the figure shown at the right.

$$\overline{ST} \cong \overline{SP} \text{ and } \overline{TO} \cong \overline{PO}$$

Can you conclude that $\triangle STO \cong \triangle SPO$? Write a proof or give a counterexample.

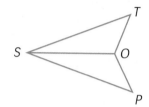

3 For the four lines shown, $\overleftrightarrow{AB} \parallel \overleftrightarrow{DE}$ and C is the midpoint of $\overline{BD}$.

   a. Prove that $\triangle ABC \cong \triangle EDC$.

   b. Using your work in Part a, explain why you also can conclude that C is the midpoint of $\overline{AE}$.

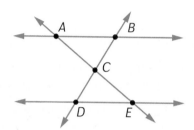

4 Hutchins Lake is a long, narrow lake. Its length is represented by $\overline{AB}$ in the diagram shown below. Andrea designed the following method to determine its length. First, she paced off and measured $\overline{AC}$ and $\overline{BC}$. Then, using a transit, she sighted $\angle ACP$ so that $m\angle ACP = m\angle ACB$. She then marked point D on $\overrightarrow{CP}$ so that $DC = BC$, and she measured $\overline{AD}$.

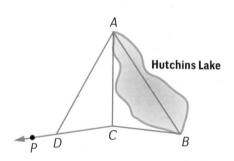

Hutchins Lake

   a. Andrea claimed $AB = AD$. Is she correct? Justify your answer.

   b. Dmitri claimed he could find the length AB of the lake without sighting $\triangle ADC$. Suppose $AC = 500$ m, $BC = 200$ m, and $m\angle ACB = 100°$. What is the length of Hutchins Lake?

5 Consider the diagram below, in which $\overrightarrow{BD}$ is the angle bisector of ABC and $\overleftrightarrow{AE}$ is perpendicular to $\overrightarrow{BD}$.

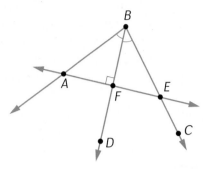

a. Write a proof for each statement.

　　i. $\angle ABF \cong \angle EBF$

　　ii. $\angle ABE$ is an isosceles triangle.

b. On a copy of the diagram, draw a line perpendicular to $\overrightarrow{BD}$ through point C and intersecting $\overleftrightarrow{AB}$. Identify all pairs of congruent and similar triangles. Justify your response.

6 To ship cylindrical objects such as plans, posters, and blueprints, one popular shipping company supplies packages in the shape of an equilateral triangular prism.

Shipping Guide
Mailing Tube

• Triangular package for plans, posters, blueprints, etc.

• Inside dimensions: 96 cm × 15 cm × 15 cm × 15 cm

a. Sketch the triangular base of the prism and the circular outline of a rolled up blueprint of largest diameter that can be shipped in this package.

b. What is the diameter of the rolled up blueprint in Part a?

7 In the diagram, $\overline{TO} \cong \overline{SP}$, $\overleftrightarrow{TO} \parallel \overleftrightarrow{SP}$, and $\overleftrightarrow{TH} \parallel \overleftrightarrow{PA}$.

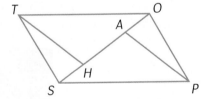

a. Prove that $\overline{TH} \cong \overline{PA}$.

b. Prove that $\angle STH \cong \angle OPA$.

8 Trapezoid $ABCD$ is an **isosceles trapezoid**—it has exactly one pair of parallel sides and the nonparallel sides are congruent.

Write an argument to prove that the base angles, $\angle D$ and $\angle C$, are congruent.

9 On a copy of the diagram in Applications Task 8, draw in diagonals $\overline{AC}$ and $\overline{BD}$. What appears to be true about the two diagonals? Write an argument that proves that your observation is correct.

10 When a ball with no spin and medium speed is banked off a flat surface, the angles at which it strikes and leaves the surface are congruent. You can use this fact and knowledge of line reflections to your advantage in games of miniature golf and billiards.

To make a hole-in-one on the miniature golf green on the left below, visualize point H', the reflected image of the hole H across line ℓ. Aim for the point P where $\overline{BH'}$ intersects ℓ.

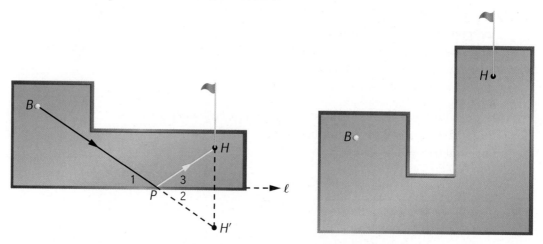

a. If you aim for point P, give reasons to justify that the ball will follow the indicated path to the hole. That is, show $\angle 1 \cong \angle 3$.

b. The diagram at the right shows another par 2 green. On a copy of that diagram, test if you can make a hole-in-one using a single line reflection as in Part a.

c. Is there a way that by finding the reflection image of the hole H across one side and then the reflection of that image across another side, that you can find a point to aim and still make a hole-in-one? Illustrate your answer. Explain why a hit golf ball will follow the path you have drawn.

11 The strategies you used in making a hole-in-one in miniature golf can also be applied in games such as pool and billiards.

a. On the billiards table on the following page, a player wanted to strike the red ball by bouncing the cue ball off three cushions as shown in the left diagram. She visualized the reflection of the red ball R across side ℓ, then visualized the reflection of the image R_1 across side m. Finally, she visualized the reflection of the image R_2 across side n. She aimed for point V where $\overline{CR_3}$ intersects side n, being careful to exert enough force and not put any spin on the ball.

Explain as precisely as you can why the cue ball will follow the indicated path and strike the red ball.

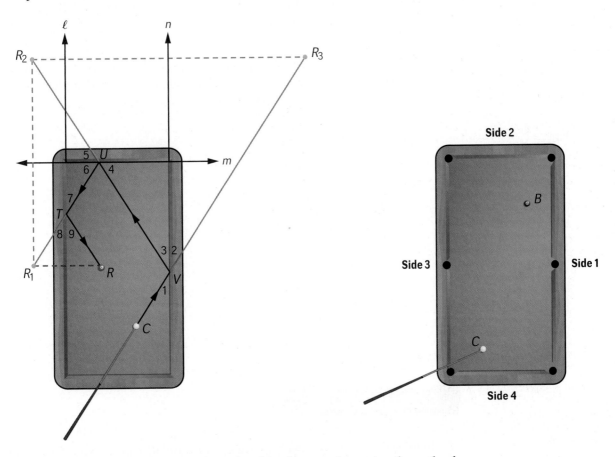

b. On copies of the pool table at the right, show how to determine the path of the cue ball *C* so that it bounces off the given side(s) and then hits ball *B*.

    i. side 1   **ii.** side 2

    iii. side 1, then side 2   **iv.** side 3, then side 2, then side 1

12 In 2007, the southeastern and southwestern regions of the United States suffered severe drought due in part to climate changes. Some communities turned to rivers as a source of water. But pumping water from a river can be expensive. Two small, rural communities along the Flint River in Georgia decided to pool their resources and build a pumping station that would pipe water to both communities.

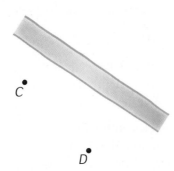

On a copy of the diagram, determine a location of the pumping station that will use the minimum amount of pipe. Then explain why the amount of pipe required is the minimum possible.

13 In each case below, $\triangle ABC \cong \triangle PQR$.

- On a copy of each pair of triangles, determine if you can map $\triangle ABC$ onto $\triangle PQR$ by a composition of line reflections. If so, find and label the appropriate line(s).

- In each case, identify the resulting composite transformation as precisely as you can. If it is a rotation, give its center and directed angle of rotation. If it is a translation, give its magnitude and direction.

a.

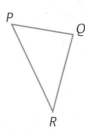

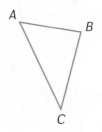

b.

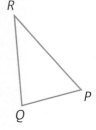

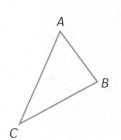

CONNECTIONS

14 In Investigation 1, you used conditions for similarity of triangles and a scale factor of 1 to establish four sets of conditions that are sufficient to prove that two triangles are congruent: SSS, SAS, ASA, and AAS.

   a. Rewrite the SSS Congruence Theorem, replacing the word "triangles" with "parallelograms." Prove or disprove that your statement about two parallelograms is a theorem.

   b. Follow the directions in Part a for the SAS Congruence Theorem.

   c. Follow the directions in Part a for the ASA Congruence Theorem.

   d. Follow the directions in Part a for the AAS Congruence Theorem.

15 Two points determine a line. Explain and provide an illustration that shows why three noncollinear points determine a circle.

16 Find the coordinates of the centroid C of $\triangle PQR$.

   a. For each median, how does the distance from the vertex to point C compare to the length of the median from that vertex?

   b. Use interactive geometry software to test if the relationship you found in Part a holds for other triangles. Write a statement summarizing your findings.

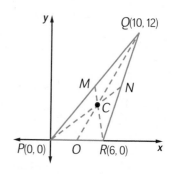

17 In Investigation 2, you saw that a cardboard triangle can be balanced on the tip of a pencil placed at the centroid of the triangle.

a. Cut a triangle from poster board or cardboard. Draw a median of the triangle. Can you balance the triangle on the edge of a ruler placed along the median?

b. Your work in Part a suggests that a median divides a triangle into two triangles of equal area. Write a proof of this statement.

18 Using interactive geometry software construction tools or a compass and straightedge:

a. construct a circle with center C.

b. construct a square inscribed in the circle.

19 In Problem 7 of Investigation 3, you likely used the Midpoint Connector Theorem for Triangles to prove the Midpoint Connector Theorem for Quadrilaterals.

If the midpoints of the sides of a quadrilateral are connected in order, the resulting quadrilateral is a parallelogram.

Consider now how you could prove that statement using coordinate methods.

a. Why was the coordinate system placed so that vertex A is at the origin and side $\overline{AD}$ is on the x-axis?

b. Why were the coordinates of vertices B, C, and D chosen to be $(2a, 2b)$, $(2c, 2d)$, and $(2e, 0)$, respectively?

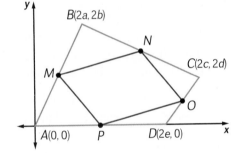

c. Write a coordinate proof of the Midpoint Connector Theorem for Quadrilaterals.

d. Describe a second strategy for proving this theorem using coordinates.

20 Use interactive geometry software or paper and pencil to explore special cases of the Midpoint Connector Theorem for Quadrilaterals on page 206.

a. Draw a square and label it $ABCD$. Find the midpoints of the sides. Connect them in order to obtain quadrilateral $EFGH$.

   i. What special kind of quadrilateral does quadrilateral $EFGH$ appear to be?

   ii. How would you justify your conjecture?

b. Draw a rectangle and label it *ABCD*. Find the midpoints of the sides. Connect them in order to obtain quadrilateral *EFGH*.

    i. What special kind of quadrilateral does quadrilateral *EFGH* appear to be?

    ii. How would you justify your conjecture?

c. Draw a rhombus and name it *ABCD*. Find the midpoints of the sides. Connect them in order to obtain quadrilateral *EFGH*.

    i. What special kind of quadrilateral does quadrilateral *EFGH* appear to be?

    ii. How would you justify your conjecture?

REFLECTIONS

21 In this unit, you have reasoned with "corresponding angles" in the contexts of similarity and congruence of triangles and in the context of parallel lines cut by a transversal. Draw sketches illustrating the different meanings of "corresponding angles."

22 Given two angles, there is a variety of conditions that would allow you to conclude, without measuring, that the angles are congruent. For example, if you know the angles are vertical angles, then you know they are congruent. What other conditions will allow you to conclude two angles are congruent? List as many as you can.

23 The Midpoint Connector Theorem for Quadrilaterals is an amazing result. It was first proved by a French mathematician, Pierre Varignon (1654–1722). Here is a more recent unexpected "point connector" result.

a. On a copy of the diagram below, continue to draw segments parallel to the sides of the triangle. What happens?

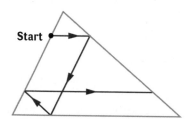

b. What was the total number of segments drawn? Can you explain why this happens?

c. Is there a "start" point for which there will be fewer segments drawn? If so, where is the point located? How many segments will there be?

24 The diagram below can be used as an aid to help prove properties of parallelograms using coordinates.

a. Determine the coordinates of vertex C.

b. Look back at Problem 3 (pages 205–206) of Investigation 3. From the list, pick two properties of parallelograms that you proved using congruent triangles. For each property, describe how you could also prove the property using coordinate methods.

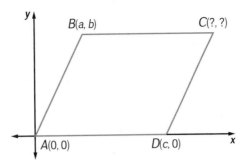

25 In Lesson 1, Investigation 3, Problem 9 (page 178), you proved that under a size transformation, a line and its image are parallel. In Lesson 2, Investigation 4, Problem 1, you proved that under a translation, a line and its image are parallel. Under a rotation of what magnitude will a line and its image be parallel? Write an argument to support your assertion.

26 In this lesson, you re-examined congruence of triangles as a special case of similarity of triangles. The following statement relates similarity to congruence in terms of a size transformation.

> *Two polygons are similar if one is congruent to a size transformation of the other.*

Is this description of similarity consistent with the definition of similarity in Lesson 1? Explain.

27 In reflecting back over this lesson, Dianna observed:

> **There is a SSS condition and a SAS condition for both congruence and similarity of triangles. But we proved only an ASA condition for congruence, why not for similarity?**

How would you respond?

EXTENSIONS

28 In the diagram at the right, $\overline{BF}$ is the perpendicular bisector of $\overline{AC}$ and $\angle 1 \cong \angle 2$. Prove that $\overline{DB} \cong \overline{BE}$.

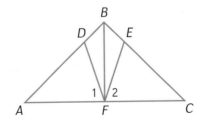

29 Use interactive geometry software or a compass and straightedge to construct a regular pentagon and a regular hexagon.

a. What appears to be true about the perpendicular bisectors of the sides?

b. Does your observation in Part a also hold for a regular quadrilateral?

c. Make a conjecture about the concurrency of the perpendicular bisectors of the sides of any regular polygon.

d. How could you prove your conjecture using the fact that the perpendicular bisector of a line segment is the set of all points equidistant from the segment's endpoints?

30 Use interactive geometry software or a compass and straightedge to construct a regular pentagon and a regular hexagon.

a. What appears to be true about the angle bisectors?

b. Does your observation in Part a also hold for a regular quadrilateral?

c. Make a conjecture about the concurrency of the angle bisectors of any regular polygon.

d. How could you prove your conjecture using the fact that the bisector of an angle is the set of all points equidistant from the sides of the angle?

31 Look back at Connections Task 14. What conditions that ensure congruence of two parallelograms could be modified with additional side or angle information to provide a test of congruence of two general quadrilaterals? Write an argument justifying your congruence condition for general quadrilaterals.

32 The two quadrilaterals below are not drawn to scale. In Parts a and b, determine if the given information is sufficient to guarantee the two quadrilaterals are congruent. Explain your reasoning. If quadrilateral *ABCD* is not congruent to quadrilateral *PQRS*, what is the least amount of additional information you would need to conclude the two quadrilaterals are congruent?

a. Is *quadrilateral ABCD* ≅ *quadrilateral PQRS* if these four conditions are satisfied?

- $\overline{AD} \cong \overline{PS}$
- $\overline{BC} \cong \overline{QR}$
- $\overline{AB}$ is parallel to $\overline{CD}$.
- $\overline{PQ}$ is parallel to $\overline{RS}$.

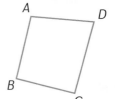
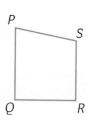

b. Is *quadrilateral ABCD* ≅ *quadrilateral PQRS* if these three conditions are satisfied?

- $\angle A \cong \angle P$
- $\angle C \cong \angle R$
- $\overline{AD} \cong \overline{PS}$

33 In Investigation 4, you proved that a triangle and its image under a line reflection or a composite of two (three) line reflections are congruent. In this task, you will justify the following striking result.

If △ABC ≅ △A'B'C', then there is a rigid transformation that maps △ABC onto △A'B'C' and this transformation is the composite of at most three line reflections.

Study the steps below to find the possible lines of reflection for a transformation that will map △ABC onto △A'B'C' by successive reflections across those lines.

Step 1

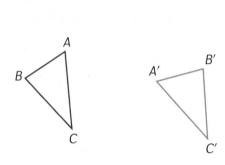

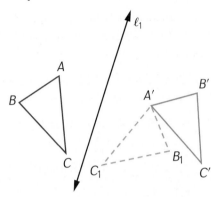

Step 2

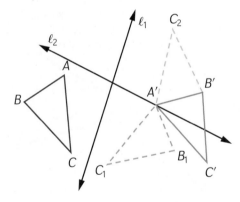

Step 3

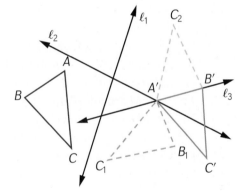

a. In Step 1, △A'B₁C₁ is the reflection image of △ABC across line ℓ₁.

- How was line ℓ₁ chosen?

- How is A'B₁ related to A'B'?

b. In Step 2, ℓ₂ was chosen to be the perpendicular bisector of $\overline{B_1B'}$.

- Why was ℓ₂ chosen that way?

- The diagram shows that point A' is on ℓ₂. Why must that be the case?

- How is A'C₂ related to A'C'?

c. In Step 3, ℓ₃ was chosen to be the perpendicular bisector of $\overline{C_2C'}$.

- Why was ℓ₃ chosen that way?

- The diagram shows that points A' and B' are on ℓ₃. Why must that be the case?

d. Why is A' the reflection image of point A across line ℓ_1, then line ℓ_2, and then line ℓ_3? Why is B' the reflection image of point B under the composition of these three line reflections? Why is C' the reflection image of point C under this composition of line reflections?

e. Suppose in Step 1, the reflection image of point B across ℓ_1 was point B'. How should ℓ_2 in Step 2 be chosen?

f. Suppose in Step 2, the reflection image of point C_1 was point C'. How should ℓ_3 in Step 3 be chosen?

g. How does your work in Parts a–f justify this fundamental theorem of rigid transformations?

34 An **equilic quadrilateral** is a quadrilateral with a pair of congruent opposite sides that, when extended, meet to form a 60° angle. The other two sides are called *bases*.

a. Quadrilateral $ABCD$ is equilic with bases $\overline{AB}$ and $\overline{CD}$ and $\overline{AD} \cong \overline{BC}$.

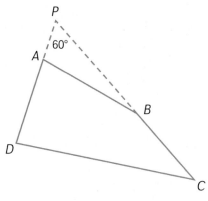

- Could base $\overline{AB}$ be parallel to base $\overline{CD}$? Explain your reasoning.

- Could $\overline{AB} \cong \overline{CD}$? Explain.

- Find $m\angle D + m\angle C$.

- Find $m\angle A + m\angle B$.

b. Suppose you have an equilic quadrilateral $ABCD$ with $\overline{AD} \cong \overline{BC}$ and diagonals $\overline{AC}$ and $\overline{BD}$. Points J, K, L, and M are midpoints of bases $\overline{AB}$ and $\overline{CD}$ and diagonals $\overline{AC}$ and $\overline{BD}$, respectively. How are the four points J, K, L, and M related? Prove you are correct.

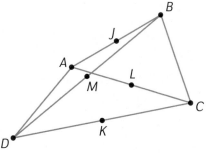

c. Draw an equilic quadrilateral $ABCD$ with $\overline{AD} \cong \overline{BC}$ and $AB < CD$. Draw an equilateral triangle on base $\overline{AB}$ that shares no points with the interior of quadrilateral $ABCD$. Label the third vertex P. How are points P, C, and D related? Prove your claim.

REVIEW

35 If $1{,}000 < x < 2{,}000$, place the following expressions in order from smallest to largest.

$$\sqrt{x} \qquad \frac{x}{5} \qquad \frac{5}{x} \qquad x^2 \qquad \frac{x+3}{5} \qquad \frac{5}{x+2} \qquad |x|$$

36 For each table of values, determine if the relationship is linear, exponential, or quadratic. Then find a rule that matches the relationship in the table.

a.

x	0	1	2	3	4	5
y	297	99	33	11	$3\frac{2}{3}$	$1\frac{2}{9}$

b.

x	0	3	6	9	12	15
y	99	115	131	147	163	179

c.

x	2	3	4	5	6	7
y	18	15.5	13	10.5	8	5.5

37 Why is the length of the perpendicular segment to a line drawn from a point not on a line the shortest distance from the point to the line?

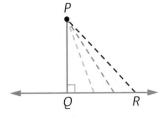

38 Find the length of the line segment that connects each pair of points. Express your answer in simplest radical form.

a. (3, 6) and (11, 10)

b. (5, −4) and (0, −1)

c. (a, b) and (a, 0)

39 Consider △XYZ with m∠Y = 90° and sin X = $\frac{3}{5}$.

a. Find m∠X.

b. Sketch two different triangles that each meet the given conditions. Identify the lengths of the sides of each triangle.

c. Are the two triangles you drew in Part b similar? Explain your reasoning.

40 Draw a segment $\overline{AB}$. Bill claims that there are two distinct lines ℓ and m such that the reflection image of $\overline{AB}$ across line ℓ is $\overline{BA}$ and the reflection image of $\overline{AB}$ across line m is the segment itself. Describe the lines ℓ and m.

41 Rewrite each expression in a simpler equivalent form.

a. $-\frac{2}{5}(14x + 12) + \frac{1}{5}(3x - 6)$

b. $\frac{8x + 10}{2} + 6x - 15$

c. $x(10 - x) - 3(x + 4) + 12$

42 On a copy of $\triangle PQR$, construct three lines ℓ, m, n so that:

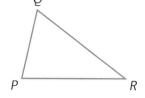

a. Point Q is the reflection image of point P across line ℓ.

b. Point R is the reflection image of point Q across line m.

c. Point P is the reflection image of point R across line n.

d. What appears to be true about lines ℓ, m, and n? Explain why this is the case.

43 The suggested serving size for a certain cereal is $1\frac{1}{4}$ cups.

a. How much cereal would you need if you wanted to have 15 servings?

b. Georgia has 25 cups of cereal. How many servings does she have?

c. A box of cereal weighs 10 ounces and contains 9 servings. What does 1 cup of cereal weigh?

d. Write a symbolic rule that describes the relationship between the weight of the cereal and the number of servings.

44 The histogram below represents the number of sit-ups that the 60 eleventh-grade students at Eisenhower High School were able to complete in one minute.

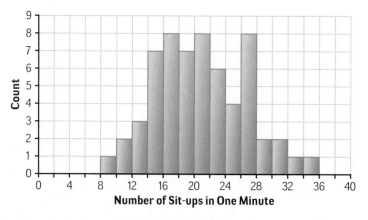

a. Jeremy did 29 sit-ups. What is his percentile ranking within this group of eleventh graders?

b. Stephanie was only at the 40th percentile. How many sit-ups was Stephanie able to do?

45 Graph the solution set to each system of inequalities. On your graphs, identify the coordinates of the points of intersection of the graphs.

a. $4x + 8y < 40$
$\quad x - 2y > 8$

b. $y \geq -3$
$\quad y \leq x^2 - 5$

46 Suppose a data set x_1, x_2, x_3, x_4, x_5, has mean $\bar{x}$ and standard deviation s.

a. If a constant c is added to each of the five values, what is the mean of the transformed values? What is the standard deviation of the transformed values?

b. If each of the five values is multiplied by a positive constant d, what is the mean of the transformed values? What is the standard deviation of the transformed values?

c. How would the data transformation in Part a affect the median? The interquartile range?

d. How would the data transformation in Part b affect the median? The interquartile range?

47 Match each of the following histograms of test scores in three World History classes, I, II, and III, to the best description of class performance.

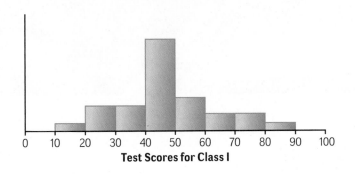

Test Scores for Class I

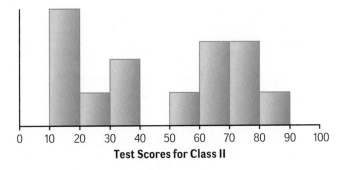

Test Scores for Class II

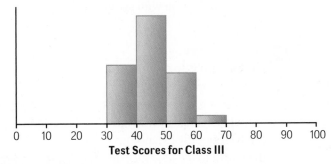

Test Scores for Class III

a. The mean of the test scores is 46, and the standard deviation is 26.

b. The mean of the test scores is 46, and the standard deviation is 8.

c. The mean of the test scores is 46, and the standard deviation is 16.

Looking Back

In this unit, you re-examined similarity and congruence as related ideas and as tools for solving problems. In Lesson 1, you used the Law of Cosines and the Law of Sines to deduce sufficient conditions to ensure similarity of triangles. You used similarity to explain the design of mechanisms and to calculate inaccessible heights and distances. You also discovered and proved important properties of size transformations.

In Lesson 2, you used similarity conditions for triangles to justify sufficient conditions for congruence of triangles that you discovered and used without proof in your previous work. You used congruence of triangles to locate special centers of triangles and to prove properties of (special) parallelograms.

You also investigated how rigid transformations provided an alternative approach to establishing sufficient conditions for congruence of triangles, and how a composite of a size transformation

and a rigid transformation could be used to prove sufficient conditions for similarity of triangles.

The tasks in this final lesson will help you review and organize your knowledge of key ideas and strategies for reasoning with similarity and congruence.

1 When a rectangular photograph is to be enlarged to fit the width of a given column of type in a newspaper or magazine, the new height can be found by extending the width to the new measure and drawing a perpendicular as shown. The point at which the perpendicular meets the extended diagonal determines the height of the enlargement. Explain as precisely as you can why this method works.

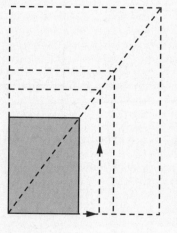

2 In the diagram below, △*ABC* is a right triangle and $\overline{BD}$ is an altitude to side $\overline{AC}$.

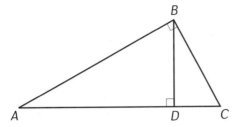

a. Prove that $(AB)^2 = (AC)(AD)$.

b. Find a similar expression for $(BC)^2$.

c. Use your work in Parts a and b to provide another proof of the Pythagorean Theorem.

3 In this unit, you completed a proof of the AA Similarity Theorem for triangles.

a. State this theorem in words and illustrate it with a diagram.

b. Inquiring minds might ask if there is an SS condition or an SA condition that would guarantee two given triangles are similar. How would you respond? Be as precise as possible in your answer.

4 In the diagram below, $\overline{PR}$ and $\overline{QS}$ are diagonals of quadrilateral *PQRS*, and intersect at point *T*. $\overline{PS} \cong \overline{QR}$, $\overline{PT} \cong \overline{QT}$, and $\overline{ST} \cong \overline{RT}$.

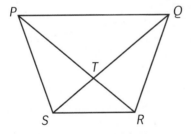

a. Identify and name eight triangles in the diagram.

b. Which pairs of the eight triangles seem to be congruent? Name each pair.

c. Using the given information, prove or disprove that each identified pair of triangles are congruent.

d. Which of the remaining pairs of triangles are similar? Provide an argument to support your answer.

e. What can you conclude about $\overline{PQ}$ and $\overline{SR}$ when you know the given information? Write a proof for your claim.

5 Civil engineers, urban planners, and design engineers are frequently confronted with "traffic center" problems. In mathematical terms, these problems often involve locating a point M that is equidistant from three or more given points or for which the sum of the distances from point M to the given points is as small as possible. In the first two parts of this task, you are to write and analyze a synthetic proof and a coordinate proof of the following theorem:

The midpoint of the hypotenuse of a right triangle is equidistant from its vertices.

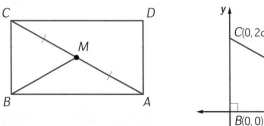
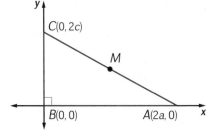

a. In the diagram on the left, $\triangle ABC$ is composed of two consecutive sides and a diagonal of rectangle $ABCD$. Using that diagram, write a synthetic proof that if M is the midpoint of $\overline{AC}$, then $MA = MB = MC$.

b. Using the diagram on the right, write a coordinate proof that if M is the midpoint of $\overline{AC}$, then $MA = MB = MC$.

c. For the proof in Part b, why were the coordinates of points A and C assigned $(2a, 0)$ and $(0, 2c)$ rather than simply $(a, 0)$ and $(0, c)$, respectively?

6 Suppose cities A, B, and C are connected by highways $\overline{AB}$, $\overline{BC}$, and $\overline{AC}$ as shown.

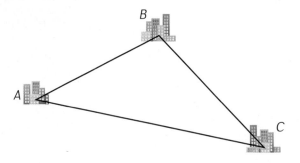

a. On a copy of $\triangle ABC$, find the location for a proposed new airport that would be the same distance from each of the existing highways.

b. Explain why your proposed location meets the constraints.

7 The display below suggests that if the midpoints of the sides of an isosceles trapezoid are connected in order, a special parallelogram is formed.

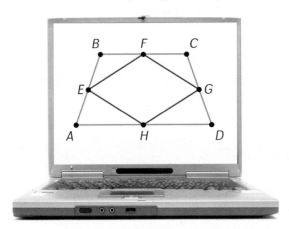

a. Make and test conjectures about the type of parallelogram that is formed.

b. Prove your conjecture using synthetic methods. Using coordinate methods.

c. Which proof method in Part b seems to be the better approach for this problem? Why?

8 In July 2006, a tsunami, triggered by an undersea earthquake, struck the coasts of India and Indonesia, destroying villages. Tidal waves damaged bridges. A new bridge connecting villages V and W positioned as shown below is to be constructed across the river.

a. If river conditions permit, why should a bridge $\overline{XY}$ be constructed perpendicular to the banks of the river?

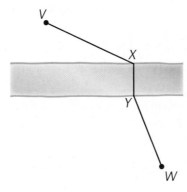

b. The diagram shows one possible location of the bridge. Write an expression for the length of the path from village V to village W.

c. On a copy of this diagram, find the image V' of point V under the translation that maps point X onto point Y. How could you use point V' to find a location for the bridge so that the path from point V to the bridge and then to point W is of minimum length? Why would that make sense?

d. Explain why your location of the bridge provides a shorter path than the path in the diagram. Why is it shorter than the path for any other possible location of the bridge?

9 The composite of a size transformation and a rigid transformation (or vice versa) is called a **similarity transformation**. Explain as precisely as you can why each statement is true or false.

a. Two triangles are *congruent* if and only if one is the image of the other under a rigid transformation or a composite of rigid transformations.

b. Two triangles are *similar* if and only if one is the image of the other under a similarity transformation.

SUMMARIZE THE MATHEMATICS

In this unit, you extended your understanding of similarity and congruence and sharpened your skills in mathematical reasoning.

a What conditions are sufficient to guarantee that two triangles are similar? Are congruent?

b How are the ideas of congruence and similarity related?

c What is the Midpoint Connector Theorem for Triangles? For Quadrilaterals? What are strategies that can be used to remember and prove each theorem?

d A circle has one center. But a triangle has more than one center. Identify and describe how to locate each center studied in this unit. Then describe an application of each.

e Identify at least two conditions on a quadrilateral that guarantees that it is a parallelogram. A rhombus. A rectangle.

f Suppose in a diagram, $\triangle X'Y'Z'$ is the image of $\triangle XYZ$ under the indicated transformation.

  i. Size transformation of magnitude k: How could you locate the center of the size transformation and determine the magnitude k?

  ii. Line reflection: How could you determine the line of reflection?

  iii. Rotation: How could you determine the center of the rotation and the directed angle of rotation?

  iv. Translation: How could you determine lines ℓ_1 and ℓ_2 so that the translation can be expressed as a composition of reflections across those lines? How are the magnitude and direction of the translation related to lines ℓ_1 and ℓ_2?

g How are congruence and similarity related to rigid transformations and size transformations?

Be prepared to explain your ideas and strategies to the class.

 **CHECK YOUR UNDERSTANDING**

Write, in outline form, a summary of the important mathematical concepts and methods developed in this unit. Organize your summary so that it can be used as a quick reference in future units and courses.

Samples and Variation

Taking political polls and manufacturing car parts do not seem similar at first glance. However, both involve processes that have variation in the outcomes. When Gallup takes a poll, different samples of voters would give slightly different estimates of the president's popularity. When Ford Motor Company builds a car, body panels will vary slightly in their dimensions even if they are made by the same machine.

In this unit, you will investigate how understanding this variability helps both pollsters and manufacturers improve their products. The statistical methods they use can be effectively applied to any area in which there is variation in the process, nearly every area of human endeavor. The essential knowledge and skills required for this work are developed in three lessons of this unit.

Steve Lindridge/Alamy

LESSONS

1 Normal Distributions

Describe characteristics of a normal distribution; compute and interpret a standardized score; and fit a normal distribution to a set of data and use it to estimate population percentages.

2 Binomial Distributions

Develop a binomial probability distribution and predict whether it will be approximately normal. Compute the expected value and standard deviation of a binomial distribution. Decide whether the data are consistent with a specified population proportion.

3 Statistical Process Control

Recognize when the mean and standard deviation change in a plot over time; use probability concepts to develop tests for product testing and compute the probability of a false alarm; and use the Central Limit Theorem.

Normal Distributions

Jet aircraft, like Boeing's latest 787 Dreamliner, are assembled using many different components. Parts for those components often come from other manufacturers from around the world. When different machines are used to manufacture the same part, the sizes of the produced parts will vary slightly. Even parts made by the same machine have slight variation in their dimensions.

Variation is inherent in the manufacturing of products whether they are made by computer-controlled machine or by hand. To better understand this phenomenon, suppose your class is working on a project making school award certificates. You have decided to outline the edge of a design on each certificate with one long piece of thin gold braid.

Certificate of Distinction

This certificate is awarded to

_____

to recognize achievement in

_____

on this day, _____.

In this lesson, you will explore connections between a normal distribution and its mean and standard deviation and how those ideas can be used in modeling the variability in common situations.

INVESTIGATION 1

Characteristics of a Normal Distribution

Many naturally occurring measurements, such as human heights or the lengths or weights of supposedly identical objects produced by machines, are **approximately normally distributed**. Their histograms are "bell-shaped," with the data clustered symmetrically about the mean and tapering off gradually on both ends, like the shape below.

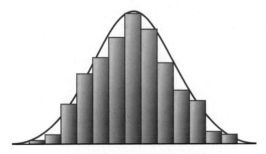

When measurements can be modeled by a distribution that is approximately normal in shape, the mean and standard deviation often are used to summarize the distribution's center and variability. As you work on the problems in this investigation, look for answers to the following question:

> *How can you use the mean and standard deviation to help you locate a measurement in a normal distribution?*

1 In Course 1, *Patterns in Data*, you estimated the mean of a distribution by finding the balance point of a histogram. You estimated the standard deviation of a normal distribution by finding the distance to the right of the mean and to the left of the mean that encloses the middle 68% (about two-thirds) of the values. Use this knowledge to help analyze the following situations.

a. The histogram below shows the political points of view of a sample of 1,271 voters in the United States. The voters were asked a series of questions to determine their political philosophy and then were rated on a scale from liberal to conservative. Estimate the mean and standard deviation of this distribution.

Political Philosophy

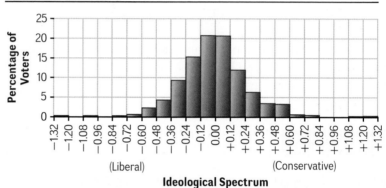

Source: Romer, Thomas, and Howard Rosenthal. 1984. Voting Models and Empirical Evidence. *American Scientist*, 72: 465–473.

b. In a science class, you may have weighed something by balancing it on a scale against a standard weight. To be sure a standard weight is reasonably accurate, its manufacturer can have it weighed at the National Institute of Standards and Technology in Washington, D.C. The accuracy of the weighing procedure at the National Institute of Standards and Technology is itself checked about once a week by weighing a known 10-gram weight, NB 10. The histogram below is based on 100 consecutive measurements of the weight of NB 10 using the same apparatus and procedure. Shown is the distribution of weighings, in micrograms below 10 grams. (A microgram is a millionth of a gram.) Estimate the mean and standard deviation of this distribution.

NB 10 Weight

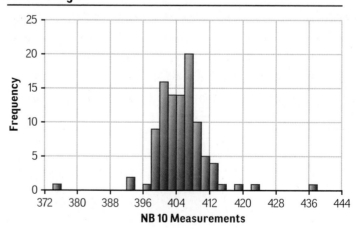

Source: Freedman, David, *et al. Statistics, 3rd edition.* New York: W. W. Norton & Co, 1998.

2 The data below and the accompanying histogram give the weights, to the nearest hundredth of a gram, of a sample of 100 new nickels.

Nickel Weights (in grams)

4.87	4.92	4.95	4.97	4.98	5.00	5.01	5.03	5.04	5.07
4.87	4.92	4.95	4.97	4.98	5.00	5.01	5.03	5.04	5.07
4.88	4.93	4.95	4.97	4.99	5.00	5.01	5.03	5.04	5.07
4.89	4.93	4.95	4.97	4.99	5.00	5.02	5.03	5.05	5.08
4.90	4.93	4.95	4.97	4.99	5.00	5.02	5.03	5.05	5.08
4.90	4.93	4.96	4.97	4.99	5.01	5.02	5.03	5.05	5.09
4.91	4.94	4.96	4.98	4.99	5.01	5.02	5.03	5.06	5.09
4.91	4.94	4.96	4.98	4.99	5.01	5.02	5.04	5.06	5.10
4.92	4.94	4.96	4.98	5.00	5.01	5.02	5.04	5.06	5.11
4.92	4.94	4.96	4.98	5.00	5.01	5.02	5.04	5.06	5.11

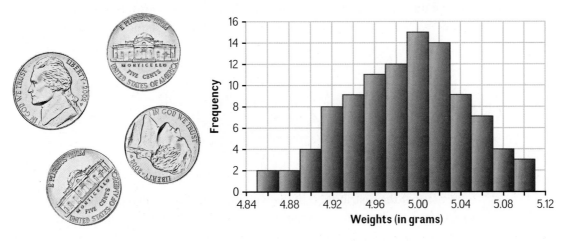

a. The mean weight of this sample is 4.9941 grams. Find the median weight from the table above. How does it compare to the mean weight?

b. Which of the following is the standard deviation?

    0.0253 grams   0.0551 grams   0.253 grams   1 gram

c. On a copy of the histogram, mark points along the horizontal axis that correspond to the mean, one standard deviation above the mean, one standard deviation below the mean, two standard deviations above the mean, two standard deviations below the mean, three standard deviations above the mean, and three standard deviations below the mean.

d. What percentage of the weights in the table above are within one standard deviation of the mean? Within two standard deviations? Within three standard deviations?

e. Suppose you weigh a randomly chosen nickel from this collection. Find the probability that its weight would be within two standard deviations of the mean.

In Problems 1 and 2, you looked at distributions of real data that had an *approximately* normal shape with mean $\bar{x}$ and standard deviation s. Each distribution was a sample taken from a larger population that is more nearly normal. In the rest of this investigation, you will think about theoretical populations that have a perfectly normal distribution.

The symbol for the *mean of a population* is μ, the lower-case Greek letter "mu." The symbol for the *standard deviation of a population* is σ, the lower case Greek letter "sigma." That is, use the symbol μ for the mean when you have an entire population or a theoretical distribution. Use the symbol $\bar{x}$ when you have a sample from a population. Similarly, use the symbol σ for the standard deviation of a population or of a theoretical distribution. Use the symbol s for the standard deviation of a sample.

All normal distributions have the same overall shape, differing only in their mean and standard deviation. Some look tall and skinny. Others look more spread out. All normal distributions, however, have certain characteristics in common. They are symmetric about the mean, 68% of the values lie within one standard deviation of the mean, 95% of the values lie within two standard deviations of the mean, and 99.7% of the values lie within three standard deviations of the mean.

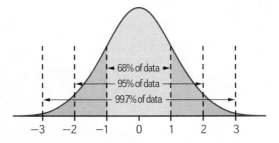

3 Suppose that the distribution of the weights of newly minted coins is a normal distribution with mean μ of 5 grams and standard deviation σ of 0.10 grams.

 a. Draw a sketch of this distribution. Then label the points on the horizontal axis that correspond to the mean, one standard deviation above and below the mean, two standard deviations above and below the mean, and three standard deviations above and below the mean.

 b. Between what two values do the middle 68% of the weights of coins lie? The middle 95% of the weights? The middle 99.7% of the weights?

 c. Illustrate your answers in Part b by shading appropriate regions in copies of your sketch.

4 Answer the following questions about normal distributions. Draw sketches illustrating your answers.

a. What percentage of the values in a normal distribution lie above the mean?

b. What percentage of the values in a normal distribution lie more than two standard deviations from the mean?

c. What percentage of the values in a normal distribution lie more than two standard deviations above the mean?

d. What percentage of the values in a normal distribution lie more than one standard deviation from the mean?

5 The weights of babies of a given age and gender are approximately normally distributed. This fact allows a doctor or nurse to use a baby's weight to find the weight percentile to which the child belongs. The table below gives information about the weights of six-month-old and twelve-month-old baby boys.

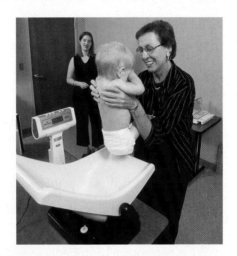

Weights of Baby Boys		
	Weight at Six Months (in pounds)	**Weight at Twelve Months** (in pounds)
Mean μ	17.25	22.50
Standard Deviation σ	2.0	2.2

Source: Tannenbaum, Peter, and Robert Arnold. *Excursions in Modern Mathematics.* Englewood Cliffs, New Jersey: Prentice Hall. 1992.

a. On separate axes with the same scales, draw sketches that represent the distribution of weights for six-month-old boys and the distribution of weights for twelve-month-old boys. How do the distributions differ?

b. About what percentage of six-month-old boys weigh between 15.25 pounds and 19.25 pounds?

c. About what percentage of twelve-month-old boys weigh more than 26.9 pounds?

d. A twelve-month-old boy who weighs 24.7 pounds is at what percentile for weight? (Recall that a value x in a distribution lies at, say, the 27th **percentile** if 27% of the values in the distribution are less than or equal to x.)

e. A six-month-old boy who weighs 21.25 pounds is at what percentile?

✔ CHECK YOUR UNDERSTANDING

Scores on the Critical Reading section of the 2011 SAT Reasoning Test are approximately normally distributed with mean μ of 497 and standard deviation σ of 114 (**Source:** professionals.collegeboard.com).

a. Sketch this distribution with a scale on the horizontal axis.

b. What percentage of students scored between 383 and 611 on the Critical Reading section of the SAT?

c. What percentage of students scored over 611 on the Critical Reading section of the SAT?

d. If you scored 383 on the Critical Reading section of the SAT, what is your percentile?

INVESTIGATION 2

Standardized Values

Often, you are interested in comparing values from two different distributions. For example: Is Sophie or her brother Pablo taller compared to others of their gender? Was your SAT score or your ACT score better? Answers to questions like these rely on your answer to the following question, which is the focus of this investigation:

How can you use standardized values to compare values from two different normal distributions?

1 Examine the table below, which gives information about the heights of young Americans aged 18 to 24. Each distribution is approximately normal.

Heights of American Young Adults (in inches)		
	Men	**Women**
Mean μ	68.5	65.5
Standard Deviation σ	2.7	2.5

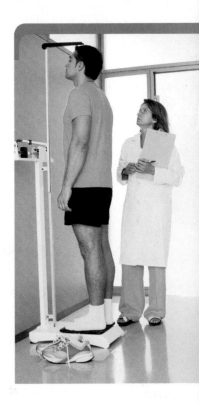

a. Sketch the two distributions. Include a scale on the horizontal axis.

b. Alexis is 3 standard deviations above average in height. How tall is she?

c. Marvin is 2.1 standard deviations below average in height. How tall is he?

d. Miguel is 74" tall. How many standard deviations above average height is he?

e. Jackie is 62" tall. How many standard deviations below average height is she?

f. Marina is 68" tall. Steve is 71" tall. Who is relatively taller for her or his gender, Marina or Steve? Explain your reasoning.

The **standardized value** tells how many standard deviations a given value lies from the mean of a distribution. For example, in Problem 1 Part b, Alexis is 3 standard deviations above average in height, so her standardized height is 3. Similarly, in Problem 1 Part c, Marvin is 2.1 standard deviations below average in height, so his standardized height is −2.1.

2 Look more generally at how standardized values are computed.

a. Refer to Problem 1, Parts d and e. Compute the standardized values for Miguel's height and for Jackie's height.

b. Write a formula for computing the standardized value z of a value x if you know the mean of the population μ and the standard deviation of the population σ.

3 Now consider how standardizing values can help you make comparisons. Refer to the table in Problem 1.

a. Find the standardized value for the height of a young woman who is 5 feet tall.

b. Find the standardized value for the height of a young man who is 5 feet 2 inches tall.

c. Is the young woman in Part a or the young man in Part b shorter, relative to his or her own gender? Explain your reasoning.

4 In an experiment about the effects of mental stress, subjects' systolic blood pressure and heart rate were measured before and after doing a stressful mental task. Their systolic blood pressure increased an average of 22.4 mm Hg (millimeters of Mercury) with a standard deviation of 2. Their heart rates increased an average of 7.6 beats per minute with a standard deviation of 0.7. Each distribution was approximately normal.

Suppose that after completing the task, Mario's blood pressure increased by 25 mm Hg and his heart rate increased by 9 beats per minute. On which measure did he increase the most, relative to the other participants?

(**Source:** "Mental Stress-Induced Increase in Blood Pressure Is Not Related to Baroreflex Sensitivity in Middle-Aged Healthy Men." Hypertension, Vol. 35, 2000.)

SUMMARIZE THE MATHEMATICS

In this investigation, you examined standardized values and their use.

a What information does a standardized value provide?

b How can standardizing values be useful?

c Kua earned a grade of 50 on a normally distributed test with mean 45 and standard deviation 10. On another normally distributed test with mean 70 and standard deviation 15, she earned a 78. On which of the two tests did she do better, relative to the others who took the tests? Explain your reasoning.

Be prepared to share your ideas and reasoning with the class.

✅ **CHECK YOUR UNDERSTANDING**

Refer to Problem 1 on page 243. Selena Gomez is 5 feet 6 inches tall. Nick Jonas is 5 feet 9 inches tall. Who is taller compared to others of their gender, Selena Gomez or Nick Jonas? Explain your reasoning.

Using Standardized Values to Find Percentiles

In the first investigation, you were able to find the percentage of values within one, two, and three standard deviations of the mean. Your work in this investigation will extend to other numbers of standard deviations and help you answer the following question:

How can you use standardized values to find the location of a value in a distribution that is normal, or approximately so?

The following table gives the proportion of values in a normal distribution that are less than the given standardized value z. By standardizing values, you can use this table for any normal distribution. If the distribution is not normal, the percentages given in the table do not necessarily hold.

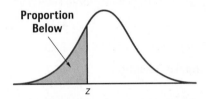

Proportion of Values in a Normal Distribution that Lie Below a Standardized Value z							
z	Proportion Below	z	Proportion Below	z	Proportion Below	z	Proportion Below
−3.5	0.0002	−1.7	0.0446	0.1	0.5398	1.9	0.9713
−3.4	0.0003	−1.6	0.0548	0.2	0.5793	2.0	0.9772
−3.3	0.0005	−1.5	0.0668	0.3	0.6179	2.1	0.9821
−3.2	0.0007	−1.4	0.0808	0.4	0.6554	2.2	0.9861
−3.1	0.0010	−1.3	0.0968	0.5	0.6915	2.3	0.9893
−3.0	0.0013	−1.2	0.1151	0.6	0.7257	2.4	0.9918
−2.9	0.0019	−1.1	0.1357	0.7	0.7580	2.5	0.9938
−2.8	0.0026	−1.0	0.1587	0.8	0.7881	2.6	0.9953
−2.7	0.0035	−0.9	0.1841	0.9	0.8159	2.7	0.9965
−2.6	0.0047	−0.8	0.2119	1.0	0.8413	2.8	0.9974
−2.5	0.0062	−0.7	0.2420	1.1	0.8643	2.9	0.9981
−2.4	0.0082	−0.6	0.2743	1.2	0.8849	3.0	0.9987
−2.3	0.0107	−0.5	0.3085	1.3	0.9032	3.1	0.9990
−2.2	0.0139	−0.4	0.3446	1.4	0.9192	3.2	0.9993
−2.1	0.0179	−0.3	0.3821	1.5	0.9332	3.3	0.9995
−2.0	0.0228	−0.2	0.4207	1.6	0.9452	3.4	0.9997
−1.9	0.0287	−0.1	0.4602	1.7	0.9554	3.5	0.9998
−1.8	0.0359	−0.0	0.5000	1.8	0.9641		

1. Use the table to help answer the following questions. In each part, draw sketches that illustrate your answers.

   a. Suppose that a value from a normal distribution is two standard deviations below the mean. What proportion of the values are below it? What proportion are above it?

   b. If a value from a normal distribution is 1.3 standard deviations above the mean, what proportion of the values are below it? Above it?

   c. Based on the table, what proportion of values are within one standard deviation of the mean? Within two standard deviations of the mean? Within three standard deviations of the mean?

2. Reproduced below is the summary table for heights of Americans aged 18 to 24. Each distribution is approximately normal.

Heights of American Young Adults (in inches)		
	Men	**Women**
Mean μ	68.5	65.5
Standard Deviation σ	2.7	2.5

   a. Miguel is 74 inches tall. What is his percentile for height? That is, what percentage of young men are the same height or shorter than Miguel?

   b. Jackie is 62 inches tall. What is her percentile for height?

   c. Abby is 5 feet 8 inches tall. What percentage of young women are between Jackie (Part b) and Abby in height?

   d. Gabriel is at the 90th percentile in height. What is his height?

   e. Yvette is at the 31st percentile in height. What is her height?

3. All 11th-grade students in Pennsylvania are tested using the Pennsylvania System of School Assessment (PSSA). The mean score on the PSSA math test was 1,372 with standard deviation 276. You may assume the distribution of scores is approximately normal. (**Source:** *Technical Report for the 2010 Pennsylvania System of School Assessment*)

   a. Draw a sketch of the distribution of these scores with a scale on the horizontal axis.

   b. What PSSA math score would be at the 50th percentile?

   c. What percentage of 11th graders scored above 1,500?

   d. Javier's PSSA score was at the 76th percentile. What was his score on this test?

SUMMARIZE THE MATHEMATICS

In this investigation, you explored the relation between standardized values and percentiles.

a How are standardized values used to find percentiles? When can you use this procedure?

b How can you find a person's score if you know his or her percentile and the mean and standard deviation of the normal distribution from which the score came?

Be prepared to share your ideas and reasoning with the class.

✔ CHECK YOUR UNDERSTANDING

Standardized values can often be used to make sense of scores on aptitude and intelligence tests.

a. Actress Brooke Shields reportedly scored 608 on the math section of the SAT. When she took the SAT, the scores were approximately normally distributed with an average on the math section of about 462 and a standard deviation of 100.

   i. How many standard deviations above average was her score?

   ii. What was Brooke Shields' percentile on the math section of the SAT?

b. The IQ scores on the Stanford-Binet intelligence test are approximately normal with mean 100 and standard deviation 15. Consider these lines from the movie *Forrest Gump*.

   (**Source:** moviequotesandmore.com; screenwriter Eric Roth)

   Mrs. Gump: Remember what I told you, Forrest. You're no different than anybody else is. Did you hear what I said, Forrest? You're the same as everybody else. You are no different.

   Principal: Your boy's … different, Miz Gump. His IQ's 75.

   Mrs. Gump: Well, we're all different, Mr. Hancock.

   i. How many standard deviations from the mean is Forrest's IQ score?

   ii. What percentage of people have an IQ higher than Forrest's IQ?

APPLICATIONS

1 Thirty-two students in a drafting class were asked to prepare a design with a perimeter of 98.4 cm. A histogram of the actual perimeters of their designs is displayed below. The mean perimeter was 98.42 cm.

Design Perimeters

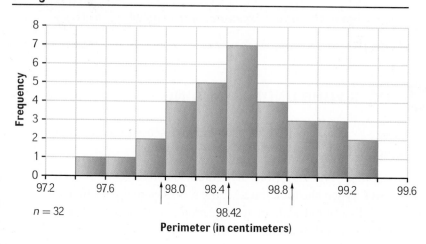

$n = 32$

Perimeter (in centimeters)

a. What might explain the variation in perimeters of the designs?

b. The arrows mark the mean and the points one standard deviation above the mean and one standard deviation below the mean. Use the marked plot to estimate the standard deviation for the class' perimeters.

c. Estimate the percentage of the perimeters that are within one standard deviation of the mean. Within two standard deviations of the mean.

d. How do the percentages in Part c compare to the percentages you would expect from a normal distribution?

2 For a chemistry experiment, students measured the time for a solute to dissolve. The experiment was repeated 50 times. The results are shown in the chart and histogram at the top of the next page.

Dissolution Time (in seconds)

4	5	6	7	8	8	8	8	9	9
9	10	10	10	10	10	10	10	11	11
11	11	11	12	12	12	12	12	12	12
12	13	13	13	13	13	14	14	14	14
14	15	15	16	16	17	17	17	19	19

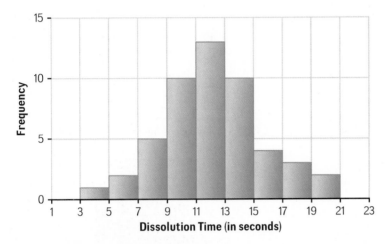

a. The mean time for the 50 experiments is 11.8 seconds. Find the median dissolution time from the table above. How does it compare to the mean dissolution time?

b. Which of the following is the best estimate of the standard deviation?

   1.32 seconds   3.32 seconds   5.32 seconds

c. On a copy of the histogram, mark points along the horizontal axis that correspond to the mean, one standard deviation above the mean, one standard deviation below the mean, two standard deviations above the mean, two standard deviations below the mean, three standard deviations above the mean, and three standard deviations below the mean.

d. What percentage of the times are within one standard deviation of the mean? Within two standard deviations? Within three standard deviations?

e. How do the percentages in Part d compare to the percentages you would expect from a normal distribution?

3 The table and histogram below give the heights of 123 women in a statistics class at Penn State University in the 1970s.

Female Students' Heights			
Height (in inches)	Frequency	Height (in inches)	Frequency
59	2	66	15
60	5	67	9
61	7	68	6
62	10	69	6
63	16	70	3
64	22	71	1
65	20	72	1

Source: Joiner, Brian L. 1975. Living histograms. *International Statistical Review 3*: 339–340.

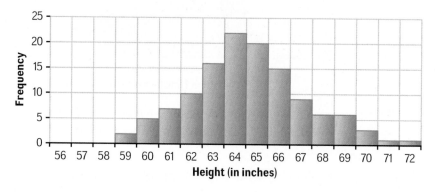

a. The mean height of the women in this sample is approximately 64.626 inches. Which of the following is the best estimate of the standard deviation?

  0.2606 inches   0.5136 inches   2.606 inches   5.136 inches

b. On a copy of the histogram, mark points along the horizontal axis that correspond to the mean, one standard deviation above the mean, one standard deviation below the mean, two standard deviations above the mean, two standard deviations below the mean, three standard deviations above the mean, and three standard deviations below the mean.

c. What percentage of the heights are within one standard deviation of the mean? Within two standard deviations? Within three standard deviations?

d. Suppose you pick a female student from the class at random. Find the probability that her height is within two standard deviations of the mean.

4 Suppose a large urban high school has 29 English classes. The number of students enrolled in each of the classes is displayed on the plot below. The outlier is a grammar class with only 7 students enrolled. One way to handle an outlier is to report a summary statistic computed both with and without it. If the two values are not very different, it is safe to report just the value that includes the outlier.

English Class Enrollment

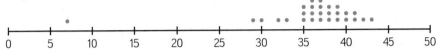

a. Justina computed the mean twice, with and without the outlier. The two means were 35.97 and 37.00. Which mean was computed with the outlier?

b. Justina also computed the standard deviation twice, with and without the outlier. The two standard deviations were 3.25 and 6.34. Which standard deviation was computed with the outlier?

c. How many standard deviations from the mean is the enrollment in the class with 43 students when the grammar class is included in computing the mean and standard deviation? When the grammar class is not included in the calculations?

d. Suppose five students drop out of each class. What will be the new mean and standard deviation if the grammar class is included in the computations?

5 The Web site www.marathonguide.com keeps track of the results of all marathons run in the United States. In 2010, in 483 marathons, 503,000 people completed the entire 26.2 miles. The mean time for men was 4:27:12 (std. dev. 1:01:18) and for women was 4:54:27 (std. dev. 1:02:16). These times were reported in the form hours:minutes:seconds. Suppose Oliver ran a marathon in 4:00:15 and Jada in 4:09:09.

a. How many standard deviations from the mean for men was Oliver's time? How many standard deviations from the mean for women was Jada's time? Who did better compared to others of their gender?

b. Write a formula that gives the number of standard deviations from the mean for women for a time of x minutes.

c. Write a formula that gives the number of standard deviations from the mean for men for a time of x minutes.

6 The length of a human pregnancy is often said to be 9 months. Actually, the length of pregnancy from conception to natural birth varies according to a distribution that is approximately normal with mean 266 days and standard deviation 16 days.

a. Draw a sketch of the distribution of pregnancy lengths. Include a scale on the horizontal axis.

b. What percentage of pregnancies last less than 250 days?

c. What percentage of pregnancies are longer than 298 days?

d. To be in the shortest 2.5% of pregnancies, what is the longest that a pregnancy can last?

e. What is the median length of pregnancy?

7 Scores on the mathematics section of the 2011 SAT Reasoning Test are approximately normally distributed with mean 514 and standard deviation 117. Scores on the mathematics part of the ACT are approximately normally distributed with mean 21.1 and standard deviation 5.3.
(**Source:** professionals.collegeboard.com, act.org)

a. Sketch graphs of the distribution of scores on each test. Include a scale on the horizontal axis.

b. What percentage of the SAT scores lie above 631? Above what ACT score would this same percentage of scores lie?

c. What ACT score is the equivalent of an SAT score of 450?

d. Find the percentile of a person who gets an SAT score of 450.

e. One of the colleges to which Eliza is applying accepts either SAT or ACT mathematics scores. Eliza scored 680 on the mathematics part of the SAT and 27 on the mathematics section of the ACT. Should she submit her SAT or ACT mathematics score to this college? Explain your reasoning.

8 Many body dimensions of adult males and females in the United States are approximately normally distributed. Approximate means and standard deviations for shoulder width are given in the table below.

U.S. Adult Shoulder Width (in inches)		
	Men	**Women**
Mean μ	17.7	16.0
Standard Deviation σ	0.85	0.85

a. What percentage of women have a shoulder width of less than 15.5 inches? Of more than 15.5 inches?

b. What percentage of men have a shoulder width between 16 and 18 inches?

c. What percentage of American women have a shoulder width more than 17.7 inches, the average shoulder width for American men?

d. What percentage of men will be uncomfortable in an airplane seat designed for people with shoulder width less than 18.5 inches? What percentage of women will be uncomfortable?

e. If you sampled 100,000 men, what is the expected number who would be uncomfortable in an airline seat designed for people with shoulder width less than 18.5 inches? If you sampled 100,000 women, what is the expected number who would be uncomfortable in an airline seat designed for people with shoulder width less than 18.5 inches?

CONNECTIONS

9 Three very large sets of data have approximately normal distributions, each with a mean of 10. Sketches of the overall shapes of the distributions are shown below. The scale on the horizontal axis is the same in each case. The standard deviation of the distribution in Figure A is 2. Estimate the standard deviations of the distributions in Figures B and C.

Figure A   **Figure B**   **Figure C**

 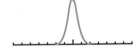 

10 How is the formula for the standard deviation $s = \sqrt{\dfrac{\Sigma(x - \bar{x})^2}{n - 1}}$ like the distance formula?

11 If a set of data is the entire population you are interested in studying, you compute the population standard deviation using the formula:

$$\sigma = \sqrt{\frac{\Sigma(x - \mu)^2}{n}}$$

If you are looking at a set of data as a sample from a larger population, you compute the standard deviation using this formula:

$$s = \sqrt{\frac{\Sigma(x - \bar{x})^2}{n - 1}}$$

a. For a given set of data, which is larger, σ or s? Explain your reasoning.

b. Does it make much difference whether you divide by n or by $n - 1$ if $n = 1{,}000$? If $n = 15$?

c. A sample tends to have less variability than the population from which it came. How does the formula for s, the standard deviation for a sample, account for this fact?

12 Suppose a normal distribution has mean 100 and standard deviation 15. Now suppose every value in the distribution is converted to a standardized value.

a. What is true about the mean of the standardized values?

b. What is true about the standard deviation of the standardized values?

13 In a normal distribution, about 5% of the values are more than two standard deviations from the mean. If a distribution is not approximately normal, is it possible to have more than 5% of the values two or more standard deviations from the mean? In this task, "standard deviation" refers to the population standard deviation σ, as defined in Task 11.

a. Make up two sets of numbers and compute the percentage that are two or more standard deviations from the mean. Your objective is to get the largest percentage that you can.

b. What is the largest percentage of numbers that you were able to find in Part a? Compare your results to those of other students completing this task.

14 The formula for Pearson's correlation that you studied in the Course 2 *Regression and Correlation* unit is

$$r = \frac{1}{n-1} \sum \left(\frac{x - \bar{x}}{s_x} \right) \left(\frac{y - \bar{y}}{s_y} \right).$$

Here, n is the sample size, $\bar{x}$ is the mean of the values of x, $\bar{y}$ is the mean of the values of y, s_x is the standard deviation of the values of x, and s_y is the standard deviation of the values of y.

a. Use this formula to find the correlation between x and y for the following pairs of numbers.

x	y
1	1
2	2
3	6

b. Explain the meaning of the correlation in the context of standardized values.

REFLECTIONS

15 Why do we say that distributions of real data are "approximately" normally distributed rather than say they are normally distributed?

16 Under what conditions will the standard deviation of a data set be equal to 0? Explain your reasoning.

17 Is it true that in all symmetric distributions, about 68% of the values are within one standard deviation of the mean? Give an example to illustrate your answer.

18 Consider the weights of the dogs in the following two groups.

- the dogs pulling a sled in a trans-Alaska dog sled race
- the dogs in a dog show, which includes various breeds of dogs

a. Which group would you expect to have the larger mean weight? Explain your reasoning.

b. Which group would you expect to have the larger standard deviation? Explain your reasoning.

19 ACT and SAT scores have an approximately normal distribution. Scores on classroom tests are sometimes assumed to have an approximately normal distribution.

a. What do teachers mean when they say they "grade on a curve"?

b. Explain how a teacher might use a normal distribution to "grade on a curve."

c. Under what circumstances would you want to be "graded on a curve"?

EXTENSIONS

20 The producers of a movie did a survey of the ages of the people attending one screening of the movie. The data are shown in the table.

a. Compute the mean and standard deviation for this sample of ages. Do this without entering each of the individual ages into a calculator or computer software. (For example, do not enter the age "14" thirty-eight times.)

b. In this distribution, what percentage of the values fall within one standard deviation of the mean? Within two standard deviations of the mean? Within three standard deviations of the mean?

c. Compare the percentages from Part b to those from a normal distribution. Explain your findings in terms of the shapes of the two distributions.

Saturday Night at the Movies	
Age (in years)	**Frequency**
12	2
13	26
14	38
15	32
16	22
17	10
18	8
19	8
20	6
21	4
22	1
23	3
27	2
32	2
40	1

21 In 1903, Karl Pearson and Alice Lee collected the heights of 1,052 mothers. Their data are summarized below. A mother who was exactly 53 inches tall was recorded in the 53–54 inches row.

Heights of Mothers (in inches)

Height	Number of Mothers	Height	Number of Mothers
52–53	1	62–63	183
53–54	1	63–64	163
54–55	1	64–65	115
55–56	2	65–66	78
56–57	7	66–67	41
57–58	18	67–68	16
58–59	34	68–69	7
59–60	80	69–70	5
60–61	135	70–71	2
61–62	163		

Source: Pearson, Karl and Alice Lee. 1903. On the Laws of Inheritance in Man. *Biometrika*: 364.

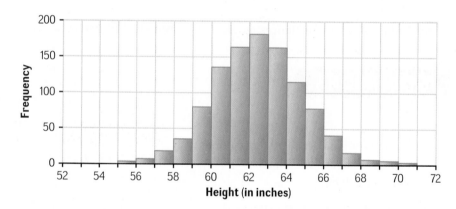

a. Is this distribution of heights approximately normal? Why or why not?

b. Collect the heights of 30 mothers. How does the distribution of your sample compare to the distribution of heights from mothers in 1903?

c. What hypothesis might you make about heights of mothers today? Design a plan that you could use to test your hypothesis.

22 The equation of the curve that has the shape of a normal distribution is:

$$y = \frac{1}{\sigma\sqrt{2\pi}} e^{-\frac{1}{2}\left(\frac{x-\mu}{\sigma}\right)^2}$$

In this formula, μ is the mean of the normal distribution, σ is the standard deviation, and the number e is approximately equal to 2.71828.

a. Use your calculator or computer software to graph the normal curve that has a mean of 0 and a standard deviation of 1.

b. Describe what happens to the curve if you increase the mean. Describe what happens if you increase the standard deviation.

c. Normal curves have two "bends," or **points of inflection**. One is on the left side of the curve where the graph changes from curved up to curved down. The other is on the right side of the curve where the graph changes from curved down to curved up. Estimate the point where the "bend" seems to occur in the curve in Part a. What relation does this point have to the mean and standard deviation?

23 Discuss whether the situations below are consistent with what you know about normal distributions and IQ tests. Recall that the mean and standard deviation for IQ tests are $\mu = 100$ and $\sigma = 15$. What could account for any inconsistencies that you see?

a. One of the largest K–12 districts in the country, educates between 700,000 and 800,000 children. In this district, there are special magnet schools for "highly gifted" children. The only way for a child in this district to be classified as highly gifted is to score 145 or above on an IQ test given by a school psychologist. Recently, at one gifted magnet school, there were 61 students in the sixth-grade class.

b. One way for a child to be identified as gifted in California is to have an IQ of 130 or above. A few years ago, a Los Angeles high school had a total enrollment of 2,830 students, of whom 410 were identified as gifted.

REVIEW

24 Suppose that the cost of filling up the gas tank of a car is a function of the number of gallons and can be found using the rule $C(g) = 2.89g$.

a. What does the 2.89 tell you about this situation?

b. What is the cost of 4.2 gallons of gas?

c. Find the value of g so that $C(g) = 19.65$. Then explain what your solution means in this context.

d. Is this relationship a direct variation, inverse variation, or neither? Explain.

25 In the diagram at the right, $\overline{BC}$ is the perpendicular bisector of $\overline{AD}$, and $AD = 20$ cm. If possible, complete each of the following tasks. If not possible, explain why not.

a. Prove that $\triangle ACB \cong DCB$.

b. Find the area of $\triangle ABD$.

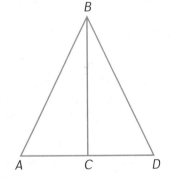

26 Solve each inequality and graph the solution on a number line. Then describe the solution using interval notation.

    a. $3(x + 7) \geq 5$
    b. $10 - 8x < 4(8 - 4x)$

    c. $2x \geq 5x - 3$
    d. $x^2 > 4$

    e. $x^2 - 5x < 0$
    f. $x - 2 \leq \sqrt{3x}$

27 Rewrite each expression in factored form.

    a. $3x^2 + 9x$
    b. $2x^2 + 9x + 4$

    c. $x^3 - 64x$
    d. $x^2 - 12x - 45$

    e. $16 - x^4$
    f. $x^4 + 10x^2 - 11$

28 Try to answer each of the following using mental computation only.

    a. What is 25% of 400?

    b. What is 60% of 80?

    c. 20 is what percent of 50?

    d. 180 is what percent of 270?

    e. 6 is 20% of what number?

    f. 14 is 2% of what number?

29 Given $\triangle XYZ$ with $m\angle X = 90°$ and $\sin Y = \frac{1}{2}$:

    a. Find a set of possible exact side lengths for the triangle. Are your side lengths the only ones possible? If so, explain why. If not, explain why not.

    b. Which of the following is the value for $\cos Z$?

$$\frac{1}{2} \qquad \frac{2}{\sqrt{3}} \qquad \sqrt{3} \qquad \frac{\sqrt{3}}{2}$$

    c. Find $m\angle Y$.

30 Rewrite each expression in simplest equivalent form.

    a. $10 - 4(5x + 1)$
    b. $(2x + 1)(3 - x) + 5(x + 3)$

    c. $\frac{6x + 9}{3}$
    d. $(4x + 5)^2$

31 Recall that two events A and B are independent if the occurrence of one of the events does not change the probability that the other occurs. For each situation, decide if it is reasonable to assume that the events are independent. Then find the probability of both events occurring.

    a. Rolling a pair of dice one time
        Event A: Getting doubles
        Event B: Getting a sum that is even

    b. Flipping a coin twice
        Event A: Getting heads on the first flip
        Event B: Getting heads on the second flip

Gregory Shamus/Getty Images Sport/Getty Images

Binomial Distributions

In the first lesson, you learned about properties of the normal distribution. This lesson is about binomial distributions. For example, suppose you flip a coin 10 times and count the number of heads. The binomial distribution for this situation gives you the probability of getting 0 heads, of getting 1 head, of getting 2 heads, and so on, up to the probability of getting 10 heads. A similar situation occurs when counting the number of successful

free throw attempts for a specific basketball player in a game or season.

Consider the case of Candace Parker who played basketball for the University of Tennessee from 2005 to 2008. She is a versatile player usually playing forward, and was listed on Tennessee's roster as forward, center, and guard. She played in 110 games and made 526 of her 738 attempted free throws.

(**Source:** chicagosports.sportsdirectinc.com)

In this lesson, you will learn how to answer questions like those above by describing the shape, center, and spread of the distribution of the possible numbers of successes in repeated samples.

INVESTIGATION 1

Shape, Center, and Spread

Repeating the same process for a fixed number of *trials* and counting the number of "successes" is a common situation in probability. For example, you might roll a pair of dice 50 times and count the number of doubles. Or, you might survey 500 randomly selected U.S. teens and count the number who play soccer. (Recall that getting a random sample of 500 U.S. teens is equivalent to putting the names of all U.S. teens in a hat and drawing out 500 at random.) These are called **binomial situations** if they have the four characteristics listed below.

- There are two possible outcomes on each trial called "success" and "failure."

- Each trial is independent of the others. That is, knowing what happened on previous trials does not change the probability of a success on the next trial. (If you take a relatively small random sample from a much larger population, you can consider the trials independent.)

- There is a fixed number of trials n.

- The probability p of a success is the same on each trial.

Your work in this investigation will help you answer the following question:

What are the shape, mean (expected value), and standard deviation of a distribution of the number of successes in a binomial situation?

1 Determine whether each of the following situations is a binomial situation by identifying:

- the two possible outcomes on each trial
- whether the trials are independent
- the number of trials
- the probability of a success on each trial

a. You flip a coin 20 times and count the number of heads.

b. You roll a six-sided die 60 times and count the number of times you get a 2.

c. About 51% of the residents of the U.S. are female. You take a randomly selected sample of 1,200 U.S. residents and count the number of females. (**Source:** www.census.gov)

d. Fifty-nine percent of children live in counties that do not meet one or more of the Primary National Ambient Air Quality Standards. You select 75 children at random from the U.S. and count the number who live in counties that do not meet one or more of these standards. (**Source:** *America's Children: Key National Indicators of Well-Being 2011.*)

e. The player with the highest career free throw percentage in NBA history is Mark Price. He made 2,135 free throws out of 2,362 attempts. Suppose Mark Price shoots 20 free throws and you count the number of times he makes his shot. (**Source:** www.nba.com/statistics/)

Recall that in probability, **expected value** means the *long-run average* or *mean* value. For example, if you flip a coin 7 times, you might get 3 heads, you might get 4 heads, and you might get more or fewer. Over many sets of 7 flips, however, the average or **expected number** of heads will be 3.5. So, you can say that if you flip a coin 7 times, the expected number of heads is 3.5.

2 In the Course 2 *Probability Distributions* unit, you learned that if the probability of a success on each trial of a binomial situation is *p*, then the expected number of successes in *n* trials is *np*.

a. What formula gives you the expected number of failures?

b. For the binomial situations in Problem 1, what are the expected number of successes and the expected number of failures?

You can use statistics software or a command on your calculator to simulate a binomial situation. For example, suppose you want to simulate flipping a coin 100 times and counting the number of heads. From the calculator Probability menu, select **randBin(**. Type in the number of trials and the probability of a success, **randBin(100,.5)** and press **ENTER**. The calculator returns the number of successes in 100 trials when the probability of a success is 0.5. The "Random Binomial" feature of simulation software like in *CPMP-Tools* operates in a similar manner.

3 Give the calculator or software command that you would use to simulate each of the binomial situations in Problem 1. Then do one run and record the number of successes.

Hisham Ibrahim/Getty Images

4 Suppose you flip a coin 100 times and count the number of heads.

a. What is the expected number of heads?

b. If everyone in your class flipped a coin 100 times and counted the number of heads, how much variability do you think there would be in the results?

c. Use the **randBin** function of your calculator or the "Random Binomial" feature of simulation software to simulate flipping a coin 100 times. Record the number of heads. Compare results with other members of your class.

d. Perform 200 runs using the software or combine calculator results with the rest of your class until you have the results from 200 runs. Make a histogram of this **approximate binomial distribution**. What is its shape?

e. Estimate the expected number of heads and the standard deviation from the histogram. How does the expected number compare to your answer from Part a?

In Problem 4, you saw a binomial distribution that was approximately normal in shape. In the next problem, you will examine whether that is the case for all binomial distributions.

5 According to the U.S. Census, about 20% of the population of the United States are children, age 14 or younger. (**Source:** *Age and Sex Composition: 2010, Census 2010 Briefs,* May 2011 at www.census.gov/prod/cen2010/briefs/c2010br-03.pdf) Suppose you take a randomly selected sample of people from the United States. The following graphs show the binomial distributions for the number of children in random samples varying in size from 5 to 100.

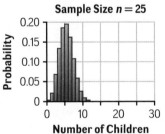

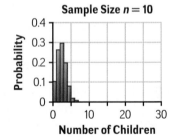

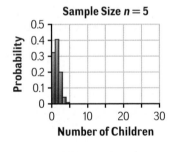

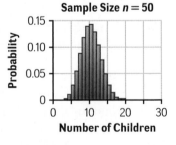

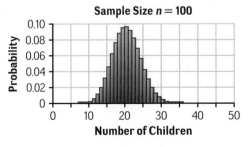

a. The histogram for a sample size of 5 has six bars, one for 0 successes, one for 1 success, one for 2 successes, one for 3 successes, one for 4 successes, and a very short bar for 5 successes.

   i. Why are there more bars as the sample size increases?

   ii. How many bars should there be for a sample size of *n*?

   iii. Why can you see only 7 bars on the histogram for a sample size of 10?

b. Use the histograms on the previous page to help answer the following questions.

   i. What happens to the shape of the distribution as the sample size increases?

   ii. What happens to the expected number of successes as the sample size increases?

   iii. What happens to the standard deviation of the number of successes as the sample size increases?

6 As you saw in Problem 5, not all binomial distributions are approximately normal. Statisticians have developed a *guideline* that you can use to decide whether a binomial distribution is approximately normal.

> *The shape of a binomial distribution will be approximately normal if the expected number of successes is 10 or more and the expected number of failures is 10 or more.*

a. Write the guideline about when a binomial distribution can be considered approximately normal using two algebraic inequalities where *n* is the number of trials and *p* is the probability of a success.

b. In each of the following situations, decide if the sample size is large enough so that the binomial distribution will be approximately normal.

   i. The five distributions in Problem 5

   ii. 100 flips of a coin and counting the number of heads

   iii. Rolling a die 50 times and counting the number of 3s

   iv. Randomly selecting 1,200 U.S. residents in Problem 1 Part c

   v. Randomly selecting 75 children in Problem 1 Part d

7 One of the reasons the standard deviation is such a useful measure of spread is that the standard deviation of many distributions has a simple formula. The **standard deviation of a binomial distribution** is given by:

$$\sigma = \sqrt{np(1-p)}$$

a. Suppose you flip a coin 100 times and count the number of heads. Use this formula to compute the standard deviation of the binomial distribution.

b. Compare your answer in Part a to the estimate from your simulation in Problem 4 Part e.

8 Suppose you plan to spin a spinner, like the one shown, 60 times and count the number of times that you get green.

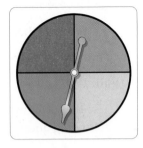

    a. If p represents the probability of getting green on a single spin, what is the value of p? What does $1 - p$ represent?

    b. In 60 spins, what is the expected number of times that you will get green?

    c. What is the standard deviation of the number of greens?

    d. Will the distribution of the number of successes be approximately normal? If so, make a sketch of this distribution, marking values of the mean and one, two, and three standard deviations from the mean on the x-axis.

    e. Would you be surprised to get 13 greens? 20 greens? Use standardized values in your explanations.

    f. In Problem 5, you learned that as the number of trials for a binomial distribution increases, the standard deviation of the number of successes increases. Suppose you increase the number of spins to 240.

        i. Compute the standard deviation of this distribution.

        ii. Describe the relationship between the standard deviation for 240 spins and the standard deviation for 60 spins.

        iii. How can you recognize this relationship by examining the formula for the standard deviation of a binomial distribution?

9 About 60% of children ages 3 to 5 are read to daily by a family member. (**Source:** *America's Children: Key National Indicators of Well-Being 2011,* page 157.) Suppose you take a random sample of 75 children this age.

    a. Describe the shape of the binomial distribution for this situation. Compute the mean and standard deviation.

    b. Using the expected number and standard deviation, describe the number of children you might get in your sample who are read to daily by a family member.

    c. Out of the 75 children in your sample, suppose you get 42 children who are read to daily by a family member. Compute the standardized value for 42 children. Use this standardized value to estimate the probability of getting 42 or fewer children who are read to daily by a family member.

    d. Out of the 75 children in your sample, suppose you get 50 children who are read to daily by a family member. Compute the standardized value for 50 children. Use this standardized value to estimate the probability of getting 50 or more children who are read to daily by a family member.

Digital Vision

SUMMARIZE THE MATHEMATICS

In this investigation, you learned to use technology tools to construct a simulated binomial distribution. You also learned formulas that give the expected value and the standard deviation of a binomial distribution.

a Describe how to use the **randBin** function of your calculator or simulation software to construct an approximate binomial distribution if you know the number of trials n and the probability of a success p.

b Write a formula for the expected number of successes in a binomial distribution. What is another term for the expected value?

c How can you tell whether a binomial distribution will be approximately normal in shape?

d How do you find the spread of a binomial distribution?

e What happens to the expected number of successes in a binomial distribution if the probability of a success remains the same and the number of trials increases? What happens to the standard deviation?

f How do standardized values help you analyze binomial situations?

Be prepared to share your ideas and reasoning with the class.

✔ CHECK YOUR UNDERSTANDING

About 66.6% of all housing units in the U.S. are owner-occupied. (**Source:** quickfacts. census.gov/qfd/states/00000.html) Suppose you randomly select 120 housing units from the U.S.

a. What is the expected number of owner-occupied units in your sample?

b. Is the binomial distribution of the number of owner-occupied units approximately normal?

c. What is the standard deviation of this binomial distribution? What does it tell you?

d. Would you be surprised to find that 65 of the housing units are owner-occupied? To find that 90 of the housing units are owner-occupied? Explain.

e. Compute a standardized value for 72 owner-occupied units. Use this standardized value to estimate the probability of getting 72 or fewer owner-occupied units.

f. Explain how to use your calculator to simulate taking a random sample of 120 housing units from the U.S. and determining how many are owner-occupied.

Binomial Distributions and Making Decisions

Sometimes things happen that make you suspicious. For example, suppose your friend keeps rolling doubles in Monopoly. Knowing about binomial distributions can help you decide whether you should look a little further into this suspicious situation. As you work on problems in this investigation, look for an answer to the following question:

How do you decide whether a given probability of success
is a plausible one for a given binomial situation?

1 Suppose that while playing Monopoly, your friend rolls doubles 9 times out of 20 rolls.

a. What is the expected number of doubles in 20 rolls?

b. What is the standard deviation of the number of doubles? Is your friend's number of doubles more than one standard deviation above the expected number? More than two?

c. It does seem like your friend is a little lucky. The following histogram shows the binomial distribution for $n = 20$ and $p = \frac{1}{6}$. This is the theoretical distribution with the probability of getting each number of doubles shown on the vertical axis. Should you be suspicious of the result of 9 doubles in 20 rolls?

20 Rolls of the Dice

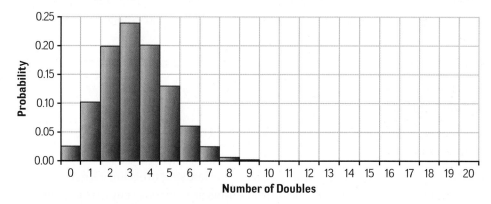

d. What would be your reaction if your friend rolled doubles all 20 times after you started counting? Use the Multiplication Rule to find the probability that 20 doubles in a row will occur just by chance.

2 Recall that a **rare event** is one that lies in the outer 5% of a distribution. In a waiting-time distribution, rare events typically must be in the upper 5% of the distribution. In binomial distributions, rare events are those in the upper 2.5% or lower 2.5% because it is unlikely to get either an unusually large number of successes or an unusually small number of successes. Use the histogram in Problem 1 to determine which of the following are rare events if you roll a pair of dice 20 times.

a. Getting doubles only once

b. Getting doubles seven times

When a distribution is approximately normal, there is an easy way to identify rare events. You learned in the previous lesson that about 95% of the values in a normal distribution lie within two standard deviations of the mean. Thus, for a binomial distribution that is approximately normal, rare events are those more than two standard deviations from the mean.

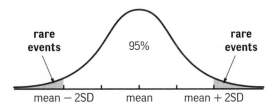

3 Suppose you roll a pair of dice 100 times and count the number of times you roll doubles. The binomial distribution for this situation appears below.

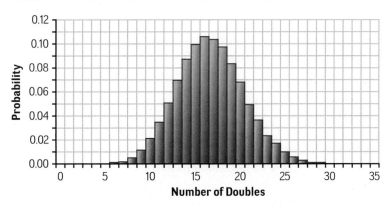

a. Is this distribution approximately normal? How does it compare to the distribution in Problem 1 on page 266?

b. Look at the tails of the histogram. Approximately what numbers of doubles would be rare events?

c. Use the histogram to estimate the expected value and standard deviation of the number of doubles in 100 rolls. Then use the formulas to calculate the expected value and standard deviation.

d. Using the values you calculated in Part c, determine which numbers of doubles would be rare events in 100 rolls of the dice. Are they the same as those you identified in Part b?

4 According to the U.S. Census, about 20% of the residents of the United States are children, age 14 or younger. (**Source:** *Age and Sex Compostion: 2010, Census 2010 Briefs, May 2011 at www.census.gov/prod/cen2010/briefs/c2010br-03.pdf*) The binomial distribution below shows the number of children in 1,000 random samples. Each sample was of size $n = 100$ U.S. residents.

Number of Children in 1,000 Random Samples of 100 U.S. Residents

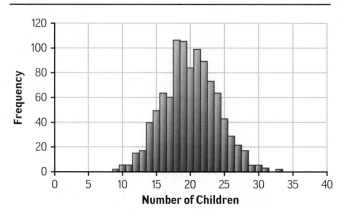

a. Is this distribution approximately normal? Calculate or estimate the mean and standard deviation.

b. Using the mean and standard deviation from Part a, determine which numbers of children would be rare events in a random sample of 100 U.S. residents.

c. Suppose you take a randomly selected sample of 100 residents of your community and find that there are 35 children.

  i. Would getting 35 children be a rare event if it is true that 20% of the residents in your community are children?

  ii. Would this sample make you doubt that 20% is the correct percentage of residents who are children? If so, the result is called *statistically significant*.

5 Look back at the process you used to analyze the result of getting 35 children in the random sample of 100 residents in the previous problem. Review the steps that you took to help you decide whether the event of 35 children was statistically significant.

Step 1. You want to decide whether it is plausible that the specified value of p is the right one for your population. State this value of p.

Step 2. For the specified value of p and your sample size (number of trials), verify that the binomial distribution is approximately normal. Compute its mean and standard deviation.

Step 3. Take a random sample from your population.

Step 4. Determine how many standard deviations the number of successes in your sample is from the mean of the binomial distribution for the specified proportion of successes in the population p and sample size n. Include a sketch of the situation.

Step 5. If the number of successes in your sample is more than two standard deviations from the mean, declare it **statistically significant**.

If you have a random sample, there are two reasons that a statistically significant result could occur.

- You have the right value of p for your population but a rare event occurred.

- You have the wrong value of p for your population.

What makes the practice of statistics challenging is that it is usually impossible to know for sure which is the correct conclusion. However, it is unlikely to get a rare event (probability only 0.05) if you have the right p for your population. Thus, you conclude that the value of p is not the right one for your population.

a. How do you verify that the distribution is approximately normal? Why did you need to determine whether or not the distribution is approximately normal?

b. How do you compute the mean and standard deviation?

c. How do you find how many standard deviations your number of successes lies from the mean?

d. What should you conclude if you have a statistically significant result? What should you conclude if you do not have a statistically significant result?

6 About 51% of the residents of the U.S. are female. Tanner takes a random sample of 1,200 people from his community and finds that 640 of the 1,200 are female. From this, he concludes that in his community, the percentage of females is greater than 51%. Use the five steps from Problem 5 to help you decide if you agree with Tanner's conclusion.

7 Fifty-nine percent of children live in counties that do not meet one or more of the Primary National Ambient Air Quality Standards. Suppose that of a randomly selected sample of 100 children hospitalized for asthma, 76 live in a county that does not meet one or more of the Primary National Ambient Air Quality Standards. Is this a statistically significant result? What conclusion should you reach? (**Source:** *America's Children: Key National Indicators of Well-Being, 2011.*)

SUMMARIZE THE MATHEMATICS

In this investigation, you learned to identify rare events in binomial situations.

a How do you determine if an outcome is a rare event in a binomial distribution that is approximately normal?

b What conclusions should you consider if you read in a newspaper article that an outcome is statistically significant?

Be prepared to share your ideas and reasoning with the class.

 CHECK YOUR UNDERSTANDING

According to a U.S. Department of Education report, about 70% of high school graduates enroll in college immediately after high school graduation. (**Source:** *The Condition of Education 2011*, nces.ed.gov/pubs2011/2011033.pdf) Suppose you take a random sample of 200 high school graduates and count the number who enroll in college immediately after graduation.

a. What is the expected number in your sample who enroll? The expected number who do not? Is this sample size large enough that the binomial distribution for this situation will be approximately normal?

b. Complete this sentence using the mean and standard deviation. If you select 200 high school graduates at random, the expected number of graduates who enroll in college immediately after high school graduation is ___, give or take ___ graduates.

c. Sketch this distribution. Include a scale on the x-axis that shows the mean and the points one and two standard deviations from the mean.

d. Suppose you want to see whether 70% is a plausible percentage for your county. You take a random sample of 200 high school graduates from your county and find that 135 enroll in college immediately after graduation. Is this statistically significant? What is your conclusion?

e. Suppose your random sample found 120 enrolled in college immediately after graduation. Is this statistically significant? What is your conclusion?

APPLICATIONS

1 About 20% of the residents of the United States are children aged 14 and younger. Suppose you take a random sample of 200 people living in the United States and count the number of children.

  a. Will the binomial distribution for the number of children in a sample of this size be approximately normal? Why or why not?

  b. Compute the mean and the standard deviation of the binomial distribution for the number of children.

  c. Make a sketch of the distribution, including a scale on the x-axis that shows the mean and one, two, and three standard deviations from the mean.

  d. Use a standardized value to estimate the probability that your sample contains 45 or more children.

  e. The number of children in one random sample of 200 residents is 1.94 standard deviations below average. How many children are in the sample?

  f. Suppose in your random sample of 200 people, 45 were children. Would you be surprised with that number of children? Explain.

2 Ty Cobb played Major League Baseball from 1905 to 1928. He holds the record for the highest career batting average (0.366). Suppose you pick 60 of his at bats at random and count the number of hits. (**Source:** baseball-almanac.com/hitting/hibavg1.shtml)

  a. Is the binomial distribution of the number of hits approximately normal?

  b. Compute the expected value and standard deviation of the binomial distribution of the number of hits in 60 at bats.

  c. Sketch this distribution. Include a scale on the x-axis that shows the mean and the points one and two standard deviations from the mean.

  d. Use a standardized value to estimate the probability that Cobb got 14 or fewer hits in this random sample of size 60.

  e. Suppose you were Cobb's coach. Would you have been concerned if Cobb got only 14 hits in his next 60 at bats?

3 A library has 25 computers, which must be reserved in advance. The library assumes that people will decide independently whether or not to show up for their reserved time. It also estimates that for each person, there is an 80% chance that he or she will show up for the reserved time.

a. If the library takes 30 reservations for each time period, what is the expected number of people who will show? Is it possible that all 30 will show?

b. How many reservations should the library take so that 25 are expected to show? Is it possible that if the library takes this many reservations, more than 25 people will show?

c. Suppose the library decides to take 30 reservations. Use the **randBin** function of your calculator to conduct 5 runs of a simulation that counts the number who show. Add your results to a copy of the frequency table and histogram below so that there is a total of 55 runs.

Computer Reservations		
Number of People Who Show	**Frequency (before)**	**Frequency (after)**
19	0	
20	1	
21	2	
22	3	
23	5	
24	8	
25	9	
26	9	
27	7	
28	4	
29	2	
30	0	
Total Number of Runs	50	55

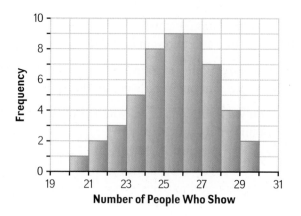

d. Describe the shape of the distribution.

e. Suppose the library takes 30 reservations. Based on the simulation, estimate the probability that more than 25 people show. Is 30 a reasonable number of reservations to take? Should the library take fewer than 30 reservations or is it reasonable to take more?

f. What assumption made in this simulation is different from the real-life situation being modeled?

4 Several years ago, a survey found that 25% of American pet owners carry pictures of their pets in their wallets. Assume this percentage is true, and you will be taking a random sample of 20 American pet owners and counting the number who carry pictures of their pets.

a. Describe how to conduct one run of a simulation using the **randBin** function of your calculator.

b. Predict the shape, mean, and standard deviation of the binomial distribution that you would get if you conduct a large number of runs.

c. Perform 10 runs of your simulation. Add your results to a copy of the following frequency table so that there is a total of 100 runs.

Number of People with Pictures of Their Pets	Frequency (before)	Frequency (after)
0	0	
1	2	
2	6	
3	12	
4	17	
5	18	
6	15	
7	10	
8	5	
9	3	
10	1	
11	1	
Total Number of Runs	90	100

d. Make a histogram of this distribution. Check your predictions in Part b.

e. Estimate the number of pet owners with pictures that would be rare events.

f. Patrick took a survey of 20 pet owners at the mall. Only one had a picture of a pet in his wallet. Does Patrick have good reason to doubt the reported figure of 25%?

5 According to the U.S. census, 66.6% of housing units are occupied by the owners. (**Source:** quickfacts.census.gov/qfd/b states/00000.html) You want to see if this percentage is plausible for your county. Suppose you take a random sample of 1,100 housing units and find that 526 are owner-occupied. Is this a statistically significant result? What conclusion should you reach? Use the five steps in Problem 5 of Investigation 2 on pages 268 and 269 to help you decide.

6 You have inherited a coin from your eccentric uncle. The coin appears a bit strange itself. To test whether the coin is fair, you toss it 150 times and count the number of heads. You get 71 heads. What conclusion should you reach?

CONNECTIONS

7 In Problem 5 (page 262) of Investigation 1, you examined the shape, center, and spread of binomial distributions with probability of success 0.2 and varying sample sizes. Use computer software like the "Binomial Distributions" custom app in *CPMP-Tools* to help generalize your findings from that problem (Part a) and to explore a related question (Part b).

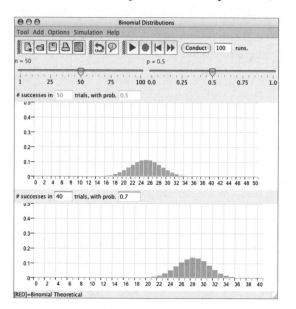

a. As *n* increases, but the probability of a success *p* remains the same, what happens to the shape, center, and spread of the binomial distribution for the number of successes?

b. As *p* increases from 0.01 to 0.99, but *n* remains the same (for example, *n* = 50), what happens to the shape, center, and spread of the binomial distribution for the number of successes?

8 Match each description of a binomial distribution in Parts a–d with its corresponding graph. It will help if you count the number of visible bars and if you compute the standard deviations from the values in Parts a–d.

    a. $p = 0.1, n = 40$

    b. $p = 0.2, n = 20$

    c. $p = 0.4, n = 10$

    d. $p = 0.8, n = 5$

Graph I

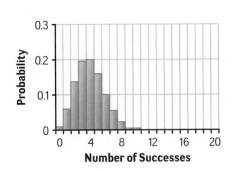

Graph II

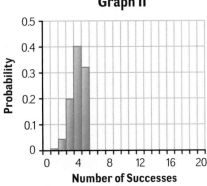

Graph III

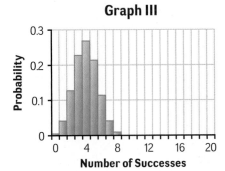

Graph IV

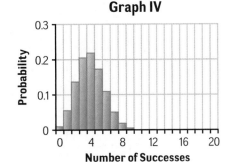

9 In Investigation 1, you examined a simple formula for the standard deviation of a binomial distribution of the number of successes with n trials and probability of success p. In this task, you will further analyze the formula to make sure it gives the values you would expect.

    a. What do you think should be the standard deviation of the binomial distribution when $p = 1$? Does the formula give this value?

    b. What do you think should be the standard deviation of the binomial distribution when $p = 0$? Does the formula give this value?

    c. If you know the standard deviation of the binomial distribution with n trials and probability of success p, what is the standard deviation of the binomial distribution with n trials and probability of success $1 - p$? Why does this make sense?

    d. For a fixed number of trials n, what value of p will make the standard deviation, $\sqrt{np(1 - p)}$, the largest?

10 In Problem 6 of Investigation 1, you found that the binomial distribution generated by rolling a die 50 times and counting the number of 3s is not approximately normal. The sample size of 50 is not large enough. What is the smallest sample size that will generate a binomial distribution that is approximately normal?

11 About 60% of the transcripts of recent high school graduates in the United States show that the student has passed a chemistry course. (**Source:** U.S. Department of Education Institute of Education Sciences, National Center for Education Statistics, High School Transcript Study (HSTS), various years, 1990–2009)

a. Describe how to use the **randBin** function of your calculator to conduct a simulation to estimate the number of students who passed a high school chemistry course in a randomly selected group of 20 high school graduates.

b. Conduct five runs of your simulation. Add your results to a copy of the frequency table and histogram below so there is a total of 400 runs.

Chemistry on Student Transcripts		
Number Who Passed Chemistry	Frequency (before)	Frequency (after)
6	3	
7	2	
8	17	
9	39	
10	51	
11	53	
12	80	
13	66	
14	39	
15	29	
16	13	
17	3	
Total Number of Runs	395	400

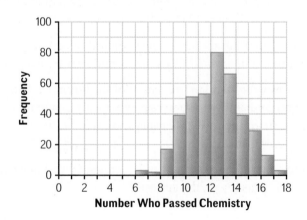

c. When 20 students are randomly selected, what is your estimate of the probability that fewer than half passed chemistry in high school?

d. Describe the shape of the distribution. According to the rule, should it be approximately normal?

e. Estimate the mean and standard deviation of this distribution from the histogram. Check your estimates using the formulas for the mean and standard deviation.

f. The histogram below shows the binomial distribution for the number of transcripts with a passing grade in chemistry when 80 transcripts are selected at random, rather than 20. Describe how the shape, center, and spread of this histogram are different from those for sample sizes of 20.

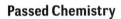

Passed Chemistry

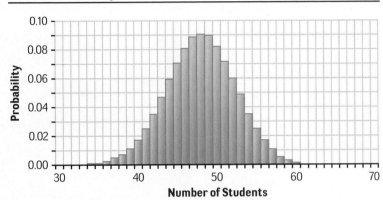

g. Is it more likely to find that fewer than half of the high school transcripts show that the student has passed a chemistry course if you take a sample of size 20 or a sample of size 80? Explain why this should be the case.

REFLECTIONS

12 The following situations are not binomial situations. For each situation, list which of the four characteristics of a binomial situation on page 260 do not hold.

a. Roll a die 35 times. On each roll, note the number that appears on top.

b. Select five students from your classroom. Note whether each is male or female. The method of selection is to write the name of each student in your classroom on a slip of paper. Select a slip at random. Do not replace that slip. Select another slip at random. Do not replace that slip either. Select another slip at random. Continue until you have five names.

c. Place four new batteries in a calculator. Count the number that need to be replaced during 200 hours of use.

d. Flip a coin. Count the number of trials until you get a head.

e. Select 25 cars at random. Count the number of cars of each color.

13 Look back at the formulas for calculating the mean and standard deviation of a binomial probability distribution (pages 261 and 263). Consider a binomial situation with probability p.

a. If you double the sample size, what happens to the mean? To the standard deviation?

b. How does the mean vary with the sample size? How does the standard deviation vary with the sample size?

c. How are your answers to Parts a and b seen in the distributions on page 262?

14 Sports writers and commentators often speak of *odds* rather than probabilities. When you roll two dice, the odds of getting 7 as the sum are 6 to 30, or 1 to 5. That is, of 36 equally likely outcomes, 6 are successes and 30 are failures. Suppose you roll a pair of dice. Give both the probability and the odds of getting each of the following events.

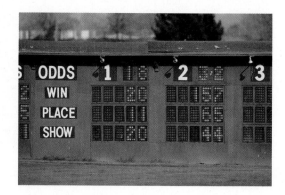

a. a sum of 2

b. a sum of 8

c. an even sum

15 Suppose a person starts flipping a coin and gets heads on each flip.

a. How many heads do you think it would take before you were suspicious that the person had a two-headed coin? Before you were confident the person had a two-headed coin?

b. For your two answers in Part a, compute the probability that a person who begins to flip a fair coin will get this many heads in a row just by chance.

16 Statistics is sometimes defined as learning about a population by taking a sample from that population. Some people say you can prove anything with statistics. The truth is more nearly the opposite. You cannot *prove* anything with statistics. Explain why the second statement is the better one.

EXTENSIONS

17 Suppose you want to estimate the percentage of teens who find "whatever" the most annoying word. As the Law of Large Numbers tells you, the more teens you ask, the closer your estimate should be to the true percentage that you would get if you could ask all teens. (However, that is true only if you randomly select the teens you ask rather than, for example, asking only teens you know.)

Suppose that you want to be at least 95% confident that your estimate is off by no more than 3%. You can use the formula below to estimate the number of teens you need to ask. This formula gives the sample size n needed in order to be 95% confident that your estimate is within your margin of error E.

$$n = \frac{1}{E^2}$$

In your survey, you want to estimate the percentage of teens who find "whatever" the most annoying word. You want to be at least 95% confident your estimated percentage is within $E = 0.03$ (3%) of the true percentage. Substituting 0.03 into the formula gives

$$n = \frac{1}{(0.03)^2} = \frac{1}{0.0009} \approx 1{,}111.$$

You need to ask approximately 1,111 teens.

Being "at least 95% confident" means that if you conduct 100 surveys with 100 different random samples, the expected number of surveys in which your estimated percentage is within E of the true percentage is 95 or more.

a. Suppose you want to estimate the probability a loaded die lands with a 6 on top and be at least 95% confident that your estimated probability is within 5% of the actual probability. How many times should you toss the die?

b. Suppose you do a simulation to estimate a proportion. How many runs do you need to estimate the proportion to within 1% with 95% confidence?

c. What is the margin of error associated with 200 runs of a simulation to estimate a proportion with 95% confidence?

d. In a national poll of 1,026 adults, 38% said that "whatever" is the most annoying word. (**Source:** Marist poll) What is the margin of error associated with a sample size of 1,026?

18 Use your graphing calculator or computer software to investigate the graph of the *sample size* function $n = \frac{1}{E^2}$ given in Extensions Task 17.

a. Should x or y represent the sample size? What does the other variable represent?

b. Graph this function.

c. Describe the shape of the graph of this function.

d. Use the trace function to estimate the sample size needed for a margin of error of 2%. Of 3%. Of 8%.

19 The Martingale is an old gambling system. At first glance, it looks like a winner. Here is how it would work for a player betting on red in roulette. The roulette wheel has 38 spaces, and 18 of these spaces are red. On the first spin of the wheel, Mr. Garcia bets $1. If red appears, he collects $2 and leaves. If he loses, he bets $2 on red on the second spin of the wheel. If red appears, he collects $4 and leaves. If he loses, he bets $4 on red on the third spin of the wheel, and so on. Mr. Garcia keeps doubling his bet until he wins.

a. If Mr. Garcia plays 76 times, what is the expected number of times he wins?

b. If Mr. Garcia wins on his first try, how much money will he be ahead? If he wins for the first time on his third try, how much money will he be ahead? If Mr. Garcia finally wins on his tenth try, how much money will he be ahead?

c. From the gambler's point of view, what are some flaws in this system? (From the gambler's point of view, there are flaws in *every* "system.")

20 Use the margin of error formula given in Extensions Task 17 to investigate the sample size needed to cut the margin of error in half.

a. What is the margin of error for a sample size of 100? What sample size would you need to cut this margin of error in half?

b. What is the margin of error with a sample size of 625? What sample size would you need to cut this margin of error in half?

c. In general, to cut a margin of error in half, how must the sample size change? Use the formula to justify your answer.

21 **Acceptance sampling** is one method that industry uses to control the quality of the parts it uses. For example, a recording company buys blank CDs from a supplier. To ensure the quality of these CDs, the recording company examines a sample of the CDs in each shipment. The company buys the shipment only if 5% or fewer of the CDs in the sample are defective. Assume that 10% of the CDs actually are defective.

a. Suppose the recording company examines a sample of 20 CDs from each shipment.

   i. Is this a binomial situation? If so, give the sample size and the probability of a "success."

   ii. Design and carry out a simulation of this situation.

   iii. What is your estimate of the probability that the shipment will be accepted?

b. Suppose the recording company examines a sample of 100 CDs from each shipment. Is a normal distribution a good model of the binomial distribution? If so, use it and a standardized value to estimate the probability that the shipment will be accepted.

"By a small sample we may judge the whole piece." Miguel de Cervantes

REVIEW

22 Solve each equation by examining its structure and then reasoning with the symbols themselves.

a. $6x - x^2 = 5$

b. $3(x + 9)^2 = 75$

c. $4(3^x) = 108$

d. $(x + 3)(x - 2) = 10$

e. $x^4 + x^2 - 12 = 0$

23 Suppose that you spin this spinner two times.

a. What is the probability that both spins land on yellow?

b. What is the probability that the first spin lands on blue and the second spin lands on red?

c. What is the probability that one spin lands on yellow and one spin lands on green?

24 Rewrite each expression in a simpler equivalent form.

a. $4x^2 - x(3 - 2x)$

b. $6 + 3(2y - 5) - 10$

c. $3(x + 2y) + 4(3x + 2y)$

d. $x(x + 1) - (x + 5)$

25 The length of a rectangle is three times its width. The area of the rectangle is 192 cm². Find the perimeter of the rectangle.

26 Determine if each statement is true or false. Explain your reasoning.

a. All rectangles are similar.

b. All circles are similar.

c. If $\triangle ABC \sim \triangle DEF$ with scale factor $\frac{3}{4}$ and $\triangle ABC$ is larger than $\triangle DEF$, then the area of $\triangle DEF$ is $\frac{3}{4}$ the area of $\triangle ABC$.

d. If $\triangle XYZ$ is a right triangle with $m\angle Y = 90°$, then $\cos X = \sin Z$.

27 What is the difference between independent events and mutually exclusive events?

28 For each pair of equations, identify to the nearest integer the intervals where $y_1 > y_2$.

a. $y_1 = 2^x$
$y_2 = 0.1x^4$

b. $y_1 = 1.2^x$
$y_2 = 0.5x^3$

29　Without using a calculator or computer graphing tool, match each function rule with the correct graph. Then use technology to check your answers. All graphs use the same scales.

a. $y = \dfrac{1}{x^2}$

b. $y = x^2 + 1$

c. $y = -x + 1$

d. $y = \dfrac{1}{x}$

e. $y = 2^x$

f. $y = x - 1$

I

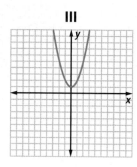

II

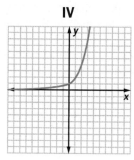

III

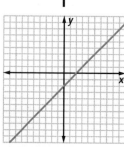

IV

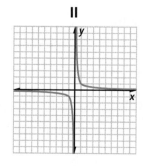

V

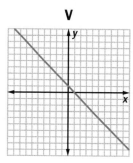

VI

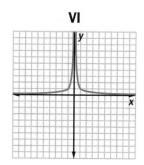

Statistical Process Control

A major West Coast metal producer received the following complaint from a customer. The customer said that the metal he had recently received had an impurity in it. (The impurity was a trace element in the metal that affected how the metal performed.) Since this particular impurity had not been a problem in the past, the metal producer had not been monitoring it. The metal producer looked up the records on metal recently shipped to the customer. The percentage of the impurity in the metal for each week's shipment is given in the table at the right.

Metal Impurity in Production					
Week Ending	Percentage	Week Ending	Percentage	Week Ending	Percentage
6/7	0.00053	7/26	0.00072	9/13	0.00219
6/14	0.00047	8/2	0.00033	9/20	0.00220
6/21	0.00043	8/9	0.00054	9/27	0.00237
6/28	0.00048	8/16	0.00046	10/4	0.00187
7/5	0.00035	8/23	0.00043	10/11	0.00269
7/12	0.00047	8/30	0.00049	10/18	0.00272
7/19	0.00066	9/6	0.00198	10/25	0.00289

The metal producer graphed the data on a **plot over time**. This is called a **run chart** when used for an industrial process. The run chart is shown below.

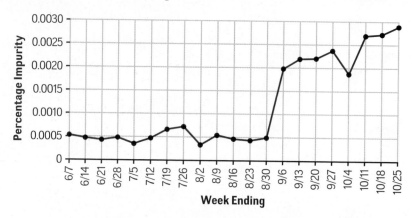

The metal producer checked with its supplier and found that the supplier had substituted a different raw material, which contained the impurity, in place of its regular raw material. The metal producer now routinely plots impurity levels of all types of impurities so that the customer will never again be the first to know.

THINK ABOUT THIS SITUATION

Think about this example of variability in industrial production.

a When do you think the metal producer first started using the different raw material?

b The percentage of impurity varies even for raw material that comes from the same source. How could you estimate the variation in the percentage of impurity in shipments from the same source?

c Companies want to stop production as soon as possible after a process goes **out of control.** In the case of the metal producer, it was obvious when that happened. Often, a process goes out of control much more gradually. What should a company look for in a run chart to indicate its process might have gone out of control? Try to list several tests or rules that would signal that the process may have gone out of control.

In this lesson, you will explore some of the methods that industry uses to identify processes that have gone out of control.

Out of Control Signals

In this investigation, you will examine what the run chart of an out-of-control process can look like. You will then explore some of the tests employed by industry to signal that a process may have gone out of control. Think about answers to the following question as you work through this investigation:

What does a run chart look like when
the process has gone out of control?

1 At Prairie Farms Dairy, a process of filling milk containers is supposed to give a distribution of weights with mean 64 ounces and standard deviation 0.2 ounces. The following four run charts come from four different machines. For each machine, 30 milk containers were filled and the number of ounces of milk measured. Two of the four machines are out of control.

a. Which two machines appear to be out of control?

b. On which of these machines did the mean change?

c. On which of these machines did the standard deviation change?

Machine 1

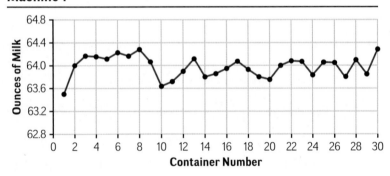

Machine 2

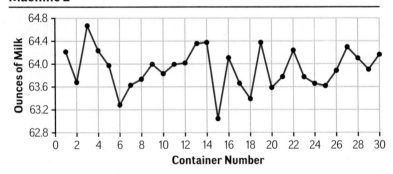

Machine 3

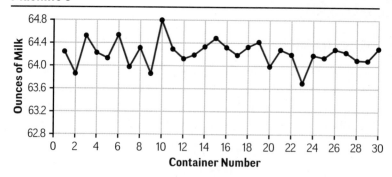

Machine 4

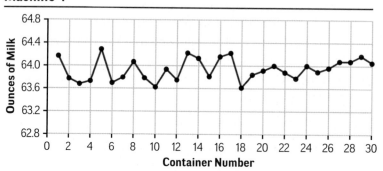

In Problem 1, you saw two ways that an out-of-control machine might behave. Since the people who monitor machines want to stop a machine as soon as possible after it has gone out of control, they have signs and patterns they look for in run charts.

2 Re-examine the data from the West Coast metal producer, reproduced below.

Metal Impurity in Production

Week Ending	Percentage	Week Ending	Percentage	Week Ending	Percentage
6/7	0.00053	7/26	0.00072	9/13	0.00219
6/14	0.00047	8/2	0.00033	9/20	0.00220
6/21	0.00043	8/9	0.00054	9/27	0.00237
6/28	0.00048	8/16	0.00046	10/4	0.00187
7/5	0.00035	8/23	0.00043	10/11	0.00269
7/12	0.00047	8/30	0.00049	10/18	0.00272
7/19	0.00066	9/6	0.00198	10/25	0.00289

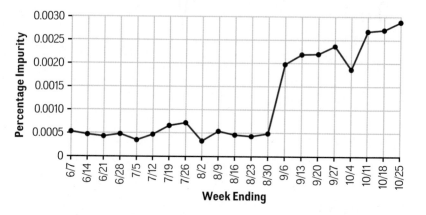

a. In the first 13 weeks of metal production, the percentage of the impurity was under control. That is, the percentage of the impurity varied a little from week to week but was of a level acceptable to the customer. Make a plot that displays the variability in these 13 percentages. Estimate the mean and standard deviation. Then compute the mean and standard deviation and compare to your estimates.

b. Now look at the percentage of impurity for the 14th week, which ended September 6. How many standard deviations from the mean computed in Part a is this percentage?

c. Assume that when the level of impurity is under control, the percentages of impurity are normally distributed with the mean and standard deviation that you calculated in Part a. If the level of impurity is under control, what is the probability of getting a percentage as high or higher than the one for September 6?

3 One commonly used test declares a process out of control when a single value is more than three standard deviations from the mean. This test assumes that the individual values are approximately normally distributed. If the process is in control, what is the probability that the next measurement will be more than three standard deviations from the mean?

4 Each of the following run charts was made from a process that was supposed to be normally distributed with a mean of 5 and a standard deviation of 1. The charts are based on displays produced using the Minitab® Statistical Software[1].

a. UCL means "upper control limit." LCL means "lower control limit." How were these limits computed?

b. Using the test of three standard deviations or more from the mean, on which of the three charts is there a point where the process should be suspected to be out of control?

[1] MINITAB® and all other trademarks and logos for the Company's products and services are the exclusive property of Minitab Inc. All other marks referenced remain the property of their respective owners. See minitab.com for more information.

Chart 1

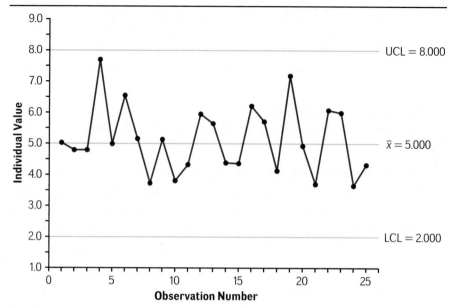

Chart 2

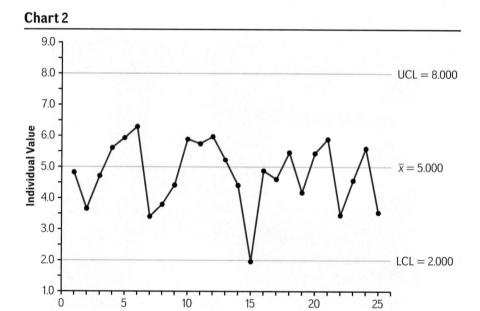

Chart 3

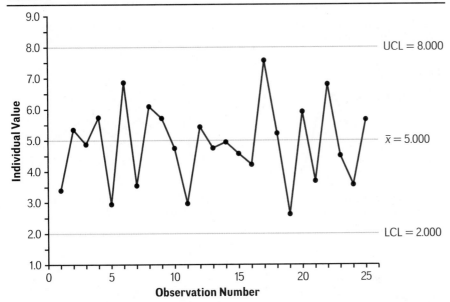

5   The process documented on the run chart below is supposed to be normally distributed with a mean of 5 and a standard deviation of 1. The small "2" below the final point, where the process was stopped, indicates that the process may have gone out of control. Why do you think Minitab has declared the process out of control?

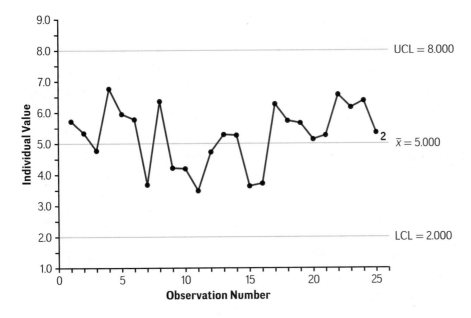

On the chart "Illustrations of Tests," which is reproduced on the next page, there are eight tests used by industry to signal that a process may have changed. The zones are marked off in standard deviations. For example, if a value falls in Zone A, it is more than two but less than three standard deviations from the mean $\bar{x}$. Each **x** marks the value at which a process is first declared out of control. Each of these tests assumes that the individual values come from a normal distribution.

Illustrations of Tests

Test 1. One observation beyond either Zone A. (Two examples are shown.)

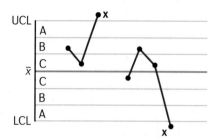

Test 2. Nine observations in a row on one half of the chart.

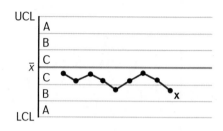

Test 3. Six observations in a row steadily increasing or decreasing. (Two examples are shown.)

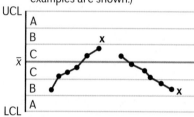

Test 4. Fourteen observations in a row alternating up or down.

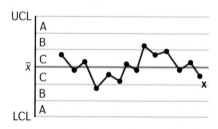

Test 5. Two out of three observations in a row on one half of the chart and in Zone A or beyond. (Three examples are shown.)

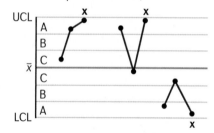

Test 6. Four out of five observations in a row on one half of the chart and in Zone B or beyond. (Two examples are shown.)

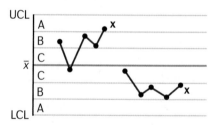

Test 7. Fifteen observations in a row within the two C zones.

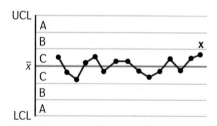

Test 8. Eight observations in a row with none in either Zone C.

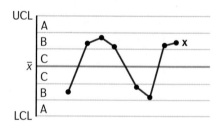

Source: *Western Electric. Statistical Quality Control Handbook.* Chicago: American Telephone and Telegraph Company, 1956.

6 Match each of the eight tests to the best description below.

   a. The observations are gradually getting larger (or smaller).

   b. One observation is very far from the mean.

   c. Two of three observations are unusually high (or low).

   d. Four of five observations are all somewhat high (or low).

   e. The mean seems to have decreased (or increased).

   f. The standard deviation seems to have decreased.

   g. The standard deviation seems to have increased.

   h. The process shows a non-random pattern that should be explained.

7 Look back at how these tests relate to your previous work in this investigation.

   a. What is the number of the test that signals when a single value is more than three standard deviations from the mean?

   b. Which test signaled the point marked "2" in Problem 5 (page 289)?

8 For each of the run charts below and on the next page, there is an **x** at the point when the process first was declared to be out of control. Give the number of the test used to decide that the process was out of control.

a.

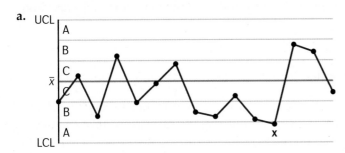

b.

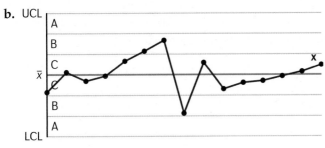

c.

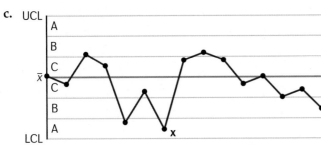

d.

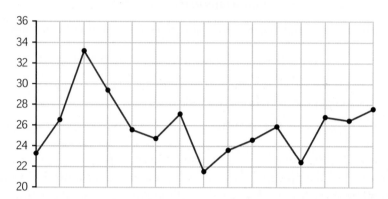

9. Here is a run chart for a process that is supposed to have a normal distribution with mean 28 and standard deviation 2.

    a. On a copy of the chart, identify and label the horizontal lines dividing the zones.

    b. When did the process first go out of control? By which test?

    c. Has the mean or the standard deviation changed?

The run charts that you have been examining are sometimes called Shewhart control charts. Dr. Walter A. Shewhart invented these charts in 1924 and developed their use while he worked for the Western Electric Company and Bell Laboratories. These charts provide a quick, visual check if a process has changed or gone "out of control." When there is change in an industrial process, the machine operator wants to know why and may have to adjust the machine.

Dr. Walter A. Shewhart

CHECK YOUR UNDERSTANDING

Examine each of the following run charts.

a. For this run chart, there is an asterisk (*) below the observation at which the statistical software warned that the process may have gone out of control. Give the number of the test used.

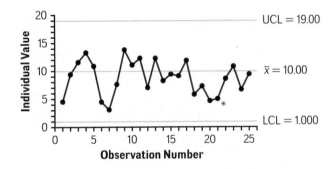

b. The process graphed on the run chart below is supposed to have a normal distribution with mean 8 ounces and standard deviation 1 ounce. Find the point at which the process should first be declared out of control. Give the number of the test used.

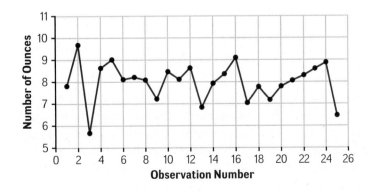

False Alarms

Even a process that is under control exhibits variation. Consequently, the eight tests on page 290 occasionally will give an out-of-control signal, called a **false alarm**, for a process that is under control. For example, if you watch a process that is in control long enough, eventually one observation will be beyond one of the A zones, or six observations in a row will be steadily increasing, and so on. The tests have been designed so that false alarms occur very rarely. If you have been monitoring a process that is under control, the probability of a false alarm on the next set of observations is very small.

As you work through this investigation, look for answers to this question:

How can you find the probability of getting a false alarm?

1. Assuming that the observations of a process are normally distributed, fill in the "Percentage of Observations in Zone" on a copy of the chart below. You can then refer to the chart when working on the remaining problems.

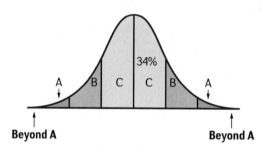

		Number of Standard Deviations from Mean	Percentage of Observations in Zone
	Beyond Zone A		
UCL		3	
	Zone A		
		2	
	Zone B		
		1	
	Zone C		
Mean		0	34%
	Zone C		
		1	
	Zone B		
		2	
	Zone A		
LCL		3	
	Beyond Zone A		

2. Suppose a machine is filling cartons of ice cream and is under control. The operator uses only Test 1 to signal that the weight of the ice cream in the cartons may have gone out of control. What is the probability of a false alarm on the very next carton of ice cream being filled? Place your result in the appropriate row of a copy of the following table. Keep your table, as you will fill out some of the remaining rows in other problems.

False Alarms	
Test	**Probability of a False Alarm on the Next Set of Observations of a Process Under Control**
1 One observation beyond either Zone A	
2 Nine observations in a row on one half of the chart	
3 Six observations in a row steadily increasing or decreasing	
4 Fourteen observations in a row alternating up and down	
5 Two out of three observations in a row on one half of the chart and in Zone A or beyond	
6 Four out of five observations in a row on one half of the chart and in Zone B or beyond	
7 Fifteen observations in a row within the two C zones	
8 Eight observations in a row with none in either Zone C	

3 Now suppose the ice cream machine operator is using only Test 7. The machine continues to stay in control.

a. What is the probability that a single observation will fall either within Zone C in the top half of the chart or within Zone C in the bottom half of the chart?

b. What is the probability that the observations from each of the next 15 cartons filled will all fall within either of the two C zones? (In this and subsequent problems, you may assume that the observations are independent, although independence of measurements is rarely attained in practice even with processes that are under control. Nevertheless, your computations will be good approximations.)

c. What is the probability that the ice cream machine operator will get a false alarm from the next 15 cartons being filled? Place your result in the appropriate row of your copy of the table, rounding to the nearest ten-thousandth.

4 Now suppose the ice cream machine operator is using only Test 2. The machine continues to stay in control.

a. What is the probability that the next nine observations are all in the bottom half of the chart?

b. What is the probability that the next nine observations are all in the top half of the chart?

c. What is the probability that the next nine observations are all in the top half of the chart or all in the bottom half of the chart?

d. What is the probability that the operator will get a false alarm from the next nine cartons being filled? Place your result in the appropriate row of your copy of the table, rounding to the nearest ten-thousandth.

5 A ceramic plate machine makes 20,000 plates in a year. It is under control. If the operator uses Test 1 only and measures every hundredth plate, what is the probability the operator will get through the year without having to stop the machine?

6 When computing the probabilities in Problems 3, 4, and 5, you used the Multiplication Rule for Independent Events. In real-life situations, the measurements will not be completely independent, even with processes that are under control. For example, when using Test 2, if the first measurement is in the bottom half of the chart, the next measurement will be slightly more likely to be in the bottom half of the chart than in the top half. Does this lack of independence mean that the probability of a false alarm when you use Test 2 in practice is greater or is less than the probability you computed in Problem 4?

The eight quality control tests were all devised so that, when you are monitoring a process that is in control, the probability of a false alarm on the next set of observations tested is about 0.005 or less. You will determine the probability of a false alarm for the remaining five tests on your chart in the On Your Own tasks.

SUMMARIZE THE MATHEMATICS

In this investigation, you examined the likelihood of a false alarm when using quality control tests.

a What is a false alarm? Why do false alarms occur occasionally if the process is under control?

b If you get an out-of-control signal, is there any way to tell for sure whether the process is out of control or whether it is a false alarm?

Be prepared to share your ideas and examples with the class.

 CHECK YOUR UNDERSTANDING

A machine operator is using the following rule as an out-of-control signal; six observations in a row in Zone B in the top half of the chart or six observations in a row in Zone B in the bottom half of the chart. Assume the machine is in control.

a. What is the probability that the next six values are all in Zone B in the top half of the chart?

b. What is the probability that the operator gets a false alarm from the next six values?

The Central Limit Theorem

The use of control charts in the previous investigations required that the individual measurements be approximately normally distributed. However, the distribution of measurements of the parts coming off an assembly line may well be skewed. What can be done in that case?

As you work through this investigation, look for answers to the following question:

What is the Central Limit Theorem and how does it allow you to use control charts even when individual measurements come from a skewed distribution?

1 Suppose that a process is under control. How would you find the probability of a false alarm on the next measurement when using Test 1 on page 290? Where did you use the assumption that the individual values are approximately normally distributed?

2 The eight tests illustrated on page 290 assume that observations come from a distribution that is approximately normal. For individual observations, this is not always the case. For example, suppose the carnival wheel below is designed to produce random digits from the set {0, 1, 2, 3, 4, 5, 6, 7, 8, 9}.

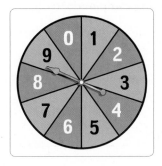

a. Suppose the wheel is operating correctly and produces digits at random.

   i. Describe the shape of the probability distribution of all possible outcomes.

   ii. What is the mean of the distribution?

   iii. What is the standard deviation of the distribution?

b. Can Test 1 ever signal that this process may be out of control? Explain. What about Test 5?

c. Since the distribution in Part a is not approximately normal, you should not use the eight tests on individual digits to check whether the wheel has gone out of control. Explore what you might do so you can use the tests.

   i. Use technology to produce 5 digits at random and find their mean.

ii. Nela repeated part i 1,400 times using simulation software. That is, the software selected 5 digits at random and found the mean $\bar{x}$. Then the process was repeated until there were 1,400 means. What is the smallest mean Nela could have observed? The largest?

iii. The histogram below shows Nela's 1,400 means. About how many times did she get a mean of 2? Of 4.2?

Means of Samples of Five Random Digits

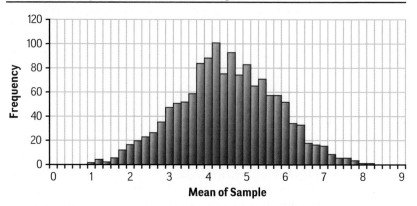

d. Use simulation software to conduct 1,500 runs of selecting 3 digits at random and calculating their means. Make a histogram of the distribution. Compare the shape and center of your distribution with Nela's.

e. Repeat Part d selecting samples of 10 digits. Of 20 digits.

f. How could you use means to meet the normality assumption of the eight tests?

3 The following plot and table show the rainfall in Los Angeles for the 129 years from 1878 through 2006. The amount varies a great deal from year to year.

Los Angeles Annual Rainfall 1878–2006

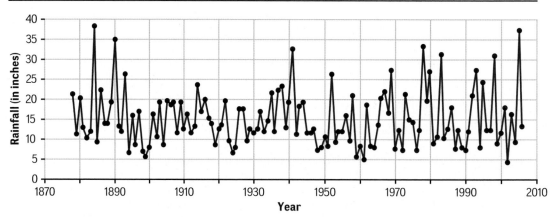

(**Source:** National Weather Service)

Year	Rainfall (in inches)	Year	Rainfall (in inches)	Year	Rainfall (in inches)
1878	21.26	1921	13.65	1964	7.93
1879	11.35	1922	19.66	1965	13.69
1880	20.34	1923	9.59	1966	20.44
1881	13.13	1924	6.67	1967	22.00
1882	10.40	1925	7.94	1968	16.58
1883	12.11	1926	17.56	1969	27.47
1884	38.18	1927	17.76	1970	7.77
1885	9.21	1928	9.77	1971	12.32
1886	22.31	1929	12.66	1972	7.17
1887	14.05	1930	11.52	1973	21.26
1888	13.87	1931	12.53	1974	14.92
1889	19.28	1932	16.95	1975	14.35
1890	34.84	1933	11.88	1976	7.22
1891	13.36	1934	14.55	1977	12.31
1892	11.85	1935	21.66	1978	33.44
1893	26.28	1936	12.07	1979	19.67
1894	6.73	1937	22.41	1980	26.98
1895	16.11	1938	23.43	1981	8.98
1896	8.51	1939	13.07	1982	10.71
1897	16.88	1940	19.21	1983	31.25
1898	7.06	1941	32.76	1984	10.43
1899	5.59	1942	11.18	1985	12.82
1900	7.91	1943	18.17	1986	17.86
1901	16.29	1944	19.22	1987	7.66
1902	10.60	1945	11.59	1988	12.48
1903	19.32	1946	11.65	1989	8.08
1904	8.72	1947	12.66	1990	7.35
1905	19.52	1948	7.22	1991	11.99
1906	18.65	1949	7.99	1992	21.00
1907	19.30	1950	10.60	1993	27.36
1908	11.72	1951	8.21	1994	8.14
1909	19.18	1952	26.21	1995	24.35
1910	12.63	1953	9.46	1996	12.46
1911	16.18	1954	11.99	1997	12.40
1912	11.60	1955	11.94	1998	31.01
1913	13.42	1956	16.00	1999	9.09
1914	23.65	1957	9.54	2000	11.57
1915	17.05	1958	21.13	2001	17.94
1916	19.92	1959	5.58	2002	4.42
1917	15.26	1960	8.18	2003	16.49
1918	13.86	1961	4.85	2004	9.25
1919	8.58	1962	18.79	2005	37.25
1920	12.52	1963	8.38	2006	13.19

a. In what year did Los Angeles receive the most rain? How many inches was this?

b. The control chart below was made using the historical mean yearly rainfall of 15.1 inches. According to this chart, rainfall was judged out of control at four points. Which years were these? What caused the out-of-control signals?

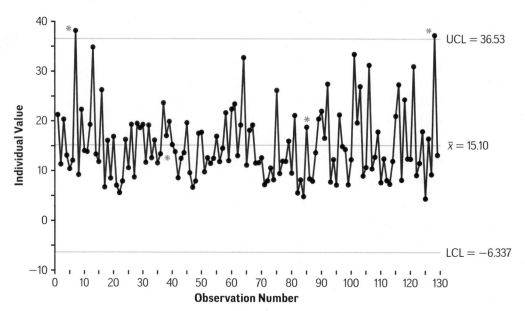

c. The histogram below shows the distribution of rainfall for the years 1878 through 2006. Describe its shape. Do these measurements meet the conditions for using the eight tests for deciding when a process is out of control?

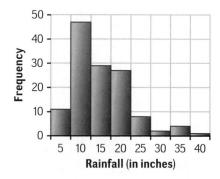

d. When individual measurements have a skewed distribution, quality control practitioners can plot the *means* of consecutive samples of measurements rather than plotting individual measurements. For example, the measurements of rainfall for the first three years were 21.26, 11.35, and 20.34 inches. The mean, $\bar{x}$, for these three years was 17.65 inches. That value is the first one plotted on the *x*-bar chart on the following page.

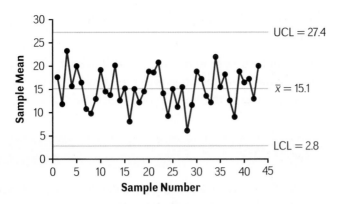

i. Using years 1881–1883, find the value of the second mean plotted.

ii. Find the value of the last mean plotted.

iii. Does any mean give an out of control signal?

iv. Note that the middle line of the x-bar chart is at 15.1, the historical mean for individual years. Explain why 15.1 is also the mean of the values on the x-bar chart.

e. The histogram below shows the distribution of the sample means for consecutive samples of size 3. Are conditions now met for using the eight tests?

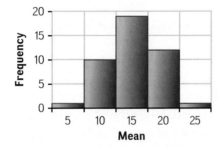

4 The two previous problems illustrated the **Central Limit Theorem**. Imagine taking repeated random samples of a fixed size from a non-normal distribution. The larger the sample size you use, the more approximately normal the distribution of sample means $\bar{x}$ tends to be. Typically, quality control practitioners use two to five measurements per sample. In general, the more skewed the distribution of individual measurements, the larger your sample size should be. Consider the strongly right-skewed population below, which has a mean of 2.

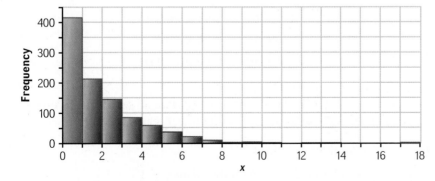

a. Using software like "Distribution of Sample Means" in *CPMP-Tools*, take 100 samples of size 1 from this skewed distribution. Describe the shape. Give the mean and standard deviation of the distribution of sample means.

b. Repeat Part a for a sample size of 2. A sample size of 5. A sample size of 10. A sample size of 20.

c. Describe what happens to the mean and standard deviation of the distribution of the sample means as the sample size increases. Compare your observations with those of your classmates.

SUMMARIZE THE MATHEMATICS

In this investigation, you saw how the Central Limit Theorem can be used in situations when you want to use a control chart, but the distribution of measurements is not approximately normal.

a What is the Central Limit Theorem?

b How is the Central Limit Theorem used in statistical process control?

c Describe how the shape, center, and spread of the distribution of sample means is related to sample size.

Be prepared to explain your ideas to the class.

 CHECK YOUR UNDERSTANDING

Your company has been successful in adjusting the machine that fills cereal boxes so that the mean weight in the boxes is 16.1 ounces. After filling several hundred boxes, it is obvious that the distribution of weights is somewhat skewed. So to ensure that you are warned if the process goes out of control, you decide to use an *x*-bar chart with samples of size 3. You set the upper control limit at 16.39 and the lower control limit at 15.81. Several days later, seven samples of the weights of three individual boxes were taken, as shown in the table.

Sample	Measurements
first	16.05, 16.27, 15.68
second	16.10, 15.80, 15.75
third	16.98, 16.12, 15.73
fourth	15.93, 15.92, 15.91
fifth	16.47, 16.04, 16.93
sixth	15.58, 15.90, 16.25
seventh	15.84, 15.60, 15.77

a. Compute the mean of each sample of size 3. Plot all seven means on an *x*-bar chart that shows the upper and lower control limits.

b. Has the process gone out of control?

c. Why were samples of size 3 used rather than size 1?

APPLICATIONS

1. The thermostat in the Simpson's apartment is set at 70° Fahrenheit. Mrs. Simpson checks the temperature in the apartment every day at noon. The table and plot below give her observations over the last 25 days. She has been satisfied with the results and felt that the process was under control. However, at noon today, the temperature in her apartment was 63°F. Should Mrs. Simpson call building maintenance? Explain why or why not, in terms of statistical process control methods.

Temperature Observations

Day	Temp (in °F)	Day	Temp (in °F)	Day	Temp (in °F)
1	72	10	70	19	70
2	71	11	68	20	67
3	67	12	65	21	69
4	72	13	68	22	68
5	72	14	67	23	72
6	71	15	72	24	71
7	73	16	73	25	70
8	72	17	71		
9	69	18	70		

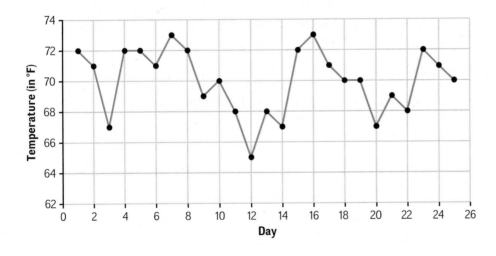

2 A person is running a machine that makes nails. The nails are supposed to have a mean length of 2 inches. But, as in all processes, the nails do not come out exactly 2 inches long each time. The machine is supposed to be set so the distribution of the lengths of the nails is normal and the standard deviation of the lengths of the nails is 0.03 inches.

a. Sketch the distribution of the lengths of the nails when the machine is under control. Mark the mean and one, two, and three standard deviations from the mean on the horizontal axis.

b. If the machine is set correctly, what percentage of the nails will be more than 2.06 inches long? Less than 1.94 inches long?

c. What percentage of the nails will be more than 2.09 inches long? Less than 1.91 inches long?

d. Suppose the machinist turns on the machine one morning after it has been cleaned. He finds these lengths in inches for the first ten nails, 2.01, 2.08, 1.97, 1.99, 1.92, 2.00, 2.03, 1.99, 1.97, and 1.95. Explain what your advice would be and why.

3 In Investigation 2, Problems 2–4, you explored the probabilities of false alarms in the context of a machine filling cartons of ice cream with the process under control. Suppose now that the operator is using only Test 8 on page 290.

a. What is the probability that the next observation is not in either Zone C?

b. What is the probability that none of the next eight observations are in either Zone C?

c. What is the probability that the operator will get a false alarm from the next eight cartons being filled? Place your result in the appropriate row of your copy of the table from Investigation 2. Round your answer to the nearest ten-thousandth.

4 The Ford Motor Company lists these four signals on its control charts.

- Any point outside of the control limits, more than three standard deviations from the mean

- Seven points in a row that are all above or all below the central line

- Seven points in a row that are either increasing or decreasing

- Any other obviously non-random pattern

(**Source:** Ford Motor Company, *Continuing Process Control and Process Capability Improvement.* December, 1987.)

a. Which of these tests is exactly the same as one of the tests (page 290) from the Western Electric handbook?

b. Assume that a manufacturing process is in control. If you are using only Ford's first test, what is the probability of a false alarm on the next observation?

c. Assume that a manufacturing process is in control. What is the probability of a false alarm on the next seven observations if you are using only Ford's second test?

d. Is Ford's second test more or less likely to produce a false alarm than the similar test from the Western Electric handbook? Explain.

e. Is Ford's third test more or less likely to produce a false alarm than the similar test from the Western Electric handbook? Explain.

5 Suppose that the ice cream machine in Investigation 2, Problem 2, is under control, and the operator is using only Test 5. You will now find the probability of a false alarm with the next three observations.

a. What is the probability that an observation will fall in Zone A at the top of the chart or beyond?

b. There are four ways that at least two of the next three observations can be in Zone A at the top of the chart or beyond. Three ways are described below. What is the fourth way?

- The first, second, and third observations are in Zone A or beyond.

- The first observation is not in Zone A or beyond. The second and third are.

- The first observation is in Zone A or beyond. The second is not, and the third is.

c. Find the probability of each of the four ways in Part b. Recall that two events are said to be **mutually exclusive** if it is impossible for both of them to occur on the same trial. Are these four ways mutually exclusive?

d. What is the probability that at least two of the next three observations will be in Zone A at the top of the chart or beyond?

e. What is the probability that at least two of the next three observations will be in Zone A at the bottom of the chart or beyond?

f. Are the events described in Parts d and e mutually exclusive?

g. What is the probability that the ice cream machine operator will get a false alarm from the next three cartons being filled? Place your result in your copy of the table from Investigation 2. Round your answer to the nearest ten-thousandth.

6 Suppose that the ice cream machine in Investigation 2, Problem 2, is under control and the operator is using only Test 6. Find the probability of a false alarm from the next five observations. Place your result in your copy of the table from Investigation 2, rounding to the nearest ten-thousandth.

7 The distribution of the lengths of widgets made by a certain process is slightly skewed. Consequently, the company uses an *x*-bar chart with samples of size 2 to determine when the process may have gone out of control. When the process is in control, the mean is 3 cm. The lower and upper control limits of the *x*-bar chart are 2.97 and 3.03. One morning, eight samples of the lengths of two individual widgets were taken, as shown in the table.

Sample	Measurements (in cm)
first	3.00, 3.02
second	3.00, 2.98
third	3.01, 2.97
fourth	3.02, 3.00
fifth	2.98, 3.00
sixth	3.00, 3.00
seventh	3.01, 3.00
eighth	2.98, 3.00

a. Compute the mean of each sample of size 2. Plot all eight means on an *x*-bar chart that shows the upper and lower control limits.

b. Has the process gone out of control?

8 At the Web site www.fueleconomy.gov, people can report their vehicles' actual gas mileage and compare it to the EPA estimate.

a. The histogram below shows the overall gas mileage reported by the first 64 owners of a 2006 Honda Civic Hybrid to make a report. The EPA estimate for the 2006 Honda Civic Hybrid was 50 miles per gallon (mpg). Does it appear that vehicles are achieving that estimate, on average?

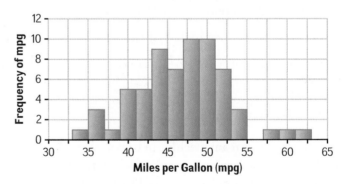

b. The following histogram shows the means $\bar{x}$ from 1,000 random samples of size 5 taken from the reported gas mileages in Part a. One of the samples had these mpg: 43.9, 53.9, 47.2, 45.0, and 46.9. Where is this sample represented on the histogram below?

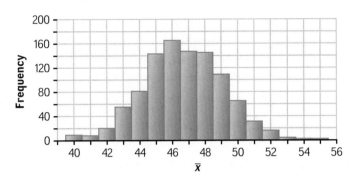

c. Describe how the shape, mean, and standard deviation of the distribution of sample means differ from that of the distribution of the individual mpgs. Is this what you would expect?

CONNECTIONS

9 Suppose you are operating a machine that fills cereal boxes. The boxes are supposed to contain 16 ounces. The machine fills the boxes so that the distribution of weights is normal and the standard deviation is 0.2 ounces. You can adjust the mean.

a. If you set the machine so that the mean is 16 ounces, what percentage of customers will get a box of cereal that contains less than 16 ounces?

b. Explain where you would recommend setting the mean and why.

c. Suppose you are buying a new box-filling machine. All else being equal, should you buy one with a standard deviation of 0.2 ounces or 0.4 ounces? Explain your reasoning.

10 A paper clip machine makes 4,000,000 paper clips each month. The machinist measures every 10,000th paper clip to be sure the machine is still set correctly. When the machine is set correctly, the measurements follow a normal distribution. The machinist uses only Test 1 (one value more than three standard deviations from the mean). Suppose the machine remains set correctly for a month. What is the expected number of times that the operator will stop production anyway?

11 Recall that two events are said to be **mutually exclusive** if it is impossible for both of them to occur on the same trial. Two events A and B are said to be **independent** if $P(A \text{ and } B) = P(A) \cdot P(B)$.

  a. What is the rule for computing $P(A \text{ or } B)$ when A and B are mutually exclusive?

  b. If you pick one value at random from a normal distribution, what is the probability that it will be more than two standard deviations above the mean or more than two standard deviations below the mean?

  c. If you pick one value at random from a normal distribution, what is the probability that it will be below the mean or more than two standard deviations above the mean?

  d. If you pick five values at random from a normal distribution, what is the probability that all five values will be above the mean or all will be below the mean?

  e. If you pick four values at random from a normal distribution, what is the probability that all four values will be more than two standard deviations above the mean?

  f. If you pick five values at random from a normal distribution, what is the probability that all five values will be more than one standard deviation above the mean or all will be more than one standard deviation below the mean?

  g. If you pick six values at random from a normal distribution, what is the probability that all six values will be below the mean?

  h. If you select two values at random from a normal distribution, what is the probability that they both are more than two standard deviations above the mean or that they are both more than two standard deviations below the mean?

12 In the Course 2 *Probability Distributions* unit, you learned that the expected waiting time for a success in a waiting-time distribution is $\frac{1}{p}$, if p is the probability of a success on any one trial.

  a. Suppose a machine operator is using only Test 1. If the process is under control, what is the expected number of items tested until Test 1 gives a false alarm? This is called the **average run length** or **ARL**.

  b. If a machine operator uses both Test 1 and Test 2, is the ARL longer or shorter than if he or she uses just Test 1? Explain.

13 Here is a distribution that is *really* non-normal. Suppose that you take random samples of a fixed size n from this distribution and compute the mean. You do this several thousand times. This is exactly like making a binomial distribution except that instead of recording the number of successes in your sample, you record the number of successes divided by n, which gives the *proportion* of successes.

Outcome	Probability
0	0.1
1	0.9

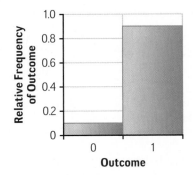

a. If the sample size is 1, what do you expect the resulting distribution to look like? Describe its shape and give its mean.

b. If the sample size is 2, what is the probability you get two 0s? Two 1s? One of each? Describe the shape and give the mean of the distribution of sample means.

c. How big does the sample size have to be before the distribution of sample means is approximately normal?

REFLECTIONS

14 A machinist is making video game tokens that are supposed to be 3.2 cm in diameter.

a. What might cause the mean diameter to change?

b. What might cause the standard deviation of the diameter to change while the mean stays the same?

15 Test 7 detects a *decrease* in variability. Why might a company want to detect a decrease in variability in the manufacturing of a product or in the processing of a service?

16 Why is it best if a machine operator does not use all eight tests but picks out just a few to use?

17 If you pick six values at random from a distribution with unknown shape, what is the probability that all six values are more than the median? Can you answer this question for the mean? Explain.

18 A false alarm occurs when a machine is in control and a test warns that it may not be.

a. What could you call a situation for which the machine is not in control and no test has given a warning?

b. The following chart has four empty cells. Two of the cells should contain the words "correct decision." Another cell should contain the words "false alarm." The fourth cell should contain your new name from Part a. Write these words in the correct cells on a copy of the chart.

		Result of Test	
		Gives Alarm	**Does Not Give Alarm**
Condition of Machine	**In Control**		
	Not in Control		

19 The *x*-bar control chart below was made from a process that was entirely under control. Two hundred samples, each of size 5, were selected from a normal distribution with mean 0. The means of successive samples are plotted on the chart. There is one out-of-control signal, when the mean for sample 11 fell below the lower control limit. Does this mean that something is wrong or is it about what you would expect to happen?

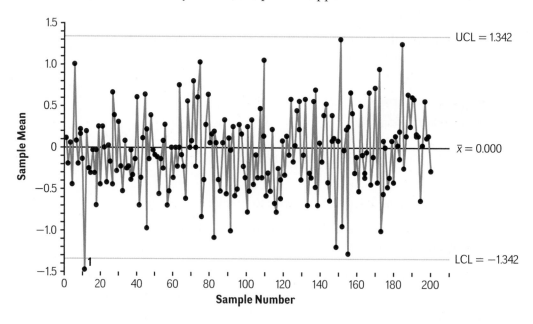

20 Look back at the distribution given in Problem 4 on page 301 of Investigation 3. Suppose 100 samples of fixed size 1, 2, 5, and 10 are taken from this population, which has mean 2. Match the sample size (1, 2, 5, or 10) with the histogram of the sample means (I, II, III, or IV).

I

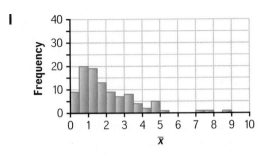

II

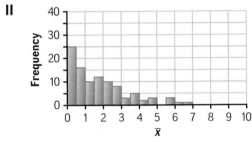

III

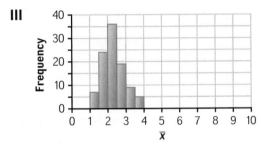

IV

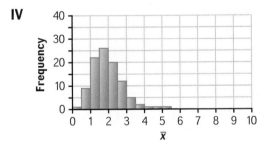

EXTENSIONS

21 Suppose that the ice cream machine in Investigation 2, Problem 2, is under control and the operator is using only Test 3 from the chart on page 290. To find the probability of a false alarm, you can use the idea of permutations (ordered arrangements).

a. In how many different orders can the digits 1, 2, and 3 be listed?

b. In how many different orders can the digits 1, 2, 3, and 4 be listed?

c. Compute 4! and 3! using the *factorial* function (!) on a calculator. Compare the calculator values of 4! and 3! to your answers in Parts a and b.

d. Use the factorial function to compute the number of different orders in which the digits 1, 2, 3, 4, 5, and 6 can be listed.

e. If the digits 1, 2, 3, 4, 5, and 6 are listed in a random order, what is the probability that they are in order from smallest to largest?

f. If the digits 1, 2, 3, 4, 5, and 6 are listed in a random order, what is the probability that they are in order from largest to smallest?

g. Are the two events described in Parts e and f mutually exclusive? If any six numbers are listed at random, what is the probability that they are in order from largest to smallest or in order from smallest to largest?

h. What is the probability that the ice cream machine operator will get a false alarm using only Test 3 with the next six cartons filled? Place your result in the appropriate row of your copy of the table from Investigation 2. Round your answer to the nearest ten-thousandth.

22 The probability of a false alarm using Test 8 is much smaller than the probability of a false alarm using any of the other tests. Describe how you could change Test 8 in order to make the probability of a false alarm closer to those of the other tests.

23 Suppose that the ice cream machine in Investigation 2, Problem 2, is under control and the operator is using only Test 4. In this task, you will use simulation to estimate the probability of a false alarm with the next fourteen observations.

a. First, describe a simulation to estimate the probability that if the digits 1, 2, 3, 4, and 5 are listed in a random order. They will alternate larger, smaller, larger, smaller, larger (for example: 5, 1, 3, 2, 4). Or, they will alternate smaller, larger, smaller, larger, smaller (for example: 4, 5, 2, 3, 1).

b. Perform your simulation 20 times. What is your estimate of the probability that the digits will alternate?

c. Make a listing of all 120 possible sequences of {1, 2, 3, 4, 5}. What is the theoretical probability that if these digits are listed at random, they will alternate?

d. If you are able to share the work with others, design and carry out a simulation to estimate the probability that if the digits 1 through 14 are listed in a random order, they will alternate larger/smaller or smaller/larger. Place your result on the appropriate row of your copy of the table from Investigation 2, rounding to the nearest ten-thousandth.

24 When all possible random samples of a given size are taken from a population with standard deviation σ, the distribution of sample means has standard deviation equal to $\frac{\sigma}{\sqrt{n}}$. This is true if the samples are taken with replacement, meaning each item drawn is replaced before selecting the next one.

a. In Problem 4 of Investigation 3 (page 301) and in other problems, you noticed that the standard deviation of the distribution of sample means is smaller than the standard deviation of the population from which the random samples were taken. How are your observations explained by the above rule?

b. The population from Problem 4 has a standard deviation of about 1.93. Compare the population standard deviation to the standard deviation of the distribution of means you found for samples of size 1, 2, 5, 10, and 20.

c. The years of rainfall in Los Angeles in the table on page 299 have mean 15.1 and standard deviation 7.1. Show how the upper and lower control limits were computed in the chart in Problem 3 Part d on page 300.

REVIEW

25 Imagine rolling a pair of tetrahedral dice.

a. Make a chart showing all possible outcomes for the sum when you roll a pair of tetrahedral dice.

b. If you roll a pair of tetrahedral dice, are the two events, getting a *sum of 5* and getting *doubles*, mutually exclusive? Explain your reasoning.

c. What is the probability that you get a sum of 5 or doubles?

Mark Steinmetz/Amanita Pictures

26 Suppose it is true that "All graduates of Rio Grande High School must successfully complete three years of high school mathematics."

  a. Write this statement in if-then form. What is the hypothesis? What is the conclusion?

  b. If Alberto is a graduate of Rio Grande High School, what can you conclude? Explain your reasoning.

  c. If Johanna was a student at Rio Grande High School and passed three years of high school mathematics, what can you conclude? Explain your reasoning.

27 Graph the solution set to each system of inequalities. Label the point of intersection of the boundary lines and the x- and y-intercepts of each line.

  a. $x \geq 3$
  $2x + y \leq 6$

  b. $4x + 3y > 30$
  $y < 2x - 5$

28 Without using technology tools, identify the x- and y-intercepts and the coordinates of the vertex of the graph of each quadratic function. Draw a sketch of the graph. Then check your work using technology.

  a. $f(x) = (x + 3)(x - 5)$

  b. $f(x) = 2x^2 - 6x$

  c. $f(x) = x^2 - 4x + 3$

29 The coordinates of three vertices of $\square ABCD$ are $A(-2, 0)$, $B(1, 4)$, and $C(13, 9)$.

  a. Find the coordinates of vertex D. Explain your reasoning, and verify that $ABCD$ is a parallelogram.

  b. Find the measures of all the angles in the parallelogram.

  c. If the diagonals of the parallelogram intersect at point E, find the length of $\overline{AE}$.

  d. Find the perimeter and area of $\square ABCD$.

30 Delaware Valley Car Rentals has rental locations in Port Jervis, Stroudsburg, and Easton. A customer who rents a car from Delaware Valley Car Rentals can return the car to any of the three locations. Company statistics show that of the cars rented in Port Jervis, 50% are returned to Port Jervis, 20% are returned to Stroudsburg, and 30% are returned to Easton. Of the cars rented in Stroudsburg, 25% are returned to Port Jervis, 40% to Stroudsburg, and 35% to Easton. Of the cars rented in Easton, 10% are returned to Port Jervis, 30% to Stroudsburg, and 60% to Easton.

  a. Represent this information in a matrix in which the rows indicate the location in which the car was rented and the columns indicate where the car was returned. Use decimal values of the percentages.

b. On one day, the company rents 20 cars in Port Jervis, 30 cars in Stroudsburg, and 45 cars in Easton. Use matrix multiplication to find the expected number of these cars that will be returned to each location.

c. Explain what the non-integer values in the matrix result might mean for this context.

31 Consider a circle with center at $(-3, 1)$ and radius 6.

a. Draw a sketch of the circle.

b. Recall that the general form for the equation of a circle is $(x - h)^2 + (y - k)^2 = r^2$, where (h, k) is the center of the circle and r is the radius. Write the equation for this circle.

c. Calculate the area of this circle.

d. Calculate the circumference of this circle.

e. Write an equation for a circle with the same center but with a circumference twice the circumference of this circle.

32 If $\cos \theta = -\frac{3}{5}$ and the terminal side of θ lies in Quadrant II, find $\sin \theta$ and $\tan \theta$.

Looking Back

You studied three basic tools in this unit: the normal distribution as a model of the variability in a distribution, the binomial distribution as a model of success/failure situations, and the control chart as a way to determine when an industrial process goes out of control.

The normal distribution is fundamental in characterizing variability for several reasons. As you learned in Lesson 1, many naturally occurring variables, such as human height, have an approximately normal distribution. As you learned in Lesson 2, if you have a large enough sample size, the binomial distribution is approximately normal in shape. As you learned in Lesson 3, the distribution of the means of random samples becomes more approximately normal as the sample size increases.

The following tasks will help you review and deepen your understanding of these basic ideas.

1 According to a U.S. Geological Survey report, the historical distribution of total rainfall for Guatemala averaged over 88 gauges at different stations during June, July, and August is approximately normal with a mean of 955 mm and a standard deviation of 257 mm. (**Source:** *June-July-August 2003 Rainfall Forecast Interpretation, Central America,* March 19, 2003)

a. Sketch this distribution. Include a scale on the horizontal axis. Mark one, two, and three standard deviations from the mean.

b. Would it be a rare event to have 500 mm of rainfall?

c. In what percentage of years is the rainfall more than 1 meter? Explain.

d. What amount of rainfall falls at the 25th percentile?

2 The claim has been made that about 57% of drivers agree that travel would be safer if the minimum driving age for new drivers was raised from 16 to 18 years old. Suppose you take a random sample of 500 drivers and ask them if they agree. (**Source:** www.cnn.com/2003/US/05/27/dangerous.driving/index.html)

a. If the claim is correct, what is the expected number in your sample who will agree? The expected number who will disagree?

b. Is this sample size large enough that the binomial distribution for this situation will be approximately normal?

c. Compute the mean and standard deviation of the binomial distribution.

d. Sketch this distribution. Include a scale on the horizontal axis that shows the mean and the points one and two standard deviations from the mean.

e. Suppose you doubt the claim of 57%. You take a random sample of 500 drivers and find only 270 people agree. Do you have statistically significant evidence that the claim is false?

3 Suppose you kept track of the gas mileage for your vehicle over a 25-week span. You recorded the data as follows.

Gas Mileage			
Week	Miles per Gallon	Week	Miles per Gallon
Feb 7	23	May 9	24
Feb 14	27	May 16	27
Feb 21	27	May 23	25
Feb 28	28	May 30	28
Mar 7	25	June 6	25
Mar 14	26	June 13	26
Mar 21	25	June 20	25
Mar 28	29.5	June 27	29
Apr 4	26	July 4	26
Apr 11	27	July 11	27
Apr 18	24	July 18	24
Apr 25	26	July 25	26
May 2	26		

a. Make a run chart for your vehicle's gas mileage. What does it tell you about the consistency of your vehicle's gas mileage?

b. Find the mean and standard deviation of the measurements of miles per gallon. Draw horizontal lines on the run chart representing the mean and one and two standard deviations from the mean. Was your vehicle's mileage unusual for any week in that time period?

c. To use the eight out-of-control tests, the data must be approximately normally distributed.

    i. Does that appear to be the case here? Explain.

    ii. How could you use the eight tests if the distribution had not been approximately normal?

d. Would you say fuel consumption of your vehicle is in control? If not, which test is violated?

e. Five weeks after July 25, you discovered that something seemed to be wrong with your vehicle, affecting its gas mileage. Write down two different sets of mileage data for those five weeks that would indicate you had a problem. Explain the test you were using and how it would apply to your data. Use a test only once.

SUMMARIZE THE MATHEMATICS

In this unit, you learned how to use the normal distribution to model situations involving chance.

a What are the characteristics of a normal distribution?

    i. What is a standardized value?

    ii. How can you find the percentile of a value in a normal distribution?

b What are the characteristics of a binomial situation?

c How can you determine the shape, mean, and standard deviation of a binomial distribution?

d How can you determine if an outcome from a random sample is statistically significant?

e How are the normal distribution, mean, and standard deviation used in quality control?

f What is a false alarm in quality control?

g Describe the Central Limit Theorem and how it is used in quality control.

Be prepared to share your responses with the class.

 CHECK YOUR UNDERSTANDING

Write, in outline form, a summary of the important mathematical concepts and methods developed in this unit. Organize your summary so that it can be used as a quick reference in future units and courses.

5

Polynomial and Rational Functions

In prior units of *Core-Plus Mathematics*, you developed understanding and skill in the use of linear, quadratic, and inverse variation functions. Those functions are members of two larger families of versatile and useful algebraic tools, the polynomial and rational functions.

In this unit, you will learn how to use polynomial and rational expressions and functions to represent and analyze a variety of quantitative patterns and relationships. The key ideas will be developed through work on problems in three lessons.

LESSONS

1 Polynomial Expressions and Functions

Use linear, quadratic, cubic, and quartic polynomial functions to represent quantitative relationships, data patterns, and graphs. Analyze the connections between symbolic expressions and graphs of polynomial functions. Combine polynomials by addition, subtraction, and multiplication.

2 Quadratic Polynomials

Extend skills in expanding and factoring quadratic expressions. Use a completing-the-square strategy to write quadratic expressions in equivalent forms and to prove the quadratic formula for solving quadratic equations.

3 Rational Expressions and Functions

Use quotients of polynomial functions to represent and analyze quantitative relationships. Analyze the connections between symbolic expressions and graphs of rational functions. Combine rational expressions by addition, subtraction, multiplication, and division.

fotog/Getty Images

Polynomial Expressions and Functions

Working for a company that designs, builds, and tests rides for amusement parks can be both fun and financially rewarding. Suppose that in such a position, your team is in charge of designing a long roller coaster. One morning, your team is handed sketches, like those at the top of the next page, that show ideas for two sections of a new roller coaster.

As team leader, your task is to find algebraic functions with graphs that match the two sketches. The functions will be useful in checking safety features of the design, like estimated speed and height at various points of the track. They will also be essential in planning manufacture of the coaster track and support frame.

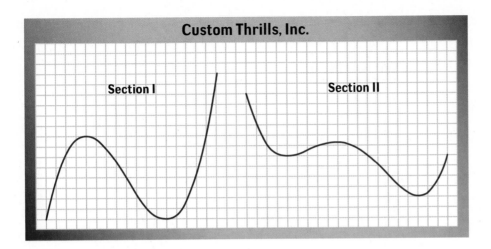

Custom Thrills, Inc.

Section I   Section II

THINK ABOUT THIS SITUATION

Study the sketches of Section I and Section II of the proposed coaster design.

a What familiar functions have graphs that match all or parts of the design sketches?

b What strategies could you use to find functions with graphs that model the sketches?

c What do you think are the key points on each sketch that should be used in finding a function model for the graph pattern?

In this lesson, you will begin study of an important class of algebraic functions called *polynomial functions*. You will learn how to use polynomial functions to model complex graphical patterns, like the roller coaster design proposals, and how to use operations on polynomials to solve problems in business situations.

INVESTIGATION 1

Modeling with Polynomial Functions

The roller coaster design sketches are not modeled well by graphs of linear, exponential, or quadratic functions. But there are other kinds of algebraic functions that do have such graphs, and there are techniques for using sample graph points to find rules for those functions. In this investigation, you should look for answers to these questions:

> *What are polynomial functions and what kinds of graphs do those functions have?*
>
> *How are rules for polynomial functions related to patterns in their graphs?*

To the nth degree The following diagram shows the graph of a function that could represent a roller coaster design based on the Section I sketch.

Section I Design

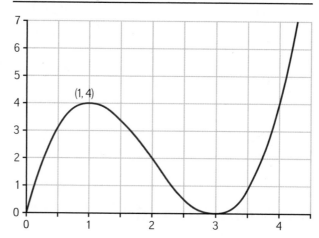

1. When you used curve-fitting tools in earlier work, all you had to do was supply coordinates of points that capture key features of the graph or data pattern and then select a function type for the model. Why would neither linear, quadratic, exponential, nor inverse variation functions provide good models for the graph pattern representing the Section I Design?

2. You might have noticed that most curve-fitting tools offer other modeling options. The one listed right after the quadratic option is usually *cubic*.

   a. What key points on the graph, beside (1, 4), do you think would be helpful in finding a cubic function that models the proposed Section I Design?

   b. Using the points you selected in Part a, apply a cubic curve-fitting routine to find a function model for the graph pattern.

   c. Compare the graph of the resulting cubic function to the shape of the Section I Design. Describe ways that the cubic function is or is not a good model of that pattern.

3. Compare results of your work toward a function model for the Section I Design with those of others in your class.

   a. Did everyone use the same points from the given graph?

   b. For those who used different points or even different numbers of points, were the resulting function models still the same?

   c. What do you think might be the minimum number of points needed to find a cubic model for a data or graph pattern?

4 As you may have noted, there are some significant differences between the design ideas for Sections I and II of the proposed roller coaster. The next diagram shows a graph of a function that could represent a roller coaster design based on the Section II sketch.

Section II Design

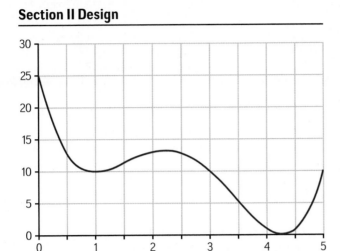

a. Find coordinates of key points outlining the shape of this graph. Then find the cubic function model for the pattern in those points and compare its graph to the shape of the proposed Section II Design.

b. You may have noticed there are other curve-fitting options on your calculator or computer software list. One such option is *quartic*. What do you think might be the minimum number of points needed to find a quartic model for a data or graph pattern?

c. Find a quartic function model for the pattern of data points you identified in Part a. Check how well its graph matches the pattern in the Section II Design.

d. Compare your result with those of others in your class. See if you can explain any differences.

Though you may not have been aware of it at the time, in earlier *Core-Plus Mathematics* units involving linear and quadratic functions, you were learning about functions from a larger class called *polynomial functions*. Cubic and quartic functions, like those that you explored in a search for models of the roller coaster designs, are also polynomial functions. A **polynomial function** is any function with a rule that can be written in the form $f(x) = a_n x^n + a_{n-1} x^{n-1} + \cdots + a_1 x + a_0$, where n is a whole number and the coefficients $a_n, a_{n-1}, a_{n-2}, \ldots, a_1, a_0$ are numbers.

The functions described by information on these calculator screens are cubic and quartic polynomials.

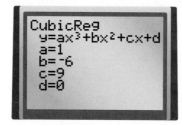

Any algebraic expression in the form $a_n x^n + a_{n-1} x^{n-1} + \cdots + a_1 x + a_0$ is called a **polynomial expression**. One of the most important characteristics of any polynomial function or expression is its degree. The **degree of a polynomial** is the greatest exponent of the variable that occurs in the expression. For example, a quadratic polynomial has degree 2, a cubic polynomial has degree 3, and a quartic polynomial has degree 4. A nonzero constant is a polynomial of degree 0.

Connecting Polynomial Expressions and Graphs From earlier work with linear, quadratic, exponential, and power functions, you know that it is often helpful to inspect a function rule and estimate the shape of its graph. In the other direction, it is helpful if you can inspect a graph and predict the kind of function rule that will model its pattern.

5 As you have seen in exploring polynomial models for roller coaster designs, one of the most interesting and important characteristics of polynomial functions is the number and location of *peaks* or *valleys* in their graphs. A peak indicates a **local maximum** value for the function, and a valley indicates a **local minimum** value for the function.

a. Graph and then estimate coordinates of local maximum and local minimum points for each polynomial function.

  i. $g(x) = -x^3 + 6x^2 - 9x$

  ii. $h(x) = -x^4 + 2x^3 + 7x^2 - 8x - 12$

b. How do the examples in Part a help to explain why the adjective "local" is used in describing function values at peaks and valleys of polynomial functions?

c. Why are the points $(0, 0)$ and $(4.25, 7)$ *not* considered local minimum or local maximum points for the cubic function you found in Problem 2 to model the Section I Design for the roller coaster?

d. Why are the points $(0, 25)$ and $(5, 10)$ *not* considered local maximum or local minimum points for the quartic function you found in Problem 4 to model the Section II Design?

6 Consider the relationship between the degree of a polynomial function and the number of local maximum and/or local minimum points on the graph of that function.

a. Give an example of a polynomial function with no local maximum or local minimum point.

b. How many local maximum and/or local minimum points can there be on the graph of a quadratic polynomial?

7 To explore the graphs of higher degree polynomial functions, it is helpful to use computer software that accepts function definitions like $f(x) = ax^3 + bx^2 + cx + d$, allows you to move "sliders" controlling the values of a, b, c, and d, and quickly produces corresponding graphs.

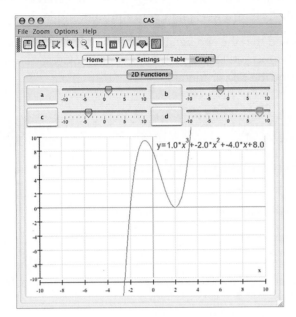

a. How many local maximum and/or local minimum points do you think there can be on the graph of a cubic polynomial function?

b. Test your conjecture in Part a by graphing cubic polynomial functions $p(x) = a_3x^3 + a_2x^2 + a_1x + a_0$ for various sets of values for the coefficients a_3, a_2, a_1, and a_0. You might start with examples like these.

$$y = x^3 - 9x$$
$$y = x^3 - 6x^2 + 12x - 3$$
$$y = -x^3 + 4x + 2$$

Then modify those examples to test other cases.

c. How many local maximum and/or local minimum points do you think there can be on the graph of a quartic polynomial?

d. Test your conjecture in Part c by graphing quartic polynomial functions $p(x) = a_4x^4 + a_3x^3 + a_2x^2 + a_1x + a_0$ for various combinations of values for the coefficients a_4, a_3, a_2, a_1, and a_0. You might start with examples like these.

$$y = x^4 - 9x^2 + 2$$
$$y = -x^4 - 8x^3 - 32x^2 - 32x + 5$$
$$y = -x^4 - 4x^3 + 4x^2 + 16x$$

Then modify those examples to test other cases.

e. What seems to be the connection between the degree of a polynomial and the number of local maximum and/or local minimum points on its graph? Test your conjecture by exploring graphs or tables of values for some polynomial functions of even higher degree like $p(x) = x^5 - 5x^3 + 4x$ and $q(x) = x^5 + x^3 + x + 4$.

8 Since calculator and computer screens provide only a small portion of a table or graph of a function $f(x)$, it is important to be able to predict the **end behavior** of the function, that is, the behavior of the function as $|x|$ gets very large.

    a. By dividing up the work, investigate the end behavior of each function in Problem 7. Share and explain your findings and look for any patterns.

    b. AJ claimed that the end behavior of any polynomial function is like that of a power function. Explain why this claim makes sense and how you would determine the power function.

SUMMARIZE THE MATHEMATICS

In this investigation, you explored some basic properties and uses of polynomial functions and expressions.

a   What kinds of expressions and functions are called polynomials?

b   What is the degree of a polynomial?

c   What are local maximum and local minimum points of a graph? How is the number of such points on the graph of a polynomial function related to the degree of the polynomial expression used in its rule?

d   How many points are needed for use of a curve-fitting tool to find the polynomial function of degree n that fits a data or graph pattern?

Be prepared to explain your ideas to the class.

 CHECK YOUR UNDERSTANDING

Use what you have learned about polynomial functions to help complete the following tasks.

    a. Consider the functions $f(x) = 2x^3 + 8x^2 + 3x - 2$ and $g(x) = x^4 - 8x^3 + 16x^2 + 4$.

        i. Identify the degree of each function.

        ii. Estimate coordinates of the local maximum and/or local minimum points on graphs of the functions.

b. Find a function whose graph models the following coaster track design.

Coaster Track Section

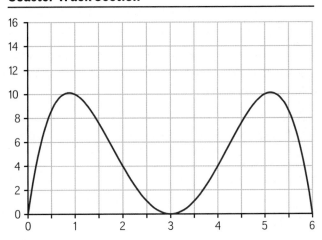

INVESTIGATION 2

Addition, Subtraction, and Zeroes

From early in your mathematical studies, you have learned how to perform arithmetic operations on numbers. As it turns out, you can do just about any operation with algebraic expressions that you can do with numbers. In this investigation, you will begin exploration of arithmetic operations on polynomials. As you work through the problems, look for answers to these questions:

> *How can the rules for polynomial functions $f(x)$ and $g(x)$ be combined to give rules for $f(x) + g(x)$ and $f(x) - g(x)$?*

> *How are the degrees of expressions being added or subtracted related to the degree of the result?*

> *How is the degree of a polynomial related to the number of zeroes for the function?*

Adding and Subtracting Polynomials When a small music venue books a popular band, like *Ice and Fire*, business prospects of the event depend on how the ticket prices are set. For example, if the ticket price is set at x dollars, income and expenses might be estimated as follows.

Ticket sale income:
$$t(x) = -25x^2 + 750x$$

Snack bar income:
$$s(x) = 7{,}500 - 250x$$

Concert operating expense:
$$c(x) = 4{,}750 - 125x$$

Snack bar operating expense:
$$b(x) = 2{,}250 - 75x$$

1 Before each show, the manager uses the functions $t(x)$ and $s(x)$ to estimate total income.

   a. Why does it make sense that each source of income—ticket sales and snack bar sales—might depend on the price set for tickets?

   b. What income should the manager expect from ticket sales alone if the ticket price is set at $12? What income from snack bar sales? What income from the two sources combined?

   c. What rule would define the function $I(x)$ that shows how combined income from ticket sales and snack bar sales depends on ticket price? Write a rule for $I(x)$ that is in simplest form for calculation.

   d. How does the degree of $I(x)$ compare to the degrees of $t(x)$ and $s(x)$?

2 The manager also uses the functions $c(x)$ and $b(x)$ to estimate total operating expenses.

   a. Why does it make sense that each source of expense—concert operation and snack bar operation—might depend on the price set for tickets?

   b. What expense should the manager expect from concert operations alone if the ticket price is set at $12? What expense from snack bar operations? What expense from the two sources combined?

   c. What rule would define the function $E(x)$ that shows how combined expense from concert and snack bar operations depends on ticket price? Write a rule for $E(x)$ that is in simplest form for calculation.

   d. How does the degree of $E(x)$ compare to the degrees of $c(x)$ and $b(x)$?

3 Consider next the function $P(x)$ defined as $P(x) = I(x) - E(x)$.

   a. What does $P(x)$ tell about business prospects for the music venue?

   b. Write two equivalent rules for $P(x)$.

   - one that shows the separate expressions for income and operating expenses

   - another that is in simplest form for calculation

   Check your work with a CAS and resolve any differences in the results.

   c. How does the degree of $P(x)$ compare to that of $I(x)$ and of $E(x)$?

   d. Compare $E(x) - I(x)$ to $I(x) - E(x)$. What caution does this result suggest in using subtraction to find the difference of quantities represented by polynomial functions?

4 In your analysis of business prospects for a concert by *Ice and Fire*, you used common sense to combine polynomials by addition and subtraction. You probably noticed a connection between the degree of a sum or difference of polynomials and the degrees of the polynomials being combined.

a. Test your ideas about the degree of the sum or difference of polynomials by finding the simplest rules for the sum and difference of each pair of functions given below. Compare your results with those of your classmates and resolve any differences.

   i. $f(x) = 3x^3 + 5x - 7$ and $g(x) = 4x^3 - 2x^2 + 4x + 3$

   ii. $f(x) = 3x^3 + 4x^2 + 5$ and $g(x) = -3x^3 - 2x^2 + 5x$

   iii. $f(x) = x^4 + 5x^3 - 7x + 5$ and $g(x) = 4x^3 - 2x^2 + 5x + 3$

   iv. $f(x) = x^4 + 5x^3 - 7x + 5$ and $g(x) = 4x^3 + 5x$

   v. $f(x) = 6x^4 + 5x^3 - 7x + 5$ and $g(x) = 6x^4 + 5x$

b. Explain in your own words.

   i. how to find the simplest rule for the sum or difference of two polynomials

   ii. how the degree of the sum or difference of two polynomials is related to the degrees of the polynomials being combined

Zeroes of Polynomial Functions In your previous work with linear and quadratic polynomial functions, you discovered that answers to many important questions are found by solving equations. You also discovered that linear and quadratic equations can be transformed to equivalent forms that look like $a_1x + a_0 = 0$ or $a_2x^2 + a_1x + a_0 = 0$, respectively.

Solving these equations requires finding values of x that are called **zeroes** of the related functions $f(x) = a_1x + a_0$ and $g(x) = a_2x^2 + a_1x + a_0$. For example, the zeroes of $g(x) = x^2 + 5x - 6$ are -6 and 1 because $g(-6) = 0$ and $g(1) = 0$. The zeroes locate the x-intercepts of the graphs of the functions.

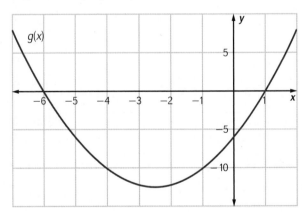

The next problems explore the possibilities for zeroes of higher degree polynomial functions.

5 The functions $f(x) = x^3 - 6x^2 + 9x$ and $g(x) = x^4 - 10x^3 + 32x^2 - 38x + 25$ could be used to model patterns in the roller coaster designs at the beginning of this lesson.

a. Graph and find the zeroes of $f(x)$.

b. Graph and find the zeroes of $g(x)$.

c. What might the zeroes represent in the roller coaster scenario?

6 Now consider possible connections between the degree of a polynomial and the number of zeroes of the related function. It might be helpful to use function graphing software that provides "sliders" to adjust coefficients and show the corresponding graphs.

a. What are the possible numbers of zeroes for linear functions? Make sketches that represent your ideas.

b. What are the possible numbers of zeroes for quadratic functions? Make sketches that represent your ideas.

c. Explore tables and graphs of several functions with rules in the form

$$f(x) = a_3x^3 + a_2x^2 + a_1x + a_0 \quad (a_3 \neq 0)$$

to discover the possible numbers of zeroes for cubic polynomial functions. Make sketches of results that illustrate your ideas.

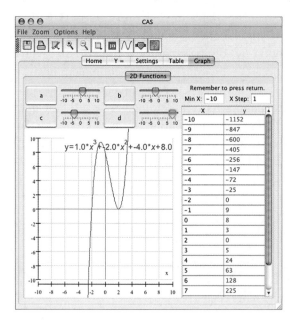

d. Explore tables and graphs of functions with rules in the form

$$g(x) = a_4x^4 + a_3x^3 + a_2x^2 + a_1x + a_0 \quad (a_4 \neq 0)$$

to discover the possible numbers of zeroes for quartic polynomial functions. Make sketches of results that illustrate your ideas.

e. How does the degree of a polynomial seem to be related to the number of zeroes of the related polynomial function? Test your conjecture by studying graphs of some polynomial functions of degrees five and six.

Remember: When checking to see if a mathematical idea is correct, you should explore many different examples to see how, if at all, some example might be constructed that disproves the conjecture. Here again, it might be helpful to use computer software that accepts function definitions like $f(x) = ax^3 + bx^2 + cx + d$ and allows you to move "sliders" controlling the values of a, b, c, and d to quickly produce corresponding graphs.

SUMMARIZE THE MATHEMATICS

In this investigation, you explored addition and subtraction of polynomials and zeroes of polynomial functions.

a What steps should one follow in finding a simplest rule for the sum or difference of two polynomials?

b How is the degree of the sum or difference of two polynomials related to the degrees of the polynomials being combined?

c What does the degree of a polynomial tell you about the possible number of zeroes for the corresponding polynomial function?

Be prepared to explain your ideas to the class.

 CHECK YOUR UNDERSTANDING

Consider the following polynomial functions.

$$f(x) = x^3 - 16x$$
$$g(x) = 16x^2 - 8x^3 + x^4$$
$$h(x) = x^4 - 8x^3 + 16x^2 + 4$$

a. For each function, (1) identify its degree, (2) sketch its graph, and (3) find its zeroes.

b. Find expressions in standard polynomial form for $f(x) + g(x)$ and $f(x) - g(x)$.

c. Find the degrees of these polynomials.

  i. $g(x) - h(x)$

  ii. $h(x) - f(x)$

Zeroes and Products of Polynomials

In earlier work with quadratic functions, you saw how both factored and expanded expressions for function rules provide useful information. For example, the two parabolas that make up the "M" in the following diagram can be created by graphing $f(x) = -x^2 + 6x$ and $g(x) = -x^2 + 14x - 40$.

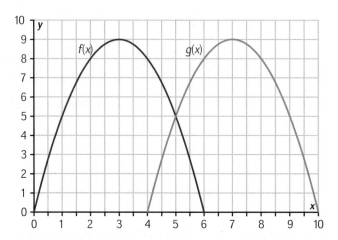

You have learned that the rules for those functions can be used to determine the location of maximum points, lines of symmetry, and y-intercepts of the graphs.

The function rules can be expressed in equivalent factored forms $f(x) = -x(x - 6)$ and $g(x) = -(x - 4)(x - 10)$. Those forms easily reveal the x-intercepts of the graphs.

Your work in this investigation will reveal strategies for working with factors and products of other polynomials. Look for answers to these questions:

How are the zeroes of a polynomial function related to its factors?

How can a product of polynomial factors be expanded to standard form?

How is the degree of a product of polynomials related to the degrees of the factors?

1 Analyze the following quadratic functions by using algebraic reasoning with the standard polynomial and factored forms of their rules.

a. $h(x) = x^2 + 4x$

b. $j(n) = -n^2 + n + 6$

c. $k(x) = (2x - 1)(x + 5)$

For each function:

i. write the rule in both standard polynomial and factored form.

ii. show how to use the factored form to find zeroes of the function and x-intercepts of its graph.

iii. show how to use the factored form and information about the *x*-intercepts to find the line of symmetry, the maximum or minimum point, and the *y*-intercept of the graph.

iv. show how to use the standard polynomial form to locate the line of symmetry, maximum or minimum point, and *y*-intercept of the graph.

2 Next, consider the function $q(x) = x(x - 3)(x + 5)$.

a. What are the zeroes of $q(x)$?

b. Use reasoning like what you apply with products of two linear factors to write a rule for $q(x)$ in standard polynomial form. Record steps in your work so that someone else could check your reasoning.

c. Identify the degree of $q(x)$. How could you have predicted that property of the polynomial before any algebraic multiplication?

d. Graph $q(x)$ and label the *x*-intercepts, *y*-intercept, and local maximum and local minimum points with their coordinates.

3 The graph below is that of a cubic polynomial $c(x)$.

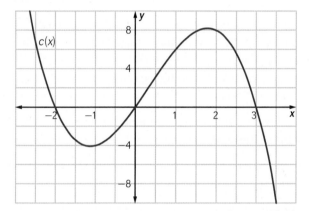

a. Use information from the graph to write a possible rule for $c(x)$. Express the rule in equivalent factored and standard polynomial forms.

b. Compare the overall shape of the graph, the local max/min points, and intercepts of the graph produced by your rule to the given graph. Adjust the rule if needed to give a better fit.

4 Look back at your work on Problems 1–3 to develop conjectures about answers to the following questions. In each case, be prepared to give other specific examples illustrating your idea.

a. How can you tell the zeroes of a polynomial function when its rule is written as a product of linear factors?

b. How can you tell the degree of a polynomial function when its rule is written as a product of linear factors?

c. Which properties of a polynomial function and its graph are shown best when the rule is written as a product of linear factors? When the rule is written in standard form?

"Advanced" Multiplication Early in your experience with variables and expressions, you learned that the Distributive Property of Multiplication over Addition guarantees correctness of statements like $3(x + 4) = 3x + 12$. More recently, you have used the distributive property to expand products of binomials like $(2x + 4)(x - 3) = (2x + 4)x - (2x + 4)3 = 2x^2 + 4x - 6x - 12 = 2x^2 - 2x - 12$.

5 Suppose that you were asked to expand the product $(2x + 1)(3x^2 + 2x + 4)$.

   a. How could you adapt the algorithm for finding the product of two binomials to find the desired product of polynomial factors? What standard polynomial form result does it give in this case?

   b. Compare the zeroes of the corresponding polynomial functions $p(x) = 2x + 1$ and $q(x) = 3x^2 + 2x + 4$ and the zeroes of the product $p(x)q(x)$.

6 When students in a *Core-Plus Mathematics* class in Denver, Colorado, were challenged to multiply a quadratic expression and a cubic expression, one group started like this.

$$(3x^2 + 4x - 10)(x^3 - 2x + 1) = 3x^5 - 6x^3 + 3x^2 + \cdots$$

   a. Do you think they were on the right track? Explain your reasoning.

   b. Based on what you know about the distributive, commutative, and associative properties, write the complete product in the form $a_nx^n + a_{n-1}x^{n-1} + a_{n-2}x^{n-2} + \cdots + a_1x + a_0$. You might choose to find the result using a CAS and then figure out why it works that way, or use your own reasoning first and check that a CAS gives the same result.

   c. What strategy would you recommend for keeping track of results in the process of multiplying two such nonlinear polynomials?

   d. Identify the degree of the product polynomial. How could you figure this out before carrying out any multiplication?

   e. The zeroes of $f(x) = 3x^2 + 4x - 10$ are approximately 1.28 and -2.61, and the zeroes of $g(x) = x^3 - 2x + 1$ are approximately 1, 0.62, and -1.62. Are those values of x also zeroes of the product function $h(x) = (3x^2 + 4x - 10)(x^3 - 2x + 1)$? Check by substituting the possible zeroes in the polynomial produced by your work on Part b.

7 Write the following products in standard polynomial form. Identify the degree of each product. Explain how that degree is related to the degrees of the factors.

   a. $(x - 4)(x^4 - 3x^2 + 2)$

   b. $(x^5 + 7)(-2x^2 + 6x - 1)$

   c. $(x^6 - 5x^5 + 3x^4 + 7x^3 - 6x^2 + 2x - 8)(x^2 + 7x + 12)$

Repeated Zeroes You may have noticed that in Problems 1–7, each first-degree polynomial function had one zero, each second-degree polynomial function had two zeroes, and each third-degree polynomial function had three zeroes. As you discovered earlier in the case of quadratic functions, this correspondence of polynomial degree and function zeroes is *not* always the case.

Jim Laser/CPMP

8 Consider the functions $r(x)$ and $s(x)$, where $r(x) = (x - 3)^2$ and $s(x) = (x + 3)^3$.

   a. Expand the expressions that define $r(x)$ and $s(x)$. Identify the degree of each polynomial.

   b. How many zeroes do $r(x)$ and $s(x)$ each have?

   c. How many zeroes does the product $p(x) = r(x)s(x)$ have?

   d. Use sketches of the respective function graphs to illustrate your responses in Parts b and c.

9 Consider the function $t(x) = (x - 3)(x + 4)^2$.

   a. Expand the expression that defines $t(x)$. Identify the degree of the resulting polynomial.

   b. What are the zeroes of $t(x)$?

   c. Sketch a graph of $t(x)$ to show that the number of zeroes and the degree of the polynomial are not the same. Explain why that happens.

SUMMARIZE THE MATHEMATICS

In this investigation, you discovered connections between factors, zeroes, and the expanded form of polynomial functions.

   a How is the degree of a product of polynomials determined by the degrees of its factors?

   b What strategies can be used to find the standard form of a product of polynomials?

   c How are the zeroes of a polynomial function related to its factors?

   d In what cases will the degree of a polynomial function not equal the number of its zeroes?

Be prepared to explain your ideas to the class.

 CHECK YOUR UNDERSTANDING

Use your understanding of polynomial multiplication to complete these tasks.

a. Rewrite the following products in standard form
$a_n x^n + a_{n-1}x^{n-1} + a_{n-2}x^{n-2} + \cdots + a_1 x + a_0$.
Identify the degree of each result. Explain how that degree is related to the degrees of the factors.

   i. $f(x) = (x^2 - 5x + 6)(x - 4)$

   ii. $g(x) = (-2x^2 + 6x - 1)(x^5 + 7)$

b. What are the zeroes of the function $h(x) = (x^2 - 5x + 6)(2x - 7)$?

c. What are the zeroes of the function $j(x) = (x + 3)^2(2x - 5)$?

d. Write rules for quadratic and cubic polynomial functions that each have zeroes at $x = 5$ and $x = -2$ only.

APPLICATIONS

1. In Parts a–d, use a curve-fitting tool and/or algebraic reasoning to find rules for functions $f(x)$, $g(x)$, $h(x)$, and $j(x)$ that model the given graph patterns. In each case, report the graph points used as the basis of your curve-fitting, the rule of the modeling function, and your reasons for choosing a model of that type.

a.
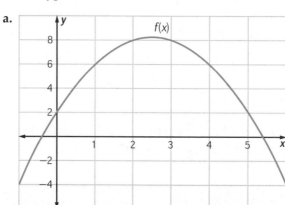

b.
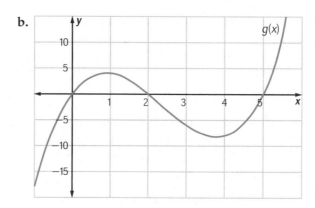

c.
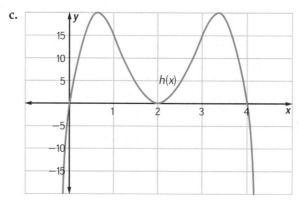

d.

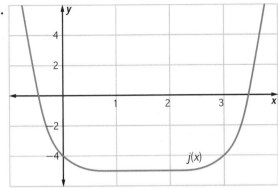

$j(x)$

2 Graph each function and then calculate or estimate coordinates of all:

- local maximum points.
- local minimum points.
- x-intercepts.
- y-intercepts.

a. $f(x) = 2x^2 + 4x + 1$

b. $g(x) = x^3 + 2x^2 + 3x + 7$

c. $h(x) = x^3 - 6x^2 + 12x - 8$

d. $s(x) = x^4 - 8x^3 + 20x^2 - 16x$

3 For each algebraic expression:

- write an equivalent expression in standard polynomial form.
- identify the degrees of the expressions being combined and the degree of the result.

a. $(2x^2 + 5x - 2) + (-2x^2 + 3x + 7)$

b. $(2x^2 + 5x - 2) - (5x + 7)$

c. $(-7x^3 + 6x^2 + 3x - 7) - (3x^4 + 7x^3 + 4x^2 - 3x + 2)$

d. $(2x^2 + 3x - 7 + 5x^5 - 3x^4 - 7x^3) + (3x^5 + 2x^4 + 2x^3 + 4x^2 + 6x + 1)$

e. $(5x^5 - 3x^4 + 2x^2 + 6x - 7) + (7x^3 - 2x^2 - 3x + 5)$

4 Without using a graph, table, or CAS, find all zeroes of these functions. Then describe the end behaviour of each function.

a. $f(x) = (x - 3)(x + 4)$

b. $g(x) = (x^2 - 9)x$

c. $h(x) = (x^2 + 5)x^2$

d. $s(t) = -(t + 5)^2$

5 For each algebraic expression:

- write an equivalent expression in standard polynomial form.

- identify the degrees of the expressions being combined and the degree of the result.

a. $(7x + 3)(x - 1)$

b. $(3x + 5)^2 x^2$

c. $(2x^2 + 3x - 7)(3x + 7)$

d. $(7x^3 - 6x + 4)(2x^2 - 7)$

e. $(7x^3 + 2)(5x^2 + 3x - 8)$

f. $(-3x^4 + 2x^2 + 6x)(7x^3 - 2x^2 + 5)$

6 *Great Lakes Supply* has been contracted to manufacture open-top rectangular storage bins for small electronics parts. The company has been supplied with 30 cm × 16 cm sheets of material. The bins are made by cutting squares of the same size from each corner of a sheet, bending up the sides, and sealing the corners.

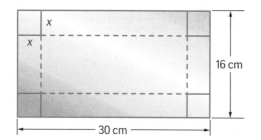
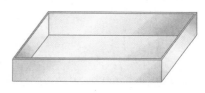

a. On centimeter graph paper, draw a rectangle 30 cm long and 16 cm wide. Cut equal-size squares from the corners of the rectangle. Fold and tape the sides as suggested above. Find the volume of your bin.

b. Write a rule expressing the volume V of a bin as a function of the length of the corner cutout x.

   i. What is the degree of the polynomial function?

   ii. What is a practical domain for the function?

   iii. Why must the graph of $V(x)$ have a local maximum point?

c. If the company is to manufacture bins with the largest possible volume, what should be the dimensions of the bins? What is the maximum volume?

7 The *Galaxy Sport and Outdoor Gear* company has a climbing wall in the middle of its store. Before the store opened for business, the owners did some market research and concluded that the daily number of climbing wall customers would be related to the price per climb x by the linear function $n(x) = 100 - 4x$.

a. According to this function, how many daily climbing wall customers will there be if the price per climb is \$10? What if the price per climb is \$15? What if the climb is offered to customers at no cost?

b. What do the numbers 100 and -4 in the rule for $n(x)$ tell about the relationship between climb price and number of customers?

c. What is a reasonable domain for $n(x)$ in this situation? That is, what values of x are plausible inputs for the function?

d. What is the range of $n(x)$ for the domain you specified in Part c? That is, what are the possible values of $n(x)$ corresponding to plausible inputs for the function?

e. If the function $I(x)$ tells how daily income from the climbing wall depends on price per climb, why is $I(x) = 100x - 4x^2$ a suitable rule or that function?

8 The function $e(x) = 2x + 150$ shows how daily operating expenses for the *Galaxy Sport* climbing wall (Task 7) depend on the price per climb x.

a. Write two algebraic rules for the function $P(x)$ that gives daily profit from the climbing wall as a function of price per climb, (1) one that shows how income and operating expense functions are used in the calculation of profit and (2) another that is in simpler equivalent form.

b. Find $P(5)$. Explain what this result tells about climbing wall profit prospects.

c. What is a reasonable domain for $P(x)$ in this problem situation?

d. What is the range of $P(x)$ for the domain you specified in Part c?

e. Write and solve an inequality that will find the climb price(s) for which *Galaxy Sport and Outdoor Gear* will not lose money on operation of the climbing wall.

f. Find the price(s) that will yield maximum daily profit from the climbing wall.

CONNECTIONS

9 Recall that graphs of quadratic polynomial functions of the form $f(x) = ax^2 + bx + c$ are called *parabolas*.

   a. A basic geometric fact is that "two points determine a line." In general, how many points are needed to determine a parabola? Explain your reasoning.

   b. Under what conditions, if any, will two points determine a parabola?

10 Evaluate each of these polynomial functions when $x = 10$.

$$f(x) = 2x^2 + 3x + 7$$
$$g(x) = 5x^3 + x^2 + 5x + 4$$
$$h(x) = 3x^4 + 7x^3 + 9x^2 + 2x + 1$$
$$j(x) = 5x^5 + 3x^4 + 6x^3 + 2x^2 + 3x + 6$$

   a. Study the set of results. Explain why things turn out as they do in every case.

   b. How does the pattern observed in Part a help to explain the arithmetic guideline that when adding numbers like 2,351.7 and 462.23, you should always "line up the decimal points"?

11 The following work illustrates one method for multiplying whole numbers.

$$
\begin{array}{r}
5{,}283 \\
\times\ 25 \\
\hline
15 \\
400 \\
1{,}000 \\
25{,}000 \\
60 \\
1{,}600 \\
4{,}000 \\
100{,}000 \\
\hline
132{,}075
\end{array}
$$

   a. What products are represented by the numbers 15, 400, 1,000, ... , 100,000?

   b. Use a similar procedure to calculate 1,789 × 64, recording all of the individual products that must be summed to get the final result.

   c. Why does the procedure give a correct result?

   d. How does this method for arithmetic multiplication relate to the procedure for multiplying two polynomials?

12 In this task, you will explore related methods for multiplication of numbers and polynomials.

a. The following table shows a method of organizing work in multiplication of numbers like 314×27.

	3	1	4
2	6	2	8
7	21	7	28

i. Why is the result of this operation equal to $6,000 + 2,300 + 150 + 28$?

ii. How are those partial products found from the table entries?

b. The next table shows a way of organizing the work involved in multiplying polynomials like $(7x^3 + 9x^2 + 2x + 1)$ and $(4x^2 + 3x + 5)$.

	7	9	2	1
4	28	36	8	4
3	21	27	6	3
5	35	45	10	5

i. What are the coefficients of each term in the result
$\_x^5 + \_x^4 + \_x^3 + \_x^2 + \_x + \_$?

ii. How are the cell entries calculated?

iii. How are those cell entries combined to produce the coefficients of each term in the result? Why does that procedure give the desired result in each case?

13 Use the fact that $x^3 + 2x^2 - 11x - 12 = (x - 3)(x + 1)(x + 4)$ to solve the following inequalities. Express your solutions with inequality notation, number line graphs, and interval notation.

a. $x^3 + 2x^2 - 11x - 12 \geq 0$

b. $x^3 + 2x^2 - 11x - 12 < 0$

14 In Investigation 3, you saw that if $(x - 3)$ is a factor of a polynomial function $q(x)$, then $q(3) = 0$.

a. Explain each step in the following proof of this general statement about polynomial functions.

If $(x - a)$ is a factor of $f(x)$, then $f(a) = 0$.

(1) If $(x - a)$ is a factor of $f(x)$, then $f(x) = (x - a)g(x)$, where $g(x)$ is a polynomial.

(2) If $f(x) = (x - a)g(x)$, then $f(a) = 0$.

b. Write the converse of the if-then statement in Part a.

c. Combine the proposition proven in Part a and its converse (which can also be proven) into a single if-and-only-if statement. That statement is called the **Factor Theorem**.

Lesson 1 | Polynomial Expressions and Functions **341**

REFLECTIONS

15 One of the attractive features of polynomial models for data or graph patterns is the fact that evaluation of a polynomial requires only repeated use of three basic operations of arithmetic—addition, subtraction, and multiplication.

a. How does that make polynomials different and easier to use than some other types of functions that you have studied?

b. How do calculators and computers make that special feature of polynomials less important than it was before such technology was available?

16 Shown below are four polynomial functions, each written in expanded standard form and as a product of linear factors.

$$f(x) = x^4 - 10x^3 + 35x^2 - 50x + 24 = (x - 1)(x - 2)(x - 3)(x - 4)$$
$$g(x) = x^4 - 7x^3 + 17x^2 - 17x + 6 = (x - 1)^2(x - 2)(x - 3)$$
$$h(x) = x^4 - 5x^3 + 9x^2 - 7x + 2 = (x - 1)^3(x - 2)$$
$$j(x) = x^4 - 4x^3 + 6x^2 - 4x + 1 = (x - 1)^4$$

a. In which form is it easier to see the zeroes of the function?

b. In which form is it easier to see the y-intercept of the graph of the polynomial function?

c. In which form is it easier to predict the end behaviour of the polynomial function?

d. How many zeroes do each of the four quartic polynomial functions have?

e. How does the collection of four functions help to explain why any polynomial function of degree n can have at most n distinct zeroes?

17 In the investigations of this lesson, you studied polynomial functions that model roller coaster designs. Some represented concerns of a concert business, while others had no accompanying context.

a. How does using contexts help make sense of polynomial properties and operations?

b. What can be learned better from study of polynomials and polynomial functions without any attached contexts?

c. In what ways are graphs of polynomial functions helpful in reasoning about the symbolic expressions of those functions?

d. In what ways do graphs of polynomial functions provide only imprecise information about critical values of those functions?

e. What are the differences among polynomial functions, expressions, and equations and why is it useful to have those separate mathematical terms defined precisely?

EXTENSIONS

18 You can work in many different ways to evaluate a polynomial function like

$$p(x) = 5x^5 + 3x^4 + 6x^3 + 2x^2 + 3x + 6$$

for some specific value of x. But to write a computer algorithm for that task, it is desirable to accomplish the work with a minimum number of instructions.

a. The following algorithm gives directions for evaluation of any polynomial. Follow the algorithm to evaluate $p(10)$. Use your calculator for the arithmetic.

Step 1. Enter the coefficient of the highest degree term.

Step 2. Multiply the result by x.

Step 3. Add the coefficient of the next lower degree term. (It might be zero.)

Step 4. If all coefficients have been used, go to Step 5. Otherwise, return to Step 2.

Step 5. Add the constant term and report the result.

b. Write this expression in equivalent standard polynomial form.

$$((((5x + 3)x + 6)x + 2)x + 3)x + 6$$

c. Use the algorithm in Part a to evaluate $f(x) = 4x^4 + 5x^3 + 7x^2 + 1x + 9$ when $x = 3$.

d. Write an expression like that in Part b that expresses the calculations involved in evaluating the polynomial $4x^4 + 5x^3 + 7x^2 + 1x + 9$ for any specific x. Then show why your alternative form is equivalent to the original.

19 A polynomial function $f(x)$ can have a *global maximum* or a *global minimum* value. A **global maximum** is a number M with the property that $f(x) \leq M$ for all values of x. A **global minimum** is a number m with the property that $f(x) \geq m$ for all values of x. For example, the quadratic function $f(x) = x^2 - 6$ has a global minimum at the point $(0, -6)$. The function $g(x) = -x^2 + 6x$ has a global maximum at $(3, 9)$.

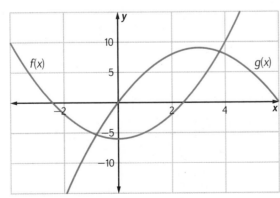

Use your experience with linear and quadratic functions and your explorations of graphs for cubic and quartic polynomial functions to answer these questions about global maximum and global minimum point possibilities. Be prepared to explain why you think your answers are correct.

a. Which polynomial functions of degree one, $p(x) = a_1x + a_0$, have global maximum or global minimum points?

b. Which polynomial functions of degree two, $p(x) = a_2x^2 + a_1x + a_0$, have global maximum or global minimum points?

c. Does the cubic polynomial function $g(x) = -x^3 + 6x^2 - 9x$ have a global maximum or global minimum?

d. Which polynomial functions of degree three, $p(x) = a_3x^3 + a_2x^2 + a_1x + a_0$, have global maximum or global minimum points?

e. Does the quartic polynomial function $h(x) = -x^4 + 2x^3 + 7x^2 - 8x - 12$ have a global maximum or global minimum?

f. Which polynomial functions of degree four, $p(x) = a_4x^4 + a_3x^3 + a_2x^2 + a_1x + a_0$, have global maximum or global minimum points?

20 When you first learned about division of whole numbers, you related that operation to multiplication by statements like "$369 \div 3 = 123$ because $3 \times 123 = 369$." You recorded the work to find a quotient in schemes that might have looked like this work.

$$
\begin{array}{r}
123 \\
3\overline{)369} \\
\underline{300} \\
69 \\
\underline{60} \\
9 \\
\underline{9}
\end{array}
$$

Apply your understanding and skill in division of whole numbers to find the quotients of polynomials in Parts a–e. It might help to organize your work in a style similar to what you learned to do with division of numbers, as shown in this example.

$$
\begin{array}{r}
x + 5 \\
x + 2\overline{)x^2 + 7x + 10} \\
\underline{x^2 + 2x} \\
5x + 10 \\
\underline{5x + 10}
\end{array}
$$

a. $(x^2 + 11x + 28) \div (x + 7)$

b. $(x^2 + 6x - 12) \div (x - 3)$

c. $(x^3 + 7x^2 + 13x + 4) \div (x + 4)$

d. $(2x^3 + x^2 - 4x - 3) \div (x + 5)$

e. $(2x^3 + 10x^2 + 16x) \div (x - 4)$

21 Look back at Extensions Task 20 and your work on Parts a–e.

a. Explain as precisely as you can why if a polynomial $p(x)$ is divided by a linear expression of the form $(x - a)$, then the remainder is a constant. That is, $p(x) = (x - a)q(x) + R$, where $q(x)$ is a polynomial function and R is a constant.

b. Write an argument proving that if a polynomial function $p(x)$ is divided by $x - a$, then the remainder is $p(a)$. This result is often referred to as the **Remainder Theorem**.

22 Look back at Connections Task 14. Use reasoning similar to that in Task 21 to prove the converse statement you wrote for Part b.

REVIEW

23 Rewrite each expression in simplest form.

a. $6x^2 - 10x + 5x^2 + x + 8$

b. $2x + 5 - (8x - 1)$

c. $3(x + 1) + 2(x^2 + 7) - 7x$

d. $(-6x^2)(9x^3)$

24 Solve each quadratic equation or inequality without the use of technology.

a. $x^2 - 4x + 4 = 0$

b. $x^2 + 6x + 8 < 0$

c. $x^2 - 4x - 5 \geq 0$

d. $x^2 + 14x + 42 = -7$

25 Laboratory tests indicate that when planted properly, 6% of a particular type of seed fail to germinate. This means that out of every 100 seeds planted according to instructions, on the average six do not sprout. The laboratory has been developing a new variety of the seed in which only 1% fail to germinate. Suppose that in an experiment, ten seeds of each of the two types are planted properly.

a. Calculate the theoretical probability that at least one seed out of the ten will fail to germinate for each variety of seed.

b. Design and carry out a simulation to estimate the chance that if ten of the seeds with the 6% germination rate are planted, at least one will fail to germinate.

c. Design and carry out a simulation to estimate the chance that if ten of the new variety of the seed are planted, at least one will fail to germinate.

d. Compare the estimates from your simulations to your calculations in Part a.

26 Write rules for quadratic functions whose graphs meet these conditions.

   a. Opens up and has two x-intercepts

   b. Opens down and has y-intercept $(0, -4)$

   c. Opens up and the only x-intercept is $(5, 0)$

   d. Has x-intercepts $(-2, 0)$ and $(6, 0)$

   e. Opens up and has line of symmetry with equation $x = 3$

27 In the diagram below, $\overleftrightarrow{AD} \parallel \overleftrightarrow{EG}$. Find $m\angle CBD$.

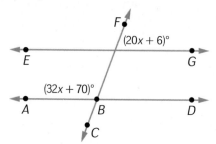

28 Rewrite each expression in equivalent simpler form with the smallest possible number under the radical sign.

   a. $\sqrt{24}$ **b.** $\sqrt{48}$

   c. $\sqrt{450}$ **d.** $\sqrt{\dfrac{8}{9}}$

   e. $\dfrac{\sqrt{12}}{2}$ **f.** $\dfrac{\sqrt{60 + 3(5)}}{6}$

   g. $\dfrac{\sqrt[3]{48}}{2}$ **h.** $\sqrt[4]{32}$

29 Write rules for linear functions whose graphs meet these conditions.

   a. Slope 0.5 and y-intercept $(0, 4)$

   b. Slope 1.6 and passing through $(2, 5)$

   c. Passing through $(-2, 4)$ and $(6, 0)$

   d. Parallel to the graph of $y = 2x + 3$ and passing through $(0, -3)$

30 Solve each of these equations by examining its algebraic form and then reasoning with the symbols themselves.

   a. $x^2 + 7 = 23$ **b.** $3x^2 - 12 = 135$

   c. $x^2 + 9 = 0$ **d.** $(x + 3)^2 = 19$

   e. $4x^3 - 49x = 0$ **f.** $x^4 - 11x^2 + 18 = 0$

31 Expand each of these expressions to an equivalent standard form quadratic expression.

   a. $(x + 5)(4x - 3)$ **b.** $5(t + 8)(4 - t)$

   c. $(x + 3)(x - 3)$ **d.** $2(s - 7)^2$

   e. $(3s + 2)^2$ **f.** $\left(y - \dfrac{5}{2}\right)^2$

32 Prove that a quadrilateral with four right angles is a rectangle.

Quadratic Polynomials

In the family of polynomial functions, the most useful are the simplest. In earlier work, you have used linear and quadratic polynomial functions to model and reason about a variety of business and scientific problems.

For example, in Lesson 1 of this unit, you analyzed business prospects for a concert promotion in which income was a quadratic function and operating expense was a linear function. Both depended on the price of tickets for the concert. The relationship of the two functions is shown below in the graph.

For concert promoters, one of the critical planning questions is estimating the break-even point(s) at which income from ticket sales and operating expenses are equal. In this case, that involves solving the equation $-25x^2 + 500x + 7,500 = 7,000 - 200x$.

Concert Income and Operating Costs

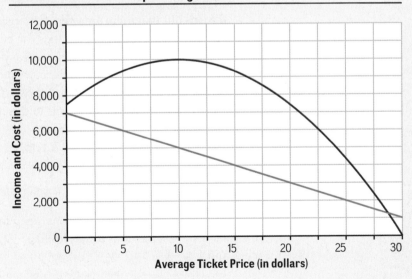

The investigations of this lesson will add to your toolkit of strategies for analyzing quadratic polynomials—writing quadratic expressions in new kinds of equivalent forms and using those expressions to solve equations and find max/min points on graphs of the corresponding functions.

INVESTIGATION **1**

Completing the Square

When problems require solving quadratic equations, the simplest cases are those like $ax^2 + b = c$ that involve no linear term. You can solve such equations easily by reasoning like this.

$$\text{If } ax^2 + b = c, \text{ then } x^2 = \frac{c - b}{a}.$$

$$\text{So, } x = \pm\sqrt{\frac{c - b}{a}}.$$

It turns out that there are ways to transform every quadratic polynomial into an equivalent expression that has the general form $a(x - h)^2 + k$. This quadratic form, called the **vertex form**, gives easy-to-read information about the shape and location of the graph of the corresponding quadratic function. It also makes solution of quadratic equations easy.

As you work on the problems in this investigation, look for answers to these questions:

How does the vertex form of a quadratic function reveal the shape and location of its graph?

How can quadratic polynomials be expressed in vertex form?

As you work on these questions, keep in mind the fact that the general form $a(x - h)^2 + k$ includes examples like $-3(x + 6)^2 - 4$ where $a = -3, h = -6,$ and $k = -4$.

Vertex Form of Quadratics To make use of quadratic polynomials in vertex form, you need to know how the parameters in those expressions are related to the shape and location of the corresponding function graphs.

1 First consider the question of how to find x- and y-intercepts on the graphs of quadratic functions with rules expressed in vertex form.

$$f(x) = a(x - h)^2 + k$$

To develop an answer to this question, you might analyze several specific examples like I–IV below and then make a conjecture. Then use algebraic reasoning to justify your conjecture. Or you might use more general algebraic reasoning to develop a conjecture and then test your ideas with the specific examples.

I. $g(x) = (x - 3)^2 - 16$ **II.** $j(x) = 3(x - 7)^2 - 12$

III. $m(x) = (x - 1)^2 + 25$ **IV.** $n(x) = -2(x + 6)^2 + 20$

a. What formula shows how to use values of a, h, and k to find coordinates for the x-intercepts on the graph of any function with rule in the form $f(x) = a(x - h)^2 + k$?

b. What formula shows how to use values of a, h, and k to find coordinates for the y-intercept on the graph of any function with rule in the form $f(x) = a(x - h)^2 + k$?

2 Consider next the question of how to locate maximum or minimum points on the graph of a quadratic function $f(x) = a(x - h)^2 + k$. Recall that this point is called the **vertex** of the graph. To develop an answer to this question, you might analyze several specific examples like I–IV below to make a conjecture. Then use algebraic reasoning to justify your conjecture. Or you might use more general algebraic reasoning to develop a conjecture and then test your ideas with the specific examples.

I. $r(x) = (x - 3)^2 + 5$ **II.** $s(x) = 3(x - 7)^2 - 4$

III. $t(x) = -2(x + 6)^2 + 3$ **IV.** $v(x) = -0.5(x + 5)^2 - 7$

a. How can you tell from the parameters a, h, and k when the graph of the function $f(x) = a(x - h)^2 + k$ will have a minimum point? When it will have a maximum point?

b. Consider the cases when the graph of the function has a minimum point.

     i. For what value of x does the expression $a(x - h)^2 + k$ take on its smallest value?

     ii. What is the y-coordinate of the corresponding point on the graph?

     iii. What are the coordinates of the minimum point when $a > 0$?

c. Now consider the cases when the function has a maximum point.

     i. For what value of x does the expression $a(x - h)^2 + k$ take on its largest possible value?

     ii. What is the y-coordinate of the corresponding point on the graph?

     iii. What are the coordinates of the maximum point when $a < 0$?

3 Your work on Problems 1 and 2 revealed a new way of thinking about max/min points and intercepts on graphs of quadratic functions. For example, the function $q(x) = x^2 - 10x + 24$ can be expressed as $q(x) = (x - 4)(x - 6)$ and also as $q(x) = (x - 5)^2 - 1$.

   a. How would you use each of the three forms to find coordinates of the y-intercept on the graph of $q(x)$? Which approach seems most helpful?

   b. How would you use each form to find coordinates of the x-intercepts on the graph of $q(x)$? Which approach seems most helpful?

   c. How would you use each form to find coordinates of the maximum or minimum point on the graph of $q(x)$? Which approach seems most helpful?

Completing the Square You know from earlier work with quadratic expressions that there are useful patterns relating factored and expanded forms of those expressions. For any number n,

$$(x + n)^2 = x^2 + 2nx + n^2 \text{ and } (x - n)^2 = x^2 - 2nx + n^2.$$

The relationship between factored and expanded forms of such *perfect square* expressions is the key to finding vertex forms for quadratic function rules. Work on Problems 4–7 will give you some valuable insight into the process of finding vertex form expressions for quadratics.

4 Use the pattern relating expanded and factored forms of perfect square polynomials (where possible) to write factored forms of the expressions in Parts a–f. Use a CAS to check that your factored forms are equivalent to the corresponding expanded forms.

   a. $x^2 + 6x + 9$ **b.** $x^2 + 10x + 25$

   c. $x^2 + 8x + 12$ **d.** $x^2 - 10x + 25$

   e. $x^2 + 5x + \dfrac{25}{4}$ **f.** $x^2 - 3x + \dfrac{9}{4}$

5 The challenge in transforming a given quadratic expression to an equivalent vertex form can be represented as a geometry problem. Study the following diagram.

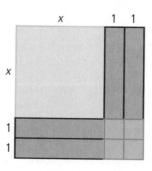

Find values of n, m, and k that make it possible to express the total area of the figure in two ways.

   a. as $(x + n)^2$

   b. as $x^2 + kx + m$

6 Now consider the expression $x^2 + 6x + 7$. The parts of this expression can be represented with an "x by x" square, six "x by 1" rectangles, and seven unit squares.

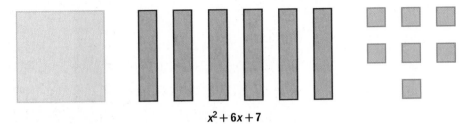

$$x^2 + 6x + 7$$

a. Why is it impossible to arrange the given squares and rectangles without overlapping to form one large square?

b. How could you produce the requested square if you were given some extra pieces or allowed to take away some pieces? What addition or subtraction would do the job most efficiently?

c. Explain how your answer to Part b illustrates the fact that
$x^2 + 6x + 7 = (x + 3)^2 - 2$.

If a quadratic expression is not a perfect square (factorable to the form $(x \pm h)^2$), it is still possible to write an equivalent expression in the form $(x \pm h)^2 \pm k$. Your work in Problem 6 gives a picture that suggests the kind of algebraic operations that are required. Problem 7 asks you to adapt that reasoning to develop a symbol manipulation procedure.

7 Study the following example of the technique called **completing the square**. Explain the choice of "9" and why it was added and subtracted in different parts of the expression.

$$
\begin{aligned}
x^2 + 6x + 11 &= (x^2 + 6x + \underline{}) + 11 \\
&= (x^2 + 6x + 9) - 9 + 11 \\
&= (x + 3)^2 + 2
\end{aligned}
$$

Use similar reasoning to write the expressions in Parts a–f in vertex form.

a. $x^2 + 6x + 5$

b. $x^2 - 6x + 10$

c. $x^2 + 8x + 19$

d. $x^2 - 8x + 12$

e. $x^2 + 10x + 27$

f. $x^2 - 3x + 1$

8 How do the results of your work on Problems 6 and 7 suggest a strategy for writing any quadratic expression $x^2 + bx + c$ in equivalent vertex form $(x \pm h)^2 \pm k$?

9 You can visualize the process of completing the square in a somewhat different way, if you focus attention on the first two terms of the given expression, $x^2 + bx$. Explain how the following proof without words shows that $x^2 + bx = \left(x + \frac{b}{2}\right)^2 - \left(\frac{b}{2}\right)^2$.

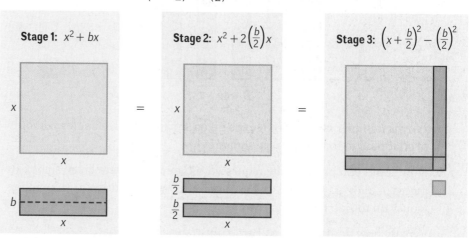

Stage 1: $x^2 + bx$

Stage 2: $x^2 + 2\left(\frac{b}{2}\right)x$

Stage 3: $\left(x + \frac{b}{2}\right)^2 - \left(\frac{b}{2}\right)^2$

SUMMARIZE THE MATHEMATICS

In this investigation, you developed skill in transforming quadratic expressions in standard form to vertex form and in using that form to analyze the graphs of corresponding quadratic functions.

a How can the vertex form of a quadratic expression like $(x - h)^2 + k$ be used to locate the max/min point and intercepts on the graph of the corresponding quadratic function?

b What are key steps in transforming $x^2 + bx + c$ to vertex form?

Be prepared to explain your ideas to the class.

 CHECK YOUR UNDERSTANDING

Use your new completing-the-square skills to complete these tasks.

a. Write the expression $x^2 - 2x - 5$ in equivalent vertex form.

b. Use the result of Part a to:

    i. solve $x^2 - 2x - 5 = 0$.

    ii. find coordinates of the minimum point on the graph of $f(x) = x^2 - 2x - 5$.

    iii. find coordinates of x- and y-intercepts on the graph of $f(x) = x^2 - 2x - 5$.

The Quadratic Formula and Complex Numbers

As you have seen in your previous studies, when problems require solving quadratic equations that cannot be factored easily, you can always turn to the *quadratic formula*. For instance, solutions for $3x^2 + 7x - 12 = 0$ are given by

$$x = \frac{-7}{2(3)} + \frac{\sqrt{7^2 - 4(3)(-12)}}{2(3)} \quad \text{and} \quad x = \frac{-7}{2(3)} - \frac{\sqrt{7^2 - 4(3)(-12)}}{2(3)}.$$

As you complete the problems in this investigation, look for answers to these questions:

How can the technique of completing the square be used to derive the quadratic formula?

How does use of the quadratic formula suggest the need for new kinds of numbers?

Proving and Using the Quadratic Formula The work that you have done to write quadratic expressions in vertex form is closely related to the quadratic formula that can be used to find solutions of any quadratic equation.

1 Consider the general form of a quadratic equation, $ax^2 + bx + c = 0$.

The solutions of this equation are given by $x = \frac{-b}{2a} + \frac{\sqrt{b^2 - 4ac}}{2a}$ and $x = \frac{-b}{2a} - \frac{\sqrt{b^2 - 4ac}}{2a}$. Explain how each step in the following derivation of the quadratic formula is justified by properties of numbers and operations.

Start: If $ax^2 + bx + c = 0$ (and $a \neq 0$),

Step 1. Then $a\left(x^2 + \frac{b}{a}x + \frac{c}{a}\right) = 0$.

Step 2. $x^2 + \frac{b}{a}x + \frac{c}{a} = 0$

Step 3. $x^2 + \frac{b}{a}x = \frac{-c}{a}$

Step 4. $x^2 + \frac{b}{a}x + \frac{b^2}{4a^2} = \frac{-c}{a} + \frac{b^2}{4a^2}$

Step 5. $x^2 + \frac{b}{a}x + \frac{b^2}{4a^2} = \frac{b^2}{4a^2} + \frac{-c}{a}$

Step 6. $\left(x + \frac{b}{2a}\right)^2 = \frac{b^2}{4a^2} + \frac{-c}{a}$

Step 7. $\left(x + \frac{b}{2a}\right)^2 = \frac{b^2}{4a^2} + \frac{-4ac}{4a^2}$

Step 8. $\left(x + \frac{b}{2a}\right)^2 = \frac{b^2 - 4ac}{4a^2}$

Step 9. $x + \frac{b}{2a} = \frac{\sqrt{b^2 - 4ac}}{2a}$ or $x + \frac{b}{2a} = -\frac{\sqrt{b^2 - 4ac}}{2a}$

Step 10. So, $x = \frac{-b}{2a} + \frac{\sqrt{b^2 - 4ac}}{2a}$ or $x = \frac{-b}{2a} - \frac{\sqrt{b^2 - 4ac}}{2a}$.

2 The quadratic formula provides a tool for solving any quadratic equation by algebraic reasoning. But you have other helpful strategies available through use of technology.

a. How could you use calculator- or computer-generated tables of function values or graphs to estimate solutions for a quadratic equation like $3x^2 + 7x - 12 = 0$?

b. How could you use a computer algebra system (CAS) to solve $3x^2 + 7x - 12 = 0$?

c. Use the CAS available to you to solve $ax^2 + bx + c = 0$ for x and compare the CAS result to the quadratic formula derived in Problem 1.

3 Use the quadratic formula to solve each of the following equations. Report your answers in exact form, using radicals where necessary rather than decimal approximations. Check each answer by substituting the solution values for x back into the original equation.

a. $3x^2 + 9x + 6 = 0$

b. $3x^2 - 6 = -3x$

c. $2x^2 + x - 10 = 0$

d. $2x^2 + 5x - 1 = 0$

e. $x^2 + 2x + 1 = 0$

f. $x^2 - 6x + 13 = 0$

4 The solutions for quadratic equations in Problem 3 included several kinds of numbers. Some could be expressed as integers. But others could only be expressed as fractions or as irrational numbers involving radicals. One of the quadratic equations appears to have no solutions.

a. At what point in use of the quadratic formula do you learn whether the equation has two distinct solutions, only one solution, or no real number solutions?

b. If the coefficients of a quadratic equation are integers or rational numbers, at what point in use of the quadratic formula do you learn whether the solution(s) will be integers, rational numbers, or irrational numbers?

Complex Numbers In work on Problems 2 and 3, you discovered that some quadratic equations do not have real number solutions. For example, when you try to solve $x^2 - 6x + 13 = 0$, the quadratic formula gives

$$x = 3 + \frac{\sqrt{-16}}{2} \quad \text{and} \quad x = 3 - \frac{\sqrt{-16}}{2}.$$

If you ask a CAS to **solve(x^2—6*x+13=0,x)**, it is likely to return the disappointing result "false."

5 Sketch a graph of the function $f(x) = x^2 - 6x + 13$. Explain how it shows that there are no real number solutions for the equation $x^2 - 6x + 13 = 0$.

The obstacle to solving $x^2 - 6x + 13 = 0$ appears with the radical $\sqrt{-16}$. Your prior experience with square roots tells you that *no* real number has -16 as its square. For thousands of years, mathematicians seemed to accept as a fact that equations like $x^2 - 6x + 13 = 0$ simply have no solutions.

6 Explain why a quadratic equation in the form $ax^2 + bx + c = 0$ has no real number solutions when the value of $b^2 - 4ac$, called the **discriminant** of the quadratic, is a negative number.

In the middle of the 16th century, Italian scholar Girolamo Cardano suggested all that was needed was a new kind of number. Cardano's idea was explored by mathematicians for several more centuries (often with strong doubts about using numbers with negative squares) until *complex numbers* became an accepted and well-understood tool for both pure and applied mathematics.

The obstacle to solving $x^2 - 6x + 13 = 0$ was removed by reasoning like this.

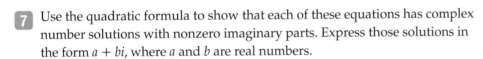

It makes sense that $\sqrt{-16}$ should equal $\sqrt{16(-1)}$.

But $\sqrt{16(-1)}$ should equal $\sqrt{16}\,\sqrt{-1}$.

So, $3 + \dfrac{\sqrt{-16}}{2}$ should equal $3 + \dfrac{4\sqrt{-1}}{2}$.

Or, the solutions for $x^2 - 6x + 13 = 0$ should be $x = 3 + 2\sqrt{-1}$ and $x = 3 - 2\sqrt{-1}$.

This kind of observation led mathematicians to develop what is now called the **complex number system** whose elements are in the form $\boldsymbol{a + bi}$, where a and b are real numbers and $i = \sqrt{-1}$. So, $3 + 2\sqrt{-1}$ is written as $3 + 2i$.

The new number system contains all real numbers (the complex numbers for which $b = 0$) and all **imaginary numbers** (the complex numbers for which $a = 0$). It also provides solutions to the problematic quadratic equations *and all other polynomial equations in the form $p(x) = 0$.* You will learn more about complex numbers in future studies.

7 Use the quadratic formula to show that each of these equations has complex number solutions with nonzero imaginary parts. Express those solutions in the form $a + bi$, where a and b are real numbers.

a. $x^2 - 10x + 29 = 0$

b. $4x^2 + 16 = 0$

c. $x^2 - 4x + 13 = 0$

d. $3x^2 - 18x + 30 = 0$

Girolamo Cardano

CHECK YOUR UNDERSTANDING

Use the quadratic formula or other reasoning to solve each of the following equations. If the solutions are real numbers, identify them as integer, noninteger rational, or irrational numbers. Write nonreal complex number solutions in *standard form a + bi.*

a. $x^2 - 6x - 7 = 0$

b. $5x^2 - 6x + 2 = 0$

c. $6x^2 - 11x - 10 = 0$

d. $5(x - 3)^2 + 6 = 11$

APPLICATIONS

1 Find coordinates of the max/min points, *x*-intercepts, and *y*-intercepts on graphs of these functions.

a. $f(x) = (x + 2)^2 - 9$

b. $g(x) = -(x - 2)^2 + 3$

c. $h(x) = (x - 5)^2 - 2$

d. $j(n) = (n + 7)^2 + 4$

2 In a *Punkin' Chunkin'* contest, the height (in feet) of shots from one pumpkin cannon is given by the function $h(t) = -16(t - 5)^2 + 425$. The height is in feet above the ground and the time is in seconds after the pumpkin leaves the cannon. Show how to use this function to answer questions about the height of the flying pumpkin.

a. At what height is the pumpkin released from the "chunker"?

b. At what time will the pumpkin hit the ground?

c. At what time does the pumpkin reach its maximum height and what is that height?

3 Consider the collection of squares and rectangles shown here.

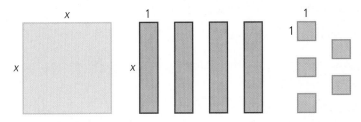

The four rectangles are congruent to each other, as are the five small squares.

a. What expression represents the total area of the ten figures?

b. Explain why the given shapes cannot be arranged to form a larger square without overlaps or gaps.

c. What is the minimal addition or subtraction of unit squares that will make it possible to arrange the new set of pieces to form a larger square?

d. Write an expression involving a binomial square to represent the area of the larger square formed by all the given pieces and the unit squares added or subtracted in Part c.

4 For each function, write the rule in vertex form and find coordinates of the max/min point on its graph.

a. $f(x) = x^2 + 12x + 11$ **b.** $g(x) = x^2 - 4x + 7$

c. $h(x) = x^2 - 18x + 74$ **d.** $j(s) = s^2 + 2s + 5$

5 Use the results of Task 4 to solve each of these equations.

a. $x^2 + 12x + 11 = 0$ **b.** $x^2 - 4x + 7 = 19$

c. $x^2 - 18x + 74 = 13$ **d.** $s^2 + 2s + 5 = 53$

6 Use the quadratic formula to solve each of these equations. If the solutions are real numbers, identify them as integer, noninteger rational, or irrational numbers. Write nonreal complex number solutions in standard form $a + bi$.

a. $2x^2 + 3x - 5 = 0$ **b.** $2x^2 + x - 3 = 0$

c. $3x^2 + x = 10$ **d.** $5x + x^2 + 10 = 0$

e. $x^2 + 9x - 10 = -24$ **f.** $3x^2 + 10 = 25$

7 Recall what you learned in Lesson 1 about factors of polynomials and zeroes of the corresponding polynomial functions. Use that knowledge to write quadratic equations in the form $ax^2 + bx + c = 0$ with two distinct solutions that:

a. are both integers.

b. include at least one fraction.

c. are both irrational numbers.

8 Use your knowledge of the quadratic formula to write quadratic equations with the indicated solutions.

a. exactly one real number solution

b. solutions that are complex numbers in the form $a + bi$, $a \neq 0$, $b \neq 0$

c. solutions that are imaginary numbers bi

9 The functions $I(x) = -25x^2 + 500x + 7{,}500$ and $E(x) = 7{,}000 - 200x$ give income and operating expenses as functions of average ticket price for a music concert. Use the quadratic formula to solve the equation $-25x^2 + 500x + 7{,}500 = 7{,}000 - 200x$ and find the ticket price(s) that will allow the promoters to break even on the event. That is, find the average ticket price that will make income equal to operating expenses.

CONNECTIONS

10 There is a useful connection between the completing the square technique and coordinate equations for circles.

a. What are the center and radius for the circle with coordinate equation $(x - 5)^2 + (y - 7)^2 = 9$?

b. What are the center and radius for the circle with coordinate equation $(x + 5)^2 + (y + 7)^2 = 4$?

c. Write the equation in Part a without parentheses in the form $x^2 + y^2 + ax + by + c = 0$.

d. Write the equation in Part b without parentheses in the form $x^2 + y^2 + ax + by + c = 0$.

e. Use what you know about completing the square to write each of the following equations in a form that gives the center and radius of the circle. State the center and radius.

  i. $x^2 + y^2 - 4x - 8y + 19 = 0$

  ii. $x^2 + y^2 + 10x + 2y - 10 = 0$

  iii. $x^2 + y^2 + 4y - 45 = 0$

  iv. $x^2 + y^2 - 8x + 6y = 0$

11 Sketch graphs of these quadratic functions. Explain how the collection of graphs show that quadratic equations can have: (1) two real number solutions, (2) one repeated real number solution, or (3) no real number solutions.

a. $f(x) = x^2 - 4x - 5$

b. $g(x) = x^2 - 4x + 4$

c. $h(x) = x^2 - 4x + 6$

12 Solve each of these equations. Then compare the forms of the given equations and the procedures used to solve them.

a. $3m + 12 = 27$

b. $\begin{bmatrix} 3 & 1 \\ 5 & 2 \end{bmatrix} \begin{bmatrix} x \\ y \end{bmatrix} + \begin{bmatrix} 2 \\ 5 \end{bmatrix} = \begin{bmatrix} 4 \\ 7 \end{bmatrix}$

c. $3n^2 + 12 = 27$

13 The complex number system was constructed in stages that began with the set **W** of whole numbers {0, 1, 2, 3, 4, …} and gradually introduced other important sets of numbers. Make a Venn diagram that illustrates the relationship among the following sets of numbers: whole numbers, integers, rational numbers, irrational numbers, real numbers, imaginary numbers, and complex numbers.

14 Each complex number can be represented in the form $a + bi$, where a and b are real numbers and $i = \sqrt{-1}$ or $i^2 = -1$. At the turn of the 19th century, three different mathematicians, Caspar Wessel (1797), Jean Robert Argand (1806), and Carl Friedrich Gauss (1811), suggested that the complex numbers could be represented by points on a coordinate plane. The complex number $a + bi$ would correspond to the point with coordinates (a, b).

   a. On a single coordinate diagram, locate and label points corresponding to these complex numbers.

$$4 + 2i \qquad -2 + 3i \qquad -4 - 3i \qquad 2 - 3i$$

   b. Use of commutative and associative properties of addition and combining "like terms" suggests a rule for addition of complex numbers, $(a + bi) + (c + di) = (a + c) + (b + d)i$. Use this rule to add the complex number $3 + 2i$ to each of the numbers in Part a.

   c. Using a second color, plot and label the points corresponding to your results in Part b. Then identify the geometric transformation that is accomplished by adding $3 + 2i$ to every complex number $a + bi$.

REFLECTIONS

15 When one student was asked to solve the quadratic equation $2x^2 + 12x = 5$, he wrote the factored form $2x(x + 6) = 5$ and concluded that the solutions are $x = 0$ and $x = -6$. What is the probable error in his thinking that led him to those incorrect answers?

16 You now know three different ways to express quadratic polynomials in equivalent forms—standard form, factored form, and vertex form.

   a. What do you see as the advantages of each form?

   b. Do you think the value added by use of each different form is equal to the effort involved in transforming the original expression to a new form?

17 The creative team of *Conklin Brothers Circus* designed a cannon to shoot a human cannonball across the arenas where they operate. They judged that the height of the "cannonball" would be a quadratic function of time and wrote three different rules for the function (time in seconds and height in feet).

$$h(t) = -16t^2 + 32t + 48$$
$$h(t) = -16(t + 1)(t - 3)$$
$$h(t) = -16(t - 1)^2 + 64$$

   a. In what different ways could you convince yourself or others that the three different rules are mathematically equivalent?

b. Which of the rules would be most helpful in answering each of these questions?

    i. What is the maximum height of the "cannonball" and when will it occur in the flight?

    ii. When would the "cannonball" hit the ground if the net collapsed while the flight was underway?

    iii. At what height does the "cannonball" leave the end of the cannon?

18 How do you decide on an approach to solving a quadratic equation? What conditions influence your strategy choice in various situations?

19 One solution of a quadratic equation $ax^2 + bx + c = 0$ is $2 + 3i$. What is the other solution? Explain your reasoning.

EXTENSIONS

20 Of the two algebraic operations, expanding a product of linear factors is generally much easier than factoring a standard form quadratic. To understand factoring, you have to develop some skill in expanding products and in reversing that process.

    Use algebraic reasoning to write each of these products in standard expanded form. Then check your results by using a CAS for the same work.

a. $(x + 4)(3x + 1)$ **b.** $(2x + 4)(x + 3)$

c. $(2x - 4)(5x + 3)$ **d.** $(x - 7)^2$

e. $(3x + 4)^2$ **f.** $(2x - 5)^2$

21 Use your experience in Task 20 to solve these quadratic equations by first writing the quadratic expression in equivalent form as a product of two linear factors.

a. $2x^2 + 7x + 3 = 0$ **b.** $5x^2 + 16x + 3 = 0$

c. $5x^2 + 17x = -6$ **d.** $6x^2 + 19x + 15 = 0$

e. $4x^2 - 12x + 9 = 0$ **f.** $9x^2 + 6x + 1 = 0$

22 Solve each of the following equations using factoring, where possible. Use a CAS to check your work.

a. $4x^2 + 12x + 9 = 0$ **b.** $9x^2 + 30x + 25 = 0$

c. $7x^2 - 18x + 11 = 0$ **d.** $9x^2 - 12x = -4$

e. $4x^2 + 10x + \frac{25}{4} = 0$ **f.** $4x^2 - 6x + \frac{9}{4} = 0$

23 When $a \neq 1$, quadratic functions in the form $f(x) = ax^2 + bx + c$ can still be transformed to equivalent vertex forms like $f(x) = m(x \pm n)^2 \pm d$.

a. Study the following example. Explain the choice of adding and subtracting 18 in two parts of the quadratic polynomial.

$$
\begin{aligned}
f(x) &= 2x^2 + 12x + 14 \\
&= 2(x^2 + 6x + \underline{\quad}) + 14 \\
&= 2(x^2 + 6x + 9) - 18 + 14 \\
&= 2(x + 3)^2 - 4
\end{aligned}
$$

b. Use similar reasoning to write each function rule below in vertex form.

   i. $f(x) = 3x^2 + 12x + 15$

   ii. $g(x) = 5x^2 + 30x + 50$

   iii. $h(x) = 2x^2 - 8x + 10$

   iv. $j(x) = 4x^2 - 20x + 7$

c. How do the results of your work in Part b suggest a strategy for writing any quadratic function $f(x) = ax^2 + bx + c$ in equivalent vertex form $f(x) = m(x \pm n)^2 \pm d$?

24 Use the Factor Theorem (Connections Task 14 in Lesson 1) to help explain why a cubic polynomial function has either zero or two nonreal complex number solutions.

25 Consider the ways that it would make sense to multiply complex numbers like $(3 + 2i)$ and $(4 + 7i)$.

a. If you think of the given complex numbers as linear expressions like $(3 + 2x)$ and $(4 + 7x)$, what would the product look like when expanded?

b. Because the letter i is the number for which $i^2 = -1$, the expression in Part a can be simplified. Write the product $(3 + 2i)(4 + 7i)$ in standard complex number form.

REVIEW

26 Express each of these sums and differences as single fractions in simplest form.

a. $\dfrac{1}{3} + \dfrac{1}{6}$ b. $\dfrac{1}{2} - \dfrac{1}{8}$

c. $\dfrac{1}{4} + \dfrac{1}{3}$ d. $\dfrac{2}{3} - \dfrac{1}{2}$

e. $\dfrac{a}{b} + \dfrac{c}{d}$

27 Consider the following numbers.

$$\sqrt{100} \qquad \frac{12}{5} \qquad -5 \qquad 2.2 \qquad \sqrt{5}$$

$$\pi \qquad 3.\overline{1} \qquad \frac{1}{6} \qquad \sqrt{2.25} \qquad \sqrt[3]{27}$$

a. Which of the listed numbers are integers?

b. Which of the listed numbers are rational numbers?

c. Which of the listed numbers are irrational numbers?

d. Place the numbers in order from smallest to largest.

28 Find equivalent expanded forms for these expressions. Then use a CAS to check your reasoning. (*Remember*: The algebraic product *mx* must generally be entered as **m*x** in a CAS.)

a. $(x + n)^2$

b. $(mx + n)^2$

c. $(x + n)(x - n)$

d. $(mx + n)(mx - n)$

29 In the diagram at the right, $\overleftrightarrow{AB} \parallel \overleftrightarrow{DE}$.

a. Prove that $\triangle ABC \sim \triangle DEC$.

b. Find the lengths of $\overline{AC}$ and $\overline{DE}$.

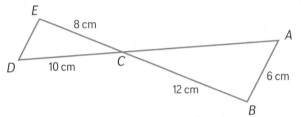

30 Without using a graphing calculator or a computer graphing tool, sketch a graph of each power function. Then check your ideas using technology.

$$f(x) = \frac{3}{x} \qquad g(x) = \frac{3}{x^2}$$

a. What is the domain of $f(x)$? Of $g(x)$?

b. What is the range of $f(x)$? Of $g(x)$?

c. Describe the end behavior of $f(x)$.

d. Describe the end behavior of $g(x)$.

31 Simplify each algebraic expression.

a. $\frac{6x - 10}{2}$ **b.** $\frac{24 - 3x}{3}$ **c.** $\frac{x - 4}{x - 4}$

32 Consider the diagram below. Point P is on the terminal side of an angle θ in standard position. Find $\sin \theta$, $\cos \theta$, and $\tan \theta$.

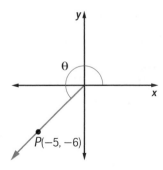

Rational Expressions and Functions

How valuable is each customer to the bottom line of a business? In Lesson 1 of this unit, you discovered how income, expenses, and profits for a concert were functions of concert ticket price. It is possible to combine those functions to see how the profit from *each* ticket is related to the ticket price, giving a measure of each customer's value to the business.

The profit-per-ticket idea might seem simple, but the resulting function has some surprising properties. For example, if you focus only on income and expenses related to the concert itself (ignoring snack bar operations), the profit-per-ticket function has a graph that looks like the one below.

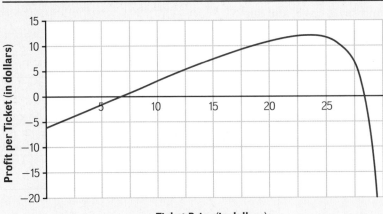

Profit per Ticket from Ticket Sales

Examine the profit-per-ticket graph on the previous page.

a How would you describe the pattern of change in profit per ticket as ticket prices approach $30?

b When the ticket price is set at $15, the business plan predicts that 375 tickets will be sold. Total profit from ticket sales is predicted to be $2,750. How would you calculate the profit per ticket sold when tickets are $15? How is your answer shown in the graph?

c How would you find a rule for calculating profit per ticket at any ticket price?

The strange properties of the profit-per-ticket function result from the fact that it is a *rational function*, the quotient of two polynomials. In this lesson, you will explore the family of rational functions and learn how to use them as mathematical models of problem situations. You will learn the connections between rules and graphs for rational functions and how to simplify and combine the expressions in those rules.

INVESTIGATION 1

Domains and Graphs of Rational Functions

Information from analysis of the concert business led to the prediction that ticket sales and profit from concert operation alone (not including snack bar operations) would be related to average ticket price x (in dollars) by these rules.

Number of tickets sold: $n(x) = 750 - 25x$

Profit from ticket sales: $P_c(x) = -25x^2 + 875x - 4,750$

As you complete the following problems, look for answers to these questions:

How can polynomials like these be combined
to give useful rational functions?

What are the important features of rational
functions and their graphs?

1 Suppose that the concert promoters want to estimate the profit per ticket for various possible ticket prices.

a. What number of tickets, total profit, and profit per ticket would be expected:

   i. if the average ticket price x is $10?

   ii. if the average ticket price x is $20?

b. Write an algebraic rule for the function $R(x)$ giving profit per ticket when the ticket price is x dollars. Express your answer as an algebraic fraction.

A function with rule that is a fraction in which both numerator and denominator are polynomials is called a **rational function**. That is, a function $f(x)$ is a rational function if and only if its rule can be written in the form $f(x) = \dfrac{p(x)}{q(x)}$, where $p(x)$ and $q(x)$ are polynomial functions and $q(x) \neq 0$.

The profit-per-ticket function $R(x)$ that you developed in Problem 1 is a rational function. Like many other rational functions, its graph has several interesting and important features. To understand those properties, it helps to begin with analysis of the functions that are the numerator and the denominator of $R(x)$.

2 Sketch graphs of $n(x) = 750 - 25x$ and $P_c(x) = -25x^2 + 875x - 4{,}750$. Use the graphs to answer the following questions.

a. What is the practical domain of $n(x)$? That is, what prices x will give predicted numbers of tickets that make sense in the concert situation?

b. Estimate the ticket price for which profit from concert operation $P_c(x)$, excluding snack bar operations, is maximized. Find that profit.

3 Now turn to analysis of the profit-per-ticket function $R(x)$.

a. Graph $R(x)$ for x between 0 and 30.

b. Estimate the ticket price for which profit per ticket is maximized.

c. Why is the profit per ticket not maximized at the same ticket price that maximizes concert profit?

4 Something unusual happens in the graph of $R(x)$ near $x = 30$. Use a calculator or computer software to examine values of $R(x)$ when x is very close to 30, say, between 29 and 31 with increments of 0.1. If using a calculator, use **Dot** rather than **Connected** mode.

a. How does your technology tool respond to requests for $R(30)$? What does the response mean? Why does it make sense in this situation?

b. Describe the pattern of change in values of $R(x)$ as x approaches very close to 30 from below. Explain the trend by referring to the values of $P_c(x)$ and $n(x)$ that are used to calculate $R(x)$.

c. Describe the pattern of change in values for $R(x)$ as x approaches very close to 30 from above. Explain the trend by referring to the values of $P_c(x)$ and $n(x)$ for such x.

d. What is the theoretical domain of $R(x)$?

Domain and Asymptotes The unusual pattern of values for the profit-per-ticket function at and on either side of $x = 30$ can be explained by analyzing the domain of $R(x)$ and values of the numerator and denominator functions used to define $R(x)$. That is the sort of analysis required to understand any rational function and its graph.

5 Study this graph of the rational function $f(x) = \dfrac{x - 1}{x^2 - 2x - 3}$.

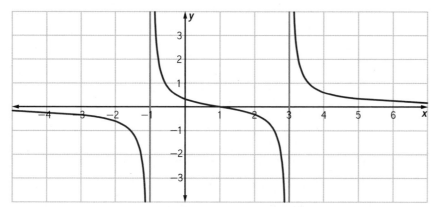

The graph has the lines $x = -1$ and $x = 3$ as **vertical asymptotes**. As values of x approach -1 from below, the graph suggests that values of $f(x)$ decrease without lower bound. As values of x approach -1 from above, the graph suggests that values of $f(x)$ increase without upper bound.

Describe the pattern of change in function values as x approaches 3.

6 Inspect values of $f(x) = \dfrac{x - 1}{x^2 - 2x - 3}$ in terms of values of the numerator $(x - 1)$ and values of the denominator $(x^2 - 2x - 3)$ to help answer these questions about the function and its graph.

a. How could you have located the x-intercept of the graph by examining the rational expression that defines $f(x)$?

b. How could you have located the two vertical asymptotes of the graph by studying the rational expression that defines $f(x)$?

c. What is the domain of $f(x)$—for what values of x can a value of $f(x)$ be calculated and for what values is it impossible to calculate $f(x)$? Explain how your answer to that question is related to your answer to the question in Part b.

d. The graph of $f(x)$ has the x-axis as a **horizontal asymptote**. As values of x decrease without lower bound and increase without upper bound, the graph of $f(x)$ gets closer and closer to the x-axis but never reaches it. How could you have located that horizontal asymptote of the graph by studying the rational expression that defines $f(x)$?

7 For each of the following rational functions, use algebraic reasoning to:

- find coordinates of all x-intercepts and the y-intercept of its graph.
- find equations of all vertical and horizontal asymptotes.
- describe the domain of the function.
- sketch a graph on which you label intercepts and asymptotes.

Then check by examining a graph of the function, and correct your responses if necessary. Be prepared to explain how to avoid any errors you made.

a. $g(x) = \dfrac{3}{x - 2}$
b. $h(x) = \dfrac{2x + 3}{x^2 - 9}$

c. $j(x) = \dfrac{2x + 1}{3x}$
d. $k(x) = \dfrac{x^2 + 1}{x + 1}$

e. $m(x) = \dfrac{6x + 1}{3x + 5}$
f. $s(x) = \dfrac{2x^2 + 2x + 3}{x + 1}$

8 What strategies for analyzing features of a rational function and its graph are suggested by your work on Problems 1–7?

Concert Profit Prospects Reprise One of the important statistics used to describe the profitability of any business venture is the ratio of profit to income. For example, if a business earns profit of $8 million on income from sales of $100 million, it would report an 8% ratio of profit to income because $8 \div 100 = 8\%$.

Now recall the profit and income functions derived in analyzing the concert business. Profit from ticket sales for the concert (excluding snack bar operations) is a function of ticket price x with rule $P_c(x) = -25x^2 + 875x - 4{,}750$. Income from concert ticket sales is also a function of ticket price with rule $t(x) = -25x^2 + 750x$.

9 Consider the function $S(x)$ that gives the ratio of profit to income from concert ticket sales as a function of ticket price.

a. Write an algebraic rule for $S(x)$.

b. Will the graph of $S(x)$ have any vertical asymptotes? If so, where?

c. Sketch a graph of $S(x)$ for domain values from 0 to 30. Label the vertical asymptotes with their equations.

d. Describe the behavior of the profit-to-income function near each of the vertical asymptotes. Then use the rule for $S(x)$ to explain why that behavior occurs.

e. Estimate coordinates of any x-intercepts for $S(x)$. Explain what the x-intercepts tell about the concert profit-to-income ratio as a function of ticket price.

f. Estimate the ticket price that would maximize the profit to income ratio. Label that point on your graph of $S(x)$.

✔ CHECK YOUR UNDERSTANDING

Use your understanding of rational functions to answer the following questions about the relationship of concert profit $P_c(x) = -25x^2 + 875x - 4{,}750$ to concert operating expenses $c(x) = 4{,}750 - 125x$.

a. Write an algebraic rule for the function $T(x)$, which gives the ratio of profit to operating expenses, that is, the number of dollars earned for every dollar spent by the concert promoters.

b. Find the location of all vertical asymptotes for $T(x)$. Describe the behavior of the function near those asymptotes. Explain why that behavior occurs.

c. What is a reasonable domain for $T(x)$ when it is being used to model conditions in the concert business?

d. At about what ticket price is the ratio of profit to operating expenses maximized?

INVESTIGATION 2

Simplifying Rational Expressions

When work on mathematical problems leads to fractions like $\frac{8}{12}$ or $\frac{12}{18}$ or $\frac{9}{15}$, you know that it is often helpful to replace each fraction by a simpler equivalent fraction. For example, $\frac{8}{12} = \frac{4 \cdot 2}{4 \cdot 3} = \frac{4}{4} \cdot \frac{2}{3} = 1 \cdot \frac{2}{3}$, so $\frac{8}{12} = \frac{2}{3}$.

As you work on the problems of this investigation, look for answers to these questions:

What principles and strategies help to simplify rational expressions?

What cautions must be observed when simplifying rational expressions?

1 Consider again the ticket sale income function $t(x) = -25x^2 + 750x$ and the function $n(x) = -25x + 750$, giving number of tickets sold. In both functions, x represents the price per ticket in dollars.

a. Write an algebraic rule for the rational function $U(x)$, giving income per ticket sold.

b. Examine the values of $U(x)$ for integer values of x from 0 to 30. Explain why the pattern in those results makes sense when you think about what the expressions in the numerator and denominator represent in the problem situation.

c. Factor both numerator and denominator of the expression defining $U(x)$. Then write the simpler rational expression that results from removing common factors.

d. How does the result in Part c relate to the pattern you observed in Part b?

e. When Rashid found the simplified expression called for in Part c, he said it is *equivalent* to the original rational expression. Flor said she did not think so, because the domain for $U(x)$ is different from that of the simplified expression. Who is correct?

2 Consider next the function $f(x) = \dfrac{x^2 + 2x - 15}{x^2 - 4x + 3}$.

a. Simplify $\dfrac{x^2 + 2x - 15}{x^2 - 4x + 3}$ by first factoring both the numerator and denominator. Let $g(x)$ be the new function defined by that simplified expression.

b. Study tables of values for $f(x)$ and $g(x)$ for integer values of x in the interval $[-10, 10]$. Find any values of x for which $f(x) \neq g(x)$. Explain why those inputs produce different outputs for the two functions.

c. Use a calculator or computer to draw graphs of $f(x)$ and $g(x)$ over the interval $[-10, 10]$. Compare the graphs.

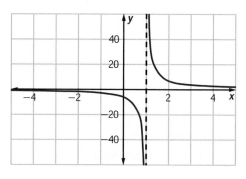

d. Now look again at tables of values for the two functions with smaller step sizes to see if you can explain why the differences observed in Part b do not seem to show up in the graphs.

3 For each of the following rational functions, use factoring to produce simpler and nearly equivalent rules. In each case, note any values of x for which the original and the simplified rules will not produce the same output values. Check your results by examining tables of values and graphs of the functions.

a. $f(x) = \dfrac{2x - 10}{x - 5}$

b. $g(x) = \dfrac{x + 3}{2x^2 + 6x}$

c. $h(x) = \dfrac{x^2 + 7x + 12}{x + 3}$

d. $j(x) = \dfrac{x^2 + 7x + 12}{x^2 + 5x + 6}$

4 In Problem 3, you examined rational functions that had zeroes in the denominator but did not have the expected vertical asymptotes at some of those domain points. Study those examples and your simplifications again to formulate an answer to this question.

In what cases will a zero of the denominator in a rational function not indicate a vertical asymptote of its graph?

5 When you are faced with the task of simplifying rational expressions, you may be tempted to make some "cancellations" that do not produce nearly equivalent results. For example, when Arlen and Tara were faced with the function $s(t) = \dfrac{t + 9}{6t + 9}$, they produced what they thought were equivalent simpler expressions for the rule.

  a. Arlen came up with $s_1(t) = \dfrac{1}{6}$. How could you show that this new function is quite different from the original? How could you help Arlen clarify his reasoning?

  b. Tara came up with $s_2(t) = \dfrac{0}{6}$. How could you show that this new function is quite different from the original? How could you help Tara clarify her reasoning?

6 To check your skill and understanding in simplifying rational expressions, follow these steps on your own.

  (1) Write an algebraic fraction in which both the numerator and the denominator are different linear expressions.

  (2) Create another linear expression and multiply both numerator and denominator by that expression. Then expand the resulting numerator and denominator into standard form quadratic polynomials.

  (3) Exchange the resulting rational expression with a classmate. See if you can find the quotient of linear expressions with which each began.

SUMMARIZE THE MATHEMATICS

In this investigation, you explored simplification of rational expressions.

  a What is meant by simplifying rational expressions?

  b For a given rational expression, what strategies can be applied to find another expression that is simpler but nearly equivalent to the original?

  c What cautions must be observed when simplifying rational expressions?

Be prepared to explain your ideas to the class.

✔ CHECK YOUR UNDERSTANDING

Use your understanding of equivalent rational expressions to complete the following tasks.

a. The function $T(x) = \dfrac{-25x^2 + 875x - 4{,}750}{-125x + 4{,}750}$ compares concert profit to operating expenses.

    i. Identify all values of x for which $T(x)$ is *undefined*.

    ii. Explain why the expression defining $T(x)$ can be simplified to $\dfrac{x^2 - 35x + 190}{5x - 190}$.

    iii. If the simplified expression in part ii were used to define a new function $K(x)$, would it have the same domain as $T(x)$?

b. Consider the expression $\dfrac{x^2 + 6x + 8}{x^2 - x - 6}$.

    i. Identify all values of x for which this expression is undefined.

    ii. Simplify the expression.

    iii. Explain why simplifying this expression does or does not produce an expression with different undefined points than the original.

INVESTIGATION 3

Adding and Subtracting Rational Expressions

In Investigation 1, you derived the profit-per-ticket function $R(x) = \dfrac{-25x^2 + 875x - 4{,}750}{-25x + 750}$ for a concert venue. As you know, the concert promoters also make profit from sales at snack bars that operate before the concert and at intermissions. Including that profit source in the business calculations requires new operations on rational expressions and functions.

As you work on the problems of this investigation, look for answers to this question:

What principles and strategies guide addition and subtraction of rational expressions?

1 The concert promoter's business analysis suggests that total profit $P_b(x)$ from sale of snack bar items will be related to ticket price x by the function $P_b(x) = -175x + 5{,}250$.

    a. Why does it make sense that profit from snack bar sales depends on ticket price?

    b. Write and simplify a rule for the function $V(x)$ that gives the profit per ticket from snack bar sales. Remember that the number of people attending the concert is related to ticket price by the function $n(x) = -25x + 750$.

    c. What are the values of $R(20)$, $V(20)$, and $R(20) + V(20)$? What do those values tell about the profit prospects for the whole concert event?

2 When the concert promoters asked their business manager to find a function $W(x)$ that shows combined profit per ticket from ticket and snack bar sales, she suggested two possibilities.

$$W_1(x) = \frac{P_c(x)}{n(x)} + \frac{P_b(x)}{n(x)} \quad \text{and} \quad W_2(x) = \frac{P_c(x) + P_b(b)}{n(x)}$$

a. Using the algebraic expressions for each function involved, you get:

$$W_1(x) = \frac{-25x^2 + 875x - 4{,}750}{-25x + 750} + \frac{-175x + 5{,}250}{-25x + 750}$$

$$\text{and} \quad W_2(x) = \frac{-25x^2 + 700x + 500}{-25x + 750}.$$

How could you convince someone that both ways of calculating total profit per ticket are correct by:

   i. reasoning about the meaning of each expression in the concert business situation?

   ii. using what you know about addition of arithmetic fractions like $\frac{2}{5} + \frac{4}{5}$?

b. Which of the two expressions for total profit per ticket would be better for:

   i. showing how each profit source contributes to total profit per ticket?

   ii. calculating total profit per ticket most efficiently?

c. Find what you believe to be the expression that is most efficient for calculating $W(x)$. Use that expression to find the profit per ticket from *both* ticket and snack bar sales when the ticket prices are $10, $15, $20, and $25.

d. Estimate the ticket price that will yield maximum profit per ticket for:

   i. the concert operation alone, $R(x)$.

   ii. the snack bar operation alone, $V(x)$.

   iii. the combination of concert and snack bar operations, $W(x)$.

3 Consider the following two rational functions.

$$f(x) = \frac{3x^2 + 5x - 2}{4x + 1} \quad \text{and} \quad g(x) = \frac{x^2 + 4x + 4}{4x + 1}$$

a. Evaluate $f(x)$, $g(x)$, and $f(x) + g(x)$ for a variety of values of x.

b. If $h(x) = f(x) + g(x)$, find what you believe to be the simplest algebraic rule for $h(x)$.

c. Use the rule from Part b to calculate $h(x)$ for the same values of x you used in Part a. Then compare the results with what you obtained in Part a.

d. If you simplify the algebraic fraction that results from adding $f(x)$ and $g(x)$, there is one value of x for which care must be exercised in work with $h(x) = f(x) + g(x)$.

    i. What is that number? (*Hint:* Examine a table of values of $h(x)$ in increments of 0.25.)

    ii. Why do you get different results from using a simplified rule for $h(x)$ and the given rules for $f(x)$ and $g(x)$?

4 Using the rules for $f(x)$ and $g(x)$ given in Problem 3, let $k(x) = f(x) - g(x)$.

    a. Find a rule in simplest form for $k(x)$.

    b. Check your answer to Part a by calculating $f(x) - g(x)$ and $k(x)$ for a variety of values of x.

    c. How can you quickly tell that the numerator and denominator of $k(x)$ do not have a common factor?

Finding Common Denominators From your experiences with arithmetic, you know that addition and subtraction of fractions become more involved if the denominators are not the same. For example, to find the sum $\frac{2}{5} + \frac{7}{10}$, you first have to replace the addends with equivalent fractions having a common denominator. In this case, you might write $\frac{2}{5}$ as $\frac{4}{10}$ to get the sum $\frac{11}{10}$.

5 The same need for common denominators arises in work with algebraic fractions. And the strategies that work in dealing with numerical fractions can be applied to algebraic fractions as well.

For example, consider $\frac{x+1}{x} + \frac{x+2}{x-1}$. Examine the result produced by a CAS shown below.

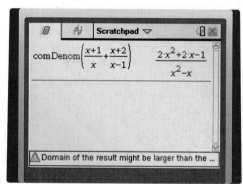

    a. Explain why the denominator of $x^2 - x$ is correct.

    b. Why is the numerator $2x^2 + 2x - 1$?

6 Consider the following questions about airline flights from Chicago to Philadelphia and back. The cities are 678 miles apart. The return trip is scheduled to take almost half an hour longer than the flight from Chicago to Philadelphia because winds in the upper atmosphere almost always flow from west to east. Suppose the speed of the airplane without wind is s miles per hour, and the speed of the wind is 30 miles per hour from west to east.

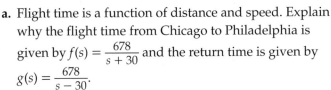

a. Flight time is a function of distance and speed. Explain why the flight time from Chicago to Philadelphia is given by $f(s) = \dfrac{678}{s + 30}$ and the return time is given by $g(s) = \dfrac{678}{s - 30}$.

b. What information is given by values of $f(s) + g(s)$?

c. Explain why the rules for $f(s)$ and $g(s)$ can be replaced by

$$f(s) = \frac{678(s - 30)}{(s + 30)(s - 30)} \text{ and } g(s) = \frac{678(s + 30)}{(s - 30)(s + 30)}.$$

Why are those symbolic manipulations a useful idea?

d. Use the results in Part c to find a rule for $f(s) + g(s)$. Simplify it if possible.

e. Use the simplified rule for $f(s) + g(s)$ to find flight time for a round trip from Chicago to Philadelphia if the airplane's speed without wind is 310 miles per hour. Compare that result to the value of $f(310) + g(310)$.

7 Adapt the reasoning used in work on Problems 5 and 6 and what you know about adding arithmetic fractions with unlike denominators to write each of the following algebraic expressions in equivalent form as a single fraction. Then simplify each result as much as possible. Check your reasoning with a CAS.

a. $\dfrac{3x + 1}{5} + \dfrac{x + 4}{x}$

b. $\dfrac{5}{x - 1} + \dfrac{x + 1}{x}$

c. $\dfrac{x + 2}{x + 3} + \dfrac{x + 1}{2x - 6}$

d. $3x + 1 + \dfrac{x + 4}{x}$

8 The following display shows how the sum of two fractions is calculated by a CAS *without* use of the **comDenom** command. Use algebraic reasoning to show that this result is equivalent to that obtained in Problem 5.

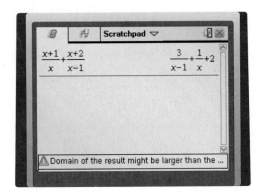

SUMMARIZE THE MATHEMATICS

In this investigation, you learned how to add and subtract rational expressions.

a Adding rational expressions is easy when the two expressions are related in a particular way. What is that relationship? If it holds, how do you add the expressions?

b If the most convenient relationship between rational expressions does *not* hold, what can you do to make addition or subtraction of the expressions possible?

Be prepared to explain your ideas to the class.

 CHECK YOUR UNDERSTANDING

Use your understanding of adding and subtracting rational expressions to find simplified rules for the sums and differences of functions in Parts a–c, where
$f(x) = \dfrac{3x - 1}{4x}$, $g(x) = \dfrac{x - 5}{4x}$, and $h(x) = \dfrac{3x}{2x + 3}$.

a. $f(x) + g(x)$

b. $f(x) - g(x)$

c. $h(x) - g(x)$

INVESTIGATION 4

Multiplying and Dividing Rational Expressions

You have learned how to simplify, add, and subtract rational expressions, and you have seen how those operations are similar to operations with numerical fractions. The similarities continue, with some limitations, to multiplication and division.

As you work on the problems of this investigation, look for answers to these questions:

If the rules for two rational functions f(x) and g(x) are given, how can you calculate rules for the product and quotient of those functions?

What cautions must be observed when simplifying products and quotients of a rational expressions?

1 Give results, in simplest form, for these products of numerical fractions.

a. $\dfrac{5}{7} \cdot \dfrac{3}{10}$

b. $\dfrac{3}{5} \cdot \dfrac{9}{6}$

c. $\dfrac{5}{8} \cdot \dfrac{12}{25}$

2 Suppose that $f(x) = \frac{x^2}{2}$ and $g(x) = \frac{12}{x}$.

a. If $h(x) = f(x) \cdot g(x)$, what do you think will be the simplest expression that gives correct values of $h(x)$ for all x?

b. Test your idea in Part a by comparing tables of values for $\left(\frac{x^2}{2}\right)\left(\frac{12}{x}\right)$ and for your proposed simpler expression to calculate $h(x)$. Use values of x in the interval $[-10, 10]$.

c. Test your idea in Part a by graphing $y = \left(\frac{x^2}{2}\right)\left(\frac{12}{x}\right)$ for values of x in the interval $[-10, 10]$ and then graphing $y = h(x)$, using your proposed simplified expression for $h(x)$.

d. What similarities and what differences between multiplication of numeric and algebraic fractions are suggested by the results of your work on Parts a–c?

3 Suppose that $f(x) = \frac{3x}{x-2}$ and $g(x) = \frac{5x-10}{x}$.

a. If $h(x) = f(x) \cdot g(x)$, what do you think will be the simplest expression that gives correct values of $h(x)$ for all x?

b. Test your idea in Part a by comparing tables of values for $\left(\frac{3x}{x-2}\right)\left(\frac{5x-10}{x}\right)$ and for your proposed simpler expression to calculate $h(x)$. Use values of x in the interval $[-10, 10]$.

c. Test your idea in Part a by graphing $y = \frac{3x}{x-2} \cdot \frac{5x-10}{x}$ for values of x in the interval $[-10, 10]$ and graphing $y = h(x)$, using your proposed simplified expression for $h(x)$.

d. What similarities and what differences between multiplication of numeric and algebraic fractions are suggested by the results of your work on Parts a–c?

4 The following work shows how Cho calculated the product $h(x) = f(x) \cdot g(x)$ in Problem 3.

$$\frac{3x}{x-2} \cdot \frac{5x-10}{x} = \frac{3x(5x-10)}{(x-2)x}$$
$$= \frac{15x^2 - 30x}{x^2 - 2x}$$
$$= \frac{15x(x-2)}{x(x-2)}$$
$$= 15$$

a. Explain how Cho could have saved time by factoring and removing identical factors before expanding the numerator and denominator expressions.

b. What restrictions on the domain of $f(x)$ and $g(x)$ has Cho ignored? What are the consequences?

5 Simplify $\dfrac{x-2}{x^2-25} \cdot \dfrac{2x^2+11x+5}{6x-12}$ by first factoring and then removing common factors before expanding the products in the numerator and denominator. Check your answer with a CAS. Be careful to explain the values of x for which the simpler expression is not equivalent to the original.

6 Suppose that $f(x) = \dfrac{6}{x}$ and $g(x) = \dfrac{2}{3x}$.

a. If $h(x) = f(x) \div g(x)$, what do think will be the simplest rule that gives correct values of $h(x)$?

b. Test your idea in Part a by comparing tables of values for $\dfrac{6}{x} \div \dfrac{2}{3x}$ and for your proposed simpler expression to calculate $h(x)$. Use values of x in the interval $[-10, 10]$.

c. Test your idea in Part a by graphing $y = \dfrac{6}{x} \div \dfrac{2}{3x}$ for values of x in the interval $[-10, 10]$ and then graphing $y = h(x)$, using your proposed simplified expression for $h(x)$.

d. What similarities and what differences between division of numeric and algebraic fractions are suggested by the results of your work on Parts a–c?

7 Suppose that $f(x) = \dfrac{2x}{x^2-9}$ and $g(x) = \dfrac{5x^2}{2x-6}$.

a. If $h(x) = f(x) \div g(x)$, what do you think will be the simplest rule that gives correct values of $h(x)$?

b. Test your idea in Part a by comparing tables of values for $\dfrac{2x}{x^2-9} \div \dfrac{5x^2}{2x-6}$ and for your proposed simpler expression to calculate $h(x)$. Use values of x in the interval $[-10, 10]$.

c. Test your idea in Part a by graphing $y = \dfrac{2x}{x^2-9} \div \dfrac{5x^2}{2x-6}$ for values of x in the interval $[-10, 10]$ and then graphing $y = h(x)$, using your proposed simplified expression for $h(x)$.

d. What similarities and what differences between division of numeric and algebraic fractions are suggested by the results of your work on Parts a–c?

8 Consider the rational expression $\dfrac{6x^2+9x+8}{3x}$.

a. Which of these expressions (if any) are equivalent to the original? How do you know?

   i. $2x + \dfrac{9x+8}{3x}$

   ii. $2x + 9x + 8$

   iii. $2x + 3 + \dfrac{8}{3x}$

b. Check your answer to Part a by comparing graphs of the functions defined by each expression in the window $-5 \le x \le 5$ and $-10 \le x \le 20$.

SUMMARIZE THE MATHEMATICS

In this investigation, you explored ways that algebraic reasoning can be used to work with multiplication and division of rational expressions and functions.

a If you are given two rational functions $f(x) = \dfrac{a(x)}{b(x)}$ and $g(x) = \dfrac{c(x)}{d(x)}$, how would you find the simplest form of the rules for these related functions?

    i. $h(x) = f(x) \cdot g(x)$

    ii. $j(x) = f(x) \div g(x)$

b How can rational functions with rules in the form $r(x) = \dfrac{ax^2 + bx + c}{x}$ be written in equivalent form as separate fractions?

Be prepared to explain your ideas to the class.

 CHECK YOUR UNDERSTANDING

Use your understanding of multiplication, division, and equivalent expressions to complete these tasks.

a. If $r(x) = \dfrac{x^2 + 3x - 10}{2x - 4}$ and $s(x) = \dfrac{3x - 6}{x - 2}$, find rules for the following expressed in simplest form.

    i. $r(x) \cdot s(x)$

    ii. $r(x) \div s(x)$

    iii. $r(x) + s(x)$

    iv. $r(x) - s(x)$

b. Write the rule $y = \dfrac{3x - 5}{2x}$ in equivalent form as a sum of two separate rational expressions.

APPLICATIONS

1 The Cannery designs and manufactures cans for packaging soups and vegetables. A&W Foods contracted with the Cannery to produce packaging for its store brand canned corn. Each can is to hold 500 cm^3 of corn. The company is interested in minimizing the cost of each can by minimizing the surface area of the can.

a. Using the information below about the volume of the can, find a rational function $h(r)$ that gives the height h of the can as a function of the radius r of the can.

Volume of a Can

$$V = \pi r^2 h$$

Surface Area of a Can

$$SA = 2\pi r^2 + 2\pi rh$$

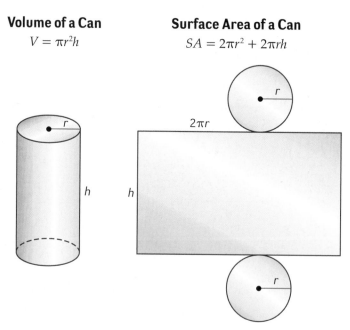

b. Based on your results from Part a, write a function $S(r)$ to calculate the surface area of the can for any given radius r.

c. Determine a practical domain for $S(r)$. Sketch a graph of the function on that domain. Identify any vertical asymptotes.

d. Determine the minimum value of $S(r)$ on this practical domain. Give the radius and height of the can with minimum surface area to the nearest tenth of a centimeter.

2 At the concert venue you have studied several times in this unit, concert attendance, snack bar income, operating expenses, and profit are all functions of the ticket price x.

- Concert attendance is given by $n(x) = 750 - 25x$.

- Snack bar income is given by $s(x) = 7,500 - 250x$.

- Snack bar operating expenses are given by $b(x) = 2,250 - 75x$.

a. Write two equivalent rules for the function $P_b(x)$ that gives snack bar profit as a function of ticket price. Give one rule that shows the separate expressions for calculating snack bar income and operating cost and another that is equivalent, yet easier to use in calculations.

b. For the rational functions described in parts i–iv:

- explain what they tell about the business situation.

- describe the theoretical and practical domains of the functions— the values of x for which a function output *can be* calculated and the values of x for which calculation of outputs makes sense in the problem situation.

- sketch graphs of each function, indicating any vertical asymptotes and enough of the graph so that its overall shape is clear.

i. $f(x) = \dfrac{s(x)}{n(x)}$ **ii.** $g(x) = \dfrac{b(x)}{n(x)}$

iii. $h(x) = \dfrac{P_b(x)}{s(x)}$ **iv.** $j(x) = \dfrac{b(x)}{s(x)}$

3 Simplify the rules for the rational functions defined in Task 2 as far as possible. In each case, check to see whether the domain of each simplified function rule is the same as that of the original.

4 When income from concert ticket and snack bar sales (Task 2) are combined, the functions that model the business situation are:

- Total income as a function of average ticket price is given by $I(x) = -25x^2 + 500x + 7,500$.

- Combined operating expenses for the concert and the snack bar is given by $E(x) = 7,000 - 200x$.

- The number of tickets sold is given by $n(x) = 750 - 25x$.

a. Write two equivalent rules for the function $P(x)$ that gives total profit as a function of ticket price. Give one rule that shows the separate expressions for calculating income and cost and another that is equivalent, yet easier to use in calculations.

b. For the rational functions described in parts i–iv:

- explain what they tell about the business situation.

- describe the theoretical and practical domains of the functions— the values of x for which a function output *can be* calculated and the values of x for which calculation of outputs makes sense in the problem situation.

- sketch graphs of each function, indicating any vertical asymptotes and enough of the graph so that its overall shape is clear.

    i. $Q(x) = \dfrac{P(x)}{I(x)}$ **ii.** $R(x) = \dfrac{E(x)}{I(x)}$

    iii. $S(x) = \dfrac{P(x)}{n(x)}$ **iv.** $T(x) = \dfrac{E(x)}{n(x)}$

5 Simplify the rules for the rational functions defined in Task 4 as far as possible. In each case, check to see whether the domain of the simplified function rule is the same as that of the original.

6 Describe the domains of these rational functions. Explain how you know that your answers are correct.

    a. $f(x) = \dfrac{4x + 1}{3x + 2}$ **b.** $g(x) = \dfrac{5x + 10}{x + 2}$

    c. $h(x) = \dfrac{x + 1}{x^2}$ **d.** $j(x) = \dfrac{x^2 + 3}{x^2 - 9}$

7 Simplify these rational expressions as much as possible. In each case, check to see whether the simplified expression is undefined for the same values of x as the original expression.

    a. $\dfrac{4x + 12}{8x + 4}$ **b.** $\dfrac{4 - 3x}{6x - 8}$

    c. $\dfrac{15x^2 + 6x}{3x^2}$ **d.** $\dfrac{4 + x^2}{16 - x^2}$

8 Write each of these sums and differences of rational expressions in equivalent form as a single algebraic fraction. Then simplify the result as much as possible.

    a. $\dfrac{4x + 13}{x - 3} + \dfrac{x + 2}{x - 3}$ **b.** $\dfrac{3x + 7}{x - 2} - \dfrac{x + 5}{x - 2}$

    c. $(x + 2) + \dfrac{x + 6}{x - 4}$ **d.** $\dfrac{3x^2 + 5x + 1}{x^2 - 25} - \dfrac{3x^2 + 4x - 4}{x^2 - 25}$

9 Write these products and quotients of rational expressions in equivalent form as single algebraic fractions. Then simplify the results as much as possible.

    a. $\dfrac{2x}{x + 2} \cdot \dfrac{3x + 6}{x^2}$ **b.** $\dfrac{2x + 6}{x + 2} \cdot \dfrac{3x + 2}{x + 3}$

    c. $\dfrac{x + 2}{x} \div \dfrac{3x + 6}{x^2}$ **d.** $\dfrac{2x}{x + 2} \div \dfrac{x^2}{2x + 4}$

10 Write each of these products and quotients of rational expressions in equivalent form as a single algebraic fraction. Then simplify the result as much as possible.

    a. $\dfrac{2x + 4}{x^2 - 6x} \cdot \dfrac{x^2 - 36}{4x + 8}$ **b.** $\dfrac{x - 3}{7x} \cdot \dfrac{3x^2}{x^2 - 2x - 3}$

    c. $\dfrac{x^2 - 4}{x^2 + 2x - 5} \div \dfrac{x + 2}{x^2 + 2x - 5}$ **d.** $\dfrac{2x^2 + 10x}{4x + 10} \div \dfrac{3x^2 + x}{9x + 3}$

11 Write each of these rational expressions as a sum or difference of two or more algebraic fractions in simplest form.

a. $\dfrac{3x + 5}{x}$

b. $\dfrac{3x^2 + 5x}{x}$

c. $\dfrac{3x^2 - 9x + 12}{6x^2}$

d. $\dfrac{3x + 21}{x + 7}$

CONNECTIONS

12 Look back at your work for Applications Task 1.

a. Record the height and radius of the can with minimum surface area in Part d.

b. Find the radius and height of the can with minimum surface area for cylindrical cans with these volumes: 236 cm³, 650 cm³, and 946 cm³.

c. Plot (*optimal radius*, *optimal height*) for each of the cans in Parts a and b.

d. What relationship seems to exist between the optimal radius and height for a given volume?

13 In any right triangle ABC with right angle at C, the *tangent* of $\angle A$ is the ratio $\dfrac{a}{b}$.

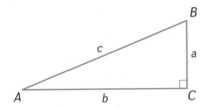

Use what you know about operations with fractions to prove that for any acute angle A,

$$\frac{\sin A}{\cos A} = \tan A.$$

14 Consider the function $r(x) = \dfrac{3(2)^x}{0.5(10)^x}$.

a. Is $r(x)$ a rational function?

b. When the function is graphed, are there any asymptotes? If so, give their location.

c. Show how the expression defining $r(x)$ can be written in a simpler form that shows the function family to which it belongs.

15 The graphs of these functions all have *horizontal asymptotes*, lines with equations in the form $y = a$.

$$f(x) = 2^x \qquad g(x) = 0.5^x + 3 \qquad h(x) = \frac{1}{x}$$

$$j(x) = \frac{1}{x^2} \qquad k(x) = \frac{8x + 5}{2x} \qquad m(x) = \frac{x^2 + 2x - 1}{x^2 + x - 2}$$

Study graphs and rules of the given functions to identify equations of their horizontal asymptotes.

16 Explain procedures that are necessary to perform these operations on arithmetic fractions.

a. Simplify a fraction like $\frac{36}{60}$.

b. Find the sum or difference of two fractions like $\frac{3}{5}$ and $\frac{7}{5}$.

c. Find the sum or difference of two fractions like $\frac{3}{5}$ and $\frac{7}{4}$.

d. Find the product of two fractions like $\frac{3}{5}$ and $\frac{7}{4}$.

e. Find the quotient of two fractions like $\frac{3}{5}$ and $\frac{7}{4}$.

17 The time required to complete a 100-mile bicycle race depends on the average speed of the rider. The function $T(s) = \frac{100}{s}$ tells time in hours if speed is in miles per hour.

a. Identify the vertical and horizontal asymptotes for the graph of $T(s)$.

b. Explain what each asymptote says about the way race time changes as average rider speed changes.

18 The intensity of sound from some source like a music speaker or a lightning strike is inversely proportional to the square of the distance from source to receiver. For example, suppose that the intensity of sound at various distances from a nearby explosion is given by the function $I(d) = \frac{100}{d^2}$, where distance is measured in meters and sound intensity in watts per square meter.

a. Identify the vertical and horizontal asymptotes for the graph of $I(d)$.

b. Explain what each asymptote says about the way received sound intensity changes as distance from the source changes.

19 When the Holiday Out hotel chain was planning to build a new property, organizers bought land for $5,000,000, and their architect made building plans that would cost $10,000,000 for the lobby, restaurant, and other common facilities. The architect also said that construction costs would average an additional $75,000 per guest room.

a. What rule defines the function that relates total construction cost C to the number n of guest rooms in the hotel?

b. What rule defines the function $A(n)$ that gives average cost per guest room as a function of the number n of guest rooms in the hotel?

c. What are the theoretical domains of the total cost and average cost functions? What are reasonable practical domains to consider for those functions?

d. Sketch a graph of the average cost per guest room function. Identify the vertical and horizontal asymptotes of that graph. Explain what the asymptotes tell about the way average cost per room changes as the number of planned guest rooms changes.

20 Rational functions arise naturally when working with similar triangles. Suppose that a long ladder is to be moved horizontally around a corner in a building with 8-foot hallways. The ladder can make it around the corner only if its length is less than all lengths L that satisfy the conditions of the diagram below.

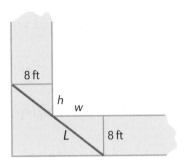

a. Using similar triangles, determine a relationship between h and w.

b. Write a function relating the distance L to the lengths h and w. Then use the relationship from Part a to write L as a function of h. (Ignore the thickness of the ladder.)

c. Determine the minimum value $L(h)$.

d. As the longest possible ladder moves through the corner, what angle is made with the walls? Explain how this angle is determined by the symmetry of the situation.

e. What is the minimum value of L if one of the hallways is 4 feet wide and the other is 8 feet wide?

REFLECTIONS

21 Why does the term *rational function* seem appropriate for any function that can be expressed as the quotient of two polynomials?

22 Suppose $f(x) = \dfrac{p(x)}{q(x)}$ is a rational function.

a. What can you conclude if $p(a) = 0$ and $q(a) \neq 0$?

b. What can you conclude if $q(a) = 0$ and $p(a) \neq 0$?

c. What can you conclude if $p(a) = 0$ and $q(a) = 0$?

23 What are some important things to remember and look for when:

a. analyzing the behavior of rational functions and their graphs?

b. simplifying rational expressions for functions?

c. adding or subtracting rational expressions and functions?

d. multiplying or dividing rational expressions and functions?

EXTENSIONS

24 Consider the function $f(x) = \dfrac{x^2 + x - 12}{x^2 - x - 6}$.

   a. What are the values of x for which $f(x)$ is undefined?

   b. Graph $f(x)$ to see whether there seem to be vertical asymptotes at the values of x for which $f(x)$ is undefined.

   c. Let $\tilde{f}(x)$ be the function resulting from simplifying the expression for $f(x)$. For how many values of x is $\tilde{f}(x)$ undefined?

   d. How many vertical asymptotes does the graph of $\tilde{f}(x)$ have? Where are they?

   e. How do the graphs of $f(x)$ and $\tilde{f}(x)$ compare?

25 The function $f(x)$ in Task 24 is *undefined* at $x = 3$. The nearly identical function $\tilde{f}(x)$ with rule obtained by "simplifying" the rational expression for $f(x)$, is *defined* at $x = 3$. However, the graph of $f(x)$ has no vertical asymptote at $x = 3$. That fact is shown by drawing a "hole" in the graph where $x = 3$.

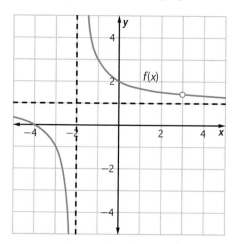

The function $f(x)$ is said to have a **removable discontinuity** at $x = 3$ because the discontinuity or break in the graph can be removed by simplifying the expression. As a result, 3 is in the domain of $\tilde{f}(x)$ but not in the domain of $f(x)$. However, the other value of x for which $f(x)$ is undefined gives rise to a vertical asymptote, a discontinuity that cannot be removed by simplification. Therefore, $f(x)$ is said to have an **essential discontinuity** at $x = -2$.

Determine the essential and removable discontinuities for each function in Parts a–d.

   a. $f(x) = \dfrac{4x + 12}{8x + 4}$ **b.** $g(x) = \dfrac{5x + 10}{x + 2}$

   c. $h(x) = \dfrac{15x^2 + 6x}{3x^2}$ **d.** $j(x) = \dfrac{x^2 + 3}{x^2 - 9}$

26 Consider $f(x) = \dfrac{20x^3 + 10x^2 - 150x}{20x^2 - 125}$.

   a. Simplify the expression for $f(x)$.

   b. Compare graphs of $f(x)$ and the function given by the simplified expression.

c. Give the values of x for which $f(x)$ is undefined. Classify each discontinuity as removable or essential. (See Task 25.)

d. What is the domain of $f(x)$?

e. What is the domain of the function derived by simplifying the expression for $f(x)$?

27 The graph of $R(x) = \dfrac{-25x^2 + 875x - 4{,}750}{750 - 25x}$ has what is called an **oblique asymptote**. The next diagram shows how the graph of $R(x)$ approaches the oblique asymptote $y = x - 5$.

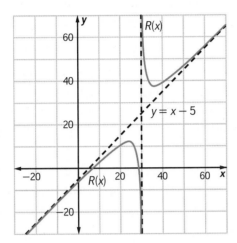

Use a spreadsheet or function table tool to examine values of $R(x)$ for very large values of x, say larger than 100, and then for "large negative" values of x.

a. Explain how your results illustrate the fact that as x approaches positive infinity, the graph of $R(x)$ gets arbitrarily close to the line $y = x - 5$, while remaining above it.

b. Explain how your results illustrate the fact that as x approaches negative infinity, the graph of $R(x)$ gets arbitrarily close to the line $y = x - 5$, while remaining below it.

28 Consider the function with rule $h(x) = 3x - 4 + \dfrac{5}{x}$.

a. What is the oblique asymptote for the graph of this function? How can you tell by analyzing values produced by the function rule when x is very large (approaches positive infinity) or very small (approaches negative infinity)?

b. How can the rule for $h(x)$ be given by an equivalent rational expression?

29 Suppose $f(x) = x - 3$ and $g(x) = \dfrac{6}{x - 2}$.

a. If you want to think of $f(x)$ as a rational function, what is its denominator?

b. What could you multiply that denominator by so it is the same as the denominator of $g(x)$?

c. Use your idea from Part b to find a rule for $f(x) - g(x)$ expressed with one simplified algebraic fraction.

d. Check your answer to Part c by graphing the simplified algebraic fraction and the function $y = (x - 3) - \dfrac{6}{x - 2}$.

e. The graph of $y = (x - 3) - \dfrac{6}{x - 2}$ has an oblique asymptote. Which form of the difference expression makes it easier to identify the equation of that asymptote, the expression $(x - 3) - \dfrac{6}{x - 2}$ or your answer to Part c? Be prepared to explain your reasoning.

30 Paco decided to spend the weekend at the local beach. He left on Friday, and his average speed was only 30 miles per hour. He took the same route home but was able to wait until Tuesday when the traffic was lighter. For the return trip, his average speed was 60 miles per hour.

a. What was Paco's average speed for the entire trip?

b. Consider the similar situation of Tory's trip to the beach. The entire trip took her two hours. For the first hour, her average speed was 60 miles per hour. During the second hour, she encountered traffic, so her average speed was 30 miles per hour for that hour of travel. What was Tory's average speed for the entire trip?

c. What information regarding Tory's trip is different from the information about Paco's trip? Which situation is in line with your definition of "average"?

REVIEW

31 Use what you know about the relationships between different units of measurement to complete the following.

a. 7.5 ft = _____ in. = _____ yd

b. 1,780 mm = _____ cm = _____ m

c. 1 yd² = _____ ft² = _____ in²

32 Rewrite each expression in a simpler equivalent form.

a. $\dfrac{108}{16}$

b. $\dfrac{3x + 12}{3}$

c. $\dfrac{10x^2 - 8}{24}$

33 Billy is standing 80 feet from the base of a building. The angle of elevation to the top of the building is 35°. If the distance from Billy's eyes to the ground is five feet, how tall is the building?

34 Consider the following statement.

The diagonals of a rhombus are perpendicular.

a. Write this statement in if-then form.

b. Is this a true statement? Explain your reasoning.

c. Write the converse of this statement.

d. Is the converse of this statement true? Explain your reasoning.

35 Solve these equations for x.

a. $2^x = 8$

b. $2^x = \frac{1}{16}$

c. $10^x = 10,000$

d. $10^x = 0.0000001$

36 Express each of these sums or differences as a single fraction in simplest form.

a. $\frac{1}{8} + \frac{5}{8}$

b. $\frac{4}{5} - \frac{2}{3}$

c. $\frac{1}{3} + \frac{5}{6}$

d. $\frac{3}{10} - \frac{7}{6}$

e. $\frac{2}{x} + \frac{3}{4}$

37 Sketch the solution set for each system of inequalities.

a. $3x + 6y \geq 12$
$y \leq -5x + 10$

b. $y > x^2$
$y < x + 2$

38 Suppose that Elizabeth rolls two regular dice and finds the sum of the numbers that are showing.

a. What is the probability that the sum is even or greater than 8?

b. What is the probability that the sum is even and greater than 8?

c. If she rolls the dice twice what is the probability that the sum is greater than 8 on both rolls?

39 Express each of these products and quotients as a single fraction in simplest form.

a. $\frac{1}{2} \cdot \frac{1}{3}$

b. $\frac{2}{3} \div \frac{4}{5}$

c. $\frac{3}{4} \cdot \frac{2}{3}$

d. $\frac{2}{3} \div \frac{1}{2}$

e. $\frac{a}{b} \cdot \frac{c}{d}$

40 In $\triangle ABC$ and $\triangle DEF$, $\angle B$ and $\angle E$ are right angles. In each case below, decide whether the given information is sufficient to conclude that $\triangle ABC \cong \triangle DEF$. If so, explain why. If not, give a counterexample.

a. $\overline{AC} \cong \overline{DF}$; $\angle C \cong \angle F$

b. $\angle A \cong \angle D$; $\angle C \cong \angle F$

c. $\overline{AC} \cong \overline{DF}$; $\overline{AB} \cong \overline{DE}$

41 Triangle ABC is an isosceles triangle with $AB = BC$.

a. If $m\angle B = 138°$, find $m\angle C$ and $m\angle A$.

b. If $AC = 16$ cm, find the perimeter and the area of $\triangle ABC$.

42 Use a compass and straightedge or interactive geometry software to complete the following constructions.

a. Draw a line segment $\overline{AB}$ and construct its perpendicular bisector. Label the midpoint of $\overline{AB}$ as M.

b. Draw a line ℓ and mark a point P on the line. Construct the line that contains P and is perpendicular to line ℓ.

c. Draw a line ℓ and mark a point P that is not on the line. Construct the line that contains P and is perpendicular to line ℓ.

C Squared Studios/Getty Images

Looking Back

The lessons of this unit extended your knowledge and skill in work with algebraic functions and expressions in three ways. First, you learned how to use polynomials that go beyond linear and quadratic functions and expressions to model data and graph patterns that are more complex than the familiar straight lines and parabolas. You discovered ways of predicting the shape of polynomial function graphs from analysis of their rules—especially the number of local maximum and local minimum points and the number of zeroes— and ways of combining polynomials by addition, subtraction, and multiplication.

Second, you revisited quadratic functions and expressions to learn about completing the square to produce vertex forms for their rules. You learned how to use the vertex form to locate maximum and minimum points on the graphs of quadratic functions and how to solve equations by algebraic reasoning. You also studied the way that completing the square can be used to prove the quadratic formula for solving equations.

Finally, you explored properties and applications of rational functions and expressions. You learned how to identify undefined points of algebraic fractions and asymptotes of their graphs. You explored simplification of algebraic fractions and the ways that cautious use of such algebraic work can provide insights into functions and the problems they model. Then you learned how to combine rational functions and expressions by addition, subtraction, multiplication, and division.

The following tasks give you an opportunity to review and to put your new knowledge of polynomial and rational functions to work in some new contexts.

1 Cassie and Cliff decided to start a company called Maryland Web Design that would help individuals build personal sites on the World Wide Web. Cassie sketched out a possible logo for their company on graph paper. Cliff thought it looked like a graph of two polynomial functions, so he drew in a pair of axes.

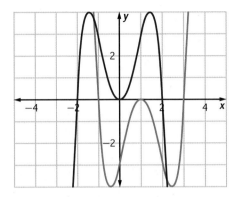

a. What is the smallest possible degree of each polynomial function?

b. Write possible functions to describe the curves Cassie drew. Use $M(x)$ to describe the curve that looks like the letter M and $W(x)$ to describe the curve that looks like the letter W.

2 Cassie and Cliff decided that their company would specialize in designing Web logs, or "blogs." Since production of any blog requires about the same amount of programming time, each customer could be charged the same price.

Cassie and Cliff also estimated that the number of customers in a year would be related to the price by $n(x) = 600 - 2x$, where x is the price charged per blog.

Production of each blog will require about $200 in programmer labor. To get started in business, they bought a computer for $2,000.

a. Write rules for the following functions, simplifying where possible.

  i. Annual income from blog designs $I(x)$ if the price is x dollars per blog

  ii. Annual production costs for blog designs $C(x)$ if the price is x dollars per blog

  iii. Annual profit from the business $P(x)$ if the price is x dollars per blog

  iv. Profit per customer $P_c(x)$ if the price is x dollars per blog

b. What would be reasonable domains for the functions in Part a?

c. For what blog prices would the following functions be maximized?

    i. Income $I(x)$

    ii. Profit $P(x)$

    iii. Profit per customer $P_c(x)$

d. What price would you recommend that Cassie and Cliff charge per blog? Explain your reasoning.

3 Find results of these algebraic operations. Express each answer in the equivalent form that would be simplest for evaluation at any specific value of x.

a. $(3x^4 + 5x^3 - 7x^2 + 9x - 1) + (2x^4 + 4x^2 + 6x + 8)$

b. $(3x^5 + 5x^3 - 7x^2 + 9x - 1) - (2x^4 + 4x^3 + 6x + 8)$

c. $(7x^2 + 9x - 1)(2x + 3)$

4 Look back at the graphs that look like an M and a W in Task 1.

a. Write and check the fit of models for the M whose function rules have these properties.

    i. A product of linear factors that has zeroes at $-2, 0$, and 2

    ii. A product of linear factors with zeroes at -2 and 2 and a repeated zero at $x = 0$

b. Write and check the fit of models for the W whose function rules have these properties.

    i. A product of linear factors that has zeroes at $-1, 1$, and 3

    ii. A product of linear factors with zeroes at -1 and 3 and a repeated zero at $x = 1$

c. Expand the products in Parts a and b that appear to give the best matches for the M and W drawings. Compare the resulting polynomials to the results produced in your work on Task 1. Reconcile any differences.

5 Express the quadratic function $f(x) = x^2 - 12x + 11$ in two ways: as the product of two linear factors and in vertex form. Then show how both forms can be used to find coordinates of these points on the graph of $y = f(x)$.

a. x-intercept(s)

b. y-intercept

c. maximum or minimum point

6 A farmer who grows wheat every summer has two kinds of expenses—*fixed costs* for equipment and taxes and *variable costs* for seed, fertilizer, and tractor fuel. Suppose that these costs combine to give total operating cost in dollars $C(x) = 25,000 + 75x$ for growing x acres of wheat.

a. What is a rule for the function $c(x)$ that gives the operating cost per acre on that farm?

b. The function giving operating cost per acre has a horizontal asymptote. Explore values of $c(x)$ as x increases to discover the equation of that asymptote.

c. Sketch a graph of $c(x)$. Explain what it shows about the way that operating cost per acre changes as the number of acres planted increases. In particular, explain what the horizontal asymptote tells about the relationship between number of acres planted and operating cost per acre.

7 Suppose that the wheat farmer in Task 6 can expect income of about $125 per acre.

a. What functions give total profit $P(x)$ and profit per acre $p(x)$ from planting x acres of wheat?

b. How many acres of wheat must the farmer plant to break even on the crop?

c. Sketch a graph of the profit per acre function $p(x)$ for $0 < x < 2,000$. What seems to be the horizontal asymptote for this graph? What does it tell about the relationship between profit per acre and number of acres planted?

8 Consider the functions $f(x) = \dfrac{2x + 7}{x^2 - 4}$ and $g(x) = \dfrac{x + 5}{x^2 - 4}$.

a. Sketch graphs of each function. Identify any asymptotes.

b. Describe the domains of the two functions.

c. Calculate expressions in simplest form for these combinations of $f(x)$ and $g(x)$.

   i. $f(x) + g(x)$

   ii. $f(x) - g(x)$

   iii. $f(x)g(x)$

   iv. $f(x) \div g(x)$

9 You know that when two integers are added, their sum is also an integer. The set of integers is said to be *closed* under the operation of addition.

a. Is the set of integers closed under the operation of subtraction? Under multiplication? Under division?

b. For which arithmetic operations is the set of polynomial expressions closed? Explain your reasoning.

c. For which arithmetic operations is the set of rational expressions closed? Explain.

SUMMARIZE THE MATHEMATICS

In this unit, you modeled problem situations using polynomial and rational functions and developed an understanding of some of the properties of those families of functions. You also learned how to add, subtract, and multiply polynomials and rational expressions.

a What kinds of functions are called:

  i. polynomial functions?

  ii. rational functions?

b Describe how you can determine:

  i. the degree of a polynomial.

  ii. local maximum or local minimum points on the graph of a polynomial function.

  iii. the zeroes of a polynomial function.

  iv. the sum or difference of two polynomials.

  v. the product of two polynomials.

c What is the vertex form of a quadratic function?

  i. What are the advantages and disadvantages of using this form?

  ii. How would you transform any quadratic expression in the form $ax^2 + bx + c$ into vertex form?

d For questions that call for solving quadratic equations $ax^2 + bx + c = 0$, how can you tell the number and kind of solutions from the value of the discriminant $b^2 - 4ac$?

e When working with rational expressions and functions, describe:

  i. when an expression will be undefined.

  ii. when the graph of a function will have a vertical asymptote.

  iii. when the graph of a function will have a horizontal asymptote.

  iv. strategies you can use to simplify the expression for a function.

  v. when two rational expressions can be added or subtracted and how you would go about performing the addition or subtraction.

  vi. how multiplication or division of two rational expressions can be performed.

Be prepared to share your examples and descriptions with the class.

 CHECK YOUR UNDERSTANDING

Write, in outline form, a summary of the important mathematical concepts and methods developed in this unit. Organize your summary so that it can be used as a quick reference in future units and courses.

6

Circles and Circular Functions

Circles are the most symmetric of all two-dimensional shapes. As a result, a circle has many interesting properties whose investigation will continue to develop your visual thinking and reasoning abilities. From another perspective, the rotation of a circle is key to the utility of wheels, circular saws, pulleys and sprockets, center-pivot irrigation, and many other mechanisms. The motion of the circle as it rotates is a special case of periodic change that can be modeled and analyzed using trigonometric or circular functions.

In this unit, you will learn about and prove properties of special lines and angles as they relate to circles. You will analyze circular motion, both how its power can be harnessed and multiplied in pulley and sprocket systems and how trigonometric functions model the motion. The key ideas will be developed through work on problems in two lessons.

LESSONS

1 Circles and Their Properties

State and prove properties of tangents to circles, chords, arcs, and central and inscribed angles. Interpret and apply these properties.

2 Circular Motion and Periodic Functions

Analyze pulley and sprocket systems in terms of their transmission factor, angular velocity, and linear velocity. Use sine and cosine functions of angles and radians to model circular motion and other periodic phenomena.

Circles and Their Properties

In Course 2 of *Core-Plus Mathematics,* you explored some properties of a circle and its center, radius, and diameter. You also learned how to determine the equation of a circle in a coordinate plane.

Circles are the most symmetric of all geometric figures. This symmetry makes them both beautiful and functional. Below, on the left, is a circle-based decorative design from the Congo region of Africa. On the right is a diagram of a circular barbeque grill.

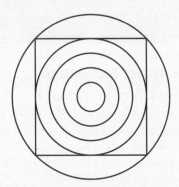

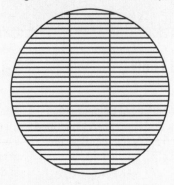

Consider the figures at the bottom of the previous page in light of your prior study of circles.

a All points in the plane of a circle with center O either lie on the circle, in its *interior*, or in its *exterior*. For any point P, how can you use the distance OP to determine whether P lies on a circle with center O and radius r; in its interior; or in its exterior?

b Suppose point P is in the interior of a circle, and line ℓ contains P. How many points of ℓ, if any, are on the circle? Use a sketch to help explain your answer.

c The five circles in the design on the bottom-left are **concentric**; that is, they share the same center. Why are these five circles similar?

d A **chord** is a line segment that joins any two distinct points on a circle. The barbeque grill shown on the previous page shows several chords. What are the longest chords in any circle? How do you think the length of a chord is related to its perpendicular distance from the center of the circle?

e In the decorative design, the sides of the square are chords of which of the five circles? What can you say about the sides of the square in relation to the second largest circle?

In this lesson, you will explore and verify important properties of special lines and angles in circles. To explore these properties, you may choose to use compass and straightedge, paper folding, measuring, or interactive geometry software. To reason about these properties, you will use congruent triangles, symmetry, coordinate methods, and trigonometric methods. The properties of circles you derive in the first lesson form a basis for the second lesson, which is the study of circular motion and how it is modeled.

INVESTIGATION **1**

Tangents to a Circle

As you learned in Course 2, lines that intersect a circle in exactly one point are said to be **tangent** to the circle. Tangent lines have great practical importance. For example, many satellites orbit Earth collecting and transmitting data that is crucial for the operation of modern communication systems. The sketch on the right at the top of the next page shows tangent lines to the circle of the equator from a satellite at point S. These tangent lines can help to determine the range of the satellite's signal.

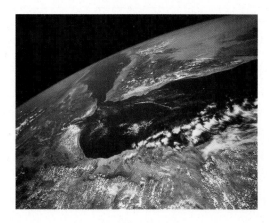

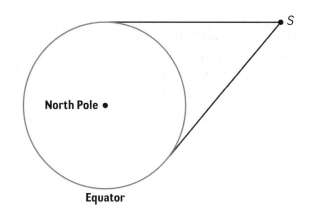

As you work on problems in this investigation, look for answers to the following question:

What are important properties of tangents to a circle,
and how can they be derived?

1 **Tangent at a Point on a Circle** Use a compass, straightedge and protractor, paper folding, or interactive geometry software to investigate properties of a line that is tangent to a circle.

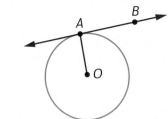

   a. Draw a circle like the one at the right with center O and a convenient radius r. Draw a tangent line touching the circle at just one point, A. Point A is called the *point of tangency*. Draw segment $\overline{OA}$. Relative to r, what is the length of $\overline{OA}$?

   b. Suppose B is any point on the tangent line other than A, the point of tangency. Relative to r, how long is $\overline{OB}$? Is B in the interior or exterior of the circle?

   c. Find the measure of the angle made by $\overline{OA}$ and the tangent line. Compare your finding with that of others. Write a conjecture about a tangent line and the radius drawn to the point of tangency.

   d. Is the converse of your conjecture in Part c also true? If you think so, write your conjecture in if-and-only-if form.

2 As with any if-and-only-if statement, proof of your conjecture in Problem 1 Part d requires an argument in both directions.

   a. One direction is the following:

      If a radius is perpendicular to a line at a point where the line intersects
      the circle, then the line is tangent to the circle at that point.

   Students at Peninsula High School wrote the following steps for a proof. Provide reasons that support each of the statements and correct any missteps.

Given: Point A on both line ℓ and circle with
     center O with $\overline{OA} \perp \ell$
Prove: Line ℓ is tangent to the circle with center O,
     that is, ℓ intersects the circle only in point A.

(1) $\overline{OA}$ is a radius. $\overline{OA} \perp \ell$ at point A.

(2) Let B be any point on line ℓ other than A.
    Draw $\overline{OB}$.

(3) △BAO is a right triangle with hypotenuse $\overline{OB}$.

(4) $OB > OA$

(5) Point B is in the exterior of the circle with center O.

(6) Line ℓ is tangent to the circle with center O.

b. What is the converse of the statement proved in Part a?

c. You are asked to prove the converse in Extensions Task 26. After your conjecture is proved in both directions, that establishes the if-and-only-if theorem. The theorem should be the same as the statement of the conjecture that you wrote for Problem 1 Part d. If necessary, adjust that statement to make it an accurate statement of the theorem. Add the theorem to your toolkit to use as needed in later problems.

3 Now refer to the figure shown at the right.

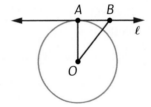

a. If line ℓ is tangent to circle O at point A, the radius of the circle is 4 inches, and $AB = 3$ inches, what is length BO? Explain why.

b. If $AB = 5$ cm, $AO = 12$ cm, and $BO = 13$ cm, why is it correct to conclude that line ℓ must be tangent to the circle at point A?

4 **Tangents from a Point not on a Circle** Examine the following algorithm for using a compass and straightedge or interactive geometry software tools to construct tangent lines to a given circle with center O from a given point P in the exterior of the circle.

Step 1. Draw $\overline{OP}$ and then construct its
    midpoint M.

Step 2. Construct the circle with center M
    and radius $\overline{OM}$. Label the points
    of intersection of the two circles A
    and B.

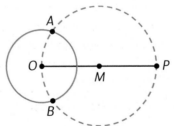

Step 3. Draw lines $\overleftrightarrow{PA}$ and $\overleftrightarrow{PB}$. These are
    the tangent lines to the circle with center O through point P.

a. On a sheet of paper, draw a large circle with center O and a point P in the exterior of the circle. Use the above algorithm to construct tangent lines to the circle with center O from point P.

b. To justify this construction, draw radii $\overline{OA}$ and $\overline{OB}$ and segments $\overline{PA}$ and $\overline{PB}$ on your copy of the figure from Part a. Provide an argument that $\overleftrightarrow{PA}$ and $\overleftrightarrow{PB}$ are tangent to the circle with center O.

5 Now suppose you are given that $\overline{PA}$ and $\overline{PB}$ are tangent to a circle centered at O. To help prove that $\overline{PA} \cong \overline{PB}$, *auxiliary line segments* $\overline{OA}, \overline{OB},$ and $\overline{OP}$ are drawn in the figure.

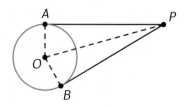

a. How could you use congruent triangles to prove that $\overline{PA} \cong \overline{PB}$?

b. How could you use the Pythagorean Theorem to show that $\overline{PA} \cong \overline{PB}$?

c. State in words the theorem you have proved about tangents drawn to a circle from an exterior point.

SUMMARIZE THE MATHEMATICS

In this investigation, you discovered and proved some important properties of tangents to a circle.

a Make a list summarizing the properties of tangents that you proved.

b Some mathematics books define a tangent to a circle with center O at point A to be a line perpendicular to radius $\overline{OA}$ at point A. With that definition, would a tangent line necessarily intersect the circle in exactly one point? Why or why not?

c If point A is any point on a circle with center O, is there always a unique line that is tangent to the circle at point A? Explain as precisely as you can.

Be prepared to compare your responses with those of others.

✔ **CHECK YOUR UNDERSTANDING**

Problems 3–5 of this investigation suggest how to determine measures related to tangents to a circle from an exterior point, such as a satellite orbiting Earth. Suppose a satellite is located in space at point S. In its view of Earth in the plane of the equator, the angle between the lines of sight at S is 50°. The radius of Earth is about 3,963 miles.

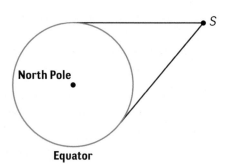

a. What is the distance from S to the horizon along the equator, that is, the length of a tangent from S to Earth's surface along the equator?

b. How high is the satellite S above Earth's surface, that is, the length of a segment S to the closest point on Earth's surface along the equator?

Chords, Arcs, and Central Angles

As noted at the beginning of this lesson, a **chord** of a circle is a line segment that joins two points of the circle. Any diameter is a chord, but chords may also be shorter than a diameter.

Automotive wheels come in a wide range of designs. One wheel option is pictured on the left below. Some students at Elmwood Park High School completed a project using the *CPMP-Tools* interactive geometry custom app "Design by Robot" to create what they thought would be a popular wheel design. A start of their design is shown below at the right.

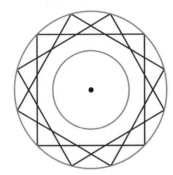

As you work on the problems of this investigation, look for answers to the following questions:

What are important properties of chords, arcs, and central angles of a circle?

How can these properties be proved and applied?

Relating Central Angles, Chords, and Arcs A **central angle** of a circle is an angle of measure less than 180° with vertex at the center of the circle and sides along radii of the circle. In the diagram below, $\angle AOB$ is a central angle and $\overline{AB}$ is the *corresponding chord*.

Each central angle separates a circle into two arcs. In the diagram, the arc $\overset{\frown}{ACB}$ that lies in the interior of the central angle is called a **minor arc**. The other arc $\overset{\frown}{ADB}$ is called a **major arc**. The simpler arc notation $\overset{\frown}{AB}$ is used to indicate the minor arc *intercepted* by the central angle $\angle AOB$.

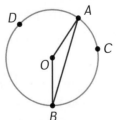

Arcs are commonly measured in degrees. The degree measure of a minor arc is equal to the measure of the corresponding central angle. The degree measure of a major arc is 360° minus the measure of the corresponding minor arc.

1 Suppose that in the diagram above m$\angle AOB = 160°$.

   a. What is m$\overset{\frown}{ACB}$, the measure of the minor arc $\overset{\frown}{ACB}$?

   b. What is m$\overset{\frown}{ADB}$, the measure of the major arc $\overset{\frown}{ADB}$?

   c. Make a copy of the sketch showing possible locations of a point X so that m$\overset{\frown}{AX} = 90°$.

2 By definition, two arcs have the same measure if and only if their central angles are congruent. But what about the corresponding chords? Consider chords $\overline{AB}$ and $\overline{CD}$ and their central angles pictured at the right.

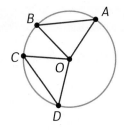

a. Suppose $AB = CD$. Prove that $m\angle AOB = m\angle COD$ and $m\overset{\frown}{AB} = m\overset{\frown}{CD}$.

b. Suppose $m\overset{\frown}{AB} = m\overset{\frown}{CD}$. Prove that $AB = CD$.

c. Summarize in an if-and-only-if statement what you have proved in Parts a and b.

3 **Relationships between Chords** Some other interesting properties of chords can be explored with compass-and-straightedge constructions or interactive geometry software.

a. Divide the work on these three constructions among your classmates so that each person completes one of the constructions. All three constructions should be done carefully, beginning with a large copy of the figure at the right.

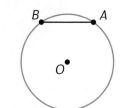

i. Construct the midpoint M of chord $\overline{AB}$. Draw line $\overleftrightarrow{OM}$.

ii. Construct line p, the perpendicular bisector of chord $\overline{AB}$.

iii. Construct line ℓ perpendicular to chord $\overline{AB}$ through point O.

b. Compare lines $\overleftrightarrow{OM}$, p, and ℓ that you and your classmates constructed. How are they alike? How are they different?

c. Based on your findings in the first two constructions, complete the following conjectures.

i. If a line contains the center of a circle and the midpoint of a chord, then _____.

ii. The perpendicular bisector of a chord of a circle contains _____.

iii. If a line through the center of a circle is perpendicular to a chord, then _____.

You are asked to prove these conjectures in Applications Task 3 in the On Your Own set.

4 The properties of chords that you explored in this investigation combined with trigonometric methods allow you to find measures of various angles and segments in circles. Suppose that a given circle has radius 6 inches.

a. What is the length of a chord that has a central angle of 115°?

b. What is the measure of the arc corresponding to a chord that is 8 inches long? What is the perpendicular distance from the center of the circle to the chord?

c. The perpendicular distance from the center of the circle to a chord is 4 inches. What is the length of the chord? What is the measure of its central angle?

SUMMARIZE THE MATHEMATICS

In this investigation, you explored and justified some interesting and useful properties of chords in the same or congruent circles.

a Describe how minor and major arcs are determined, named, and measured.

b Summarize how a chord, its minor arc, and its central angle are related.

c Complete the following if-and-only-if statement that summarizes two theorems explored in this investigation: A line through the center of a circle bisects a chord if and only if _____.

d If you know the radius of a circle and the length of a chord that is shorter than the diameter, describe how you can determine the measure of the central angle corresponding to the chord.

Be prepared to share your responses and thinking with the class.

☑ CHECK YOUR UNDERSTANDING

The design shown at the right was created using the *CPMP-Tools* "Design by Robot" custom app. The outer circle has diameter 8 inches. The chords that are shown are all of equal length.

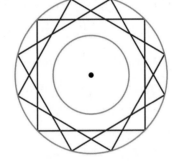

a. What is the length of each of the chords?

b. More chords can be created in the design, but none can come closer than 2 inches to the center of the circles.

  i. What is the longest possible chord that can be added to this design?

  ii. What is the measure of the central angle of one of these longest chords?

INVESTIGATION 3

Angles Inscribed in a Circle

Angle measurements and circular arcs are used as navigation aids for ships cruising in dangerous waters. For example, if a stretch of water off two lighthouses holds hazards like rocks, reefs, or shallow areas, the navigation charts might include a circle that passes through the two lighthouses and contains the hazards. It turns out that by carefully tracking the angle determined by the encompasses lighthouses and a ship, its captain can stay safely outside the hazardous area.

The key to this navigation strategy is the relationship between *inscribed angles* and arcs of circles. As you work on the problems of this investigation, look for answers to the following questions:

What is an inscribed angle in a circle?

How is the measure of an inscribed angle related to the arc it intercepts?

1 Make a drawing like that on the sketch below or use the "Lighthouse" sample sketch in *CPMP-Tools* to see how m∠ASC tells whether the ship is inside, outside, or on the circle of hazardous water conditions.

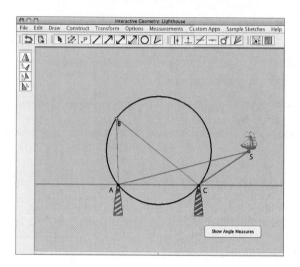

a. First find the approximate measure of $\overarc{AC}$

b. What seems to be true of m∠ASC when the ship is right on the boundary of the circle of hazardous water?

c. What seems to be true of m∠ASC when the ship is inside the circle of hazardous water?

d. What seems to be true of m∠ASC when the ship is outside the circle of hazardous water?

e. As captain of the ship, how could you use sighting instruments to keep safely outside the circle of hazardous water?

Inscribed Angles and Intercepted Arcs When the ship is located on the boundary of the region of hazardous water, the angle determined by the two lighthouses and the ship is called an **inscribed angle** in the circle. In general, if points A, B, and C are on the same circle, then those points determine three inscribed angles in that circle.

2 In the figures at the top of the next page, inscribed angle ∠ABC is said to **intercept** $\overarc{ADC}$ where $\overarc{ADC}$ consists of points A and C and all points on the circle O that lie in the interior of ∠ABC.

Explore the connection between the measures of inscribed angle ∠ABC and its intercepted arc $\overset{\frown}{ADC}$ using Parts a–c below as a guide.

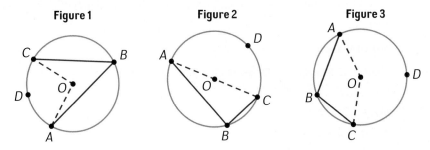

Figure 1   Figure 2   Figure 3

a. When $\overset{\frown}{ADC}$ is a semicircle, as in Figure 2, what is m∠ABC? How do the measures of ∠ABC and $\overset{\frown}{ADC}$ compare?

b. Using the "Explore Angles and Arcs" custom app or by drawing several diagrams, consider multiple instances of Figure 1 where ∠B is an acute angle on the circle. Does the same relationship between the measures of an acute inscribed angle and its intercepted arc that you noted in the case of the inscribed right angle seem to hold?

c. In Figure 3 above, ∠ABC is an obtuse inscribed angle. Its intercepted arc is *major arc* $\overset{\frown}{ADC}$. How can the measure of major arc $\overset{\frown}{ADC}$ be determined if you know m∠AOC? Use the "Explore Angles and Arcs" custom app to explore the relationship between the measure of obtuse inscribed angle ∠ABC and its intercepted arc $\overset{\frown}{ADC}$.

d. Write a conjecture about how the measure of an inscribed angle and the measure of its intercepted arc seem to be related.

3 You noted in Problem 2 Part a that $m\overset{\frown}{ADC} = 180°$ when m∠B = 90°. You also noted in Part c that when the measure of the inscribed angle is greater than 180°, you can find the measure of the major arc by subtracting the measure of the minor arc from 360°. This means that to prove the conjecture:

> *The measure of an angle inscribed in a circle is half the measure of its intercepted arc.*

it is sufficient to prove this statement for an acute angle. Consider the three cases below. In each case, ∠B is an acute angle.

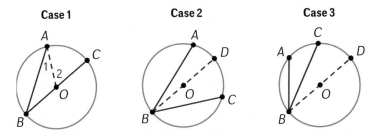

Case 1   Case 2   Case 3

a. How do these cases differ from one another?

b. A **proof by cases** strategy for establishing this conjecture is to prove it for Case 1 first. Then use that result to prove the conjecture for the other two cases. Why would carrying out this strategy successfully constitute a complete proof of the conjecture?

4 In Case 1, one side of the inscribed angle contains the center O of the circle. Study the following proof of the conjecture in Problem 3 by students at Prairie du Chien High School in Wisconsin. Give a reason why each of the statements is correct.

Proof of Case 1:

Center O lies on $\overline{BC}$. Draw radius $\overline{AO}$	(1)
In $\triangle AOB$, $\overline{AO} \cong \overline{BO}$.	(2)
$m\angle ABC = m\angle 1$	(3)
$m\angle 2 = m\angle ABC + m\angle 1$	(4)
$m\angle 2 = 2(m\angle ABC)$	(5)
$m\angle 2 = m\widehat{AC}$	(6)
$2(m\angle ABC) = m\widehat{AC}$	(7)
$m\angle ABC = \frac{1}{2}(m\widehat{AC})$	(8)

The proof of Case 1 shows that whenever one side of an inscribed angle is a diameter of the circle, its measure is half the measure of its intercepted arc. In Applications Task 8, you will be asked to prove Cases 2 and 3 to complete a proof of the **Inscribed Angle Theorem**.

Angles Intercepting the Same Arc Think back to the introduction of this investigation and the problem of assuring that a ship navigates outside a circle of hazardous water conditions. Suppose that a captain wants to steer the ship along the boundary of that circle.

5 Use the "Explore Angles and Arcs" custom app to see how, if at all, the measure of $\angle ABC$ changes as the ship moves.

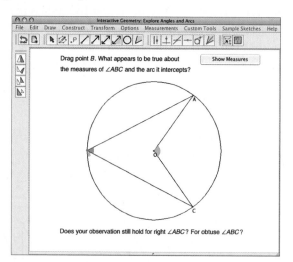

a. Keeping $m\widehat{AC}$ constant, drag the vertex B on the circle always keeping it on the same side of $\widehat{AC}$. How does the measure of $\angle B$ change as you move its vertex to various locations on the circle?

b. Now change the measure of $\overset{\frown}{AC}$ by dragging point A or point C. Then drag vertex B again. Make a conjecture about the measures of inscribed angles that intercept the same arc.

c. How could the Inscribed Angle Theorem be used to prove this result?

SUMMARIZE THE MATHEMATICS

In this investigation, you explored the connection between an inscribed angle and its intercepted arc.

a Summarize the two main properties that you considered in this investigation. How were these properties used to help a captain of a ship avoid a hazardous area?

b If $\angle ABC$ is inscribed in a semicircle, what is the measure of its intercepted arc? What is the measure of $\angle ABC$?

c A central angle and an inscribed angle intercept the same arc. How are the measures of the angles related?

Be prepared to explain your ideas and reasoning to the class.

 CHECK YOUR UNDERSTANDING

In the diagram below, $\overline{BD}$ is a diameter of the circle with center O. Points A, B, C, and D are on the circle.

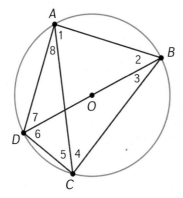

a. If $m\overset{\frown}{AB} = 100°$, find the measures of as many of the numbered angles as possible.

b. Given the two measures $m\angle 1 = 55°$ and $m\angle 2 = 50°$, find the measures of the four minor arcs $\overset{\frown}{AB}$, $\overset{\frown}{BC}$, $\overset{\frown}{CD}$, and $\overset{\frown}{DA}$.

APPLICATIONS

1 Furniture makers often round corners on tables and cabinets.

a. In the figure at the right, the lower-left corner is to be rounded, using a circle with a radius of 2 inches. Lines $\overleftrightarrow{AB}$ and $\overleftrightarrow{CB}$ will be tangent to the circle. When the corner is rounded, point B will be cut off and points A and C will lie on the circular arc. Explain how to apply tangent theorems to locate the center of the circle that rounds the corner. Sketch the rounded corner.

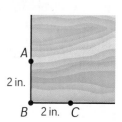

b. Determine the distance from point B to the center of the circle.

c. As shown in the figure, point B is equidistant from points A and C. When this corner is rounded as described in Part a, this illustrates a tangent theorem. Which one?

2 Prove the following direct result (**corollary**) of the theorem about tangent segments from a point exterior to a circle.

In the figure at the right, two tangent segments, $\overline{PA}$ and $\overline{PB}$, are drawn to the circle with center O from an exterior point P. What ray appears to be the bisector of $\angle APB$? Prove your answer.

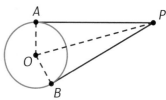

3 Suppose a circle with center O has radius 10 cm, P is a point in the exterior of the circle, and the tangent segments to the circle from point P are $\overline{AP}$ and $\overline{BP}$.

a. If point P is 18 cm from the center of the circle, determine the length of $\overline{AP}$ and the measure of $\angle APB$.

b. If m$\angle APB = 48°$, determine the lengths OP, AP, and BP.

4 Prove your three conjectures from Investigation 2, Problem 3 (page 402).

a. *If a line contains the center of a circle and the midpoint of a chord, then the line is perpendicular to the chord.*

b. *The perpendicular bisector of the chord of a circle contains the center of the circle.*

c. *If a line through the center of a circle is perpendicular to a chord, then the line contains the midpoint of the chord.*

5 At the beginning of this lesson, you explored some properties of the decorative design from the Congo region of Africa reproduced below. In this design, there are five concentric circles and square *ABCD*. The vertices of *ABCD* lie on the largest circle; that is, the square is inscribed in the largest circle. The sides of the square are tangent to the second largest circle; that is, the square is circumscribed about the second largest circle. In this task, you will continue to explore this design.

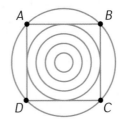

a. Suppose the radius of the smallest circle is 0.2 inches. The radii of the next three circles in increasing size are 0.2 inches more than the previous one. What is the length of a side of square *ABCD*?

b. What is the radius of the largest circle?

c. What is m$\widehat{AB}$? What is m$\widehat{ABC}$? What is the measure of major arc $\widehat{ABD}$?

d. Suppose point *E* is the point of tangency of $\overline{AB}$ to the second largest circle. Why must *E* also be the midpoint of $\overline{AB}$?

6 A broken piece of a circular plate shaped like the picture below is dug up by an archaeologist, who would like to determine the plate's original appearance and size.

a. Explain how to find the location of the center of the plate.

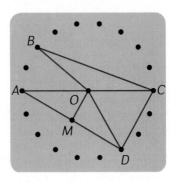

b. Once you find the center, how could you find the circumference of the circular plate?

7 Eighteen equally spaced pegs form a "circle" on the geoboard below. The measure of the diameter of circle *O* is 10 cm, and *M* is the midpoint of $\overline{AD}$.

Find the following measures:

a. m$\widehat{AB}$

b. m$\widehat{BC}$

c. *BO*

d. *BC*

e. *CD*

f. m∠*OMA*

8 Case 2 and Case 3 of the Inscribed Angle Theorem in Investigation 3, Problem 3 are shown below. In each figure, $\overline{BD}$ is a diameter. Assume these two basic properties.

Angle Addition Postulate:
If P is in the interior of ∠ABC, then m∠ABP + m∠PBC = m∠ABC.

Arc Addition Postulate:
If P is on $\widehat{AC}$, then $m\widehat{AP} + m\widehat{PC} = m\widehat{APC}$.

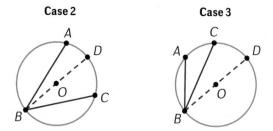

a. Prove Case 2: center *O* is in the interior of inscribed angle ∠*ABC*.

b. Prove Case 3: center *O* is in the exterior of inscribed angle ∠*ABC*.

9 As you saw in Investigation 3, to protect ships from dangerous rocks as they approach a coastline, the rocks in an area near shore are charted and a circle that contains all of them is drawn. Two lighthouses at *A* and *C* are on the circumference of the circle. The measure of any inscribed angle that intercepts $\widehat{AC}$, such as ∠*ABC* is called a *horizontal danger angle*.

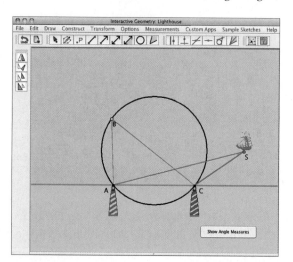

a. What is the measure of $\widehat{AC}$ in terms of the horizontal danger angle?

b. Suppose the lighthouses at *A* and *C* are 0.6 miles apart, and the horizontal danger angle is 35°. What is the radius of the hazardous water circle?

c. What is the perpendicular distance from the center of the hazardous water circle to shore?

10 A quadrilateral is inscribed in a circle if each vertex is a point on the circle. Prove the theorem:

If a quadrilateral is inscribed in a circle, then its opposite angles are supplementary.

11 Suppose △*ABC* is inscribed in a circle with diameter $\overline{BC}$ and $\overleftrightarrow{DE}$ is tangent to the circle at point *C*.

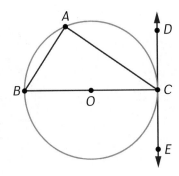

   a. Prove that m∠*A* = m∠*BCE* and m∠*B* = m∠*DCA*.

   b. ∠*DCA* and ∠*BCE* are angles formed by a chord of the circle and a tangent line at the endpoint of the chord. ∠*DCA* intercepts minor arc $\overset{\frown}{AC}$, and ∠*BCE* intercepts semicircle $\overset{\frown}{BC}$. Using Part a, what is the relationship between the measures of these angles and of their intercepted arcs? Explain your reasoning.

   c. In Part b, you justified the relationship between the measure of an angle formed by a chord and a tangent at an endpoint of the chord and its intercepted arc for acute angle ∠*DCA* and right angle ∠*BCE*. Does the same relationship hold for obtuse angle ∠*ACE* and its intercepted arc $\overset{\frown}{ABC}$? Justify your answer.

   d. Parts a, b, and c prove a theorem about the measures of angles formed by a chord of a circle and a tangent line at the endpoint of the chord. Write a statement of the theorem and add it to your toolkit.

CONNECTIONS

12 The *interior* and *exterior of a circle* are defined in terms of the circle's radius and center. Consider these ideas from a coordinate perspective.

   a. A circle with radius *r* is centered at the origin. If *P*(*x*, *y*) is any point in the coordinate plane, what conditions must *x* and *y* satisfy in each case below?

      i. *P* is on the circle.

      ii. *P* is in the interior of the circle.

      iii. *P* is in the exterior of the circle.

   b. Two overlapping circles shown at the right are centered at (0, 0) and (3, 0), respectively. They are inscribed in the rectangle as shown. What inequalities must *x* and *y* satisfy if *P*(*x*, *y*) is in each of the following regions in the diagram?

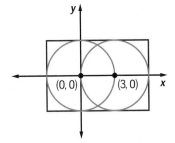

      i. Inside the rectangle

      ii. Inside the left circle but not the right one

      iii. Inside both circles

      iv. Inside the rectangle but in the exterior of both circles

13 Symmetrical circular black-and-white designs, called Lunda designs, are used in the Lunda region of Angola. The basis for some of these designs is the left-hand figure. Two completed designs are shown to the right of that figure.

a. In the figure on the left, suppose the radii are extended to the common center of the concentric circles. Twenty-four central angles of equal measures would be formed.

 i. What is the measure of each central angle?

 ii. Choose one central angle that intercepts nine different circles. What are the degree measures of these arcs?

b. In the basis figure on the left, suppose the radius of the innermost concentric circle is 1 cm. As the circles increase in size, their radii are 2 cm, 3 cm, 4 cm, and so on.

 i. What are the center and magnitude of a size transformation that maps the smallest circle to the largest circle? The second smallest circle to the third largest circle? The second largest circle to the fifth smallest circle?

 ii. What is the circumference of the largest circle? What is the length of each arc (of the largest circle) intercepted by the congruent central angles?

c. Describe the symmetries of the middle Lunda design above.

d. Describe the symmetries of the Lunda design on the right above.

e. Use an enlarged copy of the base design to create an interesting figure that has rotational symmetry of only 120° and 240°.

14 Two groups of students at Rapid City High School conjectured:

 If a line through the center of a circle is perpendicular to a chord, then it bisects the chord.

They each presented a different plan for proof.

a. Use the following proof plan of one group as a guide to write a proof for their conjecture.

Synthetic Proof Plan—Draw the circle with center O, where $\overline{OM}$ is perpendicular to chord $\overline{AB}$. Show $\triangle AMO$ and $\triangle BMO$ are congruent right triangles. So, $\overline{MA} \cong \overline{MB}$. Then consider the case when $\overline{AB}$ is a diameter of the circle.

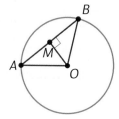

b. Use the following proof plan of the other group as a guide to help write a coordinate proof for this conjecture.

> **Coordinate Proof Plan—Draw a coordinate system. Place the center of the circle at the origin with the x-axis perpendicular to chord $\overline{AB}$. Then explain why this placement can be done for any chord. Label point $A(a, b)$ and explain why the x-coordinate of B must be a. Then, reason from the equation of a circle to find the y-coordinate of point B. Finally, show that A and B are the same distance from the x-axis and so the x-axis bisects chord $\overline{AB}$.**

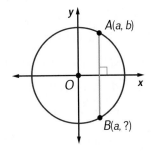

c. Which proof, the synthetic or the coordinate one, was easier for you to understand? Why?

15 *All-Star Baseball* is a classic fantasy game that simulates the batting records of current and former all-star baseball players by dividing a circular disk into *sectors* (portions of the disk bounded by two radii and an arc on the disk) representing hits of various types, walks, strikeouts, fly outs, and ground outs. The ratio of the measure of the central angle of each sector on a particular all-star player's disk to 360° is proportional to his hitting records. For example, a player who hit a double 6% of the times that he batted will have a "double" sector of (0.06)(360°), or 21.6°. When it is his turn to bat in *All-Star Baseball*, a player's disk is fitted on a random spinner. The spinner is spun, and the sector to which the arrow points when it stops determines the outcome of the player's time at bat.

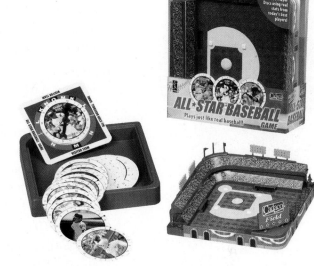

a. Would you expect a player's average hitting performance over a large number of "at-bats" in *All-Star Baseball* to approximate his actual batting record? Why or why not?

b. Babe Ruth was one of the greatest home-run hitters in baseball history. In 1927, he hit 60 home runs, the single season record for many years. He made 691 appearances at bat that year.

  i. How many degrees would Ruth's home-run sector for 1927 be in *All-Star Baseball*?

  ii. When Babe Ruth (1927 disk) comes to bat in *All-Star Baseball*, what is the probability that he will hit a home run?

c. During the 2013 season, Chris Davis hit 53 home runs, the most in major league baseball that year. On a disk based on Davis's 2013 statistics, the home run sector begins at the terminal side of 76.1° and ends at the terminal side of 103.9°.

  i. How many times did Davis come to bat in 2013?

  ii. When Chris Davis (2013 disk) comes to bat in *All-Star Baseball*, what is the probability that he will hit a home run?

Cadaco

16 A circle in a coordinate plane contains points $A(1, 6)$, $B(6, 1)$, and $C(-2, -3)$.

    a. Plot these points on graph paper.

    b. Find the coordinates of the center O of the circle. What is the radius of the circle?

    c. What is the equation of the circle?

    d. What is the length of chord $\overline{AC}$?

    e. What is the measure of $\angle AOC$? What is the measure of $\angle ABC$?

17 Water wheels are frequently used to generate energy and distribute water for irrigation. The water wheel pictured at the left operates in a river in the Guangxi Province of China.

In the diagram at the right, the 30-foot diameter water wheel has been positioned on a coordinate system so that the center of the wheel is the origin and the wheel reaches 3 feet below water level. As the wheel rotates, it takes 12 seconds for point A on the circumference of the wheel to rotate back to its starting position.

    a. What are the coordinates of the other labeled points?

    b. At its current position, how far is point A from the surface of the water?

    c. If the wheel rotates counterclockwise, through what angle does point A rotate to first reach the water at D? How many seconds does it take A to first reach the water?

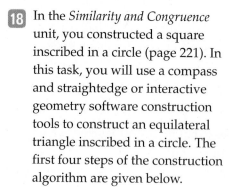

    d. The surface of the water forms a chord $\overline{DE}$. What is the measure of $\overset{\frown}{DE}$? What is the length of chord $\overline{DE}$?

    e. What is the measure of inscribed angle $\angle DAE$?

18 In the *Similarity and Congruence* unit, you constructed a square inscribed in a circle (page 221). In this task, you will use a compass and straightedge or interactive geometry software construction tools to construct an equilateral triangle inscribed in a circle. The first four steps of the construction algorithm are given below.

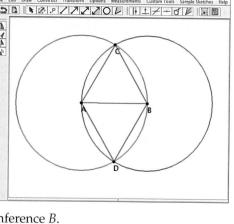

    Step 1. Construct a circle with center point A. Label a point on the circumference B.

    Step 2. Construct a congruent circle whose center is point B.

    Step 3. Mark and label the intersection points of the two circles, C and D as shown.

    Step 4. Draw $\overline{AB}$, $\overline{AC}$, $\overline{AD}$, $\overline{CB}$, and $\overline{DB}$.

Michael Townsend/Stone/Getty Images

a. You may find these questions helpful in completing the construction of an equilateral triangle inscribed in the circle with center A.

- What type of triangle is $\triangle ABC$? Explain.

- Can $\overline{CD}$ be one of the sides of an inscribed equilateral triangle? Why or why not?

b. Write and carry out the remaining steps in the construction algorithm. Justify the construction is correct.

19 Extending your work on Task 18, write an algorithm that guides the construction of a regular hexagon inscribed in a circle.

a. Construct a regular hexagon inscribed in a given circle. Justify the steps of your construction.

b. For one of your constructions in Tasks 18 and 19, describe how it could be accomplished using a method different from that outlined in the original algorithm.

REFLECTIONS

20 Illustrate and describe how to construct a line tangent to a circle at a point on the circle. This construction is based on a theorem in this lesson. Which one?

21 Summarize and illustrate with a figure the theorems that relate a chord, its midpoint, and the center of a circle. Explain why these theorems follow from the fact that a circle is symmetric about any line through its center.

22 Use your knowledge about degree measures of arcs and central angles to help answer the following questions related to linear measure and area measure. Suppose a center-pivot irrigation system waters an agricultural field as shown in the diagram.

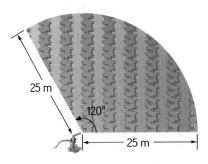

a. What is the area of the watered section of the field which is a sector of the circle with radius 25 meters?

b. If the farmer wishes to fence in this section of field, what length of fencing would be needed?

23 Arcs of circles can be measured in degrees as you did in Investigation 2 or in linear measures as when finding the circumference of a circle.

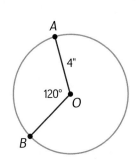

a. In the circle with center O at the right, $AO = 4"$, and $m\angle AOB = 120°$, find $m\widehat{AB}$ in degrees and the length of $\widehat{AB}$ in inches.

b. Is the following statement true or false? Explain.

Arcs that have the same degree measure also have the same linear measure.

24 A theme of many of the lessons in Course 3 has been to construct sound arguments and critique the reasoning of others. Identify a problem in Lesson 1 in which you critiqued the reasoning of a classmate. What did you and your classmate learn from this experience?

25 How does the arc intercepted by an acute, right, or obtuse central angle of a circle compare to a semicircle of the same circle? Of different circles?

EXTENSIONS

26 Prove: *If a line is tangent to a circle, then the line is perpendicular to the radius at the point of tangency.* You may use the following two facts.

- If a line contains an interior point of a circle, then it intersects the circle in two points.

- The shortest distance from a point to a line is the length of the perpendicular segment.

27 The following is a variation of an ancient Chinese problem. When the stem of a water lily is pulled up so it is vertical, the flower is 12 centimeters above the surface of the lake. When the lily is pulled to one side, keeping the stem straight, the blossom touches the water at a point 22 centimeters from where the vertical stem cuts the surface. The figure at the right is not drawn to scale; the center of the circle has been placed on the bed of the pond. What is the depth x of the water?

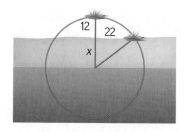

28 A familiar pattern in developing new mathematical ideas is to explore, conjecture, and then prove or disprove the conjecture.

a. If two chords of the same circle have different lengths, which is closer to the center of the circle? Explore figures like the one on the right in which one chord, $\overline{CD}$, is longer than the other, $\overline{AB}$. How do you think their perpendicular distances from the center, ON and OM, are related?

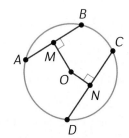

b. Using right triangles $\triangle AOM$ and $\triangle DON$, prove your conjecture.

29 Circles and similar triangles provide the foundation for a compass-and-straightedge construction of the square root of the length of any segment. Study the construction algorithm given below.

Step 1. Segments of length 1 and x are given. Draw a segment that is greater than $x + 1$. Label one endpoint A.

Step 2. On this segment, mark point B with your compass so that $AB = x$. Then mark point C so that $BC = 1$.

Step 3. Construct the midpoint of $\overline{AC}$. Label the midpoint O. Draw the circle centered at O with radius $OA = OC$.

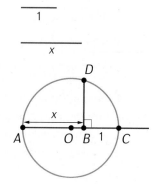

Step 4. Construct the line perpendicular to $\overline{AC}$ at point B. Label the point D at which the perpendicular line intersects the upper semicircle. Then $BD = \sqrt{x}$, as you will prove in Parts b and c below.

a. To test this construction, draw two segments, one 1 inch long and the other $x = 3$ inches long. Use Steps 1–4 to construct a segment that is $\sqrt{3}$ inches long. Measure the segment with your ruler. What is its length to the nearest sixteenth of an inch?

b. To justify this construction, draw segments $\overline{AD}$ and $\overline{CD}$. What kind of angle is $\angle ADC$?

c. Show that $\triangle ABD$ is similar to $\triangle DBC$. Explain why $\frac{x}{BD} = \frac{BD}{1}$. What is the length of $\overline{BD}$ in terms of x?

30 The theorems about angles and chords in circles can help justify interesting conjectures about products of lengths of segments that intersect circles. Consider the following instance.

Case 1   **Case 2** 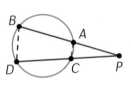   **Case 3**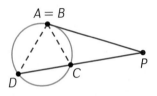

a. The "Chord Lengths in Circles" sample sketch in *CPMP-Tools* Interactive Geometry is designed to help you gather evidence that for each case above, $PA \times PB = PC \times PD$. Use the sketch to consider multiple instances of each case.

b. In each case shown above, prove that $\triangle ACP$ is similar to $\triangle DBP$.

c. Show that the result in Part a implies that $PA \times PB = PC \times PD$. To what does this equation reduce in Case 3?

31 The result described below is called Keith's Corollary, named for the high school student who discovered it (although it is not a direct result of any theorem). In the diagram, $ABCD$ is an arbitrary quadrilateral inscribed in a circle. Each of its four angles is bisected, and the bisectors intersect the circle at the points E, F, G, and H.

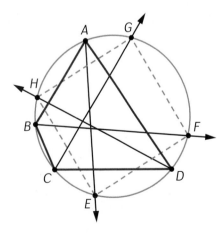

Prove that $EFGH$ is a rectangle. *Hint:* Show that the angles of $EFGH$ are right angles by using the facts that opposite angles of quadrilateral $ABCD$ are supplementary (see Applications Task 10) and that two inscribed angles intercepting the same arc have equal measure.

REVIEW

32 Write each expression in standard polynomial form.

  a. $(3x^3 + 4x^2 - 5x) + (9x^2 - 3x + 12)$

  b. $(x^5 - x^3 - 2x - 1) - (5x^5 - x^3 + 5x + 10)$

  c. $(x^2 + 3x)(x - 2)$

  d. $\dfrac{4x^2 - 2x^2 + 8x}{2x}$

33 Make a sketch of the graph of each function. Label the x- and y-intercepts. Compare each graph to the graph of $f(x) = x^2$.

  a. $f(x) = ax^2$, where $a > 1$ **b.** $f(x) = ax^2 + 2$, where $0 < a < 1$

  c. $f(x) = ax^2$, where $a < 0$ **d.** $f(x) = x^2 + b$, where $b > 0$

  e. $f(x) = 0.5x^2 + b$, where $b < 0$ **f.** $f(x) = -x^2 + b$, where $b > 0$

34 Find the perimeter and area of each figure. Each curved boundary is a portion of a circle.

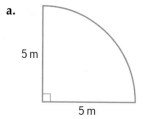

  a.

  5 m

  5 m

  b.

  6"

  6"

  3"

35 Solve each inequality.

  a. $-10x + 5 \geq 9 + 7x$ **b.** $|3a - 2| \leq 0$

  c. $x^2 + 4x > 5$ **d.** $\dfrac{-6}{x} < \dfrac{x + 1}{2}$

36 Quadrilateral $ABCD$ is a rhombus with $AB = 12$ cm and $m\angle BAD = 120°$. Determine each of the following lengths or angle measures.

  a. $m\angle ABC$ **b.** BC

  c. $m\angle BAC$ **d.** AC

  e. BD

37 Rewrite each expression in an equivalent form using only positive exponents.

  a. $3x^{-2}(x^4 + 5)$ **b.** $(4x^3y^{-2})^3$

  c. $\left(a^{-\frac{1}{2}}b^{\frac{1}{3}}\right)^6$ **d.** $\left(t^{-\frac{1}{4}}\right)\left(2t^{\frac{9}{4}}\right)$

38 Suppose the terminal side of an angle in standard position with measure θ contains the indicated point. Find $\sin \theta$, $\cos \theta$, and $\tan \theta$. Then find the measure of θ to the nearest degree.

  a. $P(3, 4)$ **b.** $P(3, -4)$

  c. $P(-3, 4)$ **d.** $P(0, -5)$

39 In the diagram below, $\overline{AB}$ is the perpendicular bisector of $\overline{CD}$.

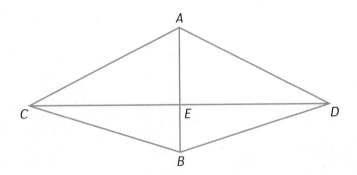

a. Is $\overline{CE} \cong \overline{DE}$? Explain your reasoning.

b. Is $\overline{AC} \cong \overline{AD}$? Explain your reasoning.

c. Is $\triangle CEB \cong \triangle DEB$? Justify your reasoning.

d. What type of quadrilateral is quadrilateral $CBDA$? Justify your response.

40 Solve each equation.

a. $2(3x) = 96$

b. $3x^2 = 96$

c. $3 + x^2 = 96$

d. $3 + 2x = 96$

e. $3(2x) = 96$

f. $(3 + x)^2 = 96$

41 Find the missing side lengths in each right triangle.

a.
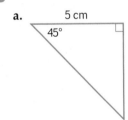

b.
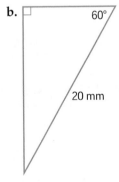

42 If $f(x) = 2x^2 + 11x - 6$, $g(x) = x^2 - 9$, and $h(x) = f(x) \cdot g(x)$, find each of the following.

a. $h(-2)$

b. The degree of $h(x)$

c. All zeroes of $h(x)$

d. The factored form of $h(x)$

e. The y-intercept of $h(x)$

f. The end behavior of $h(x)$

Circular Motion and Periodic Functions

The perfect radial symmetry of circles makes them attractive features in decorative designs and works of art. But that defining property is most useful when circular objects like disks, spheres, and cylinders are set in motion. So, it helps to understand mathematical strategies for measuring rates of circular motion and for modeling the (*time, position*) patterns that are generated.

For example, the instrument panel on a high performance car usually contains both a speedometer and a tachometer. The speedometer shows forward speed in miles or kilometers per hour; the tachometer shows revolutions per minute (rpm) by the engine crankshaft.

The investigations of this lesson develop strategies for measuring the velocity of circles rotating about axes and rolling along tangent lines. Trigonometric functions are then used to model the relationship between time and position of points on such rotating objects. Finally, those functions are used to describe the *periodic variation* of other phenomena like seasons of the year, pendulums, ocean tides, and alternating electrical current.

Angular and Linear Velocity

There are two common measures of circular motion. The angle of rotation about the center of a circle in a unit of time is called its **angular velocity**. It is commonly measured in units like revolutions per minute. The distance traveled by each point on the circle in a unit of time is called its **linear velocity**. It is commonly measured in units like miles per hour or meters per second.

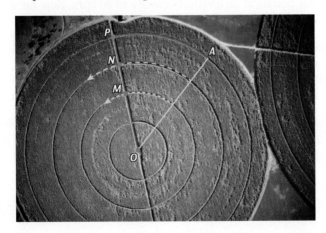

As you work on the problems in this investigation, look for answers to the following questions:

> *How are angular and linear velocity related in circular motion?*
>
> *How do mechanical systems connect the motion of driver and follower circles?*

©Michael Fay/Getty Images

Bicycle Gear Ratios One familiar example of a transmission connecting moving circles is the gear mechanism used in bicycles. For example, suppose that a mountain bike is set up so that the pedal sprocket in use has 42 teeth and the rear-wheel sprocket in use has 14 teeth of the same size.

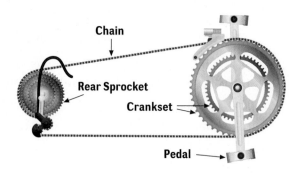

1 Consider the given "teeth per sprocket" information above.

    a. What does this information tell you about the circumferences of the two sprockets?

    b. Suppose that the rider is pedaling at an angular velocity of 80 revolutions per minute.

      i. What is the angular velocity of the rear sprocket?

      ii. What is the angular velocity of the rear wheel?

2 Suppose that the wheels on the mountain bike have a radius of 33 centimeters.

    a. How far does the bike travel for each complete revolution of the 14-tooth rear sprocket?

    b. How far does the bike travel for each complete revolution of the front pedal sprocket?

    c. If the rider pedals at an angular velocity of 80 rpm, how far will she travel in one minute?

    d. How long will the rider need to pedal at 80 rpm to travel 2 kilometers?

3 In the connection between the pedal and rear sprockets of a bicycle, the pedal sprocket is called the **driver** and the rear sprocket is called the **follower**.

    a. Write rules that express the relationships between angular velocities of the driver v_d and follower v_f when the bicycle gears are set in the following ways.

      i. Driver sprocket with 30 teeth and follower sprocket with 15 teeth

      ii. Driver sprocket with 20 teeth and follower sprocket with 30 teeth

    b. Which of the gear settings in Part a will make pedaling harder for the rider? Why?

Crankshafts and Transmissions Many internal combustion engines have pistons that move up and down, causing a cylindrical crankshaft to rotate. The crankshaft *driver* then transfers power to the *follower* wheels but also to other parts of the engine. For example, in many cars, a belt connects the crankshaft to pulleys that drive the fan and the alternator. When the engine is running, the fan cools the radiator while the alternator generates electrical current.

4. The pulley diameters from the diagram above allow us to compare angular velocities of the crankshaft, alternator, and fan pulleys. If the crankshaft makes one complete revolution:

   a. how many revolutions will the alternator pulley make?

   b. how many revolutions will the fan pulley make?

5. Write rules expressing the relationships of angular velocities for the driver v_d and follower v_f in case the crankshaft is the driver and:

   a. the alternator pulley is the follower.

   b. the fan pulley is the follower.

6. Suppose that the crankshaft (with diameter 10 cm) of the car is rotating with angular velocity of 1,500 rpm.

   a. Calculate the linear velocities in centimeters per minute of points on the edge of each pulley.

      i. crankshaft

      ii. alternator

      iii. fan

   b. Explain why the pattern of your answers in Part a makes sense if one thinks about how the belt and pulley mechanism operates.

In this investigation, you studied angular and linear velocities—important concepts in the design and analysis of connections between rotating circular objects.

 a What do angular and linear velocity measurements tell about the motion of a circular object?

b What are some common units of measurement for angular velocity? For linear velocity?

c How can you use information about the radii of two connected sprockets or pulleys to determine the relationship between their angular and linear velocities?

d If you know the angular velocity and radius of a rotating circular object, how can you determine the linear velocity of points on its circumference?

Be prepared to explain your ideas and reasoning to the class.

✔ CHECK YOUR UNDERSTANDING

Because power plants that burn fossil fuels are known to increase atmospheric carbon dioxide, many countries are turning to alternative, cleaner sources of electric power. In some places, it makes sense to construct pollution-free wind farms like that pictured below.

a. Suppose that a wind turbine has blades that are 100 feet long and an angular velocity of 2.5 rpm. What is the linear velocity of points at the tip of each blade?

b. Suppose a wind turbine connects a driver pulley with diameter 6 feet to a follower pulley with diameter 4 feet. What is the angular velocity of the follower pulley when the angular velocity of the driver is 3 rpm?

Modeling Circular Motion

The Ferris wheel was invented in 1893 as an attraction at the World Columbian Exhibition in Chicago, and it remains a popular ride at carnivals and amusement parks around the world. The wheels provide a great context for study of circular motion.

Ferris wheels are circular and rotate about the center. The spokes of the wheel are radii, and the seats are like points on the circle. The wheel has horizontal and vertical lines of symmetry through the center of rotation. This suggests a natural coordinate system for describing the circular motion. As you work on the problems in this investigation, look for an answer to the following question:

How can the coordinates of any point on a rotating circular object
be determined from the radius and angle of rotation?

Coordinates of Points on a Rotating Wheel To aid your thinking about positions on a rotating circle, it might be helpful to make a Ferris wheel model that uses a disk to represent the wheel with x- and y-coordinate axes on a fixed backboard.

Connect the disk to the coordinate axis backboard with a fastener that allows the disk to turn freely while the horizontal and vertical axes remain fixed in place.

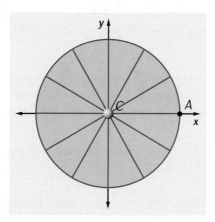

1 Imagine that your model represents a small Ferris wheel that has radius 1 decameter (about 33 feet) and that your seat is at point A when the wheel begins to turn counterclockwise about its center at point C.

   a. How does the x-coordinate of your seat change as the wheel turns?

   b. How does the y-coordinate of your seat change as the wheel turns?

2 Find angles of rotation between 0° and 360° that will take the seat from point A to the following special points.

   a. Maximum and minimum distance from the horizontal axis

   b. Maximum and minimum distance from the vertical axis

   c. Points with equal x- and y-coordinates

   d. Points with opposite x- and y-coordinates

When a circle like that modeling the Ferris wheel is placed on a rectangular coordinate grid with center at the origin $(0, 0)$, you can use what you know about geometry and trigonometry to find the x- and y-coordinates of any point on the circle.

3 Find coordinates of points that tell the location of the Ferris wheel seat that begins at point $A(1, 0)$ when the wheel undergoes the following rotations. Record the results on a sketch that shows a circle and the points with their coordinate labels.

    a. $\theta = 30°$ **b.** $\theta = 70°$ **c.** $\theta = 90°$

    d. $\theta = 120°$ **e.** $\theta = 140°$ **f.** $\theta = 180°$

    g. $\theta = 220°$ **h.** $\theta = 270°$ **i.** $\theta = 310°$

4 When the Ferris wheel has rotated through an angle of 40°, the seat that started at $A(1, 0)$ will be at about $A'(0.77, 0.64)$. Explain how the symmetry of the circle allows you to deduce the location of that seat after rotations of 140°, 220°, 320°, and some other angles as well.

5 Suppose that $P(x, y)$ is a point on the Ferris wheel model with m$\angle PCA = \theta$ in degrees.

    a. What are the coordinates x and y?

    b. How will the coordinate values be different if the radius of the circle is r decameters?

6 With your calculator or computer graphing program set in degree mode, graph the functions $\cos \theta$ and $\sin \theta$ for $0° \le \theta \le 360°$. Compare the patterns in those graphs to your ideas in Problems 1 and 2 and to the results of your work on Problem 3.

7 How will the x- and y-coordinates of your seat on the Ferris wheel change during a second complete revolution? How will those patterns be represented in graphs of the coordinate functions for $360° \le \theta \le 720°$?

SUMMARIZE THE MATHEMATICS

In this investigation, you explored the connection between circular motion and the cosine and sine functions.

a Think about the motion of a point on a wheel that starts at $(1, 0)$ and rotates counterclockwise about the center of the wheel.

    i. Describe the pattern of change in the x-coordinate of the point as the wheel makes one complete revolution.

    ii. Describe the pattern of change in the y-coordinate of the point as the wheel makes one complete revolution.

b How are the patterns of change in coordinates extended as the wheel continues to turn counterclockwise through more revolutions?

c Explain how trigonometry can be used to model circular motion.

Be prepared to share your ideas with the class.

✔️ CHECK YOUR UNDERSTANDING

At the Chelsea Community Fair, there is a Ferris wheel with a 15-meter radius. Suppose the Ferris wheel is positioned on an *x-y* coordinate system as in Problem 1.

a. Danielle is on a seat to the right of the vertical axis halfway to the highest point of the ride. Make a sketch showing the wheel, *x*- and *y*-axes through the center of the wheel, and her starting position.

b. Find the *x*- and *y*-coordinates of Danielle's seat after the wheel has rotated counterclockwise through an angle of 56°. Through an angle of 130°.

c. Write expressions that give Danielle's position relative to the *x*- and *y*-axes for any counterclockwise rotation of θ in degrees from her starting point.

d. Suppose the wheel turns at one revolution every minute. Describe in several different ways the location of Danielle's seat at the end of 50 seconds.

INVESTIGATION 3

Revolutions, Degrees, and Radians

The degree is the most familiar unit of measurement for angles and rotations. However, it is common in scientific work to measure angles and rotations in *radians*, a unit that directly connects central angles, arcs, and radii of circles.

As you work on the problems of this investigation, look for answers to the following questions:

How are angles measured in radians?

*How are radians related to other units of measure
for angles and rotations?*

Radian Measurement Every central angle in a circle intercepts an arc on the circle. The length of that arc is some fraction of the circumference of the circle. This is the key to linking length and angle measurement with radian measure.

1 To see how a *radian* is defined, collaborate with classmates to conduct the following experiment.

a. The diagram at the right represents three concentric circles with center O and radii varying from 1 to 3 inches. Points B_1, B_2, B_3, and O are collinear.

On a full-size copy of the diagram, for each circle:

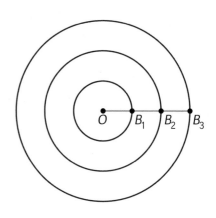

- wrap a piece of string along the circle counterclockwise from each point B_n to a second point A_n so that the length of $\overparen{A_nB_n}$ is equal to the radius of the circle with radius $\overline{OB_n}$.

- draw $\angle A_nOB_n$ and measure it in degrees.

- record the circle radius and the measure of $\angle A_nOB_n$.

Examine the resulting data and describe any interesting patterns you observe.

   b. What is the approximate degree measure of the central angle you would expect to be determined by a 15-cm arc in a circle with a radius of 15 cm?

   c. Suppose a circle with center O has radius r. What would you expect to be the approximate degree measure of $\angle AOB$ if $\overset{\frown}{AB}$ is r units in length?

A **radian** is the measure of any central angle in a circle that intercepts an arc equal in length to the radius of the circle. To use radian measure effectively in reasoning about angles and rotations, it helps to develop number sense about how radian measures are related to degrees and revolutions.

   That is the goal of the next several problems and the "Explore Radians" custom app in *CPMP-Tools*.

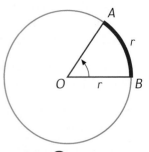

length $\overset{\frown}{AB} = r = $ radius
m$\angle AOB = 1$ radian

2 A circle with center O has radius r units. Central angle $\angle AOB$ has positive radian measure θ and intercepts $\overset{\frown}{AB}$ of length s units.

   a. If $r = 5$ and $s = 10$, determine θ. What is the degree measure of $\angle AOB$?

   b. If $r = 8$ and $s = 24$, determine θ. What is the degree measure of $\angle AOB$?

   c. If $r = 6$ and $\theta = 2.1$, determine s.

   d. Write s in terms of r and θ. If $r = 1$, how are s and θ related?

3 A formula for calculating the circumference of a circle is $C = 2\pi r$. Analyze the ways that Khadijah and Jacy used that fact to find the radian measure equivalent to 120°.

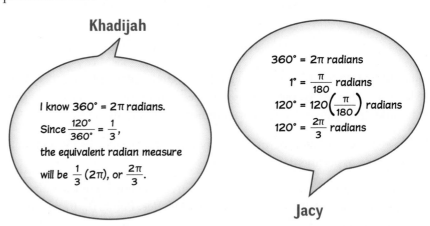

   a. How do you think Khadijah would find the radian measure equivalent to 30°?

   b. How do you think Jacy would find the radian measure equivalent to 30°?

   c. Which method, Khadijah's or Jacy's, do you find easiest to use?

d. How do you think Khadijah would reason to find the degree measure equivalent to $\frac{\pi}{4}$ radians?

e. How do you think Jacy would reason to find the degree measure equivalent to $\frac{\pi}{4}$ radians?

4 Use any method you prefer to determine the equivalent angle measures in Parts a–c.

a. Find the measures in radians and revolutions equivalent to these degree measures.

   i. 90°   **ii.** 150°   **iii.** 75°   **iv.** 210°

b. Find the measures in degrees and revolutions equivalent to these radian measures.

   i. $\frac{\pi}{3}$ **ii.** $\frac{5\pi}{4}$ **iii.** $\frac{\pi}{5}$ **iv.** $\frac{11\pi}{6}$

c. Complete a copy of the following table to show equivalent revolution, degree, and radian measurements. Save your table as a reference for later use.

Revolution/Degree/Radian Equivalents																
Revolutions	0	?	?	?	?	?	?	?	?	?	?	?	?	?	?	?	1
Degrees	0	30	?	?	90	?	135	150	?	210	?	240	270	300	315	?	360
Radians	?	?	$\frac{\pi}{4}$	$\frac{\pi}{3}$	?	$\frac{2\pi}{3}$	?	?	π	?	$\frac{5\pi}{4}$	?	?	?	?	$\frac{11\pi}{6}$	?

5 Once again, consider the information provided by an engine tachometer.

a. The tachometer of a particular SUV traveling in overdrive on a level road at a speed of 60 mph reads about 2,100 rpm. Find the equivalent angular velocity in degrees per minute and in radians per minute.

b. The idle speed of the SUV is about 1,000 rpm. Find the equivalent angular velocity in degrees per minute and in radians per minute.

c. Suppose that the SUV engine has an angular velocity of $6,000\pi$ radians per minute.

   i. What rpm reading would the tachometer show?

   ii. What is the degrees-per-minute equivalent of that angular velocity?

To further develop your understanding and skill in use of radian measures for angles and rotations, you might find it helpful to work with the "Explore Radians" custom app in *CPMP-Tools*. This software displays angles, asks you to estimate the radian and degree measures of those angles, and then allows you to check your estimates numerically and visually.

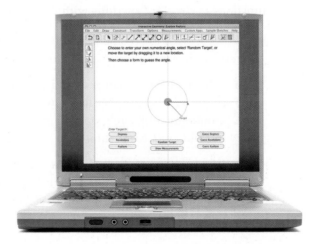

Linking Angular and Linear Motion The mathematical appeal of radian measure lies in the way that it enables use of linear measurement ideas and tools to produce meaningful measurement of angles and rotations. For example, suppose that you wanted to measure $\angle AOB$ and the only available tools were a compass and a tape measure marked off in inches.

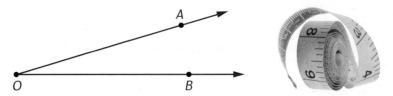

To measure the angle in radians, you could construct a circle with center O and radius equal to the unit segment on the tape measure. Then wrap the tape measure around the circle, starting where $\overrightarrow{OB}$ meets the circle and noting the point on the tape measure where $\overrightarrow{OA}$ meets the circle.

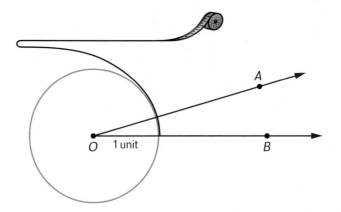

The arc length indicated on the tape measure will be the radian measure of the angle.

The mathematical description of what you have done to measure the angle in radians is to wrap a number line around a circle. If you imagine continuing that wrapping of a number line around a circle (in both directions from the original point of contact), the result is a correspondence between points on the number line and points on the circle.

If you imagine the angle and the circle located on vertical and horizontal coordinate axes, each point has x- and y-coordinates—the cosine and sine of the measure of the angle, respectively. In this way, it is possible to redefine these functions as **circular functions** with domain all real numbers.

6 Consider first the cosine function, the correspondence between points on the number line and the x-coordinates of points on the unit circle.

   a. What pattern of change would you expect in values of the cosine function as the positive number line is wrapped in a counterclockwise direction around the unit circle starting at $(1, 0)$? How would that pattern appear in a plot of $(t, cos\ t)$ values?

   b. What pattern of change would you expect in values of the cosine function as the negative real number line is wrapped in a clockwise direction around the unit circle starting at $(1, 0)$? How would that pattern appear in a plot of $(t, cos\ t)$ values?

c. Set your graphing calculator or computer software in radian mode. Graph cos t for $-2\pi \leq t \leq 2\pi$ to test your ideas in Parts a and b.

d. What would you expect to appear in the graphing window if the lower and upper bounds for t were reset to $-4\pi \leq t \leq 4\pi$?

e. Why were the suggested upper and lower bounds for the graphing windows multiples of 2π?

7 Consider next the sine function, the correspondence between points on the number line and the y-coordinates of points on the unit circle.

a. What pattern of change would you expect in values of the sine function as the positive number line is wrapped in a counterclockwise direction around the unit circle starting at $(1, 0)$? How would that pattern appear in a plot of $(t, sin\ t)$ values?

b. What pattern of change would you expect in values of the sine function as the negative real number line is wrapped in a clockwise direction around the unit circle starting at $(1, 0)$? How would that pattern appear in a plot of $(t, sin\ t)$ values?

c. Set your graphing calculator or computer software in radian mode. Graph sin t for $-2\pi \leq t \leq 2\pi$ to test your ideas in Parts a and b.

d. What would you expect to appear in the graphing window if the lower and upper bounds for t were reset to $-4\pi \leq t \leq 4\pi$?

e. In Parts c and d, the upper and lower bounds given for the graphing windows are multiples of 2π. What are advantages of this form?

SUMMARIZE THE MATHEMATICS

In this investigation, you learned how to measure angles in radians. Radian measure is useful because it connects length and angular measurement.

a Describe how to draw an angle that measures 1 radian.

b In a circle, how are the radius, the positive radian measure of a central angle, and the length of the arc intercepted by that angle related to each other?

c If an angle has measure r in radians, d in degrees, and v in revolutions what rule expresses:

    i. r as a function of d?

    ii. r as a function of v?

    iii. d as a function of r?

    iv. v as a function of r?

d How are the cosine and sine functions defined for radian measures or real numbers? How are the resulting numeric and graphic patterns of values for those circular functions similar to or different from the corresponding trigonometric functions defined for degree measure?

Be prepared to share your ideas and reasoning with the class.

✅ **CHECK YOUR UNDERSTANDING**

Suppose that one of the wind turbines shown in the Check Your Understanding of Investigation 1 has a driver pulley with 6-foot diameter attached by a belt to a follower pulley with 4-foot diameter.

a. If the driver pulley turns at 3 revolutions per minute:

   i. what is the angular velocity of the driver pulley in radians per minute?

   ii. what is the angular velocity of the follower pulley in radians per minute?

b. Suppose that economic use of the wind turbine to generate electric power requires the follower pulley to turn at a rate of 90π radians per minute. How fast, in radians per minute, must the driver pulley turn to accomplish this rate?

c. Draw sketches and use geometric reasoning to find exact values of the sine and cosine functions when the input values are the following radian measures or real numbers. Check your answers using technology or the table you completed in Problem 4.

   i. $\dfrac{3\pi}{4}$ ii. π iii. $\dfrac{7\pi}{6}$ iv. $\dfrac{5\pi}{3}$

INVESTIGATION 4

Patterns of Periodic Change

Tracking the location of seats on a spinning Ferris wheel shows how the cosine and sine functions can be used to describe rotation of circular objects. Wrapping a number line around a circle shows how the radius of the circle can be used to measure angles, arcs, and rotations in radians and how the cosine and sine can be defined as functions of real numbers. In both cases, the resulting functions had repeating graph patterns that are called *sinusoids*.

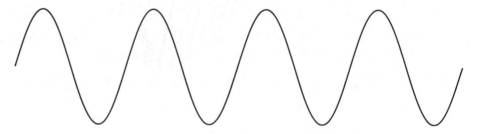

Many important variables that change over time have sinusoidal graphs.

As you work on the problems of this investigation, look for answers to the following question:

How can the cosine and sine functions be modified to represent the variety of important patterns of periodic change?

A Family of Ferris Wheels The Ferris wheel that you analyzed during work on Investigation 2 had a radius of one decameter and an unspecified angular velocity. To model the motion of wheels with different radii and to account for different rates of rotation, it is necessary to construct functions that are variations on the basic cosine and sine functions.

1 If a Ferris wheel has radius 1.5 decameters (15 meters), the functions $1.5 \cos \theta$ and $1.5 \sin \theta$ give x- and y-coordinates after rotation of θ for a seat that starts at $(1.5, 0)$. Compare the graphs of these new coordinate functions with the graphs of the basic cosine and sine functions.

   a. Find the maximum and minimum points of the graphs of $\cos \theta$ and $1.5 \cos \theta$ when:

   i. θ is measured in degrees.

   ii. θ is measured in radians.

   b. Find the θ-axis intercepts of the graphs of $\cos \theta$ and $1.5 \cos \theta$ when:

   i. θ is measured in degrees.

   ii. θ is measured in radians.

   c. Find the maximum and minimum points of the graphs of $\sin \theta$ and $1.5 \sin \theta$ when:

   i. θ is measured in degrees.

   ii. θ is measured in radians.

   d. Find the θ-axis intercepts of the graphs of $\sin \theta$ and $1.5 \sin \theta$ when:

   i. θ is measured in degrees.

   ii. θ is measured in radians.

   e. How would the maximum and minimum points and the θ-axis intercept change if the Ferris wheel being modeled had radius a and the coordinate functions were $a \cos \theta$ and $a \sin \theta$?

2 When riding a Ferris wheel, customers are probably more nervous about their height above the ground than their distance from the vertical axis of the wheel.

   Suppose that a large wheel has radius 25 meters, the center of the wheel is located 30 meters above the ground, and the wheel starts in motion when seat S is at the "3 o'clock" position.

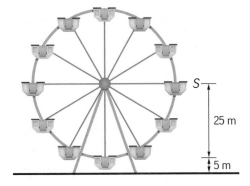

   a. Modify the sine function to get a rule for $h(\theta)$ that gives the *height* of seat S in meters after rotation of θ. Compare the graph of this height function with the graph of $\sin \theta$.

   b. Find the maximum and minimum points on the graphs of $\sin \theta$ and $h(\theta)$ when:

   i. θ is measured in degrees.

   ii. θ is measured in radians.

c. Find the θ-axis intercepts on the graphs of sin θ and $h(\theta)$.

d. How would the maximum and minimum points and the θ-axis intercept change if the Ferris wheel being modeled had radius a and its center was c meters above the ground? Why is $c > a$?

Functions with sinusoidal graphs all have patterns that repeat as values of the independent variable increase. Functions that have that kind of repeating pattern are called **periodic functions**. The **period** is the length of the shortest interval of values for the independent variable on which the repeating pattern of function values occurs. If a periodic function has maximum and minimum values, half the absolute value of the difference between those values is called its **amplitude**. In addition to the change in amplitude of sin θ used to model Ferris wheel motion, the graph of $h(\theta)$ appears to involve a shift of that graph upward. That upward shift of 30 meters is called the **y-displacement**.

3 The functions cos θ, sin θ, 1.5 cos θ, 1.5 sin θ, and $h(\theta)$ all have the same period. Find that common period and the amplitude of each function when:

a. θ is measured in degrees.

b. θ is measured in radians.

4 How are the period and the amplitude of $f(\theta) = a \cos \theta + c$ and $g(\theta) = a \sin \theta + c$ related to the values of the numbers a and c?

Period and Frequency In various problems of this lesson, you have seen how to find the location of points on a rotating circle from information about the starting point and the angle of rotation. For actual rotating wheels, it is much easier to measure elapsed *time* than the *angle* of rotation. The link between these two variables is *angular velocity*.

For constant angular velocity, the angle of rotation varies directly with the elapsed time. So, it is often convenient to describe the location of points on a wheel as a function of time. Because radian measure connects points on a number or time line axis to points on a circle with a common unit of measurement, it is customary to use radian measure in analyzing circular motion as a function of time.

5 Suppose that the height of a Ferris wheel seat changes in a pattern that can be modeled by the function $h(t) = 25 \sin t + 30$, where time is in minutes and height is in meters.

a. What are the period and amplitude of $h(t)$? What do those values tell about motion of the Ferris wheel?

b. If a seat starts out in the "3 o'clock" position, how long will it take the seat to first return to that position? At what times will it revisit that position?

c. What is the angular velocity of this wheel:

　　i. in revolutions per minute? (This is also called the **frequency** of the periodic motion.)

　　ii. in degrees per minute?

　　iii. in radians per minute?

6 Suppose that the height (in meters) of seats on different Ferris wheels changes over time (in minutes) according to the functions given below. For each function:

- find the height of the seat when motion of the wheel begins.
- find the amplitude of $h(t)$. Explain what it tells about motion of the wheel.
- find the period of $h(t)$. Explain what it tells about motion of the wheel.
- find the angular velocity of the wheel in revolutions per minute—the frequency of $h(t)$.

a. $h(t) = 15 \sin 0.5t + 17$ **b.** $h(t) = 24 \cos 2t + 27$

c. $h(t) = 12 \sin 1.5t + 13$ **d.** $h(t) = -12 \cos t + 14$

7 In Problems 5 and 6, you analyzed properties of functions with rules in the general form

$$h(t) = a \sin bt + c \quad \text{and} \quad h(t) = a \cos bt + c.$$

a. How do each of the values a, b, and c seem to affect the graphs of the functions?

b. Check your ideas by reasoning about the function rules and by graphing examples with a calculator or by using the parameter variation capability of the *CPMP-Tools* CAS. Be sure your answer accounts for both positive and negative values of a and c.

Modeling Periodic Change over Time Variations of the basic cosine and sine functions can be used to model the patterns of change in many different scientific and practical situations. The next three problems ask you to analyze familiar situations.

8 Pendulums are among the simplest but most useful examples of periodic motion. Once set in motion, the arm of the pendulum swings left and right of a vertical axis. The angle of displacement from vertical is a periodic function of time that depends on the length of the pendulum and its initial release point.

Suppose that the function $d(t) = 35 \cos 2t$ gives the displacement from vertical (in degrees) of the tire swing pendulum shown at the right as a function of time (in seconds).

a. What are the amplitude, period, and frequency of $d(t)$? What does each tell about the motion of the swing?

b. If the motion of a different swing is modeled by $f(t) = 45 \cos \pi t$, what are the amplitude, period, and frequency of $f(t)$? What does each tell about the motion of that swing?

c. Why does it make sense to use variations of the circular function $\cos t$ to model pendulum motion?

d. What function $g(t)$ would model the motion of a pendulum that is released from a displacement of 18° right of vertical and swings with a frequency of 0.25 cycles per second (a period of 4 seconds)?

9 At every location on Earth, the number of hours of daylight varies with the seasons in a predictable way. One convenient way to model that pattern of change is to measure time in days, beginning with the spring equinox (about March 21) as $t = 0$. With that frame of reference, the number of daylight hours in Boston, Massachusetts is given by $d(t) = 3.5 \sin \frac{2\pi}{365}t + 12.5$.

a. What are the amplitude, period, and frequency of $d(t)$? What do those values tell about the pattern of change in daylight during a year in Boston?

b. What are the maximum and the minimum numbers of hours of daylight in Boston? At what times in the year do they occur?

c. If the function giving the number of daylight hours in Point Barrow, Alaska, had the form $f(t) = a \sin bt + c$, how would you expect the values of a, b, and c to be related to the corresponding numbers in the rule giving daylight hours in Boston?

d. Why does it make sense that the function giving daylight hours at points on Earth should involve the circular function $\sin t$?

10 Oscilloscopes are common scientific instruments for display of the periodic variation in pressure and electronic waves that carry sound and light. For example, the function

$$V(t) = 170 \sin 120\pi t$$

gives the voltage in a standard alternating current electrical line as a function of time in seconds.

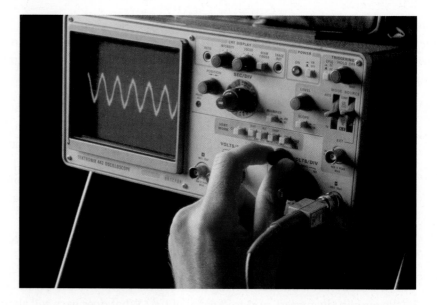

a. What maximum and minimum voltage are implied by the function $V(t)$?

b. What period and frequency are implied by the rule for $V(t)$? (That standard frequency is called **one Hertz**.)

SUMMARIZE THE MATHEMATICS

In this investigation, you explored the use of the sine and cosine functions to represent periodic phenomena.

a What is the defining property of a periodic function?

b What are the amplitude and frequency of a periodic function?

c What are the period, amplitude, and frequency of $f(t) = \sin t$? Of $g(t) = \cos t$?

d How can the values of a, b, and c be used to find the amplitude, period, and frequency of $f(t) = a \sin bt + c$ and $g(t) = a \cos bt + c$?

e What are some common examples of periodic change over time?

Be prepared to discuss your responses with the class.

✔ CHECK YOUR UNDERSTANDING

Portions of periodic graphs are shown below with windows $-4\pi \leq x \leq 4\pi$ and $-6 \leq y \leq 6$. Without using technology, match each graph to one of the given functions. (Not all function rules will be used.) In each case, explain the reason for your choice.

I. $y = 3 \sin x$ **II.** $y = 3 \cos x$ **III.** $y = 3 + \sin x$

IV. $y = -3 \sin x$ **V.** $y = -3 \cos x$ **IV.** $y = \sin 3x$

a.

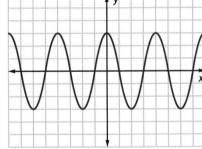

b.

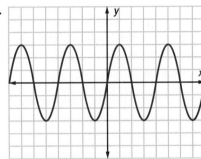

c.

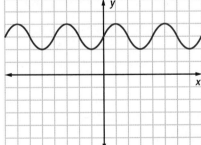

d.

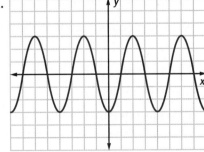

APPLICATIONS

1 Lena's mountain bike has 21 speeds. To get started, she shifts gears so that the chain connects the 42-tooth crankset with a 28-tooth rear-wheel sprocket.

   a. If Lena pedals at 40 rpm, at what rate do the rear sprocket and wheel turn?

   b. If Lena pedals at 40 rpm, how far will she travel in one minute if her bike has tires with 66-cm diameters?

   c. What is the relationship between the angular velocities of the pedal sprocket v_p and the rear wheel sprocket v_w?

2 In go-carts, the engine driver sprocket is attached to a follower sprocket on the rear axle by a belt. These sprockets can have many different diameters depending upon course demands and safety.

   a. Sketch the situation in which an engine sprocket with a 7-cm diameter drives a rear-axle sprocket with a 10-cm diameter.

   b. Suppose that the engine is turning at 1,620 rpm.

      i. Find the angular velocity of the rear axle.

      ii. Find the go-cart's speed, in cm per minute and km per hour, if the rear wheels have a diameter of 28 cm.

   c. When rounding corners, a speed of 30 km per hour or less is needed to reduce lateral sliding. What engine speed in rpm is optimal?

3 The operation of most vacuum cleaners depends on circular motion at several points in the vacuuming process. In one vacuum cleaner model, the rotating brush is 1.5 inches in diameter and is driven by a rubber belt attached to a driver pulley on the motor. The driver pulley is 0.5 inches in diameter. The belt is attached with a half twist, so that the rotating brush and the driver rotate in opposite directions.

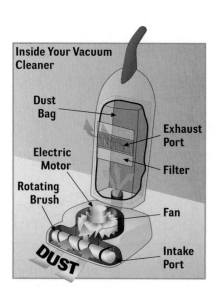

Inside Your Vacuum Cleaner

Dust Bag

Exhaust Port

Electric Motor

Filter

Rotating Brush

Fan

Intake Port

DUST

   a. The rotating brush is designed to help pull dust from the floor into the vacuum cleaner. Should the rotating brush be turning clockwise or counterclockwise? Explain.

   b. Sketch this pulley system.

   c. What are the circumferences of the driver pulley and the rotating brush?

   d. How far does a point on the edge of the rotating brush travel in one revolution of the driver?

   e. If the driver pulley is turning 600 rpm, how fast is the rotating brush turning?

   f. Write a rule that shows the relationship between angular velocities of the engine drive shaft v_D and the rotating brush v_B.

4 The crankshaft of a particular automobile engine has an angular velocity of 1,500 rpm at 30 mph. The crankshaft pulley has a diameter of 10 cm, and it is attached to an air conditioner compressor pulley with a 7-cm diameter and an alternator pulley with a 5-cm diameter.

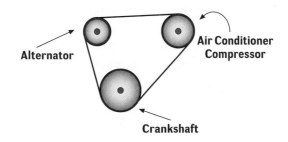

Alternator

Air Conditioner Compressor

Crankshaft

a. At what angular velocities do the compressor and alternator turn?

b. The three pulleys are connected by a 60-cm belt. At a crankshaft rate of 1,500 rpm, how many times will the belt revolve through its 60-cm length in one minute?

c. Most belts do not show significant wear until each point of the belt has traveled about 20,000 kilometers. How long can the engine run at 1,500 rpm before the belt typically would show wear?

5 One of the scarier rides at carnivals and amusement parks is provided by a long rotating arm with riders in capsules on either end. Of course, the capsules not only move up and down as the arm rotates, but they spin other ways as well.

Suppose that the arm of one such ride is 150 feet long and that you get strapped into one of the capsules when it is at ground level. Assume simple rotating motion and treat the capsule as a single point. Find your height above the ground when the arm has made the following rotations counterclockwise from its starting vertical position.

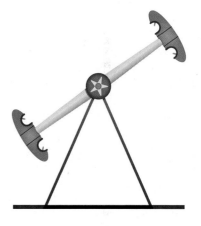

a. 20°
b. 45°

c. 85°
d. 90°

e. 120°
f. 180°

g. 270°
h. 300°

i. 340°

6 Radio direction and ranging (radar) is one of the most widely used electronic sensing tools. Most of us probably know about radar from its applications in measuring speed of baseball pitches and automobiles. But it is also an invaluable tool in navigation and weather forecasting.

In those applications, the echoes to a rotating transmitter/receiver are displayed as blips on a scope. Each blip is located by distance and angle.

To describe locations of radar blips by distance east/west and north/south of the radar device, the (*distance*, *angle*) information needs to be converted to rectangular (*x*, *y*) coordinates.

Find the (*x*, *y*) coordinates of radar blips located by the following (*distance*, *angle*) data.

a. (30 km, 40°)
b. (20 km, 160°)

c. (70 km, 210°)
d. (15 km, 230°)

e. (45 km, 270°)
f. (40 km, 310°)

India Images/Alamy

7 A rack-and-pinion gearset is used in the steering mechanism of most cars. The gearset converts the rotational motion of the steering wheel into the linear motion needed to turn the wheels, while providing a gear reduction that makes it easier to turn the wheels. As shown in the sketch below, when you turn the steering wheel, the gear (pinion) spins, moving the rack. The tie rod at the end of the rack connects to the steering arm on the spindle of each wheel.

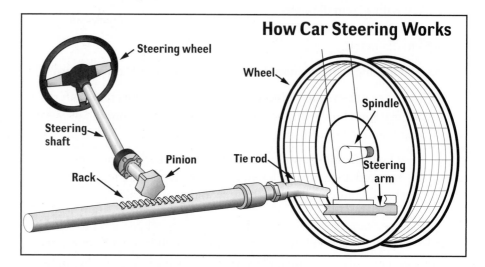

How Car Steering Works

a. The steering ratio is the ratio of how far you turn the steering wheel to how far the wheels turn. For example, a steering ratio of 18:1 means you must turn the steering wheel 18° in order to turn the front wheels 1°. A higher steering ratio such as 20:1 means you have to turn the steering wheel more to get the wheels to turn through a 1° angle, but less effort is required because of "gearing down." Would you expect lighter, sportier cars to have higher or lower steering ratios than larger cars and trucks? Explain.

b. Suppose the steering ratio of a car is 18:1.

   i. Through how many degrees will the front wheels turn if the steering wheel is turned a complete revolution?

   ii. Suppose the maximum turn of the front wheels is 75°. How many revolutions of the steering wheel are needed to make the maximum turn?

8 A circle of radius r centimeters has a circumference of $2\pi r$ centimeters.

a. Suppose a point A on the circle rotates through an angle of p radians. What is the length of the arc traversed by the point?

b. Suppose a point A on the circle rotates at p radians per minute. Find the linear velocity of the point.

c. Suppose a circle with radius 10 cm has an angular velocity of 80 radians per second. Find the linear velocity of a point A on the circle.

d. Suppose a point on a circle with radius 10 cm has linear velocity of 30π cm per second. Find the angular velocity of the point.

e. Explain how to convert an angular velocity v (in radians per second) for a circle of radius r centimeters into the linear velocity of a point on the circle.

9 Degree measure of angles is more familiar to you than radian measure. But the difference between the two is just a matter of scale. That is, $1° = \frac{\pi}{180}$, or about 0.0175 radians, and 1 radian $= \left(\frac{180}{\pi}\right)°$, or about 57.3°. The following tasks may help you better understand radians.

a. Each of the following is the radian measure of an angle in standard position. Give the quadrant in which the terminal side of each angle lies.

i. 5

ii. -5

iii. $-\frac{4\pi}{3}$

iv. $\frac{7\pi}{3}$

b. The measure of an angle in standard position is p radians, and the angle's terminal side is in the second quadrant. In what quadrant is the terminal side of an angle with each of the following radian measures? Give all possibilities and explain.

i. $p + 2\pi$

ii. $p - \frac{\pi}{2}$

iii. $p + 9\pi$

iv. $p - 2$

10 Many variations on the sine and cosine functions are also periodic functions.

a. Find the amplitude, period, and y-displacement for each of the following functions.

i. $y = 2 \cos(-x) + 3$

ii. $y = -3 \sin 0.1x + 5$

iii. $y = 12 \sin 3x - 8$

b. Suppose data from a periodic variable are fit well by a function of the form $y = a \cos bx + c$ where $a < 0$ and $b > 0$. A plot of the data suggests a y-displacement of -5, amplitude of 7, and period of 6π. Write a function rule for this data.

11 The center of a Ferris wheel in an amusement park is 7 meters above the ground and the Ferris wheel itself is 12 meters in diameter. The wheel turns counterclockwise at a constant rate and takes 20 seconds to make one complete revolution.

a. Yolanda and her friend enter their seat when it is directly below the wheel's center. Sketch a graph that you would expect to show their height above the ground during one minute at full rotational speed, starting from the entry point.

b. Make a table of values for the radian measure t and the height $h(t)$ of Yolanda's seat above the ground at 5 second intervals for the one-minute ride.

c. Use the data pairs in Part b to sketch a graph of the height $h(t)$ as a function of time t. Compare this graph to the one you drew by hand in Part a?

d. Is this function periodic? If so, what is its period? What is its amplitude?

e. Several functions that students predicted would fit the data in Part b are given below. In each case, $\frac{\pi}{10}$ is angular velocity in radians per second, t is time in seconds, and $h(t)$ is height in meters. Determine if any are good fits for the data. Explain your reasoning.

   i. $h(t) = 6 \sin \frac{\pi t}{10} + 7$

   ii. $h(t) = -6 \cos \frac{\pi t}{10} + 7$

   iii. $h(t) = -6 \sin \frac{\pi}{10} (t + 5) + 7$

   iv. $h(t) = 6 \cos \frac{\pi t}{10} + 7$

12 Suppose that you are trying to model the motion of a clock pendulum that moves as far as 5 inches to the right of vertical and swings with a period of 2 seconds.

a. Find variations of $d(t) = \cos t$ that fit the conditions for each part below.

   i. A modeling function whose values range from −5 to 5 and has a period of 2π

   ii. A modeling function that has a period of 2 and whose values range from −1 to 1

   iii. A modeling function that has a period of 2 and whose values range from −5 to 5

b. How are the numbers in the function for Part aiii related to the motion of the pendulum you are modeling?

c. Graph the function that models the motion of the clock pendulum. Identify the coordinates of the t-intercepts and minimum and maximum points of the graph.

CONNECTIONS

13 Some automobile manufacturers use an automatic, continuously-variable transmission (CVT) based on segments of cones. A simplified model is shown below. It consists of two 10-centimeter segments of right cones. The diameters of the circular ends are given. These partial cones form the basis for a *variable-drive* system, in which either partial cone can be moved laterally (left and right) along a shaft.

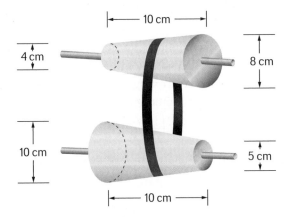

a. When two pulleys are connected by a chain or belt, their angular velocities are related. The number by which the angular velocity of the driver pulley is multiplied to get the angular velocity of the follower pulley is called the **transmission factor** from driver to follower. For example, in the figure above, the transmission factor for the top cone (driver) to the bottom cone (follower) is $\frac{6}{7.5} = 0.8$. Explain why.

b. If the upper shape is the driver, what are the maximum and minimum transmission factors?

c. Suppose the belt is halfway between the two circular ends of the upper shape. If the lower shape is permitted to move laterally, what range of transmission factors is possible?

d. Describe a position of the cones for which the transmission factor is 1.

14 A function f is called an **even function** provided $f(-x) = f(x)$ for all values of x in the domain. A function g is called an **odd function** provided $g(-x) = -g(x)$ for all values of x in the domain.

a. Is the cosine function an even function, an odd function, or neither? Explain.

b. Is the sine function an even function, an odd function, or neither? Explain.

c. Is the absolute value function $a(x) = |x|$ an even function, an odd function, or neither? Explain.

d. What is true about the graph of any even function? Of any odd function? Why does this make sense?

e. Give an example of a polynomial function that is an odd function. An even function. That is neither.

15 In earlier study of coordinate methods, you discovered that results of several special rotations can be produced by matrix multiplication. For example, for any point $P(x, y)$, the matrix product $\begin{bmatrix} -1 & 0 \\ 0 & -1 \end{bmatrix} \begin{bmatrix} x \\ y \end{bmatrix} = \begin{bmatrix} -x \\ -y \end{bmatrix}$. So, the matrix $\begin{bmatrix} -1 & 0 \\ 0 & -1 \end{bmatrix}$ represents a rotation of 180° or half-turn about the origin. In general, multiplication by a matrix in the form $\begin{bmatrix} \cos\theta & -\sin\theta \\ \sin\theta & \cos\theta \end{bmatrix}$ represents a counterclockwise rotation of θ° about the origin.

a. How does the matrix $\begin{bmatrix} -1 & 0 \\ 0 & -1 \end{bmatrix}$ for a rotation of 180° fit the general form for rotation matrices?

b. Write the matrix for a counterclockwise rotation of 30°. Then use matrix multiplication to find the images of the following special points on the unit circle and use geometric reasoning to prove that the matrix multiplication has produced the correct rotation images of the points.

   i. $(1, 0)$ ii. $\left(\dfrac{1}{2}, \dfrac{\sqrt{3}}{2}\right)$ iii. $(0, 1)$

16 An angle of 1 radian is shown in standard position below.

a. What is the length of the arc in the circle of radius 1 traversed by rotating through 1 radian?

b. What transformation maps the circle of radius 1 to that of radius 2? What transformation maps the circle of radius 3 to the circle of radius 2?

c. What are the lengths of the arcs in the circles of radii 2 and 3 traversed by rotating through an angle of 1 radian? Explain why this answer makes sense using the transformations in Part b.

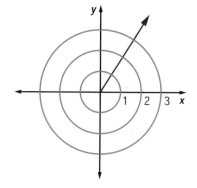

17 A **sector of a circle** is a region bounded by two radii and an arc of the circle. The two sectors AOB and COD shown at the right have central angles with equal measure, that is, m∠AOB = m∠COD.

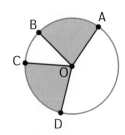

a. Use the symmetry of the circle about any line through its center O to explain why the two sectors are congruent and, therefore, have equal areas.

b. The area of a circle with radius r is πr^2. What is the area in terms of r of a sector of that circle with central angle of 120°? Of 90°? Of d°?

c. What is the area of a sector with central angle of 2 radians? Of $\dfrac{3\pi}{2}$ radians? Of θ radians?

18 In Course 2 Unit 7, *Trigonometric Methods*, you used properties of special right triangles to determine the trigonometric ratios for angles of 30°, 45°, and 60°. You also found the values of these ratios for 0° and 90°.

a. Complete the table at the right giving the sine and cosine of the radian equivalent of each of these angles.

t (in radians)	sin t	cos t
0		
	$\frac{1}{2}$	
$\frac{\pi}{4}$		$\frac{\sqrt{2}}{2}$
	$\frac{\sqrt{3}}{2}$	
$\frac{\pi}{2}$		

b. In this lesson, you learned that the sine and cosine functions can be used to model the motion of a circle with radius 1 centered at the origin. Sketch this circle and mark the points that lie on the terminal side of each of the five angles in the table. Write the coordinates of each point.

c. The terminal sides of angles measuring 0 and $\frac{\pi}{2}$ radians each lie on one of the coordinate axes. There are two other angles between 0 and 2π radians whose terminal sides lie on an axis. What are the radian measures of these angles? What are the sine and cosine of each angle?

d. In Part b, you used geometric reasoning to calculate cosine and sine of three angles with terminal side in the first quadrant. Use those results and symmetry of the circle to find the cosine and sine of nine more angles—three with terminal side in Quadrant II, three with terminal side in Quadrant III, and three with terminal side in Quadrant IV.

e. In which quadrants are the sines of angles positive? Negative? Answer the same questions for the cosine.

19 Use the language of geometric transformations to describe the relationships between the graphs of the following pairs of functions.

a. $f(x) = -\cos t$ and $g(x) = \cos t$

b. $h(x) = 5 + \sin t$ and $k(x) = \sin t$

c. $m(x) = 5\cos t$ and $n(x) = \cos t$

REFLECTIONS

20 A popular amusement park ride for young children is the carousel or merry-go-round. A typical carousel consists of a circular platform with three circles of horses to ride, an inner circle, a middle circle, and an outer circle.

Children and sometimes adults ride these horses as the carousel goes round and round. In which circle should a child ride for the slowest ride? For the fastest ride? Why?

Topix/Alamy

21 Nathan's father collects $33\frac{1}{3}$ (turntable revolutions per minute) vinyl records. Nathan often listens to classic Beatles' albums as he does his homework. After completing his homework on circular motion, he still is not comfortable with some of the ideas. He understands that a point on the circumference of the record he has been playing and a point on the edge of the record label make a complete revolution in the same amount of time. But he is not convinced that the point on the circumference of the record is traveling faster than the point on the record label.

Using mathematical ideas that you developed in Investigation 1, write a paragraph explaining to Nathan how two points on a record can move at two different speeds.

22 Karen wishes to evaluate sin 25° using her calculator. She presses
SIN 25) ENTER . The calculator displays the screen below. What indicates that there must be an error? What do you think the error is?

23 Look back at the description of the number line "winding process" on page 430 that led to the definition of the circular functions cosine and sine.

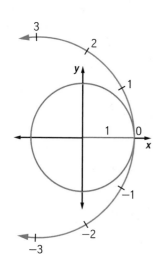

a. Explain, as precisely as you can, why this process is a function—sometimes called the *winding function*.

b. What is the domain of the winding function? The range? Explain.

c. How, if at all, is this winding function related to radian measure?

d. Based on the winding function definition of the circular functions cosine and sine, what are the domain and range of the cosine function? The sine function?

e. How are your answers for Part d similar to, and different from, those for corresponding questions about the trigonometric functions cosine and sine?

24 If you have ever ridden a Ferris wheel that was rotating at a constant rate, you know that at some positions on the wheel, you feel like you are falling or rising faster than at other points on the wheel. As the wheel rotates counterclockwise, your directed distance to the horizontal axis through the center of the wheel is $r \sin \theta$, where r is the radius of the Ferris wheel. Examine the graph of the sine function.

a. Near what values of θ in radians is the sine function increasing most rapidly? Decreasing most rapidly?

b. Where on the wheel are you located when the sine function is increasing most rapidly? Decreasing most rapidly?

c. Explain what your answers in Parts a and b may have to do with your feelings of falling or rising when you ride a Ferris wheel.

25 In this unit, as in previous units, you have engaged in important mathematical practices. Look back over Lesson 2 and describe an example of where you engaged in each of the following.

a. Made sense of a problem and persevered in solving it

b. Reasoned abstractly and quantitatively

c. Constructed a viable argument and critiqued the reasoning of others

d. Constructed a mathematical model in solving a problem

e. Used appropriate tools strategically

f. Looked for and made use of mathematical structure

EXTENSIONS

26 When two rotating circles are linked in a driver/follower relationship, the **transmission factor** from driver to follower can be expressed as the ratio of angular velocities $\frac{v_f}{v_d}$. One model of a 21-speed mountain bike has a pedal crankset of 3 sprockets with 48, 40, and 30 teeth; the bike has 7 sprockets for the rear wheel with 30, 27, 24, 21, 18, and 12 teeth.

a. What is the largest transmission factor available for this bicycle? What is the smallest?

b. Suppose on a cross-country ride, you can maintain an angular velocity of 70 rpm for the pedal sprocket. What is the fastest that you can make the rear wheel turn?

c. The radius of a mountain bike tire is about 33 cm.

   i. What is the linear velocity of the rear wheel when the pedal sprocket turns at 70 rpm and the transmission factor is greatest possible?

   ii. What is the linear velocity of the rear wheel when the pedal sprocket turns at 70 rpm and the transmission factor is least?

27 In the figure below, $\overline{AE}$ and $\overline{BD}$ are tangent to both circles. The point of intersection of the two tangents lies on the line containing the centers of the circles.

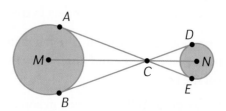

a. Draw the radii to the points of tangency in each circle. The result should be four triangles that are similar to one another. Explain why?

b. Suppose the radius of the larger circle is 8 cm, the radius of the smaller circle is 4 cm, and the distance between the centers is 30 cm. Use similar triangles to determine the length of a common tangent segment.

c. Suppose the circles represent pulleys with a belt going around the pulleys as indicated by the common tangents. Given the measures in Part c, determine the length of the belt.

28 In this task, you will explore a shape that is different in form from a circle but can serve a similar design function.

a. Make a cardboard model of this shape as follows.

Step 1. Construct an equilateral triangle of side length 6 cm.

Step 2. At each vertex, place a compass and draw the arc of a circle with radius 6 cm that gives the two other vertices. When all three pairs of vertices have been joined by these arcs, you have a rounded triangle-like figure, known as a *Reuleaux triangle* (pronounced re-low).

Step 3. Carefully cut out the Reuleaux triangle.

b. Draw a pair of parallel lines 6 cm apart, and place the edge of a ruler along one of the lines. Place your cardboard Reuleaux triangle between the lines and roll it along the edge of the ruler. Note any unusual occurrences.

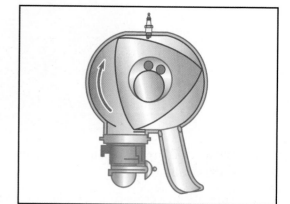

c. Conduct library or Internet research on the Wankel engine. How is it related to your model?

d. Research the concept of *shapes of constant width*. What are two other examples of such shapes?

29 An important idea in calculus involves the ratio $\frac{\sin x}{x}$ for values of x close to 0. Make a table with columns headed x and $\frac{\sin x}{x}$.

a. Using radian mode, complete one table for $x = 1, 0.5, 0.25, 0.1, 0.01, 0.001$, and 0.0001. What appears to be happening to $\frac{\sin x}{x}$ as x approaches 0 in radian mode?

b. Using degree mode, complete a second table with the same headings and same values of x. What appears to be happening to $\frac{\sin x}{x}$ as x approaches 0 in degree mode?

c. Compare your results in Parts a and b to the difference in scale of degrees and radians. Do you see any connection?

30 The electronic pattern of simple sounds, such as those you can view on an oscilloscope, are modeled by trigonometric functions of the form $y = a \sin bt$ or $y = a \cos bt$, where b is expressed in radians per second and t is time in seconds. The period of a sound wave relates to the frequency or number of waves that pass a point each second. The greater the frequency (or shorter the period) of the sound wave, the higher the pitch of the sound.

High-frequency Sound Wave

Low-frequency Sound Wave

The amplitude $|a|$ represents loudness. The greater the amplitude of the wave, the louder the sound. A good example of a simple sound is the sound produced by a piano tuning fork.

a. The middle C tuning fork oscillates at 264 cycles per second. This is called its frequency. How many radians per second is this?

b. Write a function rule that models the sound of middle C when a is 1.

c. What is the period of the function? How is it related to the frequency of 264 cycles per second?

d. Graph four cycles of the function. If you use 0 for the minimum x value, what should you use for the maximum value?

e. The C note two octaves above middle C has a frequency of 1,056 cycles per second. Model this sound with a function rule. What is the period?

f. The C note that is one octave below middle C has a frequency of 132 cycles per second. Model this sound with a function rule. Graph the function in the same viewing window with the middle C graph from Part b. What do you think is the frequency of the C note one octave above middle C?

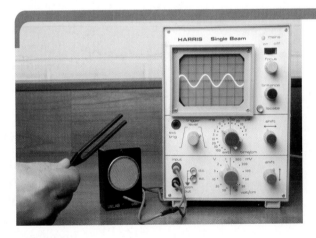

31 In this lesson, you learned that the distance from the vertical axis of a circle of radius r to a point moving counterclockwise around the circle can be described by $y = r \cos x$. Similarly, the distance from the horizontal axis of the circle to the point is modeled by $y = r \sin x$.

Another way to describe circular motion is to use *parametric equations*, which you will study more fully in Course 4. With parametric equations, a third variable t is introduced, and the x- and y-coordinates of points on the circle are written as separate functions of t.

For an introductory look at this idea for a circle of radius 5 centimeters, set your calculator or computer software to parametric mode and use degrees. Set the viewing window so that t varies from 0 to 360 in steps of 5, x varies from -9 to 9, and y varies from -6 to 6. Then in the functions list, enter $X_T = 5 \cos t$ and $Y_T = 5 \sin t$ as the first pair of functions.

a. Graph and trace values for $t = 30, 60, 90,$ and so on.

b. What physical quantity does X_T represent? What physical quantity does Y_T represent?

c. What measurement does t represent? What are the units of t?

d. How much does t change in one revolution of the circle?

e. If one revolution of the circle takes 20 seconds, what is the angular velocity of a point on the circle in units of t per second?

f. What is the linear velocity of a point on the circle in meters per second?

g. What parametric equations would represent circular motion on a circle with a radius of 3 centimeters? Enter them as the second pair of parametric equations in your calculator's function list.

h. Graph at the same time the equations for circles with radii of 3 and 5 cm. (On some calculators, you need to set the mode for simultaneous graphs.) Watch carefully.

  i. If the circles both turn with the same angular velocity, on which circle does a point have the greater linear velocity?

  ii. What is the linear velocity of the slower point?

32 AM and FM radio stations broadcast programs at various frequencies. The frequency f of a periodic function is the number of cycles per unit of time t. Radio signals broadcast by a given station transmit sound at frequencies that are multiples of 10^3 cycles per second (kilohertz, kHz) or 10^6 cycles per second (megahertz, MHz).

AM radio stations broadcast at frequencies between 520 and 1,610 kHz. An AM radio station broadcasting at 675 kHz sends out a **carrier signal** whose function rule and graph are of the form shown in the diagram below.

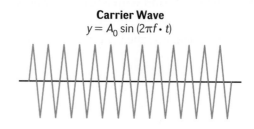

Carrier Wave
$y = A_0 \sin (2\pi f \cdot t)$

To transmit program sounds of varying frequencies, an AM station varies or "modulates" the amplitude of the carrier signal. The initials AM refer to **amplitude modulation**. An amplitude-modulated signal has a function rule and graph of the form shown below. Note that the period, and therefore, the frequency, of the carrier wave is left unchanged. The variable amplitude $A_0(t)$ is itself a sinusoidal function representing the broadcast programming.

AM Wave
$$y = A_0(t) \sin (2\pi f \cdot t)$$

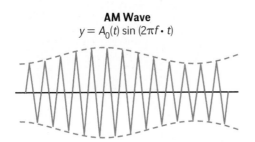

FM radio stations broadcast at frequencies between 87.8 and 108 MHz. An FM (**frequency modulation**) station transmits program sounds by varying the frequency of the carrier signal while keeping the amplitude constant. A frequency-modulated signal has a function rule and graph of the form shown below. The variable frequency $f(t)$ is itself a sinusoidal function representing the broadcast programming.

FM Wave
$$y = A_0 \sin 2\pi f(t)t$$

a. A radio station operates at a frequency of 98.5 MHz.

  i. Is the station AM or FM?

  ii. What position on your radio dial would you tune to receive the signals broadcast by this station?

b. Write a function rule for the carrier signal of a radio station that broadcasts at a frequency of 610 kHz. What is the period of the carrier wave for this station?

c. Write a function rule in terms of its variable frequency $f(t)$ for the carrier signal of a radio station that broadcasts at a frequency of 104.1 MHz. What is the period of the carrier wave for this station?

REVIEW

33 Use algebraic reasoning to find the coordinates of the maximum or minimum point, the x-intercept(s), and the y-intercept of each function.

a. $f(x) = -(x - 5)^2 + 7$

b. $g(x) = x^2 + 6x - 1$

34 Use properties of quadrilaterals to place the special types of quadrilaterals (parallelogram, rectangle, square, rhombus, kite, trapezoid, isosceles trapezoid) in the correct region of Venn diagrams with conditions given in Parts a and b below.

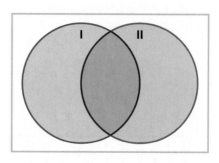

a. **I** Both pairs of opposite sides are parallel.

II Has at least one line of symmetry.

b. **I** Each diagonal divides the figure into two congruent triangles.

II Both diagonals are the same length.

35 Solve each equation for the indicated variable.

a. $C = 2\pi r$ for r

b. $A = \frac{1}{2}h(b_1 + b_2)$ for b_2

c. $E = mc^2$ for c

d. $d = \frac{m}{v}$ for v

36 Calculate these sums and differences of complex numbers. Express your answers in standard $a + bi$ form.

a. $(3 + 2i) + (5 + 8i)$

b. $(4 - 3i) + (2 + 6i)$

c. $(8 + 10i) - (3 + 2i)$

d. $(-3 + i) - (7 - i)$

37 Calculate the product of these complex numbers. Express your answers in standard $a + bi$ form.

a. $(4 + i)(2 + 3i)$

b. $(3 + 2i)(3 - 2i)$

c. $(3 + 2i)^2$

d. $(3 - 2i)^2$

38 Using the diagram at the right, determine, for each set of conditions, whether the conditions imply pairs of lines that are parallel. If so, indicate which lines are parallel.

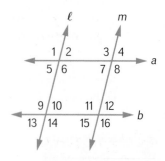

a. $m\angle 9 = m\angle 11$

b. $m\angle 13 = m\angle 7$

c. $m\angle 3 = m\angle 16$

d. $m\angle 1 + m\angle 12 = 180°$ and $m\angle 9 = m\angle 11$

39 Simplify each algebraic fraction as much as possible.

a. $\dfrac{16x + 12}{3x + 15}$

b. $\dfrac{x^2 - 9x}{x^2 - 81}$

c. $\dfrac{x^2 + 8x + 15}{x^2 + 2x - 3}$

d. $\dfrac{x - 1}{4x - 4}$

40 Use each *NOW-NEXT* rule to produce a table of values that illustrates the pattern of change from the start value through 5 stages of change.

a. $NEXT = NOW + 2.5$, starting at 3

b. $NEXT = \frac{1}{2}NOW - 50$, starting at 450

41 The number of bacteria in a wound can be modeled using the function $N(t) = 3(8^t)$, where t represents the number of hours since the wound occurred.

a. Evaluate and explain the meaning of $N(2)$.

b. How many bacteria were introduced into the wound when it occurred?

c. How long does it take for the number of bacteria in the wound to double?

d. Write a *NOW-NEXT* rule that could be used to find the number of bacteria present after any number N of complete hours.

42 Between the 2000 census and the 2010 census, the fastest growing city in the United States was Palm Coast, Florida. It had a growth rate of 92.0%. During that same decade, Youngstown, Ohio, saw a decrease in population of 18.3%. (**Source:** U.S. Census Bureau)

a. The population of Palm Coast was 49,832 in 2000. What was its approximate population in 2010?

b. The population of Youngstown, Ohio, was 82,026 in 2000. What was its approximate population in 2010?

c. Which of these *NOW-NEXT* rules could be used to help calculate the population of Youngstown, Ohio in future decades, if this rate of decline continues? All rules have a starting value of 82,026.

- $NEXT = 0.183NOW$

- $NEXT = 0.817NOW$

- $NEXT = NOW - 0.183NOW$

- $NEXT = NOW + 0.183NOW$

Looking Back

In this unit, you explored and proved important properties of special lines, segments, and angles related to circles. To reason about these properties, you drew on methods that used congruent triangles, symmetry, coordinate geometry, and trigonometry. You then built on these circle properties by embarking on a study of characteristics and applications of revolving circles. Your study of circular motion began with a study of pulleys and gears, moved to linear and angular velocity, introduced radian angle measure, and culminated with the study of the sine and cosine functions and their role in modeling aspects of circular motion and other periodic phenomena.

The following tasks will help you review, pull together, and apply what you have learned.

1. A circle with center O has diameters $\overline{BE}$ and $\overline{AH}$. Rays $\overrightarrow{CB}$ and $\overrightarrow{CD}$ are tangent to the circle at points B and D, respectively, and G is the midpoint of chord $\overline{BD}$.

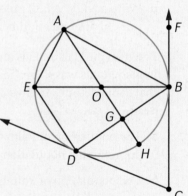

   a. Explain why m$\angle EAB$ = m$\angle EBC$ = m$\angle AGB$.

   b. Suppose m$\widehat{AED}$ = 124° and m$\angle ABF$ = 62°. Find the degree measures of $\angle AED$, $\angle BCD$, and $\widehat{DE}$. Explain your reasoning.

2. For each of the following statements, prove the statement if it is true or give a counterexample if it is false.

   a. The midpoint of the common chord of two circles that intersect in two points lies on the line through the centers of the circles.

   b. If inscribed angle $\angle ABC$ in a circle with center O measures 120°, then central angle $\angle AOC$ measures 120°.

   c. The angle bisector of any angle inscribed in a circle contains the center of the circle.

   d. If the line through the midpoints of two chords of a circle contains the center of the circle, then the chords are parallel.

3 *Wheel of Fortune* has been the most popular game show in the history of television. On the show, three contestants take turns spinning a large wheel similar to the one at the right. The result determines how much that contestant wins as she or he progresses toward solving a word, phrase, or name puzzle. The wheel is divided into 24 sectors of equal size, each corresponding to a dollar value or some other outcome.

a. If the wheel spins through 3.8 counterclockwise revolutions, what is the degree measure of the angle through which the wheel spins?

b. What is the measure of the minimum positive angle that is *coterminal* with (has the same terminal side as) the angle in Part a?

c. Suppose the wheel starts at the center of the "Bankrupt" sector and spins through 1,055°. Will it stop at "Bankrupt"? Describe the intervals of angle measures that will return the wheel to the "Bankrupt" sector.

Imagine a coordinate system superimposed on the wheel of fortune with its origin at the center of the wheel. Suppose a contestant spins the wheel releasing it at the "3 o'clock" position on the coordinate system, and the wheel completes 2.8 revolutions in 20 seconds.

d. What is the angular velocity in revolutions per second? In radians per second?

e. If the wheel is 10 feet in diameter, what is the average linear velocity in feet per second of a point on the edge of the wheel?

f. What are the coordinates of the release point when the wheel stops at the end of 20 seconds? Explain your reasoning.

4 The Thames River passes through the heart of London, England. Ships entering that part of the river need to pass under a number of bridges like the famous Tower Bridge. The height of the Tower Bridge above the river (in meters) varies over time (in hours following high tide) according to $d(t) = 12 + 3.4 \cos 0.5t$.

a. What are the maximum and minimum distances from the bridge to the river?

b. What are the period and amplitude of variation in distance from the bridge to the river?

c. How often in one day does the distance from bridge to river complete a full cycle from maximum to minimum and back to maximum?

SUMMARIZE THE MATHEMATICS

In this unit, you described and reasoned about important properties of special lines, segments, and angles related to circles. You also analyzed circular motion including linear and angular velocity, radian measure of angles, and circular functions.

a What is the relationship between a line k tangent to a circle with center O at point A and radius $\overline{OA}$?

b What is the relationship between the measure of a central angle in a circle and the measure of any inscribed angle that intercepts the same arc of the circle?

c If a radius of a circle intersects a chord (other than a diameter) of the circle at its midpoint, what are the measures of the angles formed?

d How is the angular velocity of one rotating pulley (radius r_1) related to the angular velocity of a second pulley (radius r_2) connected to it if:

    i. $r_1 = r_2$ **ii.** $r_1 > r_2$ **iii.** $r_1 < r_2$

e What does it mean to say that an angle or rotation has radian measure t? What would be the approximate corresponding measure in degrees?

f If $P(r, 0)$ is a point on a circle of radius r centered at the origin of a coordinate system, what are the coordinates of the image of P after a counterclockwise rotation of θ about the origin?

g What do the period, amplitude, and frequency of a periodic function tell about the pattern of values and graph of the function?

h When $f(t) = a \sin bt + c$ or $g(t) = a \cos bt + c$ are used to model the periodic pattern of change in some quantity, what do the values of a, b, and c tell about the situation being modeled?

Be prepared to explain your ideas to the class.

 CHECK YOUR UNDERSTANDING

Write, in outline form, a summary of the important mathematical concepts and methods developed in this unit. Organize your summary so that it can be used as a quick reference in future units and courses.

Recursion and Iteration

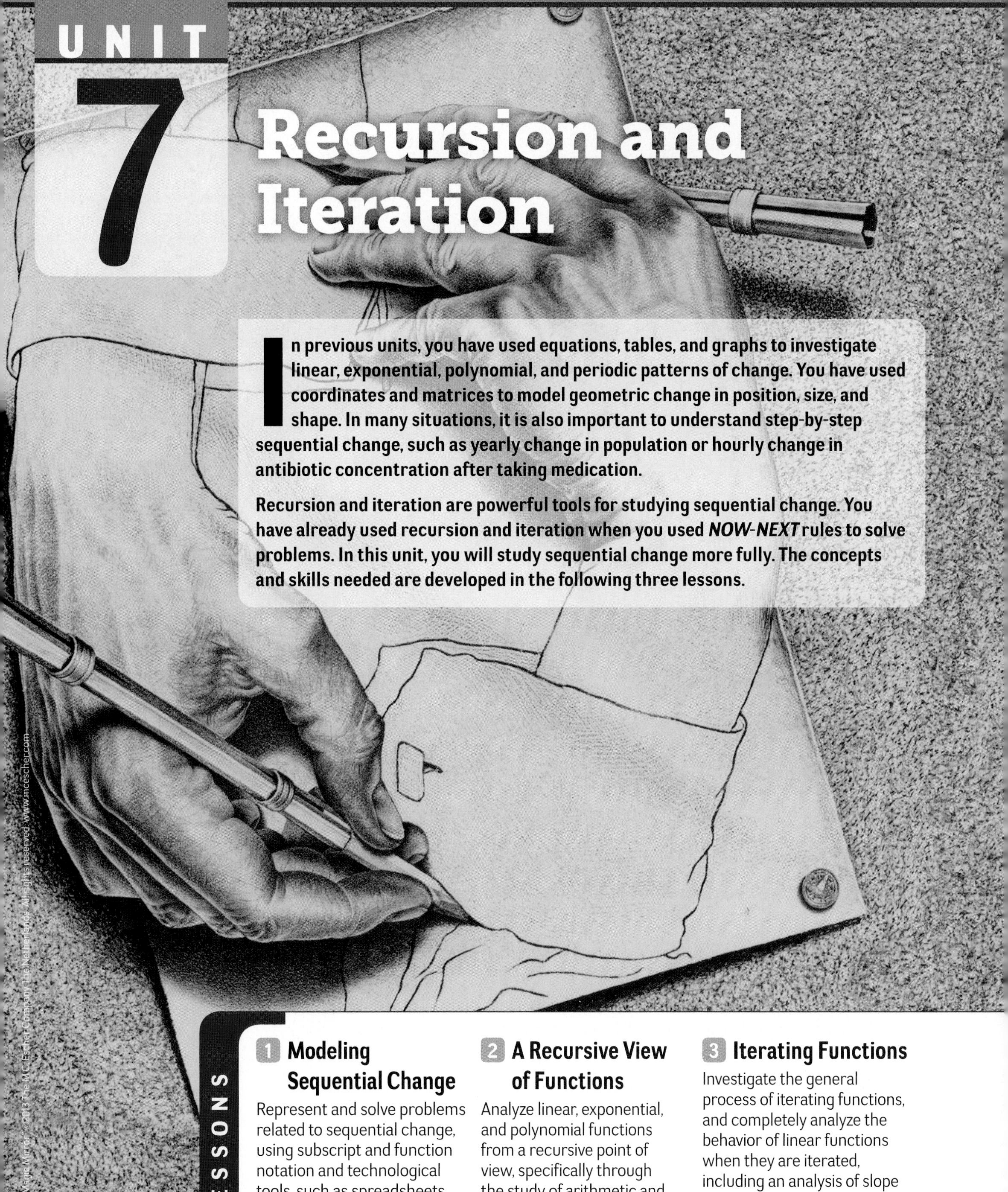

In previous units, you have used equations, tables, and graphs to investigate linear, exponential, polynomial, and periodic patterns of change. You have used coordinates and matrices to model geometric change in position, size, and shape. In many situations, it is also important to understand step-by-step sequential change, such as yearly change in population or hourly change in antibiotic concentration after taking medication.

Recursion and iteration are powerful tools for studying sequential change. You have already used recursion and iteration when you used *NOW-NEXT* rules to solve problems. In this unit, you will study sequential change more fully. The concepts and skills needed are developed in the following three lessons.

LESSONS

1 Modeling Sequential Change

Represent and solve problems related to sequential change, using subscript and function notation and technological tools, such as spreadsheets.

2 A Recursive View of Functions

Analyze linear, exponential, and polynomial functions from a recursive point of view, specifically through the study of arithmetic and geometric sequences and finite differences tables.

3 Iterating Functions

Investigate the general process of iterating functions, and completely analyze the behavior of linear functions when they are iterated, including an analysis of slope to identify attracting and repelling fixed points.

Modeling Sequential Change

Wildlife management has become an increasingly important issue as modern civilization puts greater demands on wildlife habitat. Tracking annual changes in the size of wildlife population is essential to effective management.

In previous units, you have used *NOW-NEXT* rules or formulas to model situations involving sequential change in the populations of endangered species such as timber wolves and bowhead whales. Now you will examine sequential change more fully; first in the context of a localized fish population.

U.S. Fish & Wildlife Service/Pedro Ramirez, Jr.

In this unit, you will use recursion and iteration to represent and solve problems related to sequential change, such as year-to-year change in a population or month-to-month change in the amount of money owed on a loan. You will use subscript notation and function notation. Spreadsheet software will be used to help with the analysis.

INVESTIGATION

Modeling Population Change

The first step in analyzing sequential change situations, like the fish population situation, is to build a mathematical model. As you work on the problems of this investigation, look for answers to this question:

How can you construct and use a mathematical model
to help you analyze a changing fish population?

As you have seen before, a typical first step in mathematical modeling is simplifying the problem and deciding on some reasonable assumptions. Three factors that you may have listed in the Think About This Situation discussion are initial fish population in the pond, annual growth rate of the population, and annual restocking amount, that is, the number of fish added to the pond each year. For the rest of this investigation, use just the following assumptions.

- There are 3,000 fish currently in the pond.

- 1,000 fish are added at the end of each year.

- The population decreases by 20% each year (taking into account the combined effect of all causes, including births, natural deaths, and fish being caught).

1 Using these assumptions, build a mathematical model to analyze the population growth in the pond as follows.

a. Estimate of the population after one year. Estimate the population after two years. Describe how you computed these estimates. What additional details did you assume to get the answers that you have?

b. Assume that the population decreases by 20% *before* the 1,000 new fish are added. Also assume that the population after each year is the population *after* the 1,000 new fish are added. Are these the assumptions you used to compute your answers in Part a? If not, go back and recompute using these assumptions. Explain why these assumptions are reasonable. These are the assumptions you will use for the rest of this analysis.

c. Write a formula using the words *NOW* and *NEXT* to model this situation as specified in Part b.

d. Use the formula from Part c and the last-answer feature of your calculator or computer software to find the population after seven years. Explain how the keystrokes or software features you used correspond to the words *NOW* and *NEXT* in the formula.

2 Now think about the patterns of change in the long-term population of fish in the pond.

a. Do you think the population will grow without bound? Level off? Die out? Make a quick guess about the long-term population. Compare your guess to those made by other students.

b. Determine the long-term population by continuing the work that you started in Part d of Problem 1.

c. Explain why the long-term population you have determined is reasonable. Give a general explanation in terms of the fishing pond ecology. Also, based on the assumptions above, explain mathematically why the long-term population is reasonable.

d. Does the fish population change faster around year 5 or around year 25? How can you tell?

3 What do you think will happen to the long-term population of fish if the initial population is different but all other conditions remain the same? Make an educated guess. Then check your guess by finding the long-term population for a variety of initial populations. Describe the pattern of change in long-term population as the initial population varies.

4 Investigate what happens to the long-term population if the annual restocking amount changes but all other conditions are the same as in the original assumptions. Describe as completely as you can the relationship between long-term population and restocking amount.

5 Describe what happens to the long-term population if the annual decrease rate changes but all other conditions are the same as in the original assumptions. Describe the relationship between long-term population and the annual decrease rate.

6 Now consider a situation in which the fish population shows an annual rate of *increase*.

   a. What do you think will happen to the long-term population if the population *increases* at a constant annual rate? Make a conjecture and then test it by trying at least two different annual increase rates.

   b. Write formulas using *NOW* and *NEXT* that represent your two test cases.

   c. Do you think it is reasonable to model the population of fish in a pond with an annual rate of increase? Why or why not?

SUMMARIZE THE MATHEMATICS

In this investigation, you constructed a *NOW-NEXT* formula to model and help analyze a changing fish population. Consider a population change situation modeled by *NEXT* = 0.6*NOW* + 1,500.

a Describe a situation involving a population of fish that could be modeled by this *NOW-NEXT* formula.

b What additional information is needed to be able to use this *NOW-NEXT* formula to predict the population in 3 years?

c What additional information is needed to be able to predict the long-term population? What is the long-term population in this case?

d Consider the following variations in a fish-population situation modeled by a *NOW-NEXT* formula like the one above.

   i. If the initial population doubles, what will happen to the long-term population?

   ii. If the annual restocking amount doubles, what will happen to the long-term population?

   iii. If the annual population decrease rate doubles, what will happen to the long-term population?

e How would you modify *NEXT* = 0.6*NOW* + 1,500 so that it represents a situation in which the fish population increases annually at a rate of 15%? What effect does such an increase rate have on the long-term population?

Be prepared to explain your ideas and reasoning to the class.

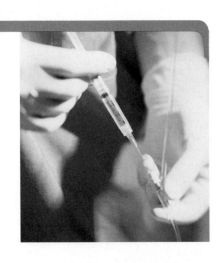

✔ CHECK YOUR UNDERSTANDING

A hospital patient is given an antibiotic to treat an infection. He is initially given a 30-mg dose and then receives another 10 mg at the end of every six-hour period thereafter. Through natural body metabolism, about 20% of the antibiotic is eliminated from his system every six hours.

a. Estimate the amount of antibiotic in his system after the first six hours and after the second six hours.

b. Write a formula using the words *NOW* and *NEXT* that models this situation.

c. Find the amount of antibiotic in the patient's system after 10 days.

d. Suppose his doctor decides to modify the prescription so that the long-term amount of antibiotic in his system will be about 25 mg. How should the prescription be modified?

INVESTIGATION 2

The Power of Notation and Technology

Situations involving sequential change, as in the fish-population problem, are sometimes called **discrete dynamical systems**. A discrete dynamical system is a situation (system) involving change (dynamical) in which the nature of the change is step-by-step (discrete). An important part of analyzing discrete dynamical systems is determining long-term behavior, as you did when you found the long-term fish population or the long-term amount of antibiotic in a patient's system.

Sequential change situations can often be analyzed using *recursion* and *iteration*. Sequential change is step-by-step change. **Recursion** is the method of describing a step in a sequential process in terms of previous steps. **Iteration** is the process of repeating the same procedure or computation over and over again.

Notation and technology can be very helpful when analyzing sequential change situations. As you work on the problems of this investigation, look for answers to these questions:

> *How can recursion and iteration be used to model and analyze sequential change situations?*
>
> *How can subscript, function, and spreadsheet notation be used in the modeling process?*
>
> *How can spreadsheets be used in the modeling process?*

1 Think about how recursion and iteration were used in your analysis of the changing fish population in Investigation 1.

a. Describe an instance where you used recursion to model or analyze the fish-population problem.

b. Describe an instance where you used iteration in the fish population problem.

Subscript and Function Notation You can use subscripts and function notation to more compactly write formulas like those in Investigation 1 that use the words *NOW* and *NEXT*. These notations can help you analyze the formulas and the situations they model. Consider the context of a changing fish population from the last investigation. The subscript notation P_n can be used to represent the population after n years. (The notation P_n is read "P sub n.") Thus, P_0 ("P sub 0") is the population after 0 years, that is, the initial population. P_1 is the population after 1 year, P_2 is the population after 2 years, and so on.

Recall that the fish-population problem is based on these three assumptions:

- There are 3,000 fish currently in the pond.

- The population decreases by 20% each year due to natural causes and fish being caught.

- 1,000 fish are added at the end of each year.

2 Find P_0, P_1, and P_2. Compare to the population values from Investigation 1. Sketch a graph of P_n versus n. Include the three values you have just found, for $n = 0$, 1, and 2. Also, indicate the shape of the graph based on all the population values from Investigation 1. A rough sketch by hand is fine.

3 The subscript notation relates closely to the way you have used the words *NOW* and *NEXT* to describe sequential change in many contexts.

a. In the context of a changing population, if P_1 is the population *NOW*, what subscript notation represents the population *NEXT* year?

b. If P_{24} is the population *NEXT* year, what subscript notation represents the population *NOW*?

c. If P_n is the population *NEXT* year, what subscript notation represents the population *NOW*?

d. If P_{n+1} is the population *NEXT* year, what subscript notation represents the population *NOW*?

e. A formula that models the annual change in fish population as described in Investigation 1 is

$$NEXT = 0.8NOW + 1{,}000, \text{ starting at } 3{,}000.$$

i. Rewrite this formula using P_n and P_{n-1} notation.

ii. Rewrite this formula using P_n and P_{n+1} notation.

4 Function notation can often be used interchangeably with subscript notation. For example, both P_0 and $P(0)$ can represent the initial population. The subscript notation is read "P sub 0" and the function notation is read "P of 0." Each notation can be used to represent the population at time 0.

a. Refer to Problem 2 above. In the context of the fish population problem, if $P(n)$ is the population after n years, what are the values of $P(0)$, $P(1)$, and $P(2)$? Calculate $P(3)$.

b. Rewrite the *NOW-NEXT* formula from Problem 3 Part e using function notation. Is there more than one way to do this? Explain.

Spreadsheet Analysis You can analyze situations that involve sequential change by using the iteration and recursion capability of a calculator or computer software. In particular, spreadsheet software can be used to analyze these situations.

5 Spreadsheets have their own notation, which is similar to subscript and function notations. Spreadsheets are organized into cells, typically labeled by letters across the top and numbers down the side. If cell **A1** contains the fish population *NOW*, and **A2** contains the population *NEXT* year, write a formula using **A1** and **A2** that shows the relationship between *NOW* and *NEXT*. (*Note:* When using formulas that reference spreadsheet cells, a * must be used for multiplication.)

6 Analyze the fish population situation using a spreadsheet, as follows.

a. A formula like that in Problem 5 does not tell you what the population is for any given year; it only shows how the population changes from year to year. To be able to compute the population for any year, you need to know the initial population. Using spreadsheet software available to you, enter the initial population of 3,000 into cell **A1**.

b. You want **A2** to show the population in the next year. So, enter "=" and your expression for **A2** from Problem 5 into cell **A2**. Press Enter or Return. The population for the next year (3,400) should appear in cell **A2**. If it does not, check with other students and resolve the problem.

c. Now you want **A3** to show the population in the next year. So, you need to enter an expression for **A3** into the **A3** cell. Then to get the next population, you would enter an expression for **A4** into the **A4** cell. And so on. Repeating this procedure is an example of iteration. The spreadsheet software will do this automatically, using the Fill Down command, as follows.

- Click and hold on cell **A2**, then drag down to cell **A50**.

- Then in the Edit menu, choose Fill Down.

- Cells **A3** through **A50** should now show the population from ear to year. If not, check with other students and resolve the problem.

d. Describe the pattern of change you see in the spreadsheet list of population values. Compare to your analysis of the fish-population situation in Investigation 1.

e. You can use the graphing capability of a spreadsheet to help analyze the situation.

    i. Use the spreadsheet software to create a scatterplot of the fish population over time. When there is only one column of data, some spreadsheet software has a default setting that will assign integers beginning with 1 as the independent variable for a scatterplot. If your spreadsheet software does not have this default setting, fill column **B** with the year numbers that correspond to the populations in column **A**.

    ii. Describe any patterns of change you see in the population graph. Compare to your analysis of the fish-population situation in Investigation 1. Be sure to describe how the long-term population trend shows up in the graph.

Compound Interest You have been using recursion and iteration, along with appropriate notation and technology tools, to analyze a population problem. Another common application of recursion and iteration is compound interest. *Compound interest* is interest that is applied to previous interest as well as to the original amount of money borrowed or invested. The original amount of money is called the *principal*. The more often the interest is applied, that is, the more often it is *compounded*, the faster the total amount of money (principal plus interest) grows. Thus, the interest rate and compounding period are crucial factors in consumer loans, such as car loans and college loans.

In the U.S., the Truth in Lending Act requires that lenders must report interest rates in a standard way, using the Annual Percentage Rate (APR), so that consumers can more easily compare and understand loans. Unfortunately, the APR itself is sometimes interpreted and computed in different ways—see Extensions Task 18. In any case, even with some clarity provided by the APR, you still need to carefully analyze the interest rate on a loan.

For example, it is common to advertise car loans with a stated *annual* interest rate. However, the interest is usually compounded *monthly*, not annually, so careful analysis is required to make sure you know what the loan will really cost. Consider the following car loan ad.

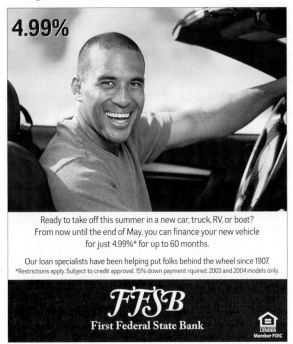

4.99%

Ready to take off this summer in a new car, truck, RV, or boat?
From now until the end of May, you can finance your new vehicle
for just 4.99%* for up to 60 months.

Our loan specialists have been helping put folks behind the wheel since 1907.
*Restrictions apply. Subject to credit approval. 15% down payment rquired. 2003 and 2004 models only.

FFSB
First Federal State Bank

LENDER
Member FDIC

7 The stated interest rate is 4.99%. Given the law about stating the APR, we will assume that this is an annual rate. The ad also implies that the loan will be computed monthly ("for up to 60 months"), so we will also assume that the interest will be compounded monthly. Finally, it is common to give the annual rate without adjusting for the effect of the monthly compounding. To carefully analyze this situation, we need to know the monthly interest rate. Under the assumptions just discussed, the monthly rate is $\frac{1}{12}$ of the annual rate. Explain why. For the annual rate shown in the ad, what is the monthly interest rate?

8 Suppose you borrow $10,000 to buy a new compact car at the interest rate shown in the ad, with repayment due in 48 monthly payments. You are told that your monthly payment will be $235. As a wise consumer, you should check to see if this payment amount is correct.

a. Figure out a method for how you can determine the amount you still owe, called the *balance*, after a given month's payment. Then find the balance of the car loan after each of the first three payments. Assume that the first payment is made exactly one month after the contract is signed. Compare your balances to those of other groups, and resolve any differences.

b. Write a formula using the words *NOW* and *NEXT* that models the month-by-month change in the balance of the loan. Write equivalent formulas using subscripts, function notation, and spreadsheet notation. Be sure to specify the initial balance.

c. Now think about whether the payment amount is correct.

- If the monthly payment of $235 is correct, what should the balance be after the 48th payment? Use the formulas from Part b and your calculator or computer to see if $235 is the correct payment for this loan.

- If the payment is incorrect, is it too high or too low? How can you tell? Experiment to find the correct monthly payment.

9 Investigate further the loan situation from Problem 8.

a. How much total interest was paid on the $10,000 loan, using the correct monthly payment?

b. Is the balance of the loan reduced faster at the beginning of the repayment period or at the end? Explain and give evidence to support your answer.

c. Suppose someone offers you a "great deal," whereby you must pay only $40 per month until the loan is paid off. Describe what happens to the repayment process in this situation. Is this such a great deal after all?

SUMMARIZE THE MATHEMATICS

Formulas involving recursion, like those you have been using in this investigation, are called recursive formulas. The recursive formulas you have studied have all been of the general form $A_n = rA_{n-1} + b$, with A_0 as the initial value. Consider a wildlife population that is modeled by the following formula and initial value.

$$U_n = 0.4U_{n-1} + 2{,}500, \ U_0 = 11{,}000$$

a What are the values for r and b in the recursive formula for the wildlife population?

b What can you say about the size of the wildlife population and how it is changing?

c Rewrite the "$U_n = \dots$" formula using the words *NOW* and *NEXT*. Write an equivalent formula using function notation.

d If the initial population is entered into cell **A8** in a spreadsheet and the population next year is entered into cell **A9**, write a formula showing the relationship between **A8** and **A9**.

e The size of the population after 3 years is 4,604. Explain how to find the population after 4 years:

    i. without using a calculator or computer.

    ii. using the last-answer feature of a calculator or computer.

    iii. using a spreadsheet.

f Describe the pattern of population change, including the long-term population trend. Sketch a graph and explain how the graph shows the trend.

g Explain why this is a sequential change situation and how it involves recursion and iteration.

Be prepared to share your responses and thinking with the entire class.

 CHECK YOUR UNDERSTANDING

Recall the situation (page 462) involving a hospital patient taking an antibiotic to treat an infection. He was initially given a 30-mg dose, and then he took another 10 mg at the end of every six hours. Through natural body metabolism, about 20% of the antibiotic was eliminated from his system every six hours.

a. Write a recursive formula in the form "$U_n = \dots$" that models this situation. Write an equivalent formula using function notation.

b. Using spreadsheet software, produce a table and graph of (n, U_n) pairs with n as the input variable.

c. Describe how the amount of antibiotic in the patient's system changes over time, including the long-term change.

For the following tasks, use spreadsheet software, *CPMP-Tools*, or a graphing calculator as needed to help you solve the problems.

1 Chlorine is used to keep swimming pools safe by controlling certain harmful microorganisms. However, chlorine is a powerful chemical, so just the right amount must be used. Too much chlorine irritates swimmers' eyes and can be hazardous to their health; too little chlorine allows the growth of microorganisms to be uncontrolled, which can be harmful.

A pool manager must measure and add chlorine regularly to keep the level just right. The chlorine is measured in parts per million (ppm) by weight. That is, one ppm of chlorine means that there is one ounce of chlorine for every million ounces of water.

Chlorine dissipates in reaction to bacteria and to the sun at a rate of about 15% of the amount present per day. The optimal concentration of chlorine in a pool is from 1 to 2 ppm, although it is safe to swim when the concentration is as high as 3 ppm.

a. Suppose you have a summer job working at a swimming pool, and one of your responsibilities is to maintain a safe concentration of chlorine in the pool. You are required to add the same amount of chlorine to the pool every day. When you take the job, you find that the concentration is 3 ppm.

How much chlorine (in parts per million) do you need to add each day in order to maintain a long-term optimal concentration? Write a recursive formula that models your optimal chlorine maintenance plan. Describe any assumptions you have made in your analysis.

b. There are three key factors in this problem: the initial concentration, the daily increase in concentration due to the amount you add, and the dissipation rate. Systematically explore changes in each of these three factors and record the corresponding effects on the long-term chlorine concentration in the swimming pool.

c. Suppose the chlorinating pellets you use are 65% active chlorine, by weight. If the pool contains 50,000 gallons of water, and water weighs 8.337 pounds per gallon, how many pounds of chlorine pellets must you add to the pool each day?

©technician/Alamy

2 Retirement is probably not something you are currently concerned about. However, working adults, even very young working adults, should have a financial plan for retirement. If you start saving early and take advantage of compound interest, then you should be in great financial shape by the time you retire. Consider twin sisters with different retirement savings plans.

Plan I: Cora begins a retirement account at age 20. She starts with $2,000 and then saves $2,000 per year at 7% interest, compounded annually, for 10 years. (*Compounded annually* means that the interest is compounded every year, once per year at the end of the year.) Then she stops contributing to the account but keeps her savings invested at the same rate.

Plan II: Miranda does not save any money in her twenties. But when she turns 30, she starts with $2,000 and then saves $2,000 per year at 7% interest, compounded annually, for 35 years.

Both sisters retire at age 65. Who do you think will have more retirement savings at age 65? Test your conjecture by determining the amount of money saved by each sister at age 65.

3 Every ten years, the United States Census Bureau conducts a complete census of the nation's population. In 2010, the census report stated that there were about 309 million residents in the United States and its territories. The population changes extensively between census reports, but it is too expensive to conduct the census more often. Annual changes can be determined using estimates like the following.

- Births will equal about 1.3% of the total population each year.

- Deaths will equal about 0.8% of the total population each year.

- Immigrants from other countries will add about 0.9 million people each year. (**Sources**: 2011 World Population Data Sheet, Population Reference Bureau, www.prb.org; www.census.gov/2010census)

a. Using the statistics above, what population is estimated for the United States in 2011? In the year 2020?

b. Write a formula using the words *NOW* and *NEXT* that represents this situation.

c. Write a recursive formula that represents this situation. Specify the initial value.

d. Produce a table and a graph that show the population estimates through the year 2030. Describe the expected long-term trend in population change over time.

e. Describe some hypothetical birth and death rates that would result in a population that levels off over time. Represent this situation with a recursive formula, a table, and a graph.

4 Commercial hunting of whales is controlled to prevent the extinction of some species. Because of the danger of extinction, scientists conduct counts of whales to monitor their population changes. A status report on the bowhead whales of Alaska estimated that the 1993 population of these whales was between 6,900 and 9,200 and that the difference between births and deaths yielded an annual growth rate of about 3.1%. No hunting of bowhead whales is allowed, except that Alaskan Inuit are allowed to take, or harvest, about 50 bowhead whales each year for their livelihood.

(**Source:** nmml.afsc.noaa.gov/CetaceanAssessment/bowhead/bmsos.htm)

a. Use 1993 as the initial year. The initial population is a range of values, rather than a single value. That is, the range of initial population values is 6,900 to 9,200. Think about the *range* of population values for later years. Using the given information about annual growth rate and harvesting by Alaskan Inuit, what is the range of population values in 1994?

b. Let LOW(n) be the lower population value in the range of values in year n.

   i. What is the initial value, that is, what is LOW(0)? What is LOW(1)? Write a recursive formula for the growth of the lower population values in the range of population values.

   ii. Similarly, let HIGH(n) be the higher population value in the range of values in year n. Write a recursive formula for the growth of the higher population values in the range of population values, and state the initial value.

c. Put the formulas in Part b together to write a recursive formula for this situation, by completing the following work. Fill in the missing steps and provide the requested explanations.

   Let $R(n)$ represent the population range in year n.

   (1) So, $R(0) = 6,900$ to $9,200$.   Explain.
   (2) $R(1) = 7,063.9$ to $9,435.2$.   Explain.
   (3) $R(2) =$ Fill in the range for $R(2)$.
   (4) $R(n) =$ Fill in the range for $R(n)$, using the formulas from Part b. Include the initial values for both the lower and higher population values.

d. Sketch a graphical representation for $R(n)$, by drawing vertical line segments for the range corresponding to each year n, with values of n on the horizontal axis and population values on the vertical axis. Use at least 4 years. Elaborate on this graphical representation by sketching graphs for n vs LOW(n) and n vs HIGH(n) on the same set of axes. Describe some patterns in the ranges $R(n)$. (See Lesson 2, Extensions Task 25 for further investigation of this situation.)

e. So far in this task, population values have been growing. Suppose that, because of some natural disaster, the initial bowhead whale population is reduced to 1,500, but growth rate and number harvested by the Inuit stay the same. Under these conditions, what happens to the long-term population?

5 Money grows when it is kept in an interest-bearing savings account. Recursive formulas can be used to analyze how the money grows due to compound interest. For this task, assume that you deposit money into a savings account and make no withdrawals.

a. Suppose you deposit $100 in a savings account that pays 4% interest, compounded annually. (*Compounded annually* means that the interest is compounded every year, once per year at the end of the year.) Write a *NOW-NEXT* formula and a recursive formula that show how the amount of money in your account changes from year to year. Find the amount of money in the account after 10 years.

b. Most savings accounts pay interest that compounds more often than annually. Suppose that you make an initial deposit of $100 into an account that pays 4% annual interest, compounded monthly.

 i. Write a recursive formula that models the month-by-month change in the amount of money in your account.

 ii. How much money is in the account after 1 month? After 2 months? After 2 years?

 iii. How much money is in the account after 10 years? Compare your answer to the answer you got in Part a. Which kind of interest is a better deal, compounding annually or compounding monthly?

c. Suppose that, in addition to the initial $100, you deposit another $50 at the end of every year, and the interest rate is 4% compounded annually.

 i. Write a recursive formula that models this situation.

 ii. How much money is in the account after 10 years?

 iii. Describe the pattern of growth of the money in the savings account.

6 In this lesson, you have investigated recursive formulas of the form $A_n = rA_{n-1} + b$. Different values for r and b yield models for different situations. Consider some of the possibilities by completing a table like the one on the next page. For each cell in the table, do the following.

• Describe a situation that could be modeled by a recursive formula with the designated values of r and b. (You may use examples from the lesson if you wish.)

• Write the recursive formula.

• Describe the long-term trend.

	$0 < r < 1$	$r > 1$
$b < 0$	A whale population is decreasing by 5% per year due to a death rate higher than the birthrate, and 50 whales are harvested each year. $A_n = 0.95A_{n-1} - 50$ Long-term trend: Extinction	
$b > 0$		

CONNECTIONS

7 In Investigation 2, you found the following recursive formula to model a fish population.

Formula I $P_n = 0.8P_{n-1} + 1,000$, with $P_0 = 3,000$

Here is another formula that represents this situation.

Formula II $P_n = -2,000(0.8^n) + 5,000$

a. You can provide an algebraic argument for why the two formulas above are equivalent (see Connections Task 11 in Lesson 2 and Extensions Task 18 in Lesson 3). For now, give an informal argument using graphs, as follows.

    i. Use geometric transformations to describe how the graph of the function indicated by Formula II is related to the graph of the exponential function $f(n) = 0.8^n$.

    ii. Compare the graph of the function given by Formula II to the graph you generated in Investigation 2 using Formula I. Describe how the graphs are similar.

    iii. Why do you think it makes sense that these two formulas could be equivalent models of the fish-population problem?

b. Verify that Formula II gives the same values for P_1 and P_5 as those found using the recursive formula.

c. Verify that Formula II yields an initial population of 3,000.

d. Use Formula II to verify the long-term population you found in Problem 2 of Investigation 1 (page 466). Explain your thinking. How is the long-term population revealed in the symbolic form of this formula?

8 In Investigation 2, there was a brief discussion of Annual Percentage Rate (APR). The APR is meant to be a standard way to state interest rates on loans so that different loans can be compared and understood more easily. A related method of stating interest rates is used for investments, like when you deposit money into a savings account.

The **Annual Percentage Yield (APY)** expresses an interest rate on an investment in a way that takes into account the effect of compounding the interest. The APY is different, and usually higher, than a "nominal" annual rate. For example, consider the table below showing interest rates for the Los Alamos National Bank in New Mexico.

Savings Account Interest Rates

as of April 3, 2008

TYPE	INTEREST RATE	APY*	MINIMUM BALANCE**	TIER BALANCE
Regular Savings Account	1.000%	1.010%	None	None
Campus Savings Account	1.000%	1.010%	None	None
Universal Savings Tier 1	1.000%	1.010%	None	$0.00 - $49,999.00
Universal Savings Tier 2	1.750%	1.770%	None	$50,000.00 - $99,999.00
Universal Savings Tier 3	2.500%	2.530%	None	$100,000.00 - $999,999.00
Universal Savings Tier 4	2.750%	2.790%	None	$1,000,000.00 and above
Money Market Tier 1	0.500%	0.500%	$1,000.00	$0.00 - $19,999.00
Money Market Tier 2	1.750%	1.770%	$1,000.00	$20,000.00 - $49,999.00
Money Market Tier 3	2.250%	2.280%	$1,000.00	$50,000.00 - $99,999.00
Money Market Tier 4	2.500%	2.530%	$1,000.00	$100,000.00 and above

The above Interest Rates are subject to change without prior notice.

*Annual Percentage Yield
**Minimum balance required to avoid periodic service charge.

Source: www.lanb.com/rates/savings.asp, retrieved April 3, 2008

This bank pays interest compounded monthly. The rates shown in the column labeled "INTEREST RATE" are sometimes called *nominal interest rates*. These rates are annual rates, and they do not take into account the effect of compounding. The APY is the nominal annual rate adjusted for the effect of monthly compounding. Here is how it works.

a. Let's think big and consider Universal Savings Tier 3. The stated nominal interest rate is 2.500%. This is an annual rate.

   i. What is the corresponding monthly rate?

   ii. Suppose you invest $1 at this monthly rate and you compound the interest for 12 months. How much will you have at the end of the year?

b. 2.500% is an annual interest rate that does *not* take into account the monthly compounding. Explain why the number you determined in Part a can be thought of as an *annual growth rate* that *does* take the monthly compounding into account.

c. Subtract 1 from the number you determined in Part a. Compare the result to the APY for Universal Savings Tier 3. Describe similarities and differences.

d. The formula for computing APY is

$$APY = \left(1 + \frac{r}{n}\right)^n - 1,$$

where r is the nominal annual rate, expressed as a decimal, and n is the number of compound periods in a year.

   i. Explain how each part of this formula relates to the numbers and computations in Parts a–c.

   ii. Use this formula to compute the APY for another type of savings account in the table on the previous page.

9 You can represent recursive formulas with subscript notation or function notation. For example, in the fish-population problem, you can think of the population P as a function of the number of years n and you can write P_n or $P(n)$ for the population after n years. In the case of the fish-population problem, what are the practical domain and practical range of the function P?

10 Think about how matrix multiplication can be used to calculate successive values generated by recursive formulas of the form $A_n = rA_{n-1} + b$. For example, consider the recursive formula for the original fish-population situation at the beginning of this lesson: $P_n = 0.8P_{n-1} + 1{,}000$, with $P_0 = 3{,}000$.

a. A first attempt at a matrix multiplication that would be equivalent to evaluating the recursive formula might use the following matrices.

$$A = \begin{bmatrix} 3{,}000 & 1 \end{bmatrix} \text{ and } B = \begin{bmatrix} 0.8 \\ 1{,}000 \end{bmatrix}$$

Compute AB and compare it to P_1.

b. AB is just a 1×1 matrix and so is not much good for repeated multiplication. Thus, you cannot use repeated multiplication of these matrices to successively evaluate the recursive formula. A better try might be to use the following matrices.

$$A = \begin{bmatrix} 3{,}000 & 1 \end{bmatrix} \text{ and } C = \begin{bmatrix} 0.8 & 0 \\ 1{,}000 & 1 \end{bmatrix}$$

Compute AC and compare it to P_1. What matrix multiplication would you use next to find P_2? To find P_3?

c. There is just one finishing touch needed. Modify the matrices so that the multiplication computes not only the successive values but also the number of the year. Consider the following matrices.

$$A = \begin{bmatrix} 0 & 3{,}000 & 1 \end{bmatrix} \text{ and } D = \begin{bmatrix} 1 & 0 & 0 \\ 0 & 0.8 & 0 \\ 1 & 1{,}000 & 1 \end{bmatrix}$$

Compute AD, AD^2, AD^3, AD^4, and AD^5. Compare the results to P_1 through P_5.

d. Use matrix multiplication to generate three successive values of the recursive formula $P_n = 1.04P_{n-1} - 350$, with $P_0 = 6{,}700$. Do so in a way that also generates the successive values of n.

e. Explain why this matrix multiplication method works to find successive values of a recursive formula.

REFLECTIONS

11 When using some graphing calculators, you have the option of graphing in "connected" mode or "dot" mode. Find out what the difference is between these two modes of graphing. Which do you think is more appropriate for the graphing you have been doing in this lesson? Why?

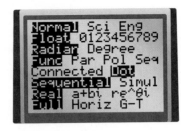

12 Consider again the fish population change modeled by $NEXT = 0.8NOW + 1,000$, from Investigation 1. You used several different formulas to represent this fish population, for example, a formula using the words *NOW* and *NEXT*, a formula with subscripts using P_n and P_{n-1}, a formula with subscripts using P_n and P_{n+1}, a formula with function notation using $A(n)$ and $A(n + 1)$, and a formula using spreadsheet notation.

Suppose that one of your classmates has not completed this task. Write a paragraph to the classmate explaining how all these formulas accurately represent the changing fish population.

13 In this lesson, you used recursive formulas to model sequential change in several situations, such as population growth, drug concentration, chlorine concentration, and money saved or owed. Describe one result of your investigations that was particularly interesting or surprising.

14 Recursive formulas are also sometimes called *difference equations*. What difference would be of interest in studying a recursive formula?

EXTENSIONS

15 These days, almost every state has a lottery, and the jackpots are often quite large. But are they really as large as they seem? Suppose you win $500,000 in a lottery (which you will, of course, donate to the mathematics department at your school). Typically, these large jackpots are not paid immediately as a lump sum. Instead, they are paid over time. For example, suppose you receive your $500,000 over 20 years at $25,000 per year. To accurately analyze how much you have *really* won in this situation, you need to include the effect of compound interest.

a. Suppose you deposit $500,000 in a bank account paying 3.5% annual interest, and you withdraw $25,000 at the end of every year. Write a recursive formula that models this situation, and calculate how much money will be in the account after 20 years.

b. Experiment to find an initial deposit that will yield a balance of $0 after 20 years.

c. The result from Part b is called the **present value** of your lottery winning. The present value is the lump sum amount that, if deposited now at 3.5% annual interest, would generate payments of $25,000 per year for 20 years. The present value is what your lottery winning is really worth, taking into account the reality of compound interest. So in this situation, how much is a $500,000 jackpot paid over 20 years really worth?

d. Find the actual value, that is, the present value, of a lottery winning of $1,000,000 that is paid at $50,000 per year for 20 years, if you can invest money at 5% annual interest.

e. What other factors besides compound interest do you think could affect the present value?

16 In this lesson, you modeled a fish population with the recursive formula $P_n = 0.8P_{n-1} + 1,000$, with $P_0 = 3,000$. You can get another modeling formula for this situation by fitting a curve to the population data and finding the equation for that curve. Use the following steps to carry out this plan.

a. Use the formula $P_n = 0.8P_{n-1} + 1,000$, with $P_0 = 3,000$, to generate the sequence of population figures for 20 years. Put these figures into a data list, say L2, on your calculator or computer. Generate another list, say L1, that contains the sequence of years 0 through 20. Produce a scatterplot of L1 and L2.

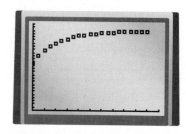

b. Modify your scatterplot by transforming the data in the lists so that the plot matches one of the standard regression models of your calculator or computer software. Find a regression equation that fits the transformed data.

c. Now, transform your regression equation so that it fits the original data.

d. Test your equation. Does the graph fit the original scatterplot? Compare the function value for $n = 20$ to what you get when you find P_{20} using the recursive formula $P_n = 0.8P_{n-1} + 1,000$, with $P_0 = 3,000$.

17 The Towers of Hanoi is a mathematical game featured in an old story about the end of the world. As the story goes, monks in a secluded temple are working on this game; and when they are finished, the world will end! The people of the world would like to know how long it will take the monks to finish the game.

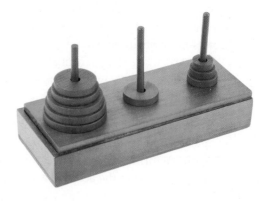

The game is played with 3 pegs mounted on a board and 64 golden discs of successively larger sizes with holes in the center. (Commercial games such as the example on the previous page have many fewer discs.) The game begins with all 64 discs stacked on one peg, from largest to smallest. The goal of the game is to move all 64 discs onto another peg, subject to these rules: (1) Only one disc may be moved at a time. (2) You are not allowed to put a larger disc on top of a smaller disc. (3) Discs may be placed temporarily on any peg, including the one on which the discs were originally stacked. (4) Eventually, the discs must be stacked from largest to smallest on a new peg.

a. You may not have 64 golden discs to play this game. But to get an idea of how it works, you can play with some homemade discs and pegs. Cut out four successively larger squares of paper to use as discs (or use different size coins for the discs). Put three large dots on a piece of paper to use as pegs mounted on a board. Play the game several times, first with just one disc stacked on the starting peg, then with two discs, then three, and so on. As you play each game, keep track of the *fewest* number of moves it takes to finish the game. Make a table listing the number of discs in the game and the fewest number of moves it takes to finish the game.

b. By thinking about strategies for how to play the game with more and more discs and by looking for patterns in the table from Part a, find a recursive formula and any other formula you can for the fewest number of moves needed to finish the game with n discs.

c. What is the fewest number of moves needed to finish a game with 64 discs? If the monks in the story move one disc every second and work nonstop, should we worry about the world ending soon? Explain.

18 The Annual Percentage Rate (APR), which you considered in Investigation 1, is not interpreted consistently in all ads and descriptions on the Internet. This may be because the law on which the APR is based is long and complicated. The APR comes from the U.S. Federal Truth in Lending Act, which is contained in Title I of the Consumer Credit Protection Act from May 29, 1968. This Act is implemented by the Truth in Lending Act Regulation, known as Regulation Z, as amended on December 20, 2001. At one place in these laws, it is stated that:

"The annual percentage rate [APR] … shall be determined … as that annual percentage rate which will yield a sum equal to the amount of the finance charge when it is applied to the unpaid balances of the amount financed … ." Furthermore, "the amount of the finance charge shall be determined as the sum of all charges … which include interest, [and other charges]." (Sections 106 and 107 of the Truth in Lending Act)

Elsewhere in the law it is stated that:

"The annual percentage rate [APR] shall be the nominal annual percentage rate determined by multiplying the unit-period rate by the number of unit-periods in a year." (Appendix J of Regulation Z)

a. Suppose a lender advertises a loan with an interest rate of 0.8% per month compounded monthly.

    i. Using the first statement from the Truth in Lending Act, which implies that the APR must be stated as an annual rate that takes into account compounding interest, what annual interest rate might the lender state?

    ii. Using the second statement above, from Regulation Z, what annual rate might the lender state?

b. Explain how the different interpretations of the APR that may be possible based on the two statements above can be described in terms of the difference between growth determined by an exponential function and growth determined by a linear function.

19 Suppose you have a recursive formula that can be used to compute the next term in a sequence from the current term. Think about how you could compute every second term or perhaps every tenth term. (For example, you might want to predict the population every decade instead of every year.)

a. Suppose $NEXT = 3NOW$ and the starting value is 1. List the first 10 terms in the sequence generated by this recursive formula and initial value. Starting with 1, list every second term within this list. Write a recursive formula that describes the list of every second term.

b. Think about every tenth term in the original sequence in Part a, that is, the sequence determined by $NEXT = 3NOW$, starting at 1. Write a new recursive formula that would generate every tenth term in the sequence.

c. Consider the sequence of numbers generated by $A_n = 3A_{n-1} + 2$, with $A_0 = 1$. List the first 11 terms of this sequence. Think about new recursive formulas that will generate every second term or every tenth term of this sequence.

    i. Let $\{S_n\}$ be the sequence of every second term, starting at 1. Explain each step in the following argument to find a recursive formula for S_n.

$$
\begin{aligned}
S_0 &= A_0 = 1 &\quad (1)\\
S_1 &= A_2 = 17 &\quad (2)\\
S_2 &= A_4 = 161 &\quad (3)
\end{aligned}
$$

$$\vdots$$

$$
\begin{aligned}
S_n &= A_{2n} &\quad (4)\\
&= 3A_{2n-1} + 2 &\quad (5)\\
&= 3(3A_{2n-2} + 2) + 2 &\quad (6)\\
&= 3^2 A_{2n-2} + 3(2) + 2 &\quad (7)\\
&= 3^2 A_{2(n-1)} + 3(2) + 2 &\quad (8)\\
&= 3^2 S_{n-1} + 3(2) + 2 &\quad (9)
\end{aligned}
$$

Thus, $S_n = 3^2 S_{n-1} + 3(2) + 2$.

ii. Let $\{T_n\}$ be the sequence of every tenth term, starting at 1. Explain each step in the following argument to find a recursive formula for T_n.

$$
\begin{aligned}
T_n &= A_{10n} & (1)\\
&= 3A_{10n-1} + 2 & (2)\\
&= 3(3A_{10n-2} + 2) + 2 & (3)\\
&= 3^2 A_{10n-2} + 3(2) + 2 & (4)\\
&= 3^2(3A_{10n-3} + 2) + 3(2) + 2 & (5)\\
&= 3^3 A_{10n-3} + 3^2(2) + 3(2) + 2 & (6)\\
&\;\;\vdots\\
&= 3^{10} A_{10n-10} + 3^9(2) + 3^8(2) + \cdots + 3(2) + 2 & (7)\\
&= 3^{10} A_{10(n-1)} + 3^9(2) + 3^8(2) + \cdots + 3(2) + 2 & (8)\\
&= 3^{10} T_{n-1} + 3^9(2) + 3^8(2) + \cdots + 3(2) + 2 & (9)
\end{aligned}
$$

Thus, $T_n = 3^{10} T_{n-1} + 3^9(2) + 3^8(2) + \cdots + 3(2) + 2$.

Use this formula to compute T_1. Compare to A_{10}, which you computed at the beginning of Part c on the previous page.

20 So far, you have used recursive formulas to predict future values in a sequential change situation. This is a natural use of recursive formulas since they typically show how to get from one term in a sequence to the next. Think about reversing the perspective and computing past terms.

a. Suppose $NEXT = 2NOW$, and one of the terms in the sequence is 460.8. What is the previous term? Describe how you found the previous term.

b. Suppose a changing population is modeled by this recursive formula.

$$P_n = 0.8P_{n-1} + 500, \text{ where } P_n \text{ is the population at year } n$$

If the population predicted by this formula in a given year is 1,460, what is the population the previous year?

c. The recursive formula in Part b can be used to compute the population in future years. Construct a new recursive formula that you could use to compute the population in past years. Use the new recursive formula to compute the population 10 years ago if the current population is 2,341.

REVIEW

21 A study of milk samples from 167,460 Holstein cows measured the percent of fat in each milk sample. The data was normally distributed with mean milk fat of 3.59% and a standard deviation of 0.71%.

a. Approximately how many of the samples had milk fat between 2.17% and 5.01%?

b. Approximately how many of the samples had less than 4.3% milk fat?

c. Approximately what percentage of the samples had more than 5.01% milk fat?

22 Perform each indicated operation. Write the polynomial in standard form.

a. $(3x^3 - 5x + 14) - (x^3 + 4x^2 - 6x)$

b. $(6 - x^3)^2(x + 1)$

c. $8(-x^2 + 7) + 5(x^4 - 5x^2 + 10)$

d. $(x^3 + 5x) - (7x^3 - 3x) + (4x^2 + 8x)$

23 Determine whether each data pattern is linear, exponential, or quadratic. Then find a function rule that matches each pattern.

a.

x	−3	−2	−1	0	1	2	3
y	3	9	27	81	243	729	2,187

b.

x	−4	−3	−2	0	2	4	7
y	1	5	9	17	25	33	45

c.

x	−2	−1	0	1	2	3	4
y	4,800	2,400	1,200	600	300	150	75

d.

x	−3	−2	−1	0	1	2	3
y	20	12	6	2	0	0	2

24 Simplify each algebraic fraction as much as possible. Then determine if the two expressions are undefined for the same values.

a. $\dfrac{12x^2 + 4x}{3x}$

b. $\dfrac{2x}{6x - 8}$

c. $\dfrac{x^2 - 9}{x + 3}$

d. $\dfrac{x^2 + 2x - 8}{x^2 - 7x + 10}$

25 Determine if each set of conditions will guarantee that quadrilateral $ABCD$ is a parallelogram. If not, draw a counterexample. If so, provide a proof.

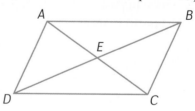

a. $\overline{CE} \cong \overline{EA}$

b. $\triangle AED \cong \triangle CEB$

c. $\overline{AC} \perp \overline{DB}$

26 Solve each exponential equation.

a. $10^x = 0.01$

b. $5(10^x) = 5,000$

c. $25(10^x) = 1,300$

d. $12(3^x) = 324$

e. $3(2^x) = 3\sqrt{2}$

LESSON
2

A Recursive View of Functions

In the previous lesson, you investigated a variety of situations that were modeled by recursive formulas, including population growth, consumer loans, and medicine dosage. In each of these situations, you examined a sequence of numbers and looked for patterns and formulas to represent the situation. In this lesson, you will take a closer look at sequences of numbers. As you do so, you will extend your understanding of recursion and iteration. In the process, you will revisit three important function families: linear, exponential, and polynomial. To begin, consider the exciting but dangerous sport of sky diving.

Imagine a sky diver jumping from a plane at a height of about 5,000 feet. Because of Earth's gravity and ignoring wind resistance, the sky diver will fall 16 feet in the first second. Thereafter, until the parachute opens, the distance fallen during each second will be 32 feet more than the distance fallen during the previous second.

a Think about the pattern of change in the distance the sky diver falls *during* each second. How would you describe the number of feet fallen during each second:

    i. using a recursive formula?

    ii. as a function of the number of seconds n?

b Now consider the pattern of change in the *total* distance fallen *after* a given number of seconds.

    i. How far has the sky diver fallen after each of the first four seconds?

    ii. Describe any patterns you see in this sequence of numbers. Compare this sequence to the sequence of distances fallen during each second.

c Think about a possible formula for the total distance fallen.

    i. How could the total distance fallen be computed from the sequence of numbers in Part a?

    ii. If you found a function formula for the total distance fallen after n seconds, what type of function do you think it will be?

    iii. How might a sky diver use a formula for total distance fallen in planning a jump?

In this lesson, you will learn about arithmetic and geometric sequences and series, and finite differences tables. In the process, you will use recursion and iteration to revisit linear, exponential, and polynomial functions.

INVESTIGATION **1**

Arithmetic and Geometric Sequences

Sequences of numbers occur in many situations. Certain sequences are particularly common and important, such as *arithmetic* and *geometric* sequences. You will learn about these sequences in this investigation.

As you work on the problems of this investigation, look for answers to these questions:

What are formulas for arithmetic and geometric sequences?

What functions correspond to arithmetic and geometric sequences?

Arithmetic Sequences As you can imagine, sky diving requires considerable training and careful advance preparation. Although the sport of bungee jumping may require little or no training, it also requires careful preparation to ensure the safety of the jumper. In Course 1, you may have explored the relationship between jumper weight and bungee cord length by conducting an experiment with rubber bands and weights.

1 In one such experiment, for each ounce of weight added to a 3-inch rubber band, the rubber band stretched about $\frac{1}{2}$ inch.

a. Describe the relationship between weight added and stretched rubber band length. What type of function represents this relationship?

b. Complete a copy of the table below.

Bungee Experiment												
Weight (in ounces)	0	1	2	3	4	5	...	50	...	99	...	*n*
Length (in inches)							...		...		...	

c. The sequence of numbers that you get for the lengths is called an *arithmetic sequence*. An **arithmetic sequence** of numbers is one in which you add the same constant to each number in the sequence to get the next number. An arithmetic sequence models **arithmetic growth**. Explain why the sequence of stretched rubber band lengths is an arithmetic sequence.

d. Which, if any, of the sequences in the Think About This Situation is an arithmetic sequence? Why?

2 There are several different formulas that can be used to represent the weight-length relationship.

a. Write a formula using the words *NOW* and *NEXT* that shows how the length changes as weight is added.

b. Write a formula for this situation that looks like

$$L_n = (\text{expression involving } L_{n-1}).$$

This is called a *recursive formula for the sequence*.

c. Use the recursive formula to predict the rubber band length when 15 ounces of weight are attached.

d. Write a formula for the weight-length situation that looks like

$$L_n = (\text{expression involving } n, \text{ but not } L_{n-1}).$$

This form can be called a *function formula for the sequence* of lengths.

  i. Explain why it makes sense to call this form a function formula.

  ii. Rewrite the formula using function notation.

e. Use the function formula in Part d to predict the rubber band length when 15 ounces of weight are attached.

f. Describe the difference between the processes of using the recursive formula and using the function formula to compute the rubber band length when 15 ounces of weight are attached.

3 Many foreign exchange students in the U.S. use international calling cards to stay in contact with family and friends at home. The details of phone cards can be quite complicated, even though the prices and plans may look simple. For example, there might be connection fees, payphone surcharges, and different schemes for rounding the number of minutes used. Suppose you have a phone card with a rate of 2¢ per minute and no other fees or charges, except for a payphone charge of 89¢ per call. When the minutes used are not an integer, the value is rounded to the next minute.

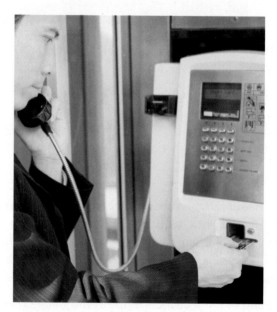

a. Suppose you make a call from a payphone using this phone card. Determine recursive and function formulas for the minute-by-minute sequence of phone card charges for this call.

b. Compare these formulas to those you found for the rubber band experiment in Problem 2. How are they similar? How are they different?

4 Now think about possible connections between arithmetic sequences and functions you have studied.

a. In each case, describe the shape of the graph of the function formula.

b. What is the slope of each graph? How does the slope appear in each of the recursive and function formulas you have been examining?

5 Suppose t_0 is the initial term of a *general* arithmetic sequence for which you add d to each term of the sequence to get the next term of the sequence.

   a. Write the first five terms of this sequence. Then find recursive and function formulas for the sequence. Compare your formulas to those of other students. Resolve any differences.

   b. LaToya explained her function formula as follows.

> **This sequence starts with t_0, and then to get to t_n you add d, n times. Thus, the general formula is $t_n = t_0 + dn$.**

   Explain how LaToya's explanation matches her formula.

   c. How would you interpret t_0, d, and n in the context of the rubber band experiment?

   d. The constant d is sometimes called the **common difference** between terms. Explain why it makes sense to call d the common difference.

   e. Suppose an arithmetic sequence $t_0, t_1, t_2, \ldots$ begins with $t_0 = 84$ and has a common difference of -6. Find function and recursive formulas for this sequence. Then find the term t_{87}.

Geometric Sequences Now consider a different type of sequence. As you investigate this new type of sequence, think about similarities to and differences from arithmetic sequences.

6 Consider the growth sequence of bacteria cells in a laboratory experiment in which 25 bacteria cells are placed in a petri dish. The lab technician finds that the number of bacteria doubles every quarter-hour.

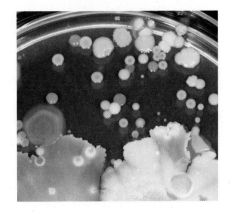

   a. Use words, graphs, tables, and algebraic rules to describe the relationship between the number of quarter-hours and the number of bacteria. What type of function represents this relationship?

   b. Complete a copy of the table below for this situation.

Bacterial Growth												
Number of Quarter-Hours	0	1	2	3	4	5	...	50	...	99	...	n
Bacteria Count							...		...		...	

   c. The sequence of numbers that you get for the bacteria count is called a *geometric sequence*. A **geometric sequence** of numbers is one in which each number in the sequence is multiplied by a constant to get the next number. Explain why the sequence of bacteria counts is a geometric sequence.

   d. Which, if any, of the sequences in the Think About This Situation (page 482) is a geometric sequence? Why?

Don Rubbelke/McGraw-Hill Education

7 If B_n represents the bacteria count after n quarter-hours, then there are several different formulas that can be used to model this situation.

    a. Write a formula using the words *NOW* and *NEXT* that shows how the bacteria count increases as time passes.

    b. Write a recursive formula beginning "$B_n = \ldots$" for the sequence of bacteria counts. Use the recursive formula to predict the bacteria count after 18 quarter-hours.

    c. Write a function formula beginning "$B_n = \ldots$" for the sequence of bacteria counts. Use the function formula to predict the bacteria count after 18 quarter-hours.

    d. Describe the difference between the processes of using the recursive formula and using the function formula to compute the bacteria count after 18 quarter-hours.

8 One of the simplest fractal patterns is the Sierpinski carpet. Starting with a solid square "carpet" one meter on a side, smaller and smaller squares are cut out of the carpet. The first two stages in forming the carpet are shown below.

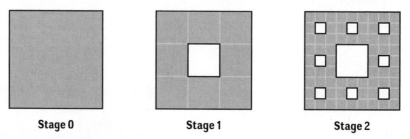

          Stage 0   **Stage 1**   **Stage 2**

    a. Find recursive and function formulas for the sequence of carpet area that remains at each stage.

    b. Compare your formulas to those you found for the bacteria present after a cut by a rusty nail, in Problem 7. How are they similar? How are they different?

9 Think about connections between geometric sequences and functions you have studied.

    a. What type of function is represented by each of the function formulas in Problems 7 and 8?

    b. The growth (or decay) modeled by a geometric sequence is called **geometric growth**. It is also called **exponential growth**. Explain why *exponential growth* is a reasonable way to describe geometric sequences and geometric growth.

10 Suppose t_0 is the initial term of a *general* geometric sequence for which each term of the sequence is multiplied by r ($r \neq 1$) to get the next term of the sequence.

    a. Write the first five terms of this sequence and then find recursive and function formulas for the sequence. Compare your formulas to those of other students. Resolve any differences.

b. Eva explained her function formula as follows.

> I started the sequence with t_0, then I multiplied by r each time to get the next terms. To get to t_n, I multiplied by r, n times. Thus, $t_n = t_0 r^n$.

Explain how Eva's explanation matches her formula.

c. What are t_0, r, and n in the situation involving bacterial growth?

d. The constant r is sometimes called the **common ratio** of terms. Explain why it makes sense to call r the common ratio.

e. Suppose a geometric sequence $t_0, t_1, t_2, \dots$ begins with $t_0 = -4$ and has a common ratio of 3. Find function and recursive formulas for this sequence and then find the term t_{17}.

Different Starting Points The sequences you have studied in this investigation have started with $n = 0$. For example, the initial term for the sequence of rubber band lengths was L_0, and the initial term for the sequence of bacteria counts was B_0. It is possible to have different starting subscripts. Consider what happens to the recursive and function formulas when the starting subscript is 1 instead of 0.

11 In the Think About This Situation, you considered the distance fallen by a sky diver. The table below shows the distance fallen in a free fall by a sky diver, assuming no wind resistance, during each of the first nine seconds of a jump.

Sky Diving Free Fall									
Time n (in seconds)	1	2	3	4	5	6	7	8	9
Distance Fallen during Second n, D_n (in feet)	16	48	80	112	144	176	208	240	272

a. Is D_n an arithmetic sequence, geometric sequence, or neither? Explain.

b. What is the starting value of n for this sequence?

c. What is a recursive formula for this sequence?

d. In the Think About This Situation, you may have figured out that a function formula for this sequence is

$$D_n = 16 + 32(n - 1).$$

Check this formula for a few values of n. Explain this formula by modifying LaToya's explanation in Problem 5 on page 485, so that the explanation fits this formula.

12 In Problem 5, you found that a function formula for an arithmetic sequence with initial term t_0 and common difference d is

$$t_n = t_0 + dn.$$

Using the form "$t_n = \dots$," write a similar function formula if the initial term is t_1, instead of t_0.

13 Consider this geometric sequence: 3, 6, 12, 24, 48, … .

    a. What is the common ratio r?

    b. Call the general term of this sequence A_n. Suppose that the starting subscript for the sequence is 0. Thus, the initial term is $A_0 = 3$. What is a recursive formula for this sequence? What is a function formula for this sequence? (Write both formulas in the form "$A_n = \ldots$.")

    c. Suppose the starting subscript is 1. Thus, $A_1 = 3$. What is a recursive formula for the sequence in this case? What is a function formula? (Write both formulas in the form "$A_n = \ldots$.")

    d. Compare the formulas in Parts b and c. Explain similarities and differences.

14 Working individually, write the first five terms for three different sequences. One sequence should be arithmetic, one should be geometric, and the third should be neither.

    a. Challenge a classmate to correctly identify your sequence type.

    b. Describe how you can tell by inspection whether a sequence is arithmetic, geometric, or neither.

    c. For the two sequences you constructed that are arithmetic or geometric, find a recursive and a function formula. For one sequence, choose the starting subscript to be 0. For the other, choose the starting subscript to be 1.

SUMMARIZE THE MATHEMATICS

In this investigation, you studied two important patterns of growth: arithmetic growth, modeled by arithmetic sequences, and geometric growth, modeled by geometric sequences. You also investigated recursive and function formulas for sequences and made connections to linear and exponential functions.

a Consider arithmetic and geometric sequences.

    i. How are arithmetic and geometric sequences different? How are they similar?

    ii. Describe one real-world situation different from those you studied in this investigation that could be modeled by an arithmetic sequence. Do the same for a geometric sequence.

    iii. What is the connection between arithmetic and geometric sequences and linear and exponential functions?

b Consider recursive and function formulas for sequences.

    i. Describe the difference between a recursive formula and a function formula for a sequence. What information do you need to know to find each type of formula?

    ii. What is one advantage and one disadvantage of a recursive formula for a sequence? What is one advantage and one disadvantage of a function formula?

c Explain why changing the starting subscript for a sequence has no effect on a recursive formula. Explain why this will cause a change in a function formula.

d In the first lesson of this unit, you investigated situations that could be modeled by recursive formulas of the form $A_n = rA_{n-1} + b$. You can think of these formulas as *combined recursive formulas*. What is the connection between such combined recursive formulas and recursive formulas for arithmetic and geometric sequences? Why does it make sense to call these formulas "combined recursive formulas?"

e You have represented sequences using recursive and function formulas. A function formula for a sequence is also called an **explicit formula** or a **closed-form formula**. Why do you think the terms "explicit" or "closed-form" are used?

Be prepared to explain your ideas to the class.

CHECK YOUR UNDERSTANDING

For each of the sequences below, do the following.

- Complete a copy of the table.

- State whether the sequence is arithmetic, geometric, or neither.

- For those sequences that are arithmetic or geometric, find a recursive formula and a function formula.

- If a sequence is neither arithmetic nor geometric, find whatever formula you can that describes the sequence.

a.

n	0	1	2	3	4	5	6	...	10	...	100	...
A_n	2	6	18	54	162							

b.

n	0	1	2	3	4	5	6	...	10	...	100	...
B_n	1	2	5	10	17							

c.

n	0	1	2	3	4	5	6	...	10	...	100	...
C_n	3	7	11	15	19							

INVESTIGATION 2

Some Sums

Sometimes it is useful to sum the numbers in a sequence, which is the focus of this investigation. As you work on the problems of this investigation, look for answers to the following questions:

What are strategies for summing the terms of arithmetic and geometric sequences?

What are formulas for these sums?

Arithmetic Sums Consider the sky diving situation once again. The table showing the distance fallen during each second is reproduced below.

Sky Diving Free Fall									
Time *n* (in seconds)	1	2	3	4	5	6	7	8	9
Distance Fallen during Second *n*, D_n (in feet)	16	48	80	112	144	176	208	240	272

1 From the previous investigation, you know that a function formula for this sequence is

$$D_n = 16 + 32(n - 1).$$

This formula gives the distance fallen *during the nth second*. Use the function formula for D_n to:

  a. verify the entries in the table for D_1 and D_5.

  b. compute D_{18}, D_{19}, and D_{20}.

2 A sky diver would be very interested in knowing the *total distance* fallen *after n seconds*.

  a. What is the total distance fallen after 3 seconds? After 5 seconds?

  b. How could you determine the total distance fallen after 20 seconds? Do not actually find this total distance yet. Just explain how you could find it.

  c. One approach to finding the sum of the first 20 terms of the sequence D_n is based on a method reportedly discovered by Carl Friedrich Gauss (1777–1855) when he was only 10 years old. Gauss, considered to be one of the greatest mathematicians of all time, noticed that the sum of the terms of an arithmetic sequence, such as $16 + 48 + 80 + \cdots + 560 + 592 + 624$, could be quickly calculated by writing the sum again in reverse order and then adding pairs of corresponding terms, as follows.

$$
\begin{array}{l}
16 + 48 + 80 + \cdots + 560 + 592 + 624 \\
\underline{624 + 592 + 560 + \cdots + 80 + 48 + 16} \\
640 +
\end{array}
$$

    i. What is the sum of each pair of terms? How many pairs are there?

    ii. What is the total distance fallen after 20 seconds?

  d. Use Gauss' method, and the function formula for D_n given in Problem 1, to determine the total distance the sky diver would fall in 30 seconds.

  e. An expert sky diver typically free-falls to about 2,000 feet above Earth's surface before pulling the rip cord on the parachute. If the altitude of the airplane was about 5,000 feet when the sky diver began the jump, how much time can the sky diver safely allow for the free-fall portion of the flight?

3 You can use Gauss' idea in Problem 2 and algebraic reasoning to derive a general formula for the sum of the terms of an arithmetic sequence with common difference d.

a. If $a_1, a_2, \dots, a_n$ is an arithmetic sequence with common difference d, explain why the sum S_n of the terms a_1 through a_n can be expressed by the formula

$$S_n = a_1 + (a_1 + d) + (a_1 + 2d) + \cdots + (a_n - 2d) + (a_n - d) + a_n.$$

b. If you rewrite S_n in reverse order and then add pairs of corresponding terms as in Problem 2, what is the sum of each pair of terms? How many pairs are there?

c. Use your answers in Part b to write a formula for S_n in terms of a_1, a_n, and n. Compare your formula to those of others. Resolve any differences.

d. Sondra developed the following formula-in-words for the sum of the terms of an arithmetic sequence.

$$S = \frac{(initial\ term + final\ term)(number\ of\ terms)}{2}$$

Explain why this formula makes sense.

4 As you discovered in the last investigation, the starting subscript used in a sequence can affect the formulas for the sequence. An important feature of the formula-in-words in Part d of Problem 3 is that it does not depend on the subscript notation used for the sequence. In contrast, the formula in Part c of Problem 3 only works if the starting subscript is 1. Consider what happens when the starting subscript is 0.

a. Suppose $a_0, a_1, \dots, a_n$ is an arithmetic sequence. Find a formula for the sum S_n of the terms a_0 through a_n. Your formula should use a_0, a_n and n.

b. Find the sum of the terms of this sequence: 7, 12, 17, 22, ..., 52.

Accumulated versus Additional Amount In modeling a situation involving sequential change, it is important to decide whether the situation involves a pattern of change in *additional* amount or in *accumulated* amount.

5 It may happen that in two different situations, you get the same sequence of numbers, but it makes sense to add the terms of the sequence in only one of the situations. Consider an epidemic that begins at Day 1 with one infected person and spreads rapidly through two counties.

a. In Adair County, a health official states, "The population of sick people triples every day." The table below can be used to represent this situation.

Day	1	2	3	4	5	6	7	8	9
Number of Sick People	1	3	9						

What is the total number of sick people at the end of Day 6?

b. In Benton County, a different health official states, "The number of new sick people triples every day." Given this statement, you could construct the same table as in Part a. In this case, what is the total number of sick people at the end of Day 6 (assuming no sick person gets well in that time)?

c. Look back at Parts a and b. In which situation do the terms in the sequence represent an *additional* amount? In which situation do the terms represent an *accumulated* amount? Explain why it makes sense to sum the terms of the sequence in one situation but not in the other.

Geometric Sums In the epidemic example from Problem 5, assume that the epidemic begins at Day 1 with one sick person and the *additional* number of sick people triples every day. Then the total number of sick people at the end of Day n is found by summing the terms in the sequence.

6 Using algebraic reasoning, you can derive a formula for quickly calculating this geometric sum, as outlined below.

a. The total number of sick people at the end of day n can be represented as

$$S_n = 1 + 3 + 3^2 + 3^3 + \cdots + 3^{n-1}.$$

Explain why the exponent on the final term is $n - 1$ and not n.

b. The formula in Part a can be used to compute the sum, but it would take many steps for a large value of n. Derive a shorter formula by carrying out the following plan.

- Create a new formula by multiplying both sides of the formula in Part a by 3.

- Subtract the formula in Part a from your new formula.

- Then solve the resulting formula for S_n.

c. Compare your work in Part b to the derivation below. Provide reasons for each step in the derivation.

$$\begin{align}
S_n &= 1 + 3 + 3^2 + 3^3 + 3^4 + \cdots + 3^{n-1} & (1) \\
3(S_n) &= 3(1 + 3 + 3^2 + 3^3 + 3^4 + \cdots + 3^{n-1}) & (2) \\
&= 3 + 3^2 + 3^3 + 3^4 + \cdots + 3^{n-1} + 3^n & (3) \\
3S_n - S_n &= 3^n - 1 & (4) \\
(3-1)S_n &= 3^n - 1 & (5) \\
S_n &= \frac{3^n - 1}{3 - 1} & (6)
\end{align}$$

d. Explain why the strategy in Part c works to create a short formula.

e. Use this formula to calculate the total number of sick people at the end of Day 6. Compare your answer to your response to Problem 5 Part b.

f. Using similar reasoning, prove that the *sum of the terms of the geometric sequence* $1, r, r^2, r^3, r^4, \ldots, r^{n-1}$, where $r \neq 1$, is

$$1 + r + r^2 + r^3 + r^4 + \cdots + r^{n-1} = \frac{r^n - 1}{r - 1}.$$

7 When summing terms in a geometric sequence, the exponent of the last term could be any positive integer. In Part f of Problem 6, the exponent of r in the last term is $n - 1$; and then the exponent of r in the numerator on the right side of the sum formula is n. Consider other possibilities.

a. Complete the formula for this sum.
$$1 + r + r^2 + r^3 + r^4 + \cdots + r^n = \ldots$$

b. Complete the formula for this sum.
$$1 + r + r^2 + r^3 + r^4 + \cdots + r^{n+1} = \ldots$$

c. It can get very confusing keeping track of when the formula should use n, $n - 1$, $n + 1$, and so on. But the pattern is always the same. State in words the relationship between the ending exponent on r in the sum in the left side of the formulas above and the exponent on r in the numerator of the right side.

8 The sums and formulas you have worked with so far apply to sequences with initial term 1. Consider sums of geometric sequences when the initial term is not 1.

a. Suppose an epidemic begins with two infected people on Day 1 and the additional number of infected people triples every day. What is the total number of sick people at the end of Day 4?

b. Suppose an ice cream store sold 22,000 ice cream cones in 2008. Based on sales data from other stores in similar locations, the manager predicts that the number of ice cream cones sold each year will increase by 5% each year. Using the formulas for the sum of a geometric sequence that you developed in Problems 6 and 7, find the total predicted number of ice cream cones sold during the period from 2008 to 2014. (*Hint*: You may find it helpful to write the sum term-by-term and then factor out 22,000.)

c. Consider a general geometric sequence with common ratio r, where $r \neq 1$, and initial term b. Find a formula for the sum, $S_n = b + br + br^2 + \cdots + br^n$. Provide an argument for why your formula is correct.

9 An expression showing the terms of a sequence added together is called a **series**. For example, $a_0 + a_1 + a_2 + \cdots + a_n$ and $B_1 + B_2 + \cdots + B_n$ are both examples of a series. A series is called an **arithmetic series** or a **geometric series**, depending on whether the sequence that generates the terms is arithmetic or geometric, respectively. (Note that the terms "sequence" and "series" have very specific meanings in mathematics, which may be different than their English meanings.)

i. Write a formula for an arithmetic series $B_1 + B_2 + \cdots + B_n$.

ii. Write a formula for a geometric series $B_1 + B_2 + \cdots + B_n$, with common ratio r.

SUMMARIZE THE MATHEMATICS

In this Investigation, you examined total change in situations involving arithmetic growth and geometric growth or decay.

a Consider the following types of sequential growth models, in which the goal is to find the total amount after a certain number of steps.

> **Model 1** The growth sequence is arithmetic, and each term in the sequence represents an accumulated amount
>
> **Model 2** The growth sequence is arithmetic, and each term in the sequence represents an additional amount
>
> **Model 3** The growth sequence is geometric, and each term in the sequence represents an accumulated amount
>
> **Model 4** The growth sequence is geometric, and each term in the sequence represents an additional amount

> **i.** Give an example of each of the four sequential growth situations.
>
> **ii.** In which situations would you sum the sequence to find the total amount after a certain number of steps?

b Suppose an arithmetic sequence has initial term a_0 and common difference d.

> **i.** Write recursive and function formulas for the sequence.
>
> **ii.** Write a formula for the sum of the terms up through a_n.

c Suppose a geometric sequence has initial term a_0 and common ratio r.

> **i.** Write recursive and function formulas for the sequence.
>
> **ii.** Write a formula for the sum of the terms up through a_n.

Be prepared to explain your thinking and formulas to the entire class.

 **CHECK YOUR UNDERSTANDING**

Find the indicated sum for each of the sequences below.

a. If $t_1 = 8$ and $d = 3$ in an arithmetic sequence, find the sum of the terms up through t_{10}.

b. If $t_0 = 1$ and $r = 2.5$ in a geometric sequence, find the sum of the terms up through t_{12}.

c. Find the sum of the first 15 terms of each sequence below.

n	1	2	3	4	5	6	7	...	15	...
A_n	2	6	18	54	162					

n	0	1	2	3	4	5	6	...	14	...
B_n	95	90	85	80	75					

d. A popular shoe store sold 5,700 pairs of athletic shoes in 2008. Projections are for a 3% increase in sales each year for several years after 2008. What is the projected total sales (number of athletic shoes sold) during the period from 2008 to 2015?

INVESTIGATION 3

Finite Differences

So far, you have been able to find a function formula for a sequence by detecting and generalizing a pattern. Sometimes that is not so easily done. In this investigation, you will learn how to use *finite differences tables* to find a function formula for certain sequences.

As you work on the problems of this investigation, look for answers to these questions:

> *How do you construct a finite differences table for a sequence?*
>
> *How can you use such a table to find a function formula for a sequence?*
>
> *What kind of function formulas can be found using finite differences tables?*

1 To learn how to use finite differences tables, start with a sample function formula. Suppose you toss a rock straight down from a high bridge. If the rock is thrown from a bridge 300 feet above the river with an initial velocity of 10 feet per second, then the distance D_n of the rock from the river after n seconds is given by the function formula

$$D_n = -16n^2 - 10n + 300.$$

An analysis of the pattern of change in the distance between the rock and the river is shown in the following table.

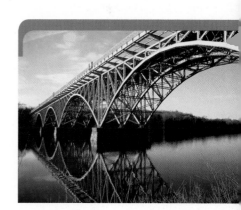

Number of Seconds, n	D_n	1st Difference	2nd Difference
0	300		
		−26	
1	274		−32
		−58	
2	216		___
		___	
3	___		___
		___	
4	___		

a. On a copy of the table, complete the D_n column.

b. The third column contains differences between consecutive terms of the sequence D_n. How was the "−58" calculated? Find the remaining entries in the "1st Difference" column.

c. The fourth column contains differences between consecutive terms of the "1st Difference" column. Verify the entry "−32" and find the remaining entries in the "2nd Difference" column.

d. A table like this is called a **finite differences table**. Describe any patterns you see in your completed table.

Fact 1 about Finite Differences It is a fact that for any function formula that is a quadratic, such as the one in Problem 1, the 2nd differences in a finite differences table will be a constant.

Conversely, if the 2nd differences are a constant, then the function formula is a quadratic. (See Extensions Task 24.) For example, consider the following counting problem involving vertex-edge graphs.

2 A complete graph is a graph in which there is exactly one edge between every pair of vertices. The diagram at the right shows a complete graph with 4 vertices. In this problem, you will investigate the number of edges E_n in a complete graph with n vertices.

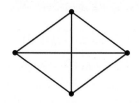

a. On a copy of a table like the one below, enter the number of edges for complete graphs with 0 to 5 vertices. Then compute the 1st and 2nd differences. Describe any patterns you see in the table.

Number of Vertices, n	E_n	1st Difference	2nd Difference
0	—		
1	—	—	—
2	—	—	—
3	—	—	—
4	—	—	—
5	—	—	

b. What pattern in the finite differences table tells you that the function formula for the sequence of edge counts must look like $E_n = an^2 + bn + c$?

c. How could you find the coefficients a, b, and c? With some classmates, brainstorm some ideas and try them out. When you find a, b, and c, or if you are having trouble finding them, go on to Problem 3.

3 One way to find a, b, and c is to set up and solve a system of three linear equations. If you already did that in Part c of Problem 2, then read through this problem, compare it to your method, and discuss any differences. If you did not set up and solve a system of three linear equations, then do so now, as follows. You can get a system of equations by letting n equal any three values, for example, 1, 2, and 3.

a. To get the first equation, suppose $n = 1$. If you substitute $n = 1$ in the equation $E_n = an^2 + bn + c$, you get $E_1 = a(1)^2 + b(1) + c = a + b + c$. Explain why one equation is $a + b + c = 0$.

b. Use similar reasoning with $n = 2$ and $n = 3$ to get the second and third equations.

c. Compare the system of three linear equations you found to the systems found by others. Resolve any differences.

d. You can solve systems of three equations like these using matrices, as you did for the case of systems of two linear equations in Course 2 Unit 2, *Matrix Methods*. Written in matrix form, the system looks like the partially completed matrix equation below. Fill in the missing entries and solve this matrix equation.

$$\begin{bmatrix} 1 & 1 & - \\ - & 2 & 1 \\ 9 & - & 1 \end{bmatrix} \begin{bmatrix} a \\ b \\ c \end{bmatrix} = \begin{bmatrix} 0 \\ 1 \\ - \end{bmatrix}$$

$$A \cdot X = C$$

e. Use the results of Part d to write the function formula for the number of edges in a complete graph with n vertices. Use this formula to check the entries in the table from Problem 2.

f. Now that you have a function formula, find a recursive formula for the sequence of edge counts. Do so by examining the table in Problem 2.

4 So far in this investigation, you have examined situations in which the function formula is a quadratic. Now consider the case of a higher-degree function formula. Suppose $A_n = 4n^3 + 2n^2 - 5n - 8$. Make a prediction about which column in the finite differences table for this sequence will be constant. Construct the finite differences table to check your conjecture.

Fact 2 about Finite Differences In general, it is possible to compute more than just 2nd or 3rd differences. If the function formula is an *r*th degree polynomial, then the *r*th differences will be constant. The converse is also true. These facts can be used to help find function formulas for certain sequences.

SUMMARIZE THE MATHEMATICS

In this investigation, you learned about finite differences tables.

a Describe how to construct a finite differences table for a sequence of numbers.

b If the 4th differences in the finite differences table for a sequence are constant, what do you think the function formula for the sequence will look like? How would you go about finding the function formula?

c In general, what kind of function formulas can be found using finite differences tables?

d The title of this lesson is "A Recursive View of Functions." Explain how this title describes the mathematics you have been studying in all three investigations of the lesson.

Be prepared to share your descriptions and thinking with the entire class.

✅ **CHECK YOUR UNDERSTANDING**

Use a finite differences table and matrices to find a function formula for the sequence below.

n	0	1	2	3	4	5	6	7	8	9	10	11
A_n	3	12	25	42	63	88	117	150	187	228	273	322

APPLICATIONS

1 For many people, a college education is a desirable and worthwhile goal. But the cost of a college education is growing every year. For the 2011–12 school year, the average cost for four-year public colleges (tuition and fees) was $8,244 for in-state students, which was up 41.3% from five years earlier. (**Source**: College Board, *Trends in College Pricing 2006, Trends in College Pricing 2011*)

   a. Assume an annual increase rate in college costs of 7% per year. Make a table showing the average cost of a year of college education (tuition and fees) for in-state students at a four-year public college for the next 5 years following 2011–12.

   b. Is the sequence of increasing costs arithmetic, geometric, or neither?

   c. Determine recursive and function formulas for the sequence of costs.

   d. Use your formulas to predict the average cost of the first year of your college education if you go to a four-year in-state public college right after you graduate from high school.

   e. Predict the average cost of four years of college at a four-year in-state public college for a child born this year.

2 Animal behavior often changes as the outside temperature changes. One curious example of this is the fact that the frequency of cricket chirps varies with the outside temperature in a very predictable way. Consider the data below for one species of cricket.

Cricket Chirps							
Temperature (in °F)	45	47	50	52	54	55	60
Frequency (in chirps/min)	20	28	41	50	57	61	80

   a. If you were to choose an arithmetic sequence or a geometric sequence as a model for the frequency sequence, which type of sequence would you choose? Why?

   b. As is often the case in mathematical modeling, the model you chose in Part a does not fit the data exactly. Nevertheless, it may be quite useful for analysis of the situation.

      i. Find a recursive formula for the frequency sequence, based on the type of sequence you chose in Part a.

      ii. Use the formula to predict the frequency of cricket chirps for a temperature of 75°F.

   c. At what temperature would you expect crickets to stop chirping?

d. You can use the relationship between the frequency of cricket chirps and the temperature as a kind of thermometer. What would you estimate the temperature to be if a cricket is chirping at 100 chirps per minute?

e. Based on your recursive formula in Part b, find a function formula for the sequence of cricket chirps. Use your function formula to find the frequency of cricket chirps for a temperature of 75°F. Compare to your answer in Part b.

f. Rewrite your function formula to express temperature as a function of frequency of cricket chirps. Use this new formula to answer the question in Part d.

3 The square Sierpinski carpet you examined in Investigation 1 is an example of a *fractal* in that small pieces of the design are similar to the design as a whole. Other fractal shapes can also be constructed using recursive procedures.

a. A *Sierpinski triangle* is constructed through a sequence of steps illustrated by the figures below.

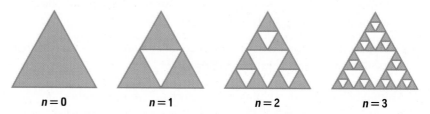

$n = 0$ $n = 1$ $n = 2$ $n = 3$

At stage $n = 0$, you construct an equilateral triangle whose sides are all of length 1 unit. In succeeding stages, you remove the "middle triangles," as shown in stages $n = 1, 2,$ and 3. This process continues indefinitely. Consider the sequence of areas of the figures at each stage. Find recursive and function formulas for the sequence of areas.

b. Another interesting fractal is the *Koch snowflake*. The procedure for constructing the Koch snowflake also begins with an equilateral triangle whose sides are of length 1 unit.

At each stage, you remove the segment that is the middle third of each side and replace it with two outward-extending segments of the same length, creating new equilateral triangles on each side as shown in the diagrams below. The process continues indefinitely.

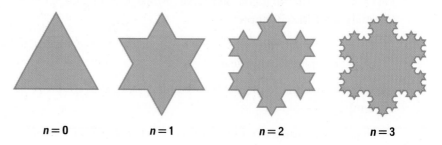

$n = 0$ $n = 1$ $n = 2$ $n = 3$

Find recursive and function formulas for the sequence of perimeters of the snowflake.

4 *Glottochronology* is the study of changes in languages. Over time, certain words in a language are no longer used. In effect, they disappear from the language. Suppose a linguist examines a list of 500 words used in a language 1,000 years ago. Let $W(n)$ be the percentage of the words in this list that are still in use n years later, given as a decimal.

It is commonly assumed that $W(n)$ is proportional to $W(n-1)$. Glottochronologists have determined that the constant of proportionality can be estimated to be about 0.99978. (Problem adapted from Sandefur, James T. *Discrete Dynamical Modeling*. Oxford, 1993, page 63.)

a. Find recursive and function formulas for this sequence.

b. According to this model, about how many of the 500 words are still in use today?

5 Irene has a sales job which pays her a monthly commission. She made $250 in her first month. Her supervisor tells her that she should be able to increase her commission income by 10% each month for the next year. For this task, assume that the supervisor's prediction is correct.

a. Find recursive and function formulas for the sequence of monthly commission income.

b. How much total commission income will Irene earn in her first ten months on the job?

6 As you complete this task, think about the defining characteristics of arithmetic and geometric sequences and how those characteristics are related to the sum of their terms. For each sequence below, determine if the sequence is arithmetic, geometric, or neither. Then find the sum of the indicated terms.

a. 13, 26, 52, 104, ... , 6,656

b. 13, 12.25, 11.5, 10.75, ... , 1

c. 1, 4, 9, 16, ... , 225

d. 5, 4, 3.2, 2.56, ... , 0.8388608

7 In this task, you will use the idea of sequential change to investigate the number of diagonals that can be drawn in a regular n-sided polygon.

a. Draw the first few regular n-gons and make a table showing the number of diagonals that can be drawn in each of them.

b. Determine a recursive formula for the number of diagonals that can be drawn in a regular n-gon.

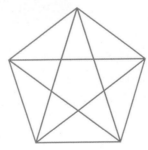

c. Use a finite differences table to find a function formula for the number of diagonals that can be drawn in a regular *n*-gon.

d. Find the number of diagonals in a regular 20-gon.

e. What other methods might you use to find a function formula for the sequence in Part a?

8 Use a finite differences table and matrices to find the function formula for the sequence given by $B_n = B_{n-1} + 3n$, with $B_0 = 2$.

CONNECTIONS

9 In Applications Task 2, you may have established that the sequence of chirping frequencies for one species of cricket is approximately an arithmetic sequence with recursive formula

$$C_n = C_{n-1} + 4$$

a. Plot the (*temperature, frequency*) data from the table on page 499. Find a regression equation that fits the data.

b. Using the equation from Part a, write a function formula for the sequence. Check that your formula generates the values given in the table for the chirping frequencies at temperatures of 47°F and 60°F.

10 In Course 2 Unit 8, *Probability Distributions*, you investigated the waiting-time distribution in the context of a modified Monopoly® game in which 36 students are in jail and a student must roll doubles to get out of jail.

a. The probability of rolling doubles is $\frac{1}{6}$. Thus, the probability of getting out of jail on any given roll of the dice is $\frac{1}{6}$. What is the probability of remaining in jail on any given roll of the dice?

(l)IT Stock Free/Alamy, (r)Mark Steinmetz/Amanita Pictures

b. Complete a copy of the following table. The waiting-time distribution refers to the first two columns, since you want to know how long a student has to wait to get out of jail. The last column provides important related information about the number of students still in jail.

Rolling Dice to Get Doubles		
Number of Rolls to Get Doubles	Expected Number of Students Released on the Given Number of Rolls	Expected Number of Students Still in Jail
1	6	30
2	5	
3		
4		
5		

c. Consider the sequence of numbers in the last column.

　　i. Is the sequence an arithmetic or geometric sequence? Why?

　　ii. Determine the recursive and function formulas for the sequence. (Use 36 as the initial term of the sequence since that is the initial number of students in jail.)

　　iii. Use the formulas to find the expected number of students left in jail after 20 rolls.

d. Sketch a histogram for the waiting-time distribution shown in the first two columns.

　　i. Write a recursive formula that shows how the height of any given bar compares to the height of the previous bar.

　　ii. What kind of sequence is the sequence of bar heights?

e. A waiting-time distribution is also called a *geometric distribution*. Explain why the use of the word "geometric" for this type of distribution seems appropriate in terms of sequences.

11 In this lesson, you found function formulas for arithmetic and geometric sequences. In Lesson 1, you studied combined recursive formulas of the form $A_n = rA_{n-1} + b$. You can now use what you know about geometric sequences and their sums to find a function formula for such combined recursive formulas.

a. Complete the following list of terms for the combined recursive formula $A_n = rA_{n-1} + b$. Examine the list for patterns so that you can write a function formula for A_n.

$A_0 = A_0$
$A_1 = rA_0 + b$
$A_2 = rA_1 + b = r(rA_0 + b) + b = r^2A_0 + rb + b$
$A_3 = r(r^2A_0 + rb + b) + b = r^3A_0 + r^2b + rb + b$
$A_4 = ?$
$A_5 = ?$
$\vdots$
$A_n = ?$

b. The expression you have for A_n at the end of Part a is a function formula, but it can be simplified. To simplify, use what you know about the sum of the terms of a geometric sequence.

c. Now reconsider the combined recursive formula that modeled the fish-population problem in Lesson 1: $A_n = 0.8A_{n-1} + 1{,}000$, with $A_0 = 3{,}000$.

   i. Use the general function formula from Part b to write a function formula that models the fish population.

   ii. Now use this formula to find A_2 and A_{10}. Choose a year far in the future, and find the long-term population. Compare your results to those you found in Lesson 1, using other methods.

d. Suppose the combined recursive formula $A_n = rA_{n-1} + b$ represents year-to-year population change. Assume that r is a positive number less than 1. Use the function formula from Part b to explain why the long-term population in this situation is $\dfrac{b}{1-r}$. Use this fact to find the long-term population in the fish-population problem, and compare it to what you had previously found.

e. Compare the function formula you obtained in Part c to the one given in Connections Task 7 of Lesson 1, page 472. Explain and resolve any differences.

12 You have seen that a sequence can be represented with function notation. There is an even stronger connection between sequences and functions. In fact, a finite sequence can be defined as a function whose domain is the set $\{1, 2, \dots, n\}$ and whose range is a set containing the terms of the sequence. For example, the sequence $7, 10, 13, \dots, 7 + 3(n-1)$ can be defined as a function, s, from the set $\{1, 2, \dots, n\}$ to the set $\{7, 10, 13, \dots, 7 + 3(n-1)\}$, where $s(1) = 7$, $s(2) = 10$, $\dots$, and $s(n) = 7 + 3(n-1)$.

a. Consider the sequence $3, 6, 12, 24, \dots, 3(2)^{n-1}$. Describe how this sequence can be defined as a function. Describe the domain and range as sets.

b. When using functions to define sequences, the domain can actually be any finite set of integers. In particular, the domain might be $\{0, 1, 2, \dots, n\}$. Give an example of a sequence that could be defined as a function whose domain is $\{0, 1, 2, \dots, n\}$.

13 A series $a_0 + a_1 + a_2 + \cdots + a_n$ can be written in a more compact form using sigma notation as $\displaystyle\sum_{i=0}^{n} a_i$. You previously used sigma notation (without subscripts) when writing a formula for standard deviation and when considering Pearson's correlation as well as in other contexts.

a. Suppose $x_1, x_2, x_3, \dots, x_n$ are sample data values. Use sigma notation with subscripts to write an expression for:

   i. the mean of this distribution.

   ii. the standard deviation of this distribution.

b. Write $\sum_{k=0}^{12} 3k$ in expanded form and then find the sum.

c. Write $\sum_{i=3}^{10} 2^i$ in expanded form and then find the sum.

d. Complete the formula below by determining the subscripts.

$$3 + 5 + 7 + 9 + 11 + 13 + 15 = \sum_{n=?}^{?} (2n+1)$$

14 Below is the sequence from the Check Your Understanding task on page 498.

n	0	1	2	3	4	5	6	7	8	9	10	11
A_n	3	12	25	42	63	88	117	150	187	228	273	322

a. Produce a scatterplot of n versus A_n using a graphing calculator or computer software. Describe the shape of the graph. Which function family would best model this data pattern?

b. Find a regression equation that seems to best fit the scatterplot. Compare this equation to the function formula you derived in the Check Your Understanding task using a finite differences table. Describe and resolve any differences in the two solutions.

c. Do you think statistical regression methods will work to find a function formula for any sequence? Explain.

15 In Investigation 2, you considered two sequences related to the free fall of a sky diver: the sequence D_n of distance fallen *during* each second and the sequence T_n of total distance fallen *after* n seconds. Continue analyzing these sequences.

a. Use the information given in the Think About this Situation on page 482 to verify that the two sequences below, D_n and T_n, are accurate models if you assume Earth's gravity and no air resistance.

Distance Fallen (in feet)		
n	D_n	T_n
1	16	16
2	48	64
3	80	144
4	112	256
5	144	400
6	176	576
7	208	784
8	240	1,024
9	272	1,296
10	304	1,600

b. Use a finite differences table to find a function formula that describes the sequence of total distance fallen.

c. Use calculator- or computer-based regression methods to find a function formula relating total distance fallen to the number of seconds fallen.

d. For the Think About This Situation questions, you wrote recursive and function formulas for the sequence D_n. Note that each term (after the first) of T_n is the result of adding the corresponding term of D_n to the previous term of T_n. Use this fact to help you find a recursive formula for T_n.

REFLECTIONS

16 Suppose you have both a recursive formula and a function formula for a sequence of numbers. Which formula would you use in each of the following situations? Why?

a. Suppose you want to find the 100th term of the sequence of numbers.

b. Suppose you want to find all of the first 100 terms of the sequence.

17 Describe one situation in your daily life or in the daily newspaper that could be modeled by a geometric sequence. Describe another situation that could be modeled by an arithmetic sequence.

18 The English economist Thomas Malthus (1766–1834) is best remembered for his assertion that *food supply grows arithmetically while population grows geometrically*. Do some research, and write a brief paper about this idea. Your essay should address the following questions.

- What is the meaning of Malthus' statement in terms of sequences you have studied?

- Do you think it is a reasonable statement?

- What are its consequences?

- Why has this statement been called "apocalyptic"?

- Have the events of the last 200 years borne out the statement?

- What can we learn from Malthus' statement even if it is not completely accurate?

19 In Problem 3 on page 497, you solved the matrix equation $AX = C$ to find the coefficients of a quadratic equation.

$$\text{You found that } A = \begin{bmatrix} 1 & 1 & 1 \\ 4 & 2 & 1 \\ 9 & 3 & 1 \end{bmatrix}.$$

Do you think this matrix will change if you work a similar problem that involves different data? Explain.

20 In this unit, you have investigated recursive formulas and sequences that can be modeled by *NOW-NEXT* formulas. This means that to find *NEXT*, you need to use only *NOW*, one step before *NEXT*. However, there are sequences for which finding *NEXT* requires using more than one step before *NEXT*. One of the most famous sequences of this type is called the *Fibonacci sequence*, named after the mathematician who first studied the sequence, Leonardo Fibonacci (c. 1175−c. 1250).

a. Here is the Fibonacci sequence: 1, 1, 2, 3, 5, 8, 13, 21, 34, 55, 89, 144, … . Using the words *NEXT*, *NOW*, and *PREVIOUS*, describe the pattern of the sequence.

b. Let F_{n+1} represent *NEXT*. Use F_{n+1}, F_n, and F_{n-1} to write a recursive formula for the Fibonacci sequence.

c. A recursive formula for a sequence similar to the Fibonacci sequence is $A_n = 2A_{n-1} - A_{n-2}$.

  i. Why can't you list the terms of this sequence?

  ii. Choose two initial values, and list the first six terms of the sequence. Then choose different initial values, and list the first six terms of the sequence. Compare the two sequences. Describe any patterns you see.

  iii. Write a recursive formula for this sequence that uses only A_n, A_{n-1}, A_1, and A_0.

d. Consider the sequence given by $A_n = A_{n-1} + A_{n-2} + 7A_{n-3}$. Choose some initial values. List the first six terms of this sequence.

e. The Fibonacci sequence has many interesting patterns and shows up in the most amazing places. In the photo at the right, the flower has 34 spirals in the counterclockwise direction and 21 in the clockwise direction. These two numbers are successive terms in the Fibonacci sequence. Sunflowers and pine cones also have spirals with numbers from the Fibonacci sequence, as do many other things in nature.

In fact, entire books and journals have been written about this sequence. Find an article or book about the Fibonacci sequence. Write a brief report on one of its patterns or applications.

21 In Investigation 2, you found formulas for the sum of a finite number of terms of a geometric sequence. A sequence can also have an infinite number of terms. It is possible to mathematically analyze the sum of an infinite number of terms. Consider this sequence.

$$\frac{1}{2}, \frac{1}{4}, \frac{1}{8}, \frac{1}{16}, \cdots, \frac{1}{2^n}, \cdots$$

a. Explain why this is a geometric sequence. Find recursive and function formulas for this sequence.

b. Find a formula for the sum of the terms up through $\frac{1}{2^n}$.

c. This sequence has infinitely many terms. Does the geometric model of this sequence shown below suggest what the infinite sum of all the terms of the sequence might be?

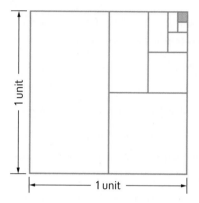

d. A mathematical analysis of infinite sums involves thinking about what happens to the sum of n terms as n gets very large. Examine the formula for the sum of terms up through $\frac{1}{2^n}$ from Part b. What happens to this formula as n gets very large? Your answer should give you the infinite sum.

e. Consider a general geometric sequence with terms t_n and common multiplier r. Suppose that $0 < r < 1$. Determine a general formula for the infinite sum by considering what happens to the finite sum formula if n is very large.

f. Use the general infinite sum formula for $0 < r < 1$ from Part e to find the infinite sum of the sequence $\frac{1}{2}, \frac{1}{4}, \frac{1}{8}, \frac{1}{16}, \cdots, \frac{1}{2^n}, \cdots$. Compare the sum to your answer for Part d.

g. Construct a geometric sequence of your choice with $0 < r < 1$. Use the general infinite sum formula from Part e to find the infinite sum.

22 Extensions Task 21 involved infinite geometric sums. Infinite sums lead to some surprising results. Consider a figure that consists of the rectangles of width 1 and height $\frac{1}{2^n}$ arranged as shown below.

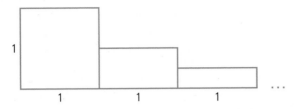

There are infinitely many rectangles that comprise this figure. Do you think it is possible for a figure to have finite area but infinite perimeter? Explain using this particular figure.

23 Oranges are one of the most popular foods during Chinese New Year. During the New Year celebration, you may find them in many homes stacked on a platter in the shape of a pyramid with a square base. The number of oranges in such a stack depends on the number of layers in the stack.

a. Complete a table like the one below.

Pyramid of Oranges							
Number of Layers in a Stack	0	1	2	3	4	5	6
Number of Oranges in a Stack	0	1	5				

b. Find a recursive formula for the sequence of number of oranges in a stack.

c. Use a finite differences table to find a function formula for the sequence of number of oranges in a stack. How many oranges are in a stack with 15 layers?

24 In Investigation 3, you used a finite differences table to find a function formula for certain sequences. You used the fact that the function formula for a sequence is a quadratic if and only if the 2nd differences in a finite differences table are constant. In this task, you will prove that fact.

a. Suppose a sequence has a quadratic function formula: $A_n = an^2 + bn + c$. Construct a finite differences table. Verify that the 2nd differences are constant. Describe any relationship you see between the constant 2nd differences and the function formula.

b. Now, conversely, suppose the 2nd differences are constant. Provide an argument for why the function formula is a quadratic.

McGraw-Hill Education

25 Recall the bowhead whale situation from Applications Task 4 in Lesson 1. A status report on the bowhead whales of Alaska estimated that the 1993 population of these whales was between 6,900 and 9,200 and that the difference between births and deaths yielded an annual growth rate of about 3.1%. No hunting of bowhead whales is allowed, except that Alaskan Inuit are allowed to take, or harvest, about 50 bowhead whales each year for their livelihood. (**Source:** www.afsc.noaa.gov/nmml/species/species_bowhead.php)

a. Use 1993 as the initial year, Year 0. The initial population is a range of values, rather than a single value. Specifically, the range of initial population values is 6,900 to 9,200. Let $L(n)$ be the lower population value in the range of values in year n. Similarly, let $H(n)$ be the higher population value in the range of values in year n. Find recursive formulas for $L(n)$ and $H(n)$.

b. Find a recursive formula and a function formula for $S(n) = H(n) - L(n)$. Describe how the size of the population range changes over time.

c. The recursive formulas for $H(n)$ and $L(n)$ that you found in Part a are combined recursive formulas of the form $A_n = rA_{n-1} + b$. Thus, as in Connections Task 11 (page 503), it is possible to find a function formula for $H(n)$ and $L(n)$ by listing the terms and summing a geometric series. This yields the following formulas.

$$L(n) = 1.031^n(6{,}900) - 50\left(\frac{1.031^n - 1}{1.031 - 1}\right)$$

$$H(n) = 1.031^n(9{,}200) - 50\left(\frac{1.031^n - 1}{1.031 - 1}\right)$$

Verify that the above two formulas can be transformed to the two equivalent formulas below.

$$L(n) = 1.031^n\left(6{,}900 - \frac{50}{0.031}\right) + \left(\frac{50}{0.031}\right)$$

$$H(n) = 1.031^n\left(9{,}200 - \frac{50}{0.031}\right) + \left(\frac{50}{0.031}\right)$$

d. In Part b, you considered the difference between the higher and lower values of the population range. Now consider the ratio of the higher value to the lower value.

i. Use the recursive formulas in Part a, along with a spreadsheet or other technology tool, to create a table of values for the ratio of higher value to lower value. Describe any patterns that you see in the table, particularly the long-term trend.

ii. Use the formulas in Part c to construct a function formula for the ratio of higher values to lower values. Estimate the formula if n is very, very large. Use the formula to predict the long-term trend in the ratios. Compare it to the trend that you observed above in part i. Explain any similarities and differences.

26 Rewrite each quadratic expression in vertex form.

a. $x^2 + 8x + 10$

b. $x^2 - 18x - 6$

c. $x^2 + 3x - 9$

27 In 2010, approximately 26% of Internet users in the U.S. age 65 or older used social networking sites. Suppose that in 2010, Nate randomly selected 150 Internet users in the U.S. who were 65 or older. (**Source:** www.pewinternet.org/ Reports/2010/Older-Adults-and-Social-Media/Report.aspx)

a. How many of these people do you expect to say that they used social networking sites?

b. Is the binomial distribution approximately normal?

c. What is the standard deviation of this distribution? What does it tell you?

d. Would you be surprised to find that 60 of the people in Nate's survey said that they used social networking sites? Explain your reasoning.

e. Use a standardized value to estimate the probability that 30 or fewer of the people in Nate's study used social networking sites.

28 Consider the statement: If quadrilateral $ABCD$ is a parallelogram, then $\overline{BD}$ divides the quadrilateral into two congruent triangles.

a. Is this a true statement? If yes, prove the statement. If not, provide a counterexample.

b. Write the converse of this statement.

c. Is the converse true? If yes, provide a proof. If not, provide a counterexample.

29 Write each product or quotient of rational expressions in equivalent form as a single algebraic fraction. Then simplify the result as much as possible.

a. $\dfrac{5x}{x + 4} \cdot \dfrac{3x + 12}{x^3}$

b. $\dfrac{x - 4}{2} \cdot \dfrac{6}{4 - x}$

c. $\dfrac{x^2 + x}{5} \div \dfrac{2x + 2}{15}$

d. $\dfrac{12x - 6}{x + 1} \div \dfrac{6x - 3}{x}$

30 Find the solution set for each inequality. Represent the solution set using a number line, inequality symbols, and interval notation.

   a. $x^2 - 5x > x + 7$ **b.** $\frac{4}{x} > 3 - x$

31 The spinner below is used at the community fair in Lennox each year. The spinner is divided into eight equal sections. Suppose Sonia buys two spins of the spinner.

   a. What is the probability that she will not win a prize?

   b. What is the probability that she will win two cookies?

   c. What is the probability that she will win a cookie and a drink?

   d. What is the probability that she will win two drinks?

32 In the triangle below, $\overline{DE} \parallel \overline{BC}$.

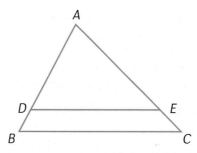

   a. Explain why $\frac{AB}{AD} = \frac{AC}{AE}$.

   b. Suppose that $AD = 10$ cm, $AB = 12$ cm, and $AE = 15$ cm. Find AC and EC.

   c. For the lengths in Part b, is it the case that $\frac{DB}{AD} = \frac{EC}{AE}$?

   d. Provide reasons for each step in the proof below that Jamal wrote to show that $\frac{DB}{AD} = \frac{EC}{AE}$ is always true given $\overline{DE} \parallel \overline{BC}$.

$$\frac{AB}{AD} = \frac{AC}{AE} \qquad (1)$$

$$\frac{AD + DB}{AD} = \frac{AE + EC}{AE} \qquad (2)$$

$$1 + \frac{DB}{AD} = 1 + \frac{EC}{AE} \qquad (3)$$

$$\frac{DB}{AD} = \frac{EC}{AE} \qquad (4)$$

   e. Suppose that $AD = 8$ ft, $DB = 2$ ft, and $AE = 18$ ft. How long is $\overline{EC}$?

33 The arch above the portion of a door shown at the right is part of a circular region bounded by a chord and part of a circle. The chord is 40 inches long and the height of the arch is 8 inches. The radius of the circle that forms $\overset{\frown}{AB}$ is 29 inches. What is the measure of $\overset{\frown}{AB}$?

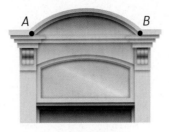

34 Determine the number and type (integer, noninteger rational, irrational, or nonreal complex) of solutions for each quadratic equation.

a. $3x^2 - 5x + 2 = 0$

b. $4x^2 + 25 = 20x$

c. $x^2 + 6x = 10$

35 As part of the preparation for the 2008 Beijing Olympics, a great observation wheel with a diameter of 198 meters was to be built. The wheel has 48 capsules which are evenly spaced around the outside of the wheel.

a. What is the distance from one capsule to the next along the outside of the wheel?

b. What is the shortest distance between two adjacent capsules?

c. If the wheel completes one revolution every 30 minutes, what is the linear velocity of the wheel in kilometers per hour?

Iterating Functions

Function iteration is a relatively new field of mathematical study, with many unanswered questions and connections to contemporary mathematical topics such as fractal geometry and chaos theory. Applications of function iteration are being discovered every day, in areas like electronic transmission of large blocks of data, computer graphics, and modeling population growth.

A function can be thought of as a machine that accepts inputs and produces outputs. For example, for the function $f(x) = x^2$, an input of 2 produces an output of 4. To begin understanding function iteration, imagine starting with a specific input, such as 2, and then sequentially feeding the outputs back into the function as new inputs.

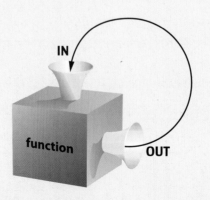

IN

function

OUT

Alfred Pasieka/Peter Arnold/Getty Images

Functions and recursive thinking are both important and unifying ideas in *Core-Plus Mathematics*. In this lesson, you will explore a connection between recursive formulas and function iteration. You will use this connection to further analyze processes of sequential change.

INVESTIGATION 1

Play It Again … and Again

The process of feeding the outputs of a function back into itself as inputs is called **iterating a function**. Imagine making a reduced copy of an image, then making a reduced copy of your copy, then making a reduced copy of that copy, and so on.

As you work on the problems of this investigation, look for answers to the following questions:

How do you iterate a function, and how can technology help?

What are connections between recursive formulas and function iteration?

What are some possibilities for long-term behavior in function iteration sequences?

Connection between Recursive Formulas and Function Iteration As you might expect, there is a close connection between iterating a function and evaluating a recursive formula.

1 Consider the rule $NEXT = 2(NOW)^2 - 5$.

   a. Use an initial value of 1 and find the next three values.

   b. Rewrite the rule as a recursive formula using U_n and U_{n-1}. Let $U_0 = 1$ and then find U_1, U_2, and U_3.

   c. Now think about iterating a function, as illustrated in the "function machine" diagram on page 514. Iterate $f(x) = 2x^2 - 5$ three times, starting with $x = 1$.

   d. Compare the sequences of numbers you got in Parts a, b, and c. Explain why the three sequences are the same, even though the representations used to generate the sequences are different—using *NOW-NEXT* in Part a, using U_n and U_{n-1} in Part b, and iterating a function in Part c. (If the three sequences you generated in Parts a, b, and c are not the same, go back and examine your work, compare to other students, and resolve any problems.)

2 Think about iterating the function $f(x) = 3x + 1$. A table that shows the iteration process is similar to function tables that you have previously used but with a new twist.

   a. Complete a table like the one below, starting with $x = 2$.

Iterating $f(x) = 3x + 1$	
x **IN**	$f(x)$ **OUT**
2	7
7	22
?	?
?	?
?	?
?	?

Start ⟶ 2

   b. How would the table be different if you started with $x = 0$?

   c. What recursive formula yields the same sequence of numbers as that generated by iterating $f(x)$?

3 Consider the recursive formula $U_n = (U_{n-1})^2 + 3U_{n-1} + 4$, with $U_0 = 1$.

    a. Compute U_1, U_2, and U_3.

    b. Rewrite the recursive formula as a rule using the words *NOW* and *NEXT*.

    c. What function can be iterated to yield the same sequence of numbers as that generated by the recursive formula? Check your answer by iterating your function, starting with an input of 1, and comparing to your answers for Part a.

Using Technology to Iterate Functions Use of technology such as calculators and spreadsheets can be very helpful when iterating functions.

4 Consider the function $g(x) = -0.7x + 6$.

    a. Complete a table like the one in Problem 2 showing the first few steps of iterating $g(x)$, starting with $x = 14$.

    b. Use a spreadsheet to iterate $g(x)$ at least 30 times, starting with 14. Describe any patterns you see in the iteration sequence.

Iterate Function.xlsx

	A	B	C	D
1	14			
2	−3.8			
3	=−0.7*A2+6			
4				
5				
6				
7				

    c. Use the last-answer feature of a calculator to iterate $g(x)$ at least 30 times, starting with 14. Make sure you get the same sequence of numbers as in Part b.

```
14
              14
-0.7Ans+6
            -3.8
            8.66
```

    d. In Parts a, b, and c, you used three methods for iterating $g(x)$: completing a table by hand, using a spreadsheet, and using the last-answer feature on a calculator. What are some advantages and disadvantages of each method for iterating a function?

    e. Use the recursion mode on your calculator or use spreadsheet software to produce a graph illustrating the iteration of $g(x)$, starting with $x = 14$. Describe how the numerical patterns that you found in Part b are illustrated in the graph.

Long-Term Behavior As with any process of sequential change, it is important to study the long-term behavior of function iteration.

5 Suppose $f(x) = \sqrt{x}$. Iterate this function, starting with $x = 256$. Describe the long-term behavior of the resulting sequence of numbers.

6 Suppose $h(x) = 1 - x$. Iterate this function. Describe the long-term behavior of the iteration sequence for each of the following starting values.

   a. Start with $x = 5$.

   b. Start with $x = -28$.

   c. Start with $x = \frac{1}{2}$.

In the next problem, you will iterate functions of the form $f(x) = rx(1 - x)$. This is a broadly useful equation in mathematics called the **logistic equation** (or **logistic map**). For different values of r, you get different functions, each of which may have a different behavior when iterated. Iterating these functions has proven to be a very useful method of modeling certain population growth situations. Also, the study of the iterated behavior of these functions has contributed to many important developments in modern mathematics over the last several decades.

7 For each of the function iterations below, use $x = 0.02$ as the starting value. Describe any patterns you see, including the long-term behavior.

   a. Iterate $f(x) = 2.7x(1 - x)$.

   b. Iterate $g(x) = 3.2x(1 - x)$.

   c. Iterate $h(x) = 3.5x(1 - x)$.

   d. Iterate $j(x) = 3.83x(1 - x)$.

   e. Iterate $k(x) = 4x(1 - x)$.

SUMMARIZE THE MATHEMATICS

In this investigation, you explored iteration of linear and nonlinear functions.

a Explain the connection between function iteration and recursive formulas.

b Consider evaluating the rule $NEXT = (NOW)^3 + 5$, starting with 2.

   i. What recursive formula (using subscript notation), when evaluated, will produce the same sequence?

   ii. What function, when iterated, will generate the same sequence?

c Describe some of the possible long-term behaviors that can occur when a function is iterated.

Be prepared to share your descriptions and thinking with the entire class.

✔ CHECK YOUR UNDERSTANDING

Function iteration can be applied to any function.

a. Iterate $g(x) = \cos x$, starting with $x = 10$. (Use radians, not degrees.) Describe the long-term behavior of the iteration sequence.

    i. What recursive formula yields the same sequence as iterating $g(x)$?

    ii. Rewrite your recursive formula using the words *NOW* and *NEXT*.

b. Choose any function not used in this investigation and several different starting points. Iterate your function using each of your starting points. Describe the long-term behavior of the iteration sequences.

INVESTIGATION 2

Iterating Linear Functions

In Investigation 1, you iterated a variety of functions, both linear and nonlinear. The iterative behavior of nonlinear functions is not yet completely understood and is currently a lively area of mathematical research. On the other hand, iteration of linear functions, which is the focus of this investigation, is well understood.

As you work on the problems of this investigation, look for answers to the following questions:

> *How can you graphically iterate a function?*
>
> *What are all the possible long-term behaviors when iterating linear functions?*
>
> *How can you use slope to predict these behaviors?*

Graphical Iteration Just as with many other ideas in mathematics, the process of function iteration can be represented visually.

To see how graphical iteration works, consider the function $f(x) = 0.5x + 2$. The graphs of $y = x$ and $y = 0.5x + 2$ are shown below on the same set of axes. Think graphically about how an input becomes an output, which then becomes the next input, which produces the next output, and so on.

Graphical Function Iteration

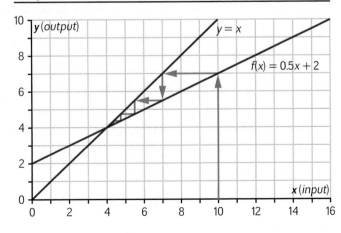

In the diagram on the previous page, 10 has been chosen as the original input. To find an input's (x value's) resulting output (y value), you go up to the graph of the function you are iterating. Next, according to the process of function iteration, the output gets put back into the function as an input. This is accomplished graphically by moving horizontally to the $y = x$ graph, since on this line the y value (the current output) is identical to the x value (the new input).

Now go vertically again to find the output associated with the new input. The process continues in this way until, in this example, you are drawn into the intersection point of the two graphs. This is the process of **graphical iteration**.

1 From this graphical perspective, what is the long-term behavior of the function $f(x) = 0.5x + 2$ when iterated? Use your calculator or a spreadsheet to iterate $f(x)$, starting with $x = 10$, and see if the numerical result matches the graphical result.

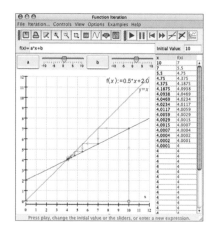

2 Analyze the process of graphical iteration as illustrated in the diagram on the previous page. If available, use a technology tool such as the "Function Iteration" custom app in *CPMP-Tools* to illustrate the process.

   a. Explain, in your own words, why you can graphically find the output of a function for a given input by moving vertically to the graph of the function.

   b. Explain, in your own words, why you can graphically turn an output into the next input by moving horizontally from the graph of the function to the graph of $y = x$.

   c. Complete a table like the one in Investigation 1 Problem 2 on page 516 that shows the first few steps of the graphical iteration illustrated above. Explain how the entries in the table correspond to the steps of the graphical iteration process.

   d. Illustrate the process of graphical iteration for this function using $x = 1$ as the original input. Describe the pattern of graphical iteration.

3 Sketch graphs and illustrate the process of graphical function iteration for the function $g(x) = -0.5x + 8$. Choose your own starting value. Compare the overall pattern of the graphical iteration to the patterns you saw for the function in Problems 1 and 2. Make a conjecture about what kinds of linear rules yield graphical iteration patterns like the one you found in this problem. You will test your conjecture later in this lesson.

Fixed Points As you have seen, sometimes when you iterate a function, you are drawn to a particular value; and if you reach that value, you never leave it. Such a value is called a *fixed point*.

4 Consider fixed points in the case of linear functions.

   a. Look back at Problem 1. What is the fixed point when iterating the function $f(x) = 0.5x + 2$?

   b. What is the fixed point when iterating the function $f(x) = -0.7x + 6$?

   c. The precise definition of a **fixed point** is a value x such that $f(x) = x$. Explain why this definition fits the previous description of a fixed point: "If you reach that value, you never leave it."

5 One method that sometimes works to find a fixed point is iterating the function, either numerically or graphically, and seeing what happens. Another method for finding a fixed point is to use the definition of a fixed point. Since the definition says that a fixed point is a value x such that $f(x) = x$, set the rule for the function equal to x and solve. Use this symbolic method to find the fixed point when iterating the function $f(x) = 0.5x + 2$. Compare your answer to your response in Problem 4 for Part a.

6 For each of the following functions, in Parts a–h, try to find a fixed point using each of these three methods.

> **Method I**   Iterate by using the last-answer feature of your calculator or spreadsheet software (numeric method).
>
> **Method II**   Iterate graphically (graphic method).
>
> **Method III**   Solve the equation $f(x) = x$ (symbolic method).

Organize your work as follows.

- Try a variety of starting values for each function.

- Keep a record of what you try, the results, and any patterns that you notice.

- Prepare a display of the graphical iteration.

Each student should use all three of the methods listed above for the functions in Parts a and b. For Parts c through h, share the workload among classmates.

a. $s(x) = 0.6x + 3$ **b.** $u(x) = 4.3x + 1$

c. $t(x) = 0.2x - 5$ **d.** $v(x) = 3x - 4$

e. $w(x) = x + 2$ **f.** $f(x) = -0.8x + 4$

g. $h(x) = -x + 2$ **h.** $k(x) = -2x + 5$

Slope and Long-Term Behavior Fixed points are examples of the long-term behavior of iterated functions. The different types of fixed points for linear functions can be completely characterized and predicted.

7 Three important characteristics to look for when iterating functions are *attracting fixed points*, *repelling fixed points*, and *cycles*. An **attracting fixed point** is a fixed point such that iteration sequences that start close to it get pulled into it. In contrast, iteration sequences move away from a **repelling fixed point**, except of course, for the sequence that begins at the fixed point. A **cycle** is a set of numbers in an iteration sequence that repeats over and over.

a. For each of the linear functions you iterated in Problem 6, decide with some classmates whether it has an attracting fixed point, a repelling fixed point, a cycle, or none of these.

b. Is there a connection between the slope of the graph of a linear function and the function's behavior when iterated? If so, explain how you could complete Part a of this problem simply by knowing the slope of each linear function's graph.

CHECK YOUR UNDERSTANDING

At the beginning of this unit, you analyzed the changing fish population in a pond. The pond has an initial population of 3,000 fish. The population decreases by 20% each year due to natural causes and fish being caught. At the end of each year, 1,000 fish are added. A rule that models this situation is

$$NEXT = 0.8NOW + 1,000, \text{ starting at } 3,000.$$

a. Rewrite this rule as a recursive formula, using subscript notation.

b. What function can be iterated to produce the same sequence of population numbers generated by the recursive formula? With what value should you start the function iteration?

c. Iterate the function in Part b and describe the long-term behavior of the iteration sequence. Compare this behavior to the long-term behavior of the fish population that you discovered in Investigation 1 of Lesson 1.

d. Graphically iterate this function, starting with $x = 3,000$.

e. Find the fixed point by solving an equation. Is the fixed point attracting or repelling? How can you tell by examining the slope of the function's graph?

f. Explain what the fixed point and its attracting or repelling property tell you about the changing fish population.

APPLICATIONS

1 Consider the three tables below. Describe similarities, differences, and connections among the three tables.

x	$f(x) = 2x + 1$
0	1
1	3
2	5
3	7
⋮	⋮

n	$U_n = 2U_{n-1} + 1,$ $U_0 = 1$
0	1
1	3
2	7
3	15
⋮	⋮

x	Iterate $f(x) = 2x + 1$
0	1
1	3
3	7
7	15
⋮	⋮

2 Iterate $f(x) = \frac{1}{x}$, using several different starting values. Describe the iterated behavior. Explain why the iterated behavior makes sense because of the nature of the function f.

3 **a.** Give an example of a function and a starting value such that the iterated sequence increases without bound.

  b. Give an example of a function and a starting value such that the iterated sequence oscillates between positive and negative values.

4 Experiment with iterating the function $f(x) = x^3 - x^2 + 1$.

  a. Describe the behavior of the iteration sequence when you iterate $f(x)$ with the following beginning values.

   i. Begin with $x = 1$.

   ii. Begin with $x = 0.8$.

   iii. Begin with $x = 1.2$.

  b. Find a fixed point for $f(x)$. Is this fixed point repelling, attracting, some combination of repelling and attracting, or none of these? Explain.

5 Experiment with iterating $g(x) = 3.7x - 3.7x^2$.

  a. Find the fixed points of $g(x)$ by writing and solving an appropriate equation.

  b. Iterate $g(x)$ starting with $x = 0.74$, which is close to a fixed point. Carefully examine the iteration sequence by listing these iterations.

  • List iterations 1 through 6.

  • List iterations 16 through 20.

  • List iterations 50 through 55.

  • List iterations 64 through 68.

c. Describe any patterns you see in the iteration sequence. Does the iteration sequence get attracted to the fixed point? Does it get steadily repelled?

A fixed point is called *repelling* if iteration sequences that begin near it get pushed away from the fixed point at some time, even if such sequences occasionally come back close to the fixed point. Is the fixed point at about 0.73 repelling?

6 Consider iterating $h(x) = 3.2x - 0.8x^2$.

a. The fixed points of $h(x)$ are repelling. Do you think you will find them by numerical iteration? Explain your reasoning.

b. Find the fixed points of $h(x)$ using symbolic reasoning.

c. Experiment with iterating $h(x)$, starting with initial values just above and below each fixed point. Carefully describe the characteristics of each fixed point.

7 Function iteration can be used to model population change. Consider, for example, the bowhead whale population described in Applications Task 4 of Lesson 1. A status report on the bowhead whales of Alaska estimated that the 1993 population of this stock was between 6,900 and 9,200 and that the difference between births and deaths yielded an annual growth rate of about 3.1%. No hunting of bowhead whales is allowed, except that Alaskan Inuit are allowed to take, or harvest, about 50 bowhead whales each year for their livelihood. (**Source:** nmml.afsc.noaa.gov/CetaceanAssessment/bowhead/bmsos.htm)

a. Write a recursive formula and a corresponding function that can be iterated to model this situation.

b. Using the low population estimate as the initial value, iterate and describe the long-term behavior of the population. Do you think this model is a good one to use for predicting the bowhead whale population hundreds of years from now? Why or why not?

c. Find the fixed point. By examining the slope of the graph of the iterated function, decide if the fixed point is attracting or repelling. Illustrate your conclusions by graphically iterating the function.

d. Write a brief analysis of the changing bowhead whale population, as described by your model. As part of your analysis, describe the role played by the fixed point. Make some long-term predictions based on different initial whale populations.

CONNECTIONS

8 Consider iteration of an arbitrary function, $y = f(x)$.

   a. Represent the process of iterating $f(x)$ by a rule using the words *NOW* and *NEXT*.

   b. Represent the process of iterating $f(x)$ by a recursive formula with subscript notation.

9 In Problem 7 of Investigation 1, you investigated the logistic equation, $f(x) = rx(1 - x)$. You found that different values of r produce different long-term iteration behavior. You can program a computer to play music that corresponds to the long-term behavior. For example, consider the BASIC program on the screen below.

```
10   volume = 50
20   duration = 0.5
30   input "r = "; r
40   x = 0.02
50   print x
60   pitch = 220*4^x
70   sound pitch,duration,volume
80   x = r*x*(1-x)
90   go to 50
```

   a. Explain the purpose of each step of this program.

   b. Enter this program, or an equivalent program, into a computer. Run the program using the values 2.7, 3.2, 3.5, 3.83, and 4.0 for r. Describe the results. Explain how the patterns in sound generated by the computer program compare to the numerical patterns that you found in Problem 7 of Investigation 1.

10 It is possible to use function iteration to solve equations. Consider the equation $x - 3 = 0.5x$.

   a. Before solving the equation by using function iteration, solve it using at least two other methods. In each case, explain your method.

   b. To solve $x - 3 = 0.5x$ by function iteration, you will make use of fixed points. Consider this equation written in an equivalent form: $x = 0.5x + 3$.

     i. Explain why this form is equivalent to the original form.

     ii. For what function does this equation define a fixed point?

MaRoDee Photography/Alamy

c. Iterate the function you identified in Part b for some initial value. Describe the long-term behavior. Explain why the observed long-term behavior gives you a solution to the original equation. Compare the solution you obtained to what you found in Part a.

d. Use the method of function iteration to solve $2x + 5 = 6x - 7$. (If at first you do not succeed, try rewriting the equation in an equivalent form.)

e. Summarize the method of function iteration to solve linear equations.

f. When you know several methods for solving a given problem, you should always think about which one is "best." In the case of solving linear equations, would you rather use one of the solution methods from Part a or use function iteration?

11 In this task, you will find and use a general formula for the fixed point of a linear function.

a. Use the algebraic method for finding fixed points to derive a general formula for the fixed point of the linear function $f(x) = rx + b$.

b. Use the general formula to find the fixed point for $h(x) = 0.75x - 4$.

c. If the fixed point is attracting, then you can find it by iteration.

  i. How can you tell from the formula for $h(x)$ that this function has an attracting fixed point?

  ii. Iterate $h(x)$ to find the fixed point. Compare it to what you found using the general formula in Part b.

12 In Lesson 1, you investigated recursive formulas of the form $A_n = rA_{n-1} + b$. Suppose such a recursive formula represents year-to-year change in some population. Assume that r is a positive number less than 1.

a. Explain why the long-term population in this situation can be found by finding the fixed point of the function $f(x) = rx + b$.

b. Use the general fixed point formula from Part a of Connections Task 11 to explain the following patterns that you previously discovered about the fish population problem in Lesson 1.

  • Changing the initial population does not change the long-term population.

  • Doubling the restocking amount doubles the long-term population.

  • Doubling the population-decrease rate cuts the long-term population in half.

13 In this unit, you have studied recursion in several contexts, for example, recursive formulas and function iteration. Recursion is sometimes described as a "self-referral" process. Explain why this is a reasonable description of recursion.

14 In this lesson, you briefly explored a famous equation in mathematics and science called the logistic equation (also called the logistic map), $f(x) = rx(1 - x)$. This equation was first studied extensively in the 1970s. Some of the first discoveries about the behavior of the iterated function came from trying to apply it as a model in biology and ecology. Its behavior turned out to be surprisingly complex and profound, giving rise to what is sometimes called *chaos theory*.

   a. Review what you found out about the iterated behavior of the logistic equation in Problem 7 of Investigation 1, page 518. Why do you think the term "chaos" has been used to describe certain long-term behavior of the logistic equation?

   b. One of the first investigators of the logistic equation was an Australian physicist and biologist named Robert May. May argued that the world would be a better place if every student was given a pocket calculator and encouraged to play with the logistic equation. What do you think May meant?

15 Obtain a copy of the book *Chaos: Making a New Science* by James Gleick (New York: Viking, 1987). Read Chapter 3, entitled "Life's Ups and Downs." This chapter is an entertaining account of some of the history of the logistic equation. Write a two-page report summarizing the chapter.

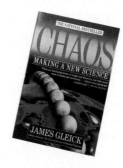

16 Chaos theory is related to certain long-term behavior of the logistic equation, which you examined in Problem 7 of Investigation 1 (page 518). Read the article on the next page honoring Edward Lorenz, who is considered the father of chaos theory.

   How does the description of chaos given in the article relate to the long-term behavior of the logistic equation? Identify specific features of chaos theory from the article and discuss how they relate to behavior of the logistic equation.

Edward Lorenz, Father of Chaos Theory and Butterfly Effect, Dies at 90

Edward Lorenz, an MIT meteorologist who tried to explain why it is so hard to make good weather forecasts and wound up unleashing a scientific revolution called chaos theory, died April 16 of cancer at his home in Cambridge. He was 90.

A professor at MIT, Lorenz was the first to recognize what is now called chaotic behavior in the mathematical modeling of weather systems. In the early 1960s, Lorenz realized that small differences in a dynamic system such as the atmosphere—or a model of the atmosphere—could trigger vast and often unsuspected results.

These observations ultimately led him to formulate what became known as the butterfly effect—a term that grew out of an academic paper he presented in 1972 entitled: "Predictability: Does the Flap of a Butterfly's Wings in Brazil Set Off a Tornado in Texas?"

Lorenz's early insights marked the beginning of a new field of study that impacted not just the field of mathematics but virtually every branch of science—biological, physical and social. ... Some scientists have since asserted that the 20th century will be remembered for three scientific revolutions—relativity, quantum mechanics and chaos.

"By showing that certain deterministic systems have formal predictability limits, Ed put the last nail in the coffin of the Cartesian universe and fomented what some have called the third scientific revolution of the 20th century, following on the heels of relativity and quantum physics," said Kerry Emanuel professor of atmospheric science at MIT. "He was also a perfect gentleman, and through his intelligence, integrity and humility set a very high standard for his and succeeding generations."

Lorenz, who was elected to the National Academy of Sciences in 1975, won numerous awards, honors and honorary degrees. In 1991, he was awarded the Kyoto Prize for basic sciences in the field of earth and planetary sciences. ... The [Kyoto Prize] committee added that Lorenz "made his boldest scientific achievement in discovering 'deterministic chaos,' a principle which has profoundly influenced a wide range of basic sciences and brought about one of the most dramatic changes in mankind's view of nature since Sir Isaac Newton."

Source: MIT News, April 16, 2008. Copyright MIT. Used with permission.

17 Explain why the fixed points for a function $f(x)$ correspond to the points of intersection of the graphs of $y = f(x)$ and $y = x$.

18 Why do you think it is sometimes said that you can never "see" a repelling fixed point?

EXTENSIONS

19 The recursive formula for a geometric sequence looks like a combined recursive formula of the form $A_n = rA_{n-1} + b$ without the added b. This connection can be used to find a function formula for such combined recursive formulas. The strategy involves building from a function formula for a geometric sequence.

a. What is the function formula for a geometric sequence with recursive formula $A_n = rA_{n-1}$ and initial value A_0?

b. As the next step, think about the long-term behavior of $A_n = r^n A_0$. Compare it to the long-term behavior of $A_n = rA_{n-1} + b$. Consider the situation when $|r| < 1$.

   i. Explain why $A_n = rA_{n-1} + b$ has an attractive fixed point for its long-term behavior.

   ii. What is the long-term behavior of $A_n = r^n A_0$?

c. Now modify the formula $A_n = r^n A_0$ so that it has the same long-term behavior as $A_n = rA_{n-1} + b$, as follows. Begin by adding the fixed point, denoted FIX, to the function formula $A_n = r^n A_0$ so that the new function will have the same long-term behavior as $A_n = rA_{n-1} + b$. Explain why $A_n = r^n A_0 + FIX$ has long-term behavior converging to FIX, if $|r| < 1$.

d. Finally, modify the formula $A_n = r^n A_0 + FIX$ so that it has the same initial value A_0 as $A_n = rA_{n-1} + b$.

   i. Explain why the initial value of $A_n = r^n A_0 + FIX$ is equal to $A_0 + FIX$.

   ii. Explain why the initial value of $A_n = r^n(A_0 - FIX) + FIX$ is equal to A_0.

e. As you explained in Part d, the function formula $A_n = r^n(A_0 - FIX) + FIX$ has the same long-term behavior and the same initial value as the combined recursive formula $A_n = rA_{n-1} + b$. Use this new type of formula to find A_5 for $A_n = 2A_{n-1} + 1$, with $A_0 = 3$. Then compute A_5 by successively evaluating $A_n = 2A_{n-1} + 1$, and compare your two results.

f. Compare the function formula in Part e to the function formula you derived in Connections Task 11 of Lesson 2 (page 503). Resolve any apparent differences.

20 In Connections Task 10, you investigated the method of function iteration to solve linear equations. Investigate if this method will work for quadratic equations. Consider the quadratic equation $2x^2 + 5x = 3$.

a. Find a function that you could iterate in order to use the method of function iteration to solve this equation. Iterate with several different initial values. Describe the results in each case.

b. Solve this equation using another method. How many solutions are there?

c. What properties of fixed points allow you to either find or not find a solution when using the method of function iteration?

d. Use your calculator or computer software to help you sketch graphs of the iterated function and $y = x$ on the same set of axes. (Your graph may look different than the one shown here, depending on which function you found to iterate in Part a.) Locate the fixed points on this graph. At each of the fixed points, visualize a line drawn tangent to the graph of the iterated function at the fixed point.

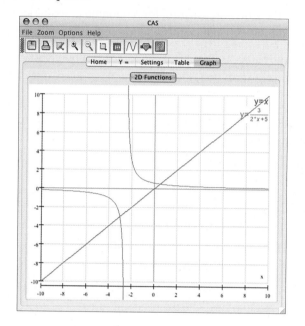

Estimate the slope of each of these tangent lines. For which fixed point is the absolute value of the slope of the tangent line greater than 1? For which tangent line is the absolute value of the slope less than 1?

e. Recall the connection between iterating linear functions and slope. Explain how the slope of the tangent line at a fixed point tells you if the fixed point is attracting or repelling. Explain what attracting or repelling fixed points have to do with solving nonlinear equations using the method of function iteration.

21 Although it is relatively easy to iterate linear functions graphically by hand with a fair degree of accuracy, it is quite difficult to iterate nonlinear functions graphically. This is because, as you have seen, the shape of a graphical iteration is determined by the slope of the graph of the iterated function. And while lines have constant slopes, graphs of nonlinear functions have changing slopes. Thus, it is usually necessary to use a computer or graphing calculator to accurately iterate nonlinear functions graphically. Graphical iteration capability is built into many calculators and computer graphing packages.

a. Consult a manual, if necessary, to find how to use your calculator or computer software to iterate graphically. Practice by graphically iterating $f(x) = -0.8x + 6$. You should get a graph that looks like the one at the right.

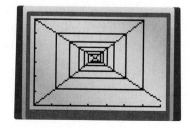

b. In Applications Task 6, you were asked to iterate the function $h(x) = 3.2x - 0.8x^2$. Use a graphing calculator or computer software to iterate $h(x)$ graphically, starting with $x = 2.6$. Compare the graphical iteration pattern to the numerical iteration results.

c. In Applications Task 4, you were asked to iterate the function $f(x) = x^3 - x^2 + 1$. Graphically iterate $f(x)$ to illustrate each of your results in Applications Task 4.

22 In this lesson, you iterated algebraic functions. It is also possible to iterate geometric transformations. As an example, play the following "Chaos Game." (Algorithm originally described in Barnsley, M., *Fractals Everywhere*, Academic Press, 1988.)

a. On a clean sheet of paper, draw the vertices of a large triangle. Any type of triangle will work; but for your first time playing the game, use an isosceles triangle. Label the vertices with the numbers 1, 2, and 3.

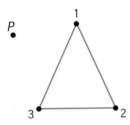

b. Start with a point anywhere on the sheet of paper. This is your initial input. Randomly choose one of the vertices (for example, use a random number generator to choose one of the numbers 1, 2, or 3). Mark a new point one-half of the distance between your input and that vertex. This is your first output and also your new input. Then randomly choose another vertex. Mark the next point, half the distance from the new input to that vertex. Repeat this process until you have plotted six points.

c. The goal of the Chaos Game is to see what happens in the long term. What do you think the pattern of plotted points will look like if you plot 300 points? Make a conjecture.

d. Program a calculator or computer to play the Chaos Game, or find such a program on the Internet, and then carry out several hundred iterations. Since you are interested only in the long-term behavior, you might carry out the first ten iterations without plotting the resulting points and then plot all points thereafter.

e. Repeat the game for several other initial points. Do you think you will always get the same resulting figure? Try it.

f. The figure that results from the Chaos Game is an example of a familiar *fractal*. One of the most important characteristics of fractals is that they are *self-similar*, which means that if you zoom in, you keep seeing figures just like the original figure. What is the scale factor of successively smaller triangles in the fractal that you produced?

g. Give a geometrical explanation for why the Chaos Game will always generate a Sierpinski triangle. (For more about Sierpinski triangles, see Applications Task 3 from Lesson 2 on page 500.)

REVIEW

23 Find the zeroes of each function.

    a. $f(x) = (x - 5)(x + 3)(2x - 1)$

    b. $g(x) = (x^2 - 7x + 12)(x - 6)$

    c. $h(x) = x^2 + 7x + 3$

24 In the figure below, $\overline{AB} \parallel \overline{CD}$, $\overline{BC} \parallel \overline{DE}$, and C is the midpoint of $\overline{AE}$. Prove that $\triangle ABC \cong \triangle CDE$.

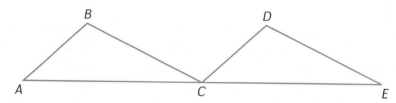

25 Use symbolic reasoning to determine if each statement is true or false.

    a. $(x + 4)^2 - 3 = x^2 + 13$

    b. $(x + 1)^3 = x^3 + 3x^2 + 3x + 1$

    c. $3(3^x)(9^x) = 9^{2x}$

    d. $\sqrt{a^2 + b^2} = a + b$

26 Sketch a graph of each function. Then state the domain and range of the function.

    a. $f(x) = -\dfrac{7}{3}x - 5$ **b.** $g(x) = x^2 + 6$

    c. $h(x) = \dfrac{6}{x}$ **d.** $j(x) = 3(2^x)$

    e. $k(x) = |x| - 2$ **f.** $p(x) = |x + 3|$

27 If a process is in control and data collected from the process are approximately normally distributed, what is the probability that the next reading on a control chart lies outside the control limits?

28 Find each product without using your calculator.

    a. $\begin{bmatrix} 3 & 5 \end{bmatrix} \begin{bmatrix} 6 \\ -2 \end{bmatrix}$

    b. $\begin{bmatrix} 1 & -3 \end{bmatrix} \begin{bmatrix} 0 & 4 \\ 1 & -2 \end{bmatrix}$

    c. $\begin{bmatrix} 2 & 1 \\ 5 & 3 \end{bmatrix} \begin{bmatrix} 3 & -1 \\ -5 & 2 \end{bmatrix}$

Looking Back

In this unit, you have investigated sequential change in a variety of contexts using the tools of recursion and iteration. You have extended the idea of using *NOW-NEXT* formulas to model sequential change; you have studied recursive formulas, function iteration, and sequences; and you have made connections to previous work with linear, exponential, and polynomial functions. In this final lesson, you will pull together and review the key ideas in the unit.

1 In Lesson 1, you modeled a variety of situations with combined recursive formulas of the form $A_n = rA_{n-1} + b$. In Lesson 3, you iterated linear functions. These two topics are closely connected. In this task, you will summarize key features of that connection.

By using different values for r and b in the combined recursive formula $A_n = rA_{n-1} + b$, you can build models for different situations. Four different possibilities are indicated by the table below. One recursive formula has already been entered into the table. If you completed Applications Task 6 on page 471, you already have a start on this task.

Four Different Versions of the Recursive Formula $A_n = rA_{n-1} + b$		
	$0 < r < 1$	$r > 1$
$b < 0$		
$b > 0$	$A_n = 0.8A_{n-1} + 1{,}000$	

Choose *one* of the empty table cells. In a copy of the table, write an appropriate recursive formula in the cell and then analyze the recursive formula and its use as a mathematical model. Title your work with the particular recursive formula on which you are reporting. Organize your analysis of the recursive formula as follows.

a. Briefly describe a real-world situation that can be modeled by the recursive formula along with a chosen initial value.

b. Rewrite the recursive formula as a *NOW-NEXT* formula.

c. Write a linear function that can be iterated to yield the same sequence as the successive values of the recursive formula. Choose an initial value.

   i. Iterate the function, and describe the long-term behavior.

   ii. Find the fixed point. Decide whether it is attracting, repelling, or neither. Explain in terms of slope.

   iii. Sketch a graph showing graphical iteration of the function for the initial value previously chosen.

d. Sketch a graph of A_n versus n, using the same initial value that you chose in Part c.

e. Describe the long-term behavior of the real-world situation being modeled, for different initial values. Refer to the fixed point and its properties, but keep your description in the context of the particular situation being modeled.

2 Many irregular shapes found in the natural world can be modeled by fractals. Study the first few stages of the fractal tree shown below.

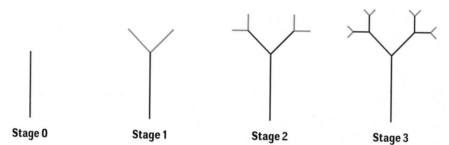

Stage 0   Stage 1   Stage 2   Stage 3

a. Write the number of new branches at each stage for the first several stages. Then write recursive and function formulas that describe this sequence. Use one of the formulas to predict the number of *new* branches at Stage 12. Check your prediction using the other formula.

b. Find the *total* number of branches at Stage 12.

c. Suppose that the length of the initial branch is 1 unit and that the branches at each successive stage of the fractal tree are half the length of the branches at the previous stage.

   i. Write the total length of all the branches at each stage for the first several stages.

   ii. Find the total length of all the branches at Stage 15.

3 When you attend a movie, concert, or theater production, you may notice that the number of seats in a row increases as you move from the front of the theater to the back. This permits the seats in consecutive rows to be offset from one another so that you have a less-obstructed view of the stage.

The center section of the orchestra level of Shaw Auditorium is arranged so that there are 42 seats in the first row, 44 seats in the second row, 46 seats in the third row, and so on for a total of 25 rows.

a. Determine the number of seats in the last row in two different ways. Compare your result and your methods to those of other students. Resolve any differences.

b. Determine the total number of seats in the center section of the orchestra level in at least two different ways. One method should involve using a rule showing the total number of seats as a function of number of rows n.

4 Amy was investigating the maximum number of regions into which a plane is separated by n lines, no two of which are parallel and no three of which intersect at a common point. For example, the diagram below shows the maximum number of regions for 3 lines.

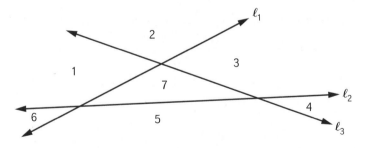

Amy gathered the data shown in the table below.

Number of Lines	0	1	2	3	4
Number of Regions	1	2	4	7	11

a. Verify the entries in the table.

b. Find a function formula for the maximum number of regions formed by n lines, using the method of finite differences.

c. Write a recursive formula for this relationship.

5 Find recursive and function formulas for each of the sequences below. Then find the sum of the terms up through the term with subscript 15.

a.

n	0	1	2	3	4	5	6	...	15	...
L_n	600	300	150	75	37.5			...		...

b.

n	1	2	3	4	5	6	7	...	15	...
P_n	−3	2	7	12	17			...		...

SUMMARIZE THE MATHEMATICS

In this unit, you used recursion and iteration to represent and solve problems involving sequential change. In the process, you investigated several new topics and revisited and extended your understanding of several familiar topics. When asked to make a list of topics covered in this unit, Derartu produced the following list.

- Recursive formulas
- Population growth
- Compound interest
- Arithmetic sequences and sums
- Geometric sequences and sums
- Finite differences

- Linear functions
- Exponential functions
- Polynomial functions
- Function iteration
- Fixed points

a Examine each topic to see if you agree that it should be on the list. Be prepared to discuss any questions that you have about what a particular topic is or why it is on the list. Add any other topics that you think are missing.

b Ray's list had just one item on it: combined recursive formulas. Explain why Ray's list provides a reasonable summary of the unit.

c Which of the topics on Derartu's list are connected in some way to combined recursive formulas of the form $A_n = rA_{n-1} + b$? Explain your reasoning.

Be prepared to share your analysis of the unit topics with the entire class.

✔ CHECK YOUR UNDERSTANDING

Write, in outline form, a summary of the important mathematical concepts and methods developed in this unit. Organize your summary so that it can be used as a quick reference in future units and courses.

UNIT

UNIT
8
Inverse Functions

In earlier units, you learned how polynomial, rational, exponential, and trigonometric functions can be useful tools for modeling the relationships between dependent and independent variables in a wide variety of problem situations. In many of those problems, the key challenge is finding values of the independent variable that lead to required values of the dependent variable.

For example, if you know the rate at which an endangered animal population is recovering, you certainly want to find the time when that population is likely to reach a specified safe level. Solution of problems like that is helped by calculation of the inverse for the function relating population to time.

Work on the investigations of this unit will develop your understanding and skill in use of inverse functions, especially the square root, logarithm, and inverse trigonometric functions. Key ideas are encountered in three lessons.

©Royalty-Free/Corbis

LESSONS

1 What Is An Inverse Function?

Discover conditions that guarantee existence of inverse functions, develop strategies for finding rules of inverse functions, and use inverse functions to solve problems of coding and decoding information.

2 Common Logarithms and Their Properties

Develop the definition and important properties of inverses for exponential functions and use those logarithmic functions to solve exponential equations.

3 Inverse Trigonometric Functions

Develop definitions and important properties for inverses of sine, cosine, and tangent functions and use those inverse functions to solve trigonometric equations.

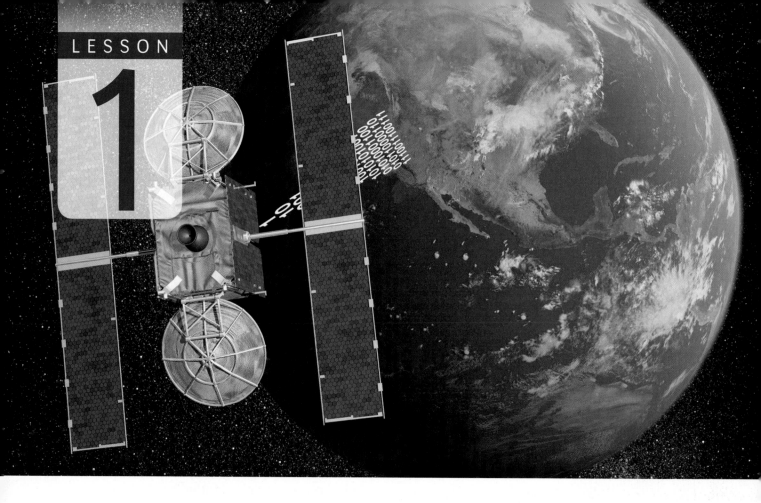

What Is An Inverse Function?

We live in a world of instant information. Numbers, text, and graphic images stream around the globe at every hour of the day on every day of the year. For most data transmission media, information must be translated into numerical form and then into electronic signals. When the signal reaches its destination, it must be translated back into the intended text, graphic, or numerical information format.

Cell phones are one of the most popular communication devices of the electronic information era. Like standard landline telephones, they permit spoken conversations across great distances. They can also be used to send *text messages*. When text messaging first became popular, cell phone texters had to use a traditional telephone keypad that had several letters associated with each number button. When the phone was set in text message mode, pressing number buttons sent letters to the intended receiver.

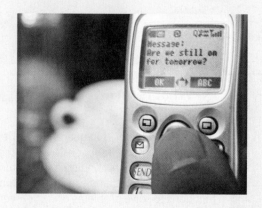

(t)Radius Images/Alamy, (b)Nick Koudis/Getty Images

THINK ABOUT THIS SITUATION

The challenge of sending clear text messages with a basic cell phone is finding a way to translate letters into numbers so that the receiver can translate those numbers back into the intended letters. One procedure for text message coding with a basic cell phone, called *predictive text*, asks the sender to simply press numbers that correspond to desired letters. Since each number button represents several different letters, it is not clear how accurate text messaging could occur with such a phone.

a How do you think a cell phone using the predictive text procedure determines intended words from number sequences?

b Suppose that a text message is entered by pressing **4, 6, 6, 3, 2, 2, 5, 5**. What do you think was the intended message?

In this lesson, you will learn how mathematical functions and their inverses are used to make accurate coding and decoding of information possible and how function inverses are used to solve other important mathematical problems.

Coding and Decoding Messages

Using the predictive text procedure, the keypad of a basic cell phone assigns each letter a numerical code. However, this coding function cannot be reversed reliably to find the intended letters from the numbers that were transmitted. As you work on the problems of this investigation, look for an answer to the following question:

> *What properties of a mathematical function f make it possible to find a related function that reverses the domain, range, and individual assignments of f?*

1 Suppose that *f* is the function that uses a basic cell phone keypad to assign number codes to letters of a text message using the predictive text procedure.

a. What number sequence would be used to send the message **GOOD CALL**?

b. What different messages could result from attempts to decode the number sequence produced in Part a?

c. What is the problem with predictive text coding that can lead to unclear message transmission?

d. What changes in the coding process could avoid the problems revealed in your answers to Parts b and c?

2 To avoid the problems that can occur when several letters are assigned the same numerical code on a cell phone keypad, information systems, like those used by computers, employ more than ten numbers to code the letters of the alphabet and important symbols. For example, a coding scheme might make the assignments in the next table.

Symbol	0	1	⋯	9	A	B	C	⋯	X	Y	Z	.	—
Code	0	1	⋯	9	10	11	12	⋯	33	34	35	36	37

a. Using this coding scheme, what message is implied by the sequence of code numbers **22, 14, 14, 29, 37, 22, 14, 37, 10, 29, 37, 9**?

b. Why does this coding scheme involve very low risk of incorrect message transmission?

c. If you want a message to be sent so that only the intended recipient can read it, why will a simple code like that shown in the table *not* be effective?

To make messages difficult for data spies to decode, the first step of letter-to-number coding is often followed by use of further *encryption* algorithms. The aim is to disguise decoding clues like known letter frequencies and patterns of repeated letters in the English language. For example, the letter "E" has been found to occur roughly 13% of the time in English text.

3 Using the basic coding scheme from Problem 2, the message **I WILL BE THERE** is translated into the number sequence **18, 37, 32, 18, 21, 21, 37, 11, 14, 37, 29, 17, 14, 27, 14**. Suppose that the basic symbols-to-numbers coding is followed by application of the linear function $f(x) = x + 16$ to each code number.

a. What number sequence gives the encrypted message now?

b. What decoding formula would you give to the recipient of your messages?

c. Why does this encryption method still fail to disguise decoding clues like known letter frequencies?

4 Because *shift encryption* algorithms like $f(x) = x + 16$ are easy to decode, it is tempting to try more complex encryption functions.

    a. Study the patterns of code assignments made by the following functions, and explain why they also have weaknesses as encryption algorithms.

        i. $g(x) = 2x + 1$

        ii. $h(x) = x^2$

        iii. $k(x) = 38x - x^2$

    b. See if you can devise an encryption algorithm of your own that does not have the flaws of the examples in Part a.

5 Where possible, describe in words and find algebraic expressions for strategies that could be used to *decode* messages that have been encrypted by the following functions.

    a. $f(x) = x + 7$

    b. $g(x) = 2x + 1$

    c. $h(x) = x^2$

    d. $k(x) = 38x - x^2$

If an encryption algorithm assigns the same code number to every occurrence of a symbol, it can be broken easily. For example, if the letter "E" is always assigned the code number **29**, code breakers will use the fact that "E" is the most common letter in English to guess that **29** stands for that letter. An encryption scheme that uses matrices avoids this problem of simpler coding algorithms.

6 One simple matrix coding scheme begins by arranging the message number codes in the form of 1×2 matrix blocks. For example, the sequence **2, 0, 11, 11, 7, 14, 12, 4** becomes:

$$M_1 = \begin{bmatrix} 2 & 0 \end{bmatrix} \quad M_2 = \begin{bmatrix} 11 & 11 \end{bmatrix} \quad M_3 = \begin{bmatrix} 7 & 14 \end{bmatrix} \quad M_4 = \begin{bmatrix} 12 & 4 \end{bmatrix}$$

Next, each of those message blocks is multiplied by a coding matrix, like $C = \begin{bmatrix} 2 & 1 \\ 5 & 3 \end{bmatrix}$.

    a. Find the encrypted form of each message block: M_1C, M_2C, M_3C, M_4C. Then rewrite the message number sequence in its newly encrypted form.

    b. What do you notice about the resulting pattern of numbers in the encrypted message sequence? How does this strategy appear to solve the problem of coding clues given by known letter frequencies?

    c. What decoding directions would you give to the message recipient?

SUMMARIZE THE MATHEMATICS

In this investigation, you compared several encryption procedures to reveal properties of schemes that provide accurate and difficult-to-break message-coding methods. The methods you examined were:

(1) the assignment of number codes to letters by cell phone buttons.

(2) the assignment of number codes to digits, letters, periods, and spaces as in $0 \rightarrow 0, 1 \rightarrow 1, \cdots, A \rightarrow 10, B \rightarrow 11, \cdots, Z \rightarrow 35, . \rightarrow 36, - \rightarrow 37$.

(3) the algorithm that encrypted numbers with the function $f(x) = x + 16$.

(4) the algorithm that encrypted numbers with the function $g(x) = 2x + 1$.

(5) the algorithm that encrypted numbers with the function $h(x) = x^2$.

(6) the algorithm that encrypted numbers with the function $k(x) = 38x - x^2$.

(7) the algorithm that encrypted number blocks with multiplication by the matrix
$$C = \begin{bmatrix} 2 & 1 \\ 5 & 3 \end{bmatrix}.$$

a Which of the coding and encryption functions assigns code numbers to message letters and numbers in ways that can always be decoded to accurately retrieve the intended message? For the methods that do *not* decode accurately, explain why not.

b What properties of a coding function *f* will guarantee that messages encrypted by that function can always be decoded accurately?

Be prepared to explain your ideas to the class.

 CHECK YOUR UNDERSTANDING

The mathematical challenge of decoding messages is similar to the problem that occurs in many other situations. For example, the student government at Rosa Parks High School was planning a fundraising project that involved selling pizza at girls' and boys' basketball games. Market research among students and parents revealed two functions relating *sales* and *profit* to *price*:

Sales: $n(x) = 250 - 80x$, where x is price in dollars per slice and $n(x)$ is number of slices sold.

Profit: $P(x) = -80x^2 + 290x - 125$, where x is price in dollars per slice and $P(x)$ is profit.

Study tables and graphs of these two functions to explain the answer to each question below.

a. Is it possible to find the number of slices sold if you know the price per slice?

b. Is it possible to find the price per slice if you know the number of slices sold?

c. Is it possible to find the profit from the project if you know the price per slice?

d. Is it possible to find the price per slice if you know the profit from the project?

Finding and Using Inverse Functions

A mathematical function f sets up a correspondence between two sets so that each element of the domain D is assigned exactly one image in the range R. If another function g makes assignments in the opposite direction so that when $f(x) = y$ then $g(y) = x$, we say that g is the **inverse** of f.

The inverse relationship between two functions is usually indicated with the notation $g = f^{-1}$ or $g(x) = f^{-1}(x)$. The notation $f^{-1}(x)$ is read "f inverse of x." In this context, the exponent "-1" *does not mean* the reciprocal "one over $f(x)$." Inverses are often useful in solving problems, but there are many functions that do not have inverses. You have seen examples in your work on the coding problems of Investigation 1.

For functions with numeric domains and ranges, it is usually very helpful to describe the rule of assignment with an algebraic expression. It is also useful to find such rules for inverse functions.

As you work on the problems of this investigation, look for answers to the following questions:

> *Which familiar types of functions have inverses?*
>
> *How can rules for inverse functions be derived?*

Which Functions Have Inverses? It is often helpful to think about inverse functions by using *arrow diagrams* and *coordinate graphs*.

For example, the next arrow diagrams show the patterns of assignments by two familiar functions.

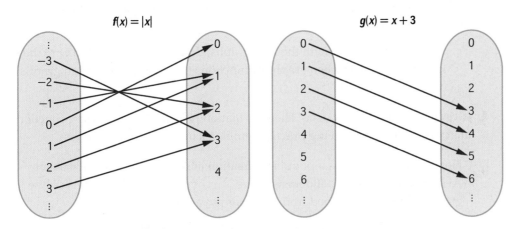

1. Consider the two functions and the given (incomplete) arrow diagrams.

    a. Does $f(x) = |x|$ have an inverse? How does the diagram show reasons for your answer?

    b. Does $g(x) = x + 3$ have an inverse? How does the diagram show reasons for your answer?

    c. How can you use an arrow diagram like those shown to decide whether a function does or does not have an inverse?

2 Since functions are often represented by coordinate graphs, it helps to be able to inspect such graphs to see if inverses exist. Consider the graphs of functions given below.

Graph I $f(x) = 2x + 1$

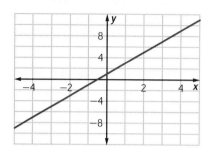

Graph II $f(x) = \dfrac{1}{x^2}$

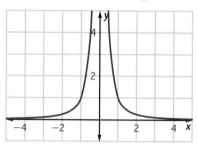

Graph III $f(x) = (x - 3)^2$

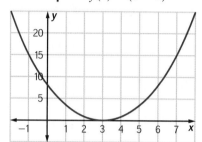

Graph IV $f(x) = x^3 - 4x$

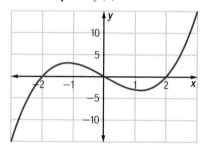

a. Which of the graphs show functions that have inverses?

b. For each function that does not have an inverse, see if you can describe a restricted (reduced) domain for the given algebraic rule so that the resulting function does have an inverse.

c. Based on your work with the four graphs, describe what you think are the key differences between graphs of functions that have inverses and graphs of those that do not.

Applications of Inverse Functions The information provided by an inverse function is often very helpful for solving practical problems in science or in business.

3 Hot air balloon flying is a popular recreation in several southwestern states. At some gatherings of balloonists, there is a competition to hit targets on the ground with bags dropped from balloons at various altitudes.

The balloons drift toward and over the target at various speeds. To have the best chance that their drop bags will hit the target, balloonists need to know how fast the bags will fall toward the ground. Principles of physics predict that the velocity in feet per second will be a function of time in seconds with rule $v(t) = 32t$. The distance fallen in feet will also be a function of time in seconds with rule $d(t) = 16t^2$.

a. How long will it take the drop bag to reach velocity of:

    i. 48 feet per second?

    ii. 100 feet per second?

    iii. 150 feet per second?

b. What rule can be used to calculate values for the inverse of the velocity function—to find the time when any given velocity is reached?

c. How long will it take the drop bag to reach the ground from altitudes of:

    i. 144 feet?

    ii. 400 feet?

    iii. 500 feet?

d. What rule can be used to calculate values of the inverse of the distance function—to find the time it takes for the drop bag to fall any given distance?

4 The daily profit of the Texas Theatre movie house is a function of the number of paying customers with rule $P(n) = 5n - 750$.

a. What numbers of paying customers will be required for the theater to have:

    i. a profit of $100?

    ii. a profit of $1,250?

    iii. a loss of $175?

b. What rule can be used to calculate values for the inverse of the profit function—to find the number of paying customers required to reach any particular profit target?

Finding Rules for Inverse Functions The situations in Problems 3 and 4 illustrate the main reason for being concerned about inverse functions: We often know a function that shows how to calculate values of one variable y from known values of another variable x. But problems in such situations often require discovering the value(s) of x that will produce specific target values of y.

If the function f shows how to use values of x to calculate values of y, then f^{-1} shows how to use values of y to calculate corresponding values of x. The challenge is using the rule for f to find the rule for f^{-1}.

5 For each of the following linear relationships between y and x, describe in words the calculations required to find the value of x corresponding to any given value of y.

   a. $y = g(x)$ and $g(x) = 3x$

   b. $y = h(x)$ and $h(x) = 3x + 5$

   c. $y = j(x)$ and $j(x) = 3x - 7$

   d. $y = k(x)$ and $k(x) = \frac{3}{5}x + 4$

   e. $y = s(x)$ and $s(x) = mx + b$ $(m \neq 0)$

When analysis of the relationship between two variables begins with the equation $y = f(x)$, it probably seems natural to express the inverse relationship with the equation $x = f^{-1}(y)$ and to write the rule for the inverse function using y as the independent variable.

   However, the convention in mathematical practice is to write the rule for f^{-1} in terms of x as well—with the understanding that the letter x stands for the independent variable in the rules for both f and f^{-1}.

6 Use this notational convention to write algebraic rules for each inverse function in Problem 5.

   a. If $g(x) = 3x$, then $g^{-1}(x) = \ldots$.

   b. If $h(x) = 3x + 5$, then $h^{-1}(x) = \ldots$.

   c. If $j(x) = 3x - 7$, then $j^{-1}(x) = \ldots$.

   d. If $k(x) = \frac{3}{5}x + 4$, then $k^{-1}(x) = \ldots$.

   e. If $s(x) = mx + b$ $(m \neq 0)$, then $s^{-1}(x) = \ldots$.

7 Finding inverses (when they exist) for nonlinear functions is generally more challenging than for linear functions. Study the following strategies for finding the inverse of $f(x) = \frac{5}{x - 2}$ to see if you agree that they are both correct.

Strategy I	**Strategy II**
If $\qquad y = \dfrac{5}{x - 2}$	If $\qquad y = \dfrac{5}{x - 2}$
Then $\quad y(x - 2) = 5$	Swap the roles of y and x to get
Then $\qquad x - 2 = \dfrac{5}{y}$	$x = \dfrac{5}{y - 2}$
Then $\qquad x = \dfrac{5}{y} + 2$	Then $\quad y - 2 = \dfrac{5}{x}$
So, $\qquad f^{-1}(x) = \dfrac{5}{x} + 2.$	Then $\quad f^{-1}(x) = \dfrac{5}{x} + 2.$

Adapt one strategy or the other to find rules for inverses of the following functions.

   a. If $g(x) = \frac{7}{x} + 4$, then $g^{-1}(x) = \ldots$.

   b. If $h(x) = \frac{7}{x + 4}$, then $h^{-1}(x) = \ldots$.

   c. If $j(x) = \frac{7}{x}$, then $j^{-1}(x) = \ldots$.

   d. If $k(x) = x^2$ $(x \geq 0)$, then $k^{-1}(x) = \ldots$.

8 In some situations, you will need to find the inverse of a function that is defined only by a table of values or a graph.

a. Suppose that assignments of the function $f(x)$ are as shown in the following table. Make a similar table that shows the assignments made by $f^{-1}(x)$. Then describe the domain and range of $f(x)$ and $f^{-1}(x)$.

x	−4	−3	−2	−1	0	1	2	3	4
f(x)	2	1	0	−1	−2	−3	−4	−5	−6

b. Suppose that the assignments of the function $g(x)$ are as shown on the graph below.

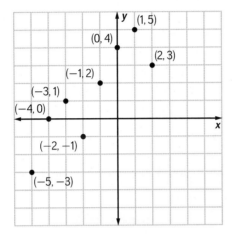

i. On a copy of that graph, plot points that represent assignments made by $g^{-1}(x)$. Then describe the domain and range of $g(x)$ and $g^{-1}(x)$.

ii. Draw line segments connecting each point (a, b) on the graph of $g(x)$ to the corresponding point (b, a) on the graph of $g^{-1}(x)$. Then describe the pattern that seems to relate corresponding points on the graphs of $g(x)$ and $g^{-1}(x)$.

SUMMARIZE THE MATHEMATICS

In this investigation, you explored properties of functions that guarantee existence of inverses and strategies for finding rules for those inverses. You also explored how the graphs of a function and its inverse are related geometrically.

a What patterns in an arrow diagram or coordinate graph for a function indicate that the function does or does not have an inverse?

b What strategies are helpful in finding the rule for f^{-1} when you know the rule for f?

c What geometric pattern relates graphs of functions and their inverses?

Be prepared to explain your ideas to the class.

✔️ CHECK YOUR UNDERSTANDING

Use your understanding of functions and their inverses to complete the following tasks.

a. Draw arrows on a copy of the next diagram to illustrate image variable assignments made by the function $f(x) = -x$. Then explain how the pattern of those arrows suggests that the function does or does not have an inverse.

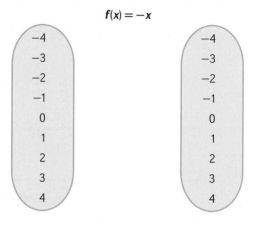

b. Which of the graphs below show functions that have inverses? Justify your response.

Graph I $f(x) = 0.5^x - 4$

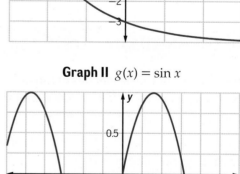

Graph II $g(x) = \sin x$

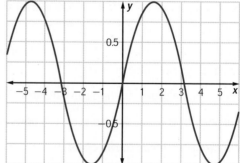

c. Find algebraic rules for the inverses of the following functions.

   i. $f(x) = 0.5x - 2$

   ii. $g(x) = \dfrac{0.5}{x} - 2$

APPLICATIONS

1 Suppose that you want to send the message **MATH QUIZ TODAY** as a text message using the predictive text procedure and a basic cell phone.

  a. What sequence of numbers would you press on the cell phone keypad?

  b. What different message might actually be received?

2 Suppose that a message was first coded by assigning **0, 1, … , 37** to numbers, letters, periods, and spaces as in Investigation 1 and then encrypted by the function $f(x) = x + 9$.

  a. If the received message is given in encrypted form by **37, 23, 23, 46, 43, 33, 39, 46, 11, 46, 32, 27, 38, 23,** what was the message that was sent?

  b. What algebraic expression would decode any message that had been encrypted by $f(x) = x + 9$?

3 Suppose that a message was first coded by assigning **0, 1, … , 37** to numbers, letters, periods, and spaces as in Investigation 1 and then encrypted by the function $f(x) = 2x$.

  a. If the received message is given in encrypted form by **38, 48, 36, 46, 74, 58, 34, 28, 74, 22, 20, 46, 26,** what was the message that was sent?

  b. What algebraic expression would decode any message that had been encrypted by $f(x) = 2x$?

  c. Why do the encrypted message numbers make the decoding rule easy to guess?

4 Suppose that a message was first coded by assigning **0, 1, … , 37** to numbers, letters, periods, and spaces as in Investigation 1 and then encrypted by the function $f(x) = 3x + 2$.

  a. If the received message is given in encrypted form by **86, 44, 71, 41, 113, 32, 113, 77, 56, 38, 89, 92, 83, 44,** what was the message that was sent?

  b. What algebraic expression would decode any message that had been encrypted by $f(x) = 3x + 2$?

5 Suppose that a message was first coded by assigning **0, 1, ... , 37** to numbers, letters, periods, and spaces as in Investigation 1. Study patterns in code numbers assigned by the following encryption functions to see which would allow decoding of any message without confusion.

For those that are suitable, give a rule for decoding. For those that are not, explain why not.

a. $f(x) = 37 - x$

b. $g(x) = x^2 - 38x + 361$

c. $h(x) = |x - 37|$

d. $k(x) = 1.5x + 0.5$

6 Which of these graphs represent functions that have inverses? Be prepared to justify each answer.

Graph I $y = 10 - x$

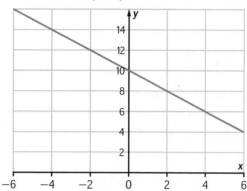

Graph II $y = 10 - |x|$

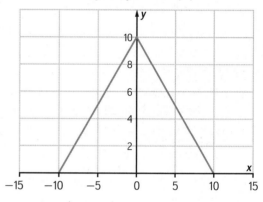

Graph III $y = x^3$

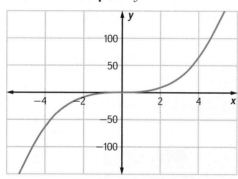

Graph IV $y = \sqrt{x}$

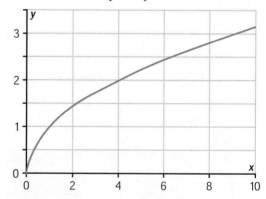

7 Find rules for the inverses of the following functions.

a. $f(x) = 4x - 5$

b. $g(x) = 8x^2$ (domain $x \geq 0$)

c. $h(x) = \dfrac{5}{x}$ (domain $x \neq 0$)

d. $k(x) = -5x + 7$

8 The following statements describe situations where functions relate two quantitative variables. For each situation:

- if possible, give a rule for the function that is described.

- determine whether the given function has an inverse.

- if an inverse exists, give a rule for that inverse function (if possible) and explain what it tells about the variables of the situation.

  a. If regular gasoline is selling for $3.95 per gallon, the price of any particular purchase *p* is a function of the number of gallons of gasoline *g* in that purchase.

  b. If a school assigns 20 students to each mathematics class, the number of mathematics classes *M* is a function of the number of mathematics students *s* in that school.

  c. The area of a square *A* is a function of the length of each side *s*.

  d. The number of hours of daylight *d* at any spot on Earth is a function of the time of year *t*.

CONNECTIONS

9 The U.S. Postal Service uses two kinds of zip codes. One short form uses five digits, like 20906. The longer form uses nine digits, like 20742-3053.

  a. How many five-digit zip codes are possible?

  b. How many nine-digit zip codes are possible?

  c. How do your answers to Parts a and b explain the fact that every possible U.S. mail address has a unique nine-digit zip code but not a unique five-digit code?

10 U.S. Social Security numbers all use nine-digits, generally written in three groups like 987-65-4321.

  a. How many U.S. citizens can be assigned social security numbers with the current system so that each person has a unique code?

  b. No social security numbers begin with the three digits 666. By how much does that restriction reduce the number of available unique social security code numbers?

  c. When social security numbers were first issued, the beginning three digits of each number indicated the area office (one of eight across the U.S.) from which it was issued. How many distinct social security numbers could be issued from one such area?

11 Consider the geometric transformation with coordinate rule $(x, y) \rightarrow (x + 3, y + 2)$.

a. What kind of transformation is defined by that rule?

b. What is the rule for the inverse of that transformation?

12 The arithmetic mean is a statistical function that assigns a single number to any set of numbers $x_1, x_2, \ldots, x_n$ by the general rule $\bar{x} = \dfrac{\Sigma x_i}{n}$. Does this function have an inverse? Explain.

13 The formula $P = 2L + 2W$ assigns a numerical perimeter to every rectangle with length L and width W. Does this function have an inverse? Explain.

14 The following graphs show pairs of functions that are inverses of each other and the line $y = x$. For each pair of functions:

- find two points (a, b) and (c, d) on one graph and show that the points (b, a) and (d, c) are on the other graph.

- explain why the transformation $(x, y) \rightarrow (y, x)$ maps every point of one graph onto a point of the other graph.

a. $f(x) = x^2$ and $f^{-1}(x) = \sqrt{x}, x \geq 0$

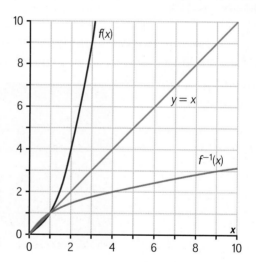

b. $g(x) = 1.5x$ and $g^{-1}(x) = \dfrac{2}{3}x$

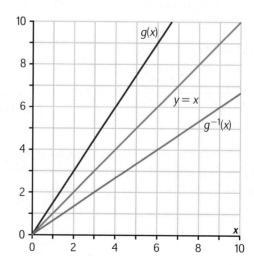

15 Calculators and computers use the *floor* and *ceiling* functions in work with data. The functions are defined as follows.

- The **floor function** $f(x) = \lfloor x \rfloor$ assigns to any number x the greatest integer less than or equal to x. For example, $\lfloor 3.1415 \rfloor = 3$.
- The **ceiling function** $c(x) = \lceil x \rceil$ assigns to any number x the smallest integer greater than or equal to x. For example, $\lceil 2.71828 \rceil = 3$.

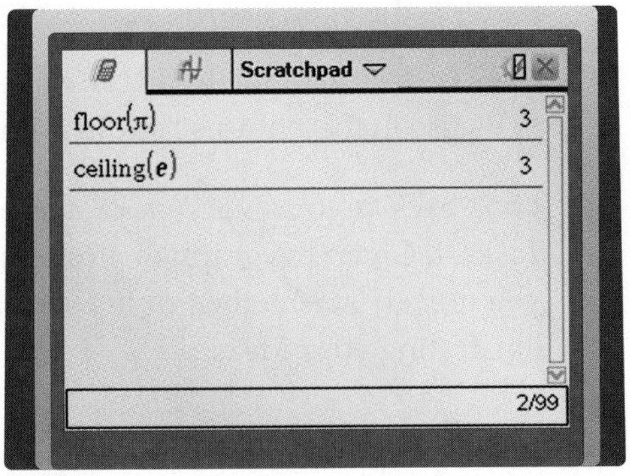

a. Explain why it is or is not possible to find the value of x when you know the value of $f(x)$.

b. Explain why it is or is not possible to find the value of x when you know the value of $c(x)$.

16 Rounding is a function used often in calculator or computer work with numeric data.

a. If $r_2(x)$ rounds every number to two decimal places, find these results from using that function.

i. $r_2(3.141)$ **ii.** $r_2(2.718)$ **iii.** $r_2(2.435)$

b. Explain why it is or is not possible to find the value of x when you know the value of $r_2(x)$.

REFLECTIONS

17 In what kinds of everyday situations do you use systems that code and decode information? Explain why that coding is required and the properties that useful coding schemes must have.

18 Many computer systems require entry of passwords before allowing users to access sites like email servers and other private sources of information. What are some criteria that make sense when assigning passwords?

19 When Brianna looked up "inverse function" on the Internet, she found a sentence that stated, "A function has an inverse if and only if it is a one-to-one function." She wondered what the phrase "one-to-one" means.

Does it mean that for each value of x there is exactly one paired value of y? Or does it mean that for each value of y there is exactly one paired value of x?

Based on your investigation of inverse functions, what do you think the one-to-one condition tells about a function?

Getty Images/SW Productions

20 The word "inverse" is used in several different ways in algebra. For example, we say that -7 is the *additive inverse* of 7 because $-7 + 7 = 0$. Similarly, we say that $\frac{7}{2}$ is the *multiplicative inverse* of $\frac{2}{7}$ because $\left(\frac{7}{2}\right)\left(\frac{2}{7}\right) = 1$.

How is the use of the word "inverse" in the phrase "inverse function" similar to its use in the phrases "additive inverse" and "multiplicative inverse"?

21 Look back at your work on Problem 8 of Investigation 1 and Connections Task 14. Given the graph of a function that has an inverse, how could you quickly sketch the graph of the inverse function? Give an example illustrating your ideas.

22 Here are two ways to think about finding the rule for an inverse function.

Brandon's Strategy: If I know the rule for $f(x)$ as an equation relating y and x, I simply swap the symbols y and x and then solve the resulting equation for y. For example, if $f(x) = 3x + 5$ or $y = 3x + 5$, then I swap y and x to get $x = 3y + 5$ and solve for y to get the rule $y = \frac{x-5}{3}$ for the inverse.

Luisa's Strategy: If I know the rule for $f(x)$ as an equation relating y and x, I solve that equation for x in terms of y and then use that equation to write the rule for the inverse of f. For example, if $f(x) = 3x + 5$ or $y = 3x + 5$, I solve for x to get $x = \frac{y-5}{3}$, that tells me the inverse function has rule $f^{-1}(x) = \frac{x-5}{3}$ or $y = \frac{x-5}{3}$.

a. Will either or both of these strategies always give the correct inverse function rule?

b. Which of the two strategies (or some other strategy of your own) makes most sense to you as a way of thinking about inverse functions and their rules?

EXTENSIONS

As you used the linear function and matrix encryption algorithms for coding, you probably noticed that the code numbers get quite large quickly. Mathematicians have devised a kind of remainder arithmetic that solves that problem. The idea is to apply the sort of counting involved in calculating with time—hours in a day or days in a week.

If you want to find the day of the week 23 days from today, you notice that 23 days is really just 2 days more than 3 weeks and count on 2 days from today. To apply this idea to simplifying code numbers, you take any number calculated with an encryption algorithm and reduce it by taking out multiples of 38. That is, you count from 0 to 37 and then start over from 0 again.

For example, suppose that you are using the encryption function $f(x) = 2x + 1$.

Step 1. The letter T is assigned the code number 29 and then $f(29) = 59$.

Step 2. If we start counting at 0 when reaching 38, the code number 59 is reduced to 21.

This method using a function and remainder arithmetic results in code numbers no larger than 37, but it must be used with some care as you will see in Extensions Tasks 23–25.

23 Simplify the result of the arithmetic of the following calculations based on the idea above.

a. $23 + 34$

b. $2(23) + 15$

c. $5(23) + 4$

d. $9(4) + 3$

e. $9(8) + 3$

f. 13^2

24 Consider the message **I WILL BE THERE AT 1** that translates first into the following number sequence.

$$18, 37, 32, 18, 21, 21, 37, 11, 14, 37, 29, 17, 14, 27, 14, 37, 10, 29, 37, 1$$

a. Use the coding function $f(x) = 2x + 1$ and remainder arithmetic that starts over at 38 to find the encrypted number sequence carrying this message.

b. What problem with this encryption algorithm is revealed by coding the given message?

25 Now explore encryption by remainder arithmetic that starts over at 38 with the function $g(x) = 5x + 1$.

a. Encrypt the following message number sequence.

$$18, 37, 32, 18, 21, 21, 37, 11, 14, 37, 29, 17, 14, 27, 14, 37, 10, 29, 37, 1$$

b. Compare the result of Part a to the encryption result using $f(x) = 2x + 1$. Show that the problem encountered with that function does not occur with $g(x) = 5x + 1$, at least for this message.

c. Complete a copy of the following table to show the encrypted form of each number from 0 to 37. Explain how the pattern of those results shows that $g(x)$ does not assign the same code number to any pair of distinct letters.

Symbol	0	1	...	9	A	B	...	Z	.	—
Code Number	0	1	...	9	10	11	...	35	36	37
Encrypted	1	6	...	8	13	18	...	24	29	34

d. Apply the function $d(x) = 23x + 15$ with remainder arithmetic to each number that results from encryption by $g(x)$ in Part c. What does the pattern of results in that table suggest about the relationship of $g(x)$ and $d(x)$?

26 You have seen that without domain restrictions, quadratic functions never have inverses. Explore that same question for the case of cubic polynomials.

a. Show by analysis of the graph of $y = x^3$, that this simplest possible cubic has an inverse, the cube root function. Then sketch a graph of both the cube and cube root functions on the same coordinate diagram for $-5 \leq x \leq 5$ and $-5 \leq y \leq 5$.

b. Give other examples (where possible) of cubic polynomials that do have inverses and some that do not.

c. The general form of a cubic polynomial is $a_3x^3 + a_2x^2 + a_1x + a_0$. Use algebraic reasoning and a graphing tool to search for patterns of coefficients in cubic functions that have inverses and patterns of coefficients in cubic functions that do not have inverses.

27 A function $f(x)$ is said to be **one-to-one** or **injective** if whenever $f(a) = f(b)$, we can conclude that $a = b$.

a. Provide justifications for each step in the following proof that the function $g(x) = 3x + 5$ is one-to-one.

Step 1. Whenever $g(a) = g(b)$, we can conclude that $3a + 5 = 3b + 5$.

Step 2. The equation in Step 1 implies that $3a = 3b$.

Step 3. The equation in Step 2 implies that $a = b$.

Step 4. Therefore, $g(x)$ is one-to-one.

b. Adapt the argument in Part a to prove that any linear function $h(x) = mx + n$, $m \neq 0$, is one-to-one.

c. Prove that $q(x) = x^2$ (and in fact, any quadratic polynomial function) is not one-to-one.

d. Explain why any exponential function is one-to-one.

e. Explain why the sine and cosine functions are not one-to-one.

f. Explain why a function that is *not* one-to-one does not have an inverse.

g. Explain why a function that is one-to-one does have an inverse.

28 Any 2 × 2 matrix defines a transformation of points in the coordinate plane that maps points in the following general form.

$$\begin{bmatrix} a & b \\ c & d \end{bmatrix} \begin{bmatrix} x \\ y \end{bmatrix} = \begin{bmatrix} ax + by \\ cx + dy \end{bmatrix}$$

a. Show that the matrices $\begin{bmatrix} 1 & 2 \\ 3 & 5 \end{bmatrix}$ and $\begin{bmatrix} -5 & 2 \\ 3 & -1 \end{bmatrix}$ are inverses of each other.

b. Evaluate $\begin{bmatrix} -5 & 2 \\ 3 & -1 \end{bmatrix} \begin{bmatrix} x \\ y \end{bmatrix}$.

c. Multiply the result from Part b by $\begin{bmatrix} 1 & 2 \\ 3 & 5 \end{bmatrix}$.

29 The simplest 2×2 coding matrices are those in the form $A = \begin{bmatrix} a & b \\ c & d \end{bmatrix}$

for which $ad - bc = 1$. In those cases, the inverse of A can be found using

the formula: $\begin{bmatrix} a & b \\ c & d \end{bmatrix}^{-1} = \begin{bmatrix} d & -b \\ -c & a \end{bmatrix}$.

a. Use the given formula to find the inverse of $\begin{bmatrix} 4 & 1 \\ 7 & 2 \end{bmatrix}$.

b. Prove that the given formula will always produce the inverse of

such a matrix by showing that $\begin{bmatrix} a & b \\ c & d \end{bmatrix} \cdot \begin{bmatrix} a & b \\ c & d \end{bmatrix}^{-1} = \begin{bmatrix} 1 & 0 \\ 0 & 1 \end{bmatrix}$,

the 2×2 identity matrix.

REVIEW

30 Use properties of exponents to write each of these expressions in a different equivalent form.

a. $(x^5)(x^2)$

b. $(x^5) \div (x^2)$

c. $(x^5)^2$

d. x^{-5}

e. $\left(a^{\frac{1}{5}}\right)^{10}$

f. $t^{-\frac{1}{2}} t^{\frac{5}{2}}$

31 In the figure at the right, $\overline{AB} \parallel \overline{CD}$.

a. Prove that $\triangle ABE \sim \triangle DCE$.

b. If $BC = 6$ cm, $CE = 9$ cm, $CD = 8$ cm, and $DE = 12$ cm, find AB, AE, and AD.

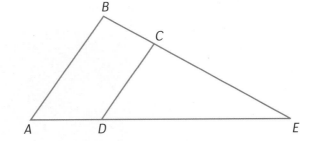

32 Consider quadrilateral $ABCD$ represented by the matrix $\begin{bmatrix} -3 & 1 & -1 & -5 \\ 5 & 4 & 0 & -1 \end{bmatrix}$.

a. Find the image of quadrilateral $ABCD$ reflected across the line $y = x$. Call the image $A'B'C'D'$.

b. Using a coordinate grid, draw quadrilateral $ABCD$, its image $A'B'C'D'$, and the line $y = x$.

c. How is the line $y = x$ related to each segment connecting an image point of the quadrilateral with its preimage point?

33 Solve these equations for x.

a. $10,000 = 10^x$

b. $100,000 = 10^{2x+1}$

c. $15 = 2^x + 7$

d. $2^{3x} = 8^x$

e. $2^{x^2+5x+6} = 1$

34 Suppose that Alvin and Tia are each starting a new walking program. On the first day, they will each walk 1,000 yards. Alvin will increase his distance by 100 yards each day. Tia will increase her distance by 6% each day.

   a. Is the sequence of each person's daily walking distance arithmetic, geometric, or neither? Explain your reasoning.

   b. Who will walk further on the 20th day? By how much?

   c. Who will have walked further over the entire 20-day period? By how much?

   d. Write a function rule that will give the distance each person walked on day n.

35 Write these expressions in equivalent form as products of linear factors.

   a. $x^2 + 7x$ **b.** $x^2 + 7x + 12$

   c. $x^2 + 7x - 8$ **d.** $x^2 - 49$

   e. $x^2 - 6x + 9$ **f.** $3x^2 - 2x - 8$

36 Matthew is riding a bike with wheels that have a 24-centimeter radius. Suppose that he is able to pedal at a steady rate so that the wheels make 120 rotations per minute.

   a. Find his angular velocity in both degrees per minute and radians per minute.

   b. How many meters will he travel in 5 minutes of riding at this speed?

37 Describe the largest possible domains for functions with these rules.

   a. $f(x) = \sqrt{5 - x}$ **b.** $g(x) = \dfrac{3x}{x - 5}$

   c. $h(x) = \dfrac{3x + 5}{x^2 + 5x + 6}$ **d.** $k(x) = 10^x$

38 Solutions of the following equations are not integers. For each equation, find the two consecutive integers between which the solution lies.

   a. $2^x = 36$ **b.** $5^x = 100$

   c. $7^x = \dfrac{1}{50}$ **d.** $10^x = 0.0035$

   e. $10^{2x} = 3,000$ **f.** $11^{x - 2} = \dfrac{1}{2}$

Common Logarithms and Their Properties

At several points in your study of algebra and functions, you have discovered relationships between variables that are well-modeled by exponential functions. The rules of those functions have the form $f(x) = a(b^x)$, where a and b are positive numbers ($b \neq 1$). Answering questions about these functions often requires solving equations in which the unknown quantity is the exponent. For example:

- If a count shows 50 bacteria in a lab dish at the start of an experiment and that number is predicted to double every hour, then to estimate the time when there will be 10,000 bacteria in the dish, you need to solve the equation $50(2^t) = 10,000$.

- The Washington Nationals baseball team was purchased in 2006 for 450 million dollars. To find how long it will take for the value of this investment to reach $1 billion, if it increases at the fairly conservative rate of 5% per year, you need to solve the equation $450(1.05^t) = 1,000$.

• If 500 mg of a medicine enters a hospital patient's bloodstream at noon and decays exponentially at a rate of 15% per hour, the amount remaining active in the patient's blood t hours later will be given by $d(t) = 500(0.85^t)$. To find the time when only 25 mg of the original amount remains active, you need to solve $25 = 500(0.85^t)$.

THINK ABOUT THIS SITUATION

Solving equations like those in the three exponential growth and decay situations is a kind of inverse problem. In each case, you know the output of a function but need to find the corresponding input.

a How could you estimate solutions for the given equations by exploring patterns in tables and graphs of function values?

b How would you approach solution of each equation by algebraic reasoning?

In this lesson, you will extend your understanding and skill in using *logarithms*—a mathematical idea that helps in solving equations like those in the given examples and in many similar problems involving exponential functions.

INVESTIGATION 1

Common Logarithms Revisited

In the *Nonlinear Functions and Equations* unit of Course 2, you learned how to use common logarithms to answer questions very similar to those in the preceding Think About This Situation. The key difference is that all problems in your earlier work involved only exponential functions with base 10 like $B(t) = 50(10^{0.3t})$, $V(t) = 450(10^{0.02t})$, and $s(t) = 200(10^{-0.046t})$.

To deal with those kinds of problems, you learned that mathematicians have developed procedures for finding exponents.

If $10^x = y$, then x is called the **base 10 logarithm** of y.

This definition of base 10 or *common* logarithms is usually expressed in function notation as:

$\log_{10} y$ or simply $\log y$, read "log base 10 of y" or "log of y"

Graphing calculators have a ☐ LOG key that automatically finds the required exponent values.

Work on the problems of this investigation will review and extend your understanding of logarithms to consider this question:

How can common logarithms be used to solve equations involving exponential functions with base 10?

1 Without use of a calculator or computer **log** function, find each of the following logarithms. Be prepared to explain how you know that your answers are correct.

   a. $\log 10^{2.5}$ **b.** $\log 100$ **c.** $\log 10{,}000$

   d. $\log 0.01$ **e.** $\log(-0.001)$ **f.** $\log(10^2 \cdot 10^5)$

2 Without use of a calculator or computer **log** function, find the consecutive integers just below and just above the exact values of these logarithms. Be prepared to explain how you know that your answers are correct.

   a. $\log 25$ **b.** $\log 314$

   c. $\log 3.14$ **d.** $\log 0.005$

3 Use what you know about logarithms to help find exact solutions for these equations.

   a. $10^x = 100$ **b.** $10^{x+2} = 100$ **c.** $10^{3x+1} = 100$

   d. $3(10)^{x+3} = 300$ **e.** $2(10)^x = 600$ **f.** $10^{2x} = 500$

   g. $10^{3x+1} = 43$ **h.** $42(10)^{3x+2} = 840$ **i.** $5(10)^{x+3} + 9 = 44$

4 Answer these questions related to the Think About This Situation on the previous page by using the helpful translation of the functions into exponential form with base 10.

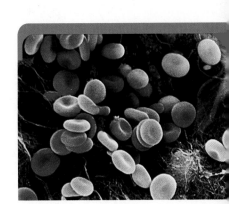

   a. Suppose that the number of bacteria in a lab dish at any time t hours after the start of an experiment is given by the function $B(t) = 50(10^{0.3t})$. Show how to use logarithms to find the time when there will be 10,000 bacteria in the lab dish.

   b. Suppose that the value (in millions of dollars) of the Washington Nationals baseball team at any time t years after 2006 is given by the function $V(t) = 450(10^{0.02t})$. Show how to use logarithms to find the time when that investment will be worth \$1 billion.

   c. Suppose that the amount of medicine (in mg) active in a patient's blood at any time t hours after an injection is given by $m(t) = 500(10^{-0.071t})$. Show how to use logarithms to find the time when only 25 mg of the medicine remain active.

5 The nature of the function $y = \log x$ makes it difficult to graph without the use of technology.

   a. Complete entries in the following table of sample values for the function.

x	0.2	0.4	0.6	0.8	1.0	1.5	2.0	3.0	4.0	5.0	6.0
log x											

   b. Plot the sample of $(x, \log x)$ values on a coordinate graph with suitable scales on the axes.

   c. Explain what the pattern of those points tells about the rate at which the log x function increases as values of x increase.

   d. Compare the graph of $y = \log x$ to that of its inverse $y = 10^x$.

CHECK YOUR UNDERSTANDING

Use your understanding of common logarithms to help complete the following tasks.

a. Find these common (base 10) logarithms without using technology.

    i. log 1,000,000   **ii.** log 0.001   **iii.** log $10^{3.2}$

b. Use the function $y = 10^x$, but not the **log** function of technology, to estimate each of these logarithms to the nearest integer. Explain how you arrived at your answers.

    i. log 85   **ii.** log 850   **iii.** log 8.5

c. In 2007, the U.S. Department of Interior removed grizzly bears from the endangered species list in Yellowstone National Park. The population of grizzly bears in the park at any time t years after 2007 can be estimated by the function $P(t) = 500(10^{0.02t})$. Use logarithms to find the time when this model predicts a grizzly population of 750.

U.S. Fish & Wildlife Service/Larry Aumiller

Covering All the Bases

Logarithms are useful for solving equations in the form $k(10^{ax+b}) = c$. However, many of the functions that you have used to model exponential growth and decay have not used 10 as the base. On the other hand, it is not too hard to transform any exponential expression in the form b^x into an equivalent expression with base 10. As you work on the problems of this investigation, look for an answer to the following question:

How can common logarithms help in finding
solutions of all exponential equations?

1 Use common logarithms to express each of the following numbers as a power of 10. For example, $15 = 10^k$ when $k \approx 1.176$.

   a. $2 = 10^k$ when $k = \dots$ **b.** $5 = 10^k$ when $k = \dots$

   c. $1.0114 = 10^k$ when $k = \dots$ **d.** $25 = 10^k$ when $k = \dots$

   e. $250 = 10^k$ when $k = \dots$ **f.** $0.003 = 10^k$ when $k = \dots$

2 Use your results from Problem 1 and what you know about properties of exponents to show how each of these exponential expressions can be written in equivalent form like $(10^k)^x$ and then 10^{kx}.

   a. 2^x **b.** 5^x **c.** 1.0114^x

3 Use your results from Problems 1 and 2 to solve these exponential equations. Check each solution. Be prepared to explain your solution strategy.

   a. $2^x = 3.5$ **b.** $5(2^n) = 35$ **c.** $5(2^t) + 20 = 125$

   d. $5^p = 48$ **e.** $3(5^r) + 12 = 60$ **f.** $300(0.9^v) = 60$

4 Use the strategies you have developed from work on Problems 1–3 to solve these problems about exponential growth and decay. In each case, it will probably be helpful to write an exponential growth or decay function with the base suggested by problem conditions. Next, transform the rule for that function into an equivalent form with exponential base 10. Then solve the related equation.

   a. The world population at the beginning of 2012 was 7.0 billion and growing exponentially at a rate of 1.2% per year. In what year will the population be double what it was in 2012?

   b. If 500 mg of a medicine enters a hospital patient's bloodstream at noon and decays exponentially at a rate of 15% per hour, when will only 10% of the original amount be active in the patient's body?

   c. If the average rent for a two-bedroom apartment in Kalamazoo, Michigan is currently $750 per month and increasing at a rate of 8% per year, in how many years will average rent for such apartments reach $1,000 per month?

✓ CHECK YOUR UNDERSTANDING

Use logarithms and other algebraic reasoning as needed to complete these tasks.

a. Solve each equation.

   i. $3^x = 243$ **ii.** $8(1.5^x) = 200$ **iii.** $8x^2 + 3 = 35$

b. The population of Nigeria in 2011 was about 162 million and growing exponentially at a rate of about 2.5% per year. (**Source:** 2011 World Population Data sheet)

   i. What function $P(t)$ will give the population of Nigeria in millions in year $2011 + t$, assuming that the growth rate stays at 2.5% per year?

   ii. According to current trends, when is the Nigerian population predicted to reach 200 million? Explain how to estimate the answer to this question with a table or a graph of $P(t)$. Show how to find an "exact" answer using logarithms and other algebraic reasoning.

INVESTIGATION 3

Properties of Logarithms

To use logarithms in reasoning about problems involving exponential functions, it helps to understand key features and properties of the function $f(x) = \log_{10} x$. As you work on the problems of this investigation, look for answers to these questions:

> *What are the important patterns in tables and graphs*
> *for the common logarithm function?*

> *How can properties of common logarithms be used to write*
> *algebraic expressions in useful equivalent forms?*

Recall the relationship between logarithms and exponents:

$$10^r = s \text{ if and only if } r = \log s.$$

1 Consider the domain and range of the function $g(x) = 10^x$.

   a. What is the domain of $g(x) = 10^x$; that is, for what values of x can you calculate 10^x?

   b. What is the range of $g(x) = 10^x$; that is, what are the possible values of 10^x?

2 What are the domain and range of $f(x) = \log x$?

For both the domain and range, explain:

- how your answer is shown by patterns in tables of values for $f(x) = \log x$.

- how your answer is shown by patterns in the graph of $f(x) = \log x$.

- how your answer can be explained logically, using the relationship of the logarithmic function $f(x) = \log x$ and the exponential function $g(x) = 10^x$.

3 Sketch graphs of the logarithmic function $f(x) = \log x$ and the exponential function $g(x) = 10^x$. Explain how the shape and relationship of those graphs illustrates the fact that the two functions are inverses of each other.

4 The logarithmic function $f(x) = \log x$ is the *inverse* of the exponential function $g(x) = 10^x$. Complete the following sentences to illustrate that relationship between the two functions. Be prepared to justify your claims.

   a. $10^3 = 1{,}000$, so $\log 1{,}000 =$ ___.

   b. $10^{1.5} \approx 31.6$, so $\log 31.6 \approx$ ___.

   c. $\log 0.01 = -2$, so $10^{-2} =$ ___.

   d. $\log 125 \approx 2.097$, so $10^{2.097} \approx$ ___.

5 Use what you know about logarithms to evaluate these expressions.

   a. $\log 10^6$ **b.** $10^{\log 100}$ **c.** $\log 10^{0.0124}$

   d. $10^{\log 8.5}$ **e.** $\log 10^w$ **f.** $10^{\log z}$

6 The close connection between logarithmic and exponential functions can be used to derive some useful rules for working with algebraic expressions involving logarithms. Use what you know about exponents to write each of these expressions in a different but equivalent form.

   a. $10^m 10^n$ **b.** $10^m \div 10^n$ **c.** 10^0

   d. $(10^m)^n$ **e.** $1 \div 10^m$ **f.** $10^{(m-n)}$

7 Which of the following claims are true? Explain each response.

   a. $\log (10^m 10^n) = m + n$ **b.** $\log (10^m \div 10^n) = m - n$

   c. $\log (10^m + 10^n) = m + n$ **d.** $\log ((10^m)^n) = mn$

   e. $\log (1 \div 10^m) = -m$ **f.** $\log 1 = 0$

8 You have seen that every positive number can be written as a power of 10. For example,

$$5 = 10^{\log 5}, \text{ or } 5 \approx 10^{0.7}$$
$$\text{and } 0.2 = 10^{\log 0.2}, \text{ or } 0.2 \approx 10^{-0.7}.$$

There are several other properties of the logarithmic function that are useful in transforming expressions to equivalent forms. Complete the sentences that describe each property below. For Part a, provide a reason for each step in the justification. Then adapt the reasoning in Part a to justify the reasoning in Parts b–e.

a. For any positive numbers s and t, $\log st = \log s + \log t$.

This property states that the **logarithm of a product** *of two numbers is equal to …*
This is true because:

$$st = 10^{\log s} \cdot 10^{\log t} \qquad (1)$$
$$st = 10^{\log s + \log t} \qquad (2)$$
$$\log st = \log s + \log t \qquad (3)$$

b. For any positive numbers s and t, $\log \frac{s}{t} = \log s - \log t$.

This property states that the **logarithm of a quotient** *of two numbers is equal to …*
This is true because …

c. For any positive number s and any number t, $\log s^t = t \log s$.

This property states that the **logarithm of a power** *of a number is equal to …*
This is true because …

d. Another way to see why the property in Part c holds is to look at an example like $\log 5^3 = \log (5 \cdot 5 \cdot 5)$, which by the _____ property is equal to …

e. For any positive number t, $\log \frac{1}{t} = -\log t$.

This property states that the **logarithm of the reciprocal** *of a number is equal to …*
This is true because … (*Hint:* Recall that $\frac{1}{t} = t^{-1}$.)

9 Use the facts that $\log 2 \approx 0.3$ and $\log 5 \approx 0.7$ and the properties of logarithms that you proved in Problem 8 to estimate the values of these expressions without the use of technology.

a. $\log 4$ **b.** $\log 20$ **c.** $\log 8$

d. $\log 25$ **e.** $\log \frac{5}{4}$ **f.** $\log 625$

10 As with properties of exponents, there are some common errors when people use the properties of logarithms. Find the errors in the following calculations. Explain why you think the error occurs and how you would help someone see the error of his or her thinking.

a. $\log (12 + 17) = \log 12 + \log 17$ **b.** $\log 0 = 0$

c. $\log 5^3 = \log 15$ **d.** $\log (7 \times 5) = (\log 7)(\log 5)$

e. $\dfrac{\log 20}{\log 2} = \log 10$

11 Properties of logarithms can be used to solve exponential equations in a different way than what you developed in Investigation 2.

a. Explain how properties of logarithms are used in this sample equation solution. Then adapt the reasoning to solve the equations in Parts b–e.

$$\text{If } 5^{2x+3} = 48, \text{ then } \log 5^{2x+3} = \log 48. \qquad (1)$$

$$(2x + 3) \log 5 = \log 48 \qquad (2)$$

$$2x + 3 = \frac{\log 48}{\log 5} \qquad (3)$$

$$2x = \frac{\log 48}{\log 5} - 3 \qquad (4)$$

$$x = \left(\frac{\log 48}{\log 5} - 3 \right) \div 2 \qquad (5)$$

b. $3^x = 25$ **c.** $5^{x+4} = 35$

d. $1.5^{2x+1} = 12$ **e.** $7(3^{0.5x}) = 42$

SUMMARIZE THE MATHEMATICS

In this investigation, you learned properties of logarithms that can be used to write logarithmic and exponential expressions in useful equivalent forms.

a Why are the functions $f(x) = 10^x$ and $g(x) = \log x$ inverses of each other?

b Rewrite each of these expressions in an equivalent form.

   i. $\log (pn)$ **ii.** $\log (p \div n)$ **iii.** $\log (n^p)$

c Explain how properties of logarithms can be used to solve equations like $a(b^{kx}) = c$.

Be prepared to share your ideas and reasoning with the class.

✔ CHECK YOUR UNDERSTANDING

Use your understanding of properties of logarithms to help complete these tasks.

a. Use properties of logarithms to write simpler expressions that are equivalent to these forms.

   i. $\log (10^5 10^3)$ **ii.** $\log (10^5 \div 10^3)$ **iii.** $\log (10^5)^3$

b. Use properties of logarithms to write expressions equivalent to these in what seem to you to be simplest possible forms.

   i. $\log x^5 x^3$ **ii.** $\log (x^5 \div x^3)$ **iii.** $\log m - \log n$

   iv. $\log m + \log n$ **v.** $m \log p + n \log p$ **vi.** $\log 10^3 + \log m^3$

c. Use properties of logarithms to solve $25(1.5^{3x}) = 1,000$.

APPLICATIONS

1 Find these common (base 10) logarithms without using technology.

a. log 100,000   **b.** log 0.001   **c.** $\log 10^{4.75}$

2 Use symbolic reasoning and the definition of logarithms to solve the following equations.

a. $\log x = 2$ **b.** $10^{x-5} = 60$ **c.** $5(10)^{2x} = 60$

3 Use what you know about logarithms to solve these equations.

a. $10^x = 49$ **b.** $10^x = 3{,}000$ **c.** $10^{3x} = 75$

d. $10^{2x+1} = 123$ **e.** $15(10)^{5x+3} = 1{,}200$ **f.** $5(10)^{x+3} + 12 = 47$

4 The income required for a family of four to stay out of poverty in the United States at some time in the future is predicted by the function $I(t) = 22{,}350(10^{0.07t})$, where t is the time in years after 2011. Use logarithms to find the time when the poverty level income will be \$25,000. (**Source:** aspe.hhs.gov/poverty/11poverty.shtml)

5 Radioactive iodine is a dangerous by-product of nuclear explosions, but it decays rather rapidly. Suppose that the function $R(t) = 6(10^{-0.038t})$ gives the amount in a test sample remaining t days after an experiment begins. Use logarithms to find the half-life of the substance.

6 Use the definition of a common logarithm to express each of the following numbers as a power of 10.

a. $0.7 = 10^k$ when $k = \ldots$

b. $7 = 10^k$ when $k = \ldots$

c. $70 = 10^k$ when $k = \ldots$

d. $700 = 10^k$ when $k = \ldots$

7 Use your results from Applications Task 6 and what you know about properties of exponents to show how each of these exponential expressions can be written in equivalent form like $(10^k)^x$ and then 10^{kx}.

a. 0.7^x **b.** 7^x

c. 70^x **d.** 700^x

8 Use your results from Applications Tasks 6 and 7 to solve these exponential equations. Check each solution. Be prepared to explain your solution strategy.

a. $7^x = 3.5$

b. $5(7^n) = 35$

c. $5(7^t) + 20 = 125$

d. $70^p = 48$

e. $3(7^r) + 120 = 600$

f. $300(0.7^v) = 60$

9 A vacuum pump attached to a chamber removes 5% of the gas in the chamber with each pump cycle.

a. What function shows the percent of gas remaining in the chamber after n pump cycles?

b. How many full cycles are needed before at least 99% of the gas is removed?

10 A light filter lets 40% of the light that hits it pass through to the other side.

a. What function shows the fraction of light intensity allowed through by n filters?

b. How many filters are needed to reduce the light intensity to:

  i. 16% of the original intensity?

  ii. less than 5% of the original intensity?

  iii. less than 1% of the original intensity?

11 Use the facts that log 20 ≈ 1.3 and log 16 ≈ 1.2 and the properties of logarithms to find approximate decimal values for each of these logarithms—without the use of technology.

a. log 320

b. log 1.25

c. log 400

12 Use properties of logarithms to write the following expressions in different equivalent forms.

a. $\log 3x$

b. $\log 5x^3$

c. $\log\left(\frac{7x}{5y}\right)$

d. $\log\left(\frac{1}{3x}\right)$

e. $\log 7 + \log x$

f. $3\log y$

g. $\log x - \log 3y$

h. $\log 7x^3y^2$

i. $\log\left(\frac{7+x}{49-x^2}\right)$

13 Use what you know about logarithms to solve these equations without rewriting the exponential expressions in equivalent base 10 form.

a. $2^x = 10,000$

b. $3^{x+3} = 10,000$

c. $4^{2x+3} = 100,000$

d. $5(2)^x + 7 = 42$

e. $5(3)^{x+1} = 2,500$

f. $1.8^{3x} = 75$

CONNECTIONS

14 The time it takes a computer program to run increases as the number of inputs increases. Three different companies wrote three different programs, A, B, and C, to perform the same calculation. The time, in milliseconds, it takes to run each of the programs when given n inputs is given by the three functions below.

$$A(n) = 10{,}000 + 2 \log n$$
$$B(n) = 100 + 4n^2$$
$$C(n) = (0.00003)(10^n)$$

a. Which program, A, B, or C, takes the least amount of time for 1 input?

b. Which program is most efficient for 300 inputs?

c. Which program is most efficient for 100,000 inputs?

15 Suppose a house was purchased five years ago for $200,000. It just sold for $265,000. Assume that this pattern of growth has been exponential at a constant annual percent rate of change.

a. What function will give the price at any time t years from now? (*Hints*: What equation relates the two house values, the five-year time interval of appreciation in value, and the percent rate of increase in value? You can estimate solutions to equations like $R^5 = k$ by studying tables or graphs of $y = x^5$.)

b. What will be the price of the house next year?

c. When will the price of the house be $1 million?

16 By measuring the decay of radioactive carbon-14, scientists can estimate the age of the remains of living things. Carbon-14 decays at a rate of 0.0121% per year (or retains 99.9879% of its radioactivity).

a. Write a rule for the function that gives the amount of a 5-mg lab sample remaining t years after its mass is first measured.

b. Find the half-life of carbon-14.

17 If $y = 5(3^x)$, then a table of sample (x, y) values for this function will look like the one below.

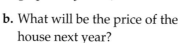

x	−2	−1	0	1	2	3	4	5	6	7
y	$\frac{5}{9}$	$\frac{5}{3}$	5	15	45	135	405	1,215	3,645	10,935

a. Add a row to a copy of this table with values $z = \log y$. For example, if $x = 2$, then $y = 45$ and $z \approx 1.65$.

b. Explain how the table pattern shows that z is a linear function of x.

c. Use properties of logarithms to write $\log 5(3^x)$ in an equivalent form that allows you to write a rule relating z to x in the form $z = mx + b$. Then compare the values produced by this linear function to those in row three of your table. Explain why the similarity occurs.

18 A large number like 2,364,700 is written in scientific notation as 2.3647×10^6 and a small number like 0.000045382 as 4.5382×10^{-5}.

a. Write each of the following numbers in scientific notation with five significant digits (rounding appropriately where necessary to meet this condition).

   i. 47,265

   ii. 584.73

   iii. 97,485,302

   iv. 0.00235148

b. Suppose the only calculator that you had was one that could multiply and divide numbers between 1 and 10 but no others. Explain how you could still use this calculator to find these products.

   i. $584.73 \times 97,485,302$

   ii. $47,265 \times 0.002351$

   iii. $47,265 \div 584.73$

19 Logarithms were invented, in part, to help with multiplication and division of very large and very small numbers in the time long before electronic calculators. When combined with use of scientific notation, they provided a powerful tool for the sort of arithmetic required by calculations in astronomy, chemistry, and physics.

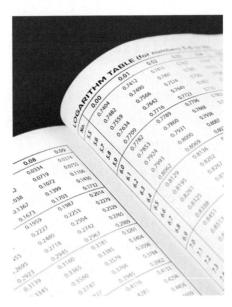

   Mathematicians in the early seventeenth century produced tables of logarithms with great precision. Use what you know about scientific notation, logarithms, and the given logarithm values to justify the following reasoning.

a. Show how the facts that $\log 1.2345 \approx 0.09149$, $\log 6.7890 \approx 0.83181$, and $\log 8.3810 \approx 0.92330$ imply that $12{,}345 \times 67{,}890 \approx 838{,}100{,}000$.

b. Show how the facts given in Part a imply that $1{,}234{,}500 \times 67{,}890{,}000 \approx 83{,}810{,}000{,}000{,}000$.

c. Use a graphing calculator to check the calculations in Parts a and b, and record the calculator outputs. Explain how the forms of those outputs tell the same story as your work with logarithms and scientific notation.

20 The common logarithm function has the very useful special property that for any positive numbers a and b, $\log ab = \log a + \log b$. Mathematicians say that the logarithm function satisfies the general functional equation $f(ab) = f(a) + f(b)$. Which, if any, of the following familiar functions satisfy that functional equation? Give proofs of those that do and counterexamples for those that do not.

    a. $f(x) = 3x$ (To begin, consider: Does $3(ab) = 3a + 3b$ for all pairs (a, b)?)

    b. $f(x) = x + 5$

    c. $f(x) = x^2$

21 Another interesting type of functional equation asks whether $f(ab) = f(a)f(b)$ for all pairs (a, b). Which, if any, of the following functions satisfy that multiplicative functional equation? Give proofs of those that do and counterexamples for those that do not.

    a. $f(x) = 3x$

    b. $f(x) = x + 5$

    c. $f(x) = x^2$

    d. $f(x) = \frac{1}{x}, x \neq 0$

22 Another interesting type of functional equation asks if $f(a + b) = f(a) + f(b)$ for all pairs (a, b). Which, if any, of the functions in Task 21 satisfy that additive functional equation?

REFLECTIONS

23 Suppose that n is a positive number.

    a. If $0 < \log n < 1$, what can you say about n?

    b. If $5 < \log n < 6$, what can you say about n?

    c. If $p < \log n < p + 1$, where p is a positive integer, what can you say about n?

24 Students learning about logarithms often find it helpful to deal with the new idea by frequent reminders that *logarithms are really just exponents*.

    a. Why is it correct to say that $\log 20 \approx 1.3010$ is an exponent?

    b. How does the phrase "logarithms are really just exponents" help in finding:

        i. $\log 1{,}000$

        ii. $\log 0.001$

25 The logarithmic function $f(x) = \log x$ is the inverse of the exponential function $g(x) = 10^x$. How can you use this fact and a graph of $g(x) = 10^x$ to quickly sketch the graph of $f(x) = \log x$ for $0 < x < 10$?

26 For $x > 1$, values of the logarithmic function $f(x) = \log x$ increase very slowly while values of the exponential function $g(x) = 10^x$ increase very rapidly. What is it about the definitions of those two functions that causes this contrasting behavior?

27 Explain why it is the case that $\log 1 = 0$.

28 With the introduction of logarithmic functions, you are now able to solve exponential equations using algebraic reasoning. Solve the equations below using each of the listed strategies. Then compare your solutions and the ease with which each method produces solutions or accurate estimates of solutions.

- Algebraic reasoning

- Approximation using function graphs

- Approximation using tables of function values

a. $100 = 4.5x - 885$

b. $3x^2 + x + 12 = 14$

c. $3(1.2^t) = 14$

29 What are the advantages of finding solutions to exponential equations by estimation using tables and graphs of exponential functions versus using algebraic reasoning with logarithms?

EXTENSIONS

30 Use properties of exponents and logarithms to solve the following equations.

a. $\log 10^x = 4$

b. $2^{2x+2} = 8^{x+2}$

31 Recall that a prime number n is an integer greater than 1 that has only 1 and n as divisors. The first eight primes are 2, 3, 5, 7, 11, 13, 17, and 19. Mathematicians have proven that the number of primes less than or equal to n is approximated by $\dfrac{0.4343n}{\log n}$. This formula is an amazing discovery since the primes appear irregularly among the natural numbers.

a. Count the actual number of primes less than or equal to n to complete a copy of the table below. Plot the resulting (n, *number of primes* $\leq n$) data.

n	10	25	40	55	70	85	100	115	130	145
Number of Primes $\leq n$	4				19	23	25	30	31	34

b. Graph $P(n) = \dfrac{0.4343n}{\log n}$, $0 < n \leq 150$. How well does this function model the counts in Part a?

c. Use the function $P(n)$ to estimate the number of primes less than or equal to 1,000. Less than or equal to 1,000,000. Less than or equal to 10^{18}.

d. According to this $P(n)$, about what percent of the numbers up to 10^6 are prime? Up to 10^{18}?

32 When you go to the movies, the number of frames that are displayed per second affects the "smoothness" of the perceived motion on the screen. If the frames are displayed slowly, our minds perceive the images as separate pictures rather than fluid motion.

However, as the frequency of the images increases, the perceived gap between the images decreases and the motion appears fluid. The frequency f at which we stop seeing a flickering image and start perceiving motion is given by the equation $f = K \log S$, where K is a constant and S is the brightness of the image being projected.

a. S is inversely proportional to the square of the observer's distance from the screen. What would be the effect on f if the distance to the screen is cut in half? What if the distance to the screen is doubled?

b. If the image is being projected at a slow frequency and you perceive a flicker, where should you move in the theater: closer to the screen or farther from the screen?

c. Suppose the show is sold out and you cannot move your seat. What could you do to reduce the flickering of the image on-screen?

33 It is possible to define logarithms for any positive base b as follows.

$$\log_b m = n \text{ if and only if } m = b^n$$

a. Justify each step in the following proof that $\log_b m = \dfrac{\log_{10} m}{\log_{10} b}$.

Suppose that $m = b^n$.

Then $\log_{10} m = n \log_{10} b$.　　　　　(1)

$\log_{10} m = (\log_b m)(\log_{10} b)$　　　　(2)

$\log_b m = \dfrac{\log_{10} m}{\log_{10} b}$　　　　　(3)

b. Use the above **change of base formula** to evaluate each of the following:

　　i. $\log_2 25$　　　　　　　　　　**ii.** $\log_5 40$

c. Sketch the graph of $y = \log_2 x$.

d. How does the graph of $y = \log_2 x$ compare to that of $y = \log x$?

REVIEW

34 Consider the function $f(x) = (x^2 - 4x)(x + 3)$.

a. Determine the value of $f(-2)$.

b. Solve $f(x) = 0$.

c. Rewrite $f(x)$ in standard polynomial form. Identify the degree of $f(x)$.

d. For what values of x is $f(x) < 0$?

e. Estimate the coordinates of all local maximum and minimum points of $f(x)$.

f. Describe the end behavior of $f(x)$.

35 In the diagram at the right, the circle with center O is inscribed in $\triangle ABC$. The radius of the circle is 6, and $m\angle DAE = 52°$. Determine each of the following measures.

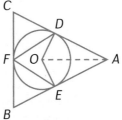

a. $m\angle DOE$ **b.** $m\widehat{DE}$

c. $m\angle DFE$ **d.** AD

36 Write each product or sum of rational expressions in equivalent form as a single algebraic fraction. Then simplify the result as much as possible.

a. $\dfrac{x^3}{x+2} \cdot \dfrac{x+4}{x}$

b. $\dfrac{3x}{x+4} \cdot \dfrac{x^2-16}{x^2}$

c. $\dfrac{6}{x+3} + \dfrac{4x+6}{x+3}$

d. $\dfrac{x}{5} + \dfrac{x+1}{10}$

37 Complete the following for each recursive definition of a sequence.

• Write the first five terms of the sequence.

• Determine if the sequence is arithmetic, geometric, or neither.

• If it is arithmetic or geometric, write a function formula for a_n.

• If it is arithmetic or geometric, find S_{12}.

a. $a_n = a_{n-1} - 3$, $a_0 = 50$

b. $a_n = 2a_{n-1} + 6$, $a_0 = 12$

c. $a_n = \dfrac{a_{n-1}}{2}$, $a_0 = 256$

38 How is the shape of an exponential function graph related to the base of the exponent?

39 Plans for the installation of a communication tower call for attaching a 200 feet long support wire to a point that is 125 feet above the ground. What is the degree measure of the angle formed by the support wire and the ground?

40 Write each of the following radicals in equivalent form with smallest possible numbers and simplest possible expressions under the radical sign.

 a. $\sqrt{121}$ **b.** $\sqrt{48}$

 c. $\sqrt{25x^2}$, $x > 0$ **d.** $\sqrt{5s^2}$, $s > 0$

 e. $\sqrt{4w^3}$, $w > 0$ **f.** $\sqrt{\dfrac{9}{4}x^2}$, $x > 0$

 g. $\sqrt[3]{-64a^4}$ **h.** $\sqrt{18a^4\,b^3}$, $a > 0$, $b > 0$

41 Recall that angles can be measured in both degrees and radians, with the two measurement scales related by the fact that $180° = \pi$ radians or $360° = 2\pi$ radians. Complete a copy of the following table showing degree and radian equivalents for some important angles.

Degrees	0		45	60		120		150	180
Radians		$\dfrac{\pi}{6}$			$\dfrac{\pi}{2}$		$\dfrac{3\pi}{4}$		

42 The sketch below shows how cosine of 30° and sine of 30° (or $\dfrac{\pi}{6}$ radians) determine coordinates of a point on the circle of radius 1 centered at the origin.

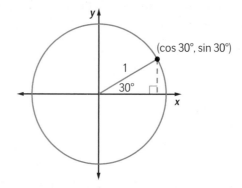

 a. Give exact values of the coordinates.

 b. Make a similar sketch showing the location and numerical coordinates of these points on the unit circle.

  i. $\left(\cos\dfrac{3\pi}{4}, \sin\dfrac{3\pi}{4}\right)$ **ii.** $\left(\cos\dfrac{\pi}{3}, \sin\dfrac{\pi}{3}\right)$

  iii. $\left(\cos\dfrac{3\pi}{2}, \sin\dfrac{3\pi}{2}\right)$ **iv.** $\left(\cos\dfrac{7\pi}{4}, \sin\dfrac{7\pi}{4}\right)$

  v. $\left(\cos\dfrac{7\pi}{6}, \sin\dfrac{7\pi}{6}\right)$

43 Without using technology, sketch a graph of each function over the interval $[-2\pi, 2\pi]$. Then give the period and the amplitude of each function. Use technology to check your work.

 a. $y = \sin x$

 b. $y = 2\sin x$

 c. $y = \sin 2x$

Inverse Trigonometric Functions

In Unit 6, you explored properties of the circular functions sine and cosine and used them to model patterns of change in periodic real-world phenomena. For example, the picture above shows Durdle Door in Dorset, England. Because of ocean tides, the water's depth d in feet under the cliff arch is a periodic function of time t in hours after noon.

- On one particular day, the depth function is

$$d(t) = 11 + 3 \sin 0.5t.$$

To find the time(s) between noon and midnight when the depth of the water is 13 feet, you need to solve

$$11 + 3 \sin 0.5t = 13.$$

- On a later date, the depth function is

$$d(t) = 11 - 3 \cos 0.5t$$

To find the time(s) between noon and midnight when the depth of the water is 10 feet, you need to solve

$$11 - 3 \cos 0.5t = 10.$$

Solving equations like those that ask for times when tidal water has a specified depth is an inverse problem. You know the output of the function, and you need to find the corresponding input.

a What is the minimum depth of the water on the two particular days? The maximum depth?

b How could you estimate solutions for the given equations by exploring patterns in tables and graphs of function values?

c How would you approach solution of each equation by using algebraic reasoning and properties of the sine and cosine functions?

d What role could the **sin⁻¹** or **cos⁻¹** calculator commands have in solving these equations?

In this lesson, you will extend your study of inverse functions to consider inverses of the sine, cosine, and tangent functions. The inverse circular functions will allow you to find complete solutions of trigonometric equations like those involved in predicting water depth under the cliff arch at Durdle Door.

INVESTIGATION 1

The Ups and Downs of the Sine

There are many situations exhibiting periodic change that are modeled well by variations of the sine function. Answering questions about those situations often requires solving equations involving that function.

As you work on the problems of this investigation, look for answers to these questions:

How is the inverse of the sine function defined?

How can the inverse sine function be used to solve trigonometric equations?

Defining the Inverse Sine Function The next diagram shows a partial graph of the sine function, with scale on the x-axis in radians from -2π to 4π or from about -6.3 to 12.6.

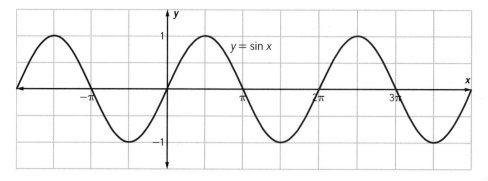

1 Find coordinates of points on the given graph that represent solutions for the following equations. Use what you know about the sine function to find exact values for the coordinates, where possible.

　Scan the given graph, a technology-generated table of $y = \sin x$, or a technology-produced graph of $y = \sin x$ to get good estimates of the coordinates as a check on your trigonometric reasoning.

a. $\sin x = 0.5$

b. $\sin x = -1$

c. $\sin x = 1$

d. $\sin x = -0.5$

2 Suppose that you were able to inspect a complete graph of $y = \sin x$.

a. What are the domain and range of $y = \sin x$?

b. For what values of k will the graph show solutions to the equation $\sin x = k$?

c. How many solutions will there be to the equation $\sin x = k$ for each value of k?

d. What do your answers to Parts a–c say about the possibility of defining an inverse for the sine function?

3 Examine the portion of the graph of $y = \sin x$ where $-\frac{\pi}{2} \leq x \leq \frac{\pi}{2}$.

a. How many solutions does $\sin x = 0.5$ have on the interval $\left[-\frac{\pi}{2}, \frac{\pi}{2}\right]$?

b. How many solutions in the interval $\left[-\frac{\pi}{2}, \frac{\pi}{2}\right]$ does $\sin x = k$ have for other values of $-1 \leq k \leq 1$?

c. Explain why $y = \sin x$ has an inverse, when x is restricted to the interval $\left[-\frac{\pi}{2}, \frac{\pi}{2}\right]$?

d. The width of the interval $\left[-\frac{\pi}{2}, \frac{\pi}{2}\right]$ is π units. Give two other intervals of width π on which $y = \sin x$ has an inverse.

4 Mathematical convention defines the **inverse sine** function as follows.

$$\sin^{-1} k = x \text{ if } \sin x = k \text{ and } -\frac{\pi}{2} \leq x \leq \frac{\pi}{2}$$

The inverse sine function is also called the *arcsine* function and written arcsin x.

a. Why do you think the definition focuses on the interval $-\frac{\pi}{2} \leq x \leq \frac{\pi}{2}$ to define values of the inverse sine function, rather than another interval of length π such as one of those you described in Part d of Problem 3?

b. What are the domain and the range of the inverse sine function?

c. The graph below shows the function $y = \sin x$ on a square window of $-1.57 \le x \le 1.57$ and $-1.57 \le y \le 1.57$. On a copy of this graph, sketch the graph of $y = \sin^{-1} x$. Check your sketch using technology that displays both functions on a square window.

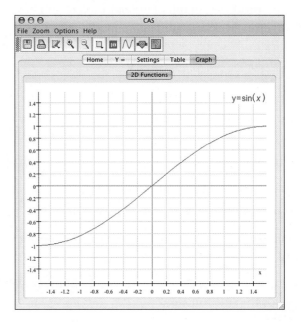

5 Use algebraic reasoning and **sin**$^{-1}$ on a graphing calculator (**2nd** **SIN**) or the inverse sine function of a computer algebra system to find one solution for each of these trigonometric equations.

a. $\sin x = 0.9$

b. $\sin x = -0.8$

c. $5 \sin x + 9 = 12$

d. $11 + 3 \sin 0.5t = 13$

6 The introduction to this lesson proposed two models for predicting the water depth under the cliff arch at Durdle Door in Dorset, England between noon and midnight. One model was

$$d(t) = 11 + 3 \sin 0.5t,$$

where water depth is in feet and time t is in hours after noon. Use that model to analyze the pattern of change in water depth over time. Check your solutions graphically and numerically.

a. What is the depth of the water at 7:00 P.M. on the day that depth tracking begins?

b. What is the maximum depth of the water and at what time will it first occur?

c. At approximately what time(s) between noon and midnight will the water depth be 13 feet?

d. Write an inequality for the time period that the depth is more than 13 feet. Use your solution in Part c to find the solutions of this inequality.

7 Graph $y = \sin^{-1} x$ in a window that shows the full domain and range of the function. Then use the graph to answer these questions.

   a. What are the coordinates of the point corresponding to a solution of $\sin x = 0.5$?

   b. What are the coordinates of the point corresponding to a solution of $\sin x = -0.8$?

   c. What are the x- and y-intercepts of the graph of $y = \sin^{-1} x$?

   d. Where is the function $y = \sin^{-1} x$ increasing? Where is it decreasing?

   e. Where are the values of $y = \sin^{-1} x$ changing most rapidly? Where are the values changing most slowly?

8 Some consequences of the definition of the inverse sine function are important to emphasize.

   a. Consider the expression $\sin (\sin^{-1} k)$.

      i. What are the values of the expression for $-1 \le k \le 1$?

      ii. What can you say about the expression if $k > 1$ or if $k < -1$?

   b. Consider next the expression $\sin^{-1} (\sin x)$.

      i. What are the values of the expression for $-\frac{\pi}{2} \le x \le \frac{\pi}{2}$?

      ii. What can you say about the expression if $x > \frac{\pi}{2}$ or $x < -\frac{\pi}{2}$?

Finding All Solutions of Trigonometric Equations Your work on Problems 1 and 2 showed that trigonometric equations like $\sin x = k$ have infinitely many solutions as long as $-1 \le k \le 1$. The inverse sine function identifies one primary solution for each such equation, but there are certainly times when the other solutions are of interest.

   For example, if the depth of water under the cliff arch at Durdle Door on one day is given by the function $d(t) = 11 + 3 \sin 0.5t$, the maximum depth of 14 feet occurs at about 3 P.M. But high tides occur at different times on the following days. It would be helpful to be able to use $d(t)$ to predict those other times of maximum water depth.

9 The equation $\sin x = 0.5$ has one solution at $x = \frac{\pi}{6}$ (or $x \approx 0.52$). But studying the graph of $y = \sin x$ reveals other solutions as well.

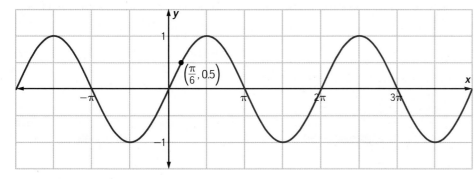

   a. How does the fact that the period of $\sin x$ is 2π make it possible to identify infinitely many solutions of $\sin x = 0.5$ other than $x = \frac{\pi}{6}$? Which of those other solutions are shown on the given graph?

b. The graph of $y = \sin x$ suggests that there are other solutions to the equation $\sin x = 0.5$ that *do not* differ from $x = \frac{\pi}{6}$ by a multiple of 2π.

   i. What is the exact radian value of the solution nearest to $x = \frac{\pi}{6}$?

   ii. What other solutions shown on the graph differ from that in part i by a multiple of 2π?

c. To find solutions to the equation $\sin x = 0.5$, other than $\frac{\pi}{6}$, students at Providence High School reasoned as follows.

> **Since the graph of $y = \sin x$ is symmetric about the line $x = \frac{\pi}{2}$, another solution is $\frac{\pi}{3}$ units to the right of $\frac{\pi}{2}$.**

Why is the second solution $\frac{\pi}{3}$ units to the right of $\frac{\pi}{2}$? What is that solution?

d. Explain why $\frac{\pi}{6} + 2\pi n$, for any integer n, describes one set of solutions. Write a similar expression representing the remaining solutions.

10 Give formulas that describe *all* solutions to the following equations from Problem 5.

a. $\sin x = 0.9$

b. $\sin x = -0.8$

Inverse Sine of Degrees and Radians In situations where trigonometric functions are used to analyze properties of triangles, the angles are often measured in degrees, not radians. The inverse trigonometric functions can be applied to degree measures as well as radians.

For example, $30° = \frac{\pi}{6}$; so, $\sin 30° = \sin \frac{\pi}{6} = 0.5$. It follows that $\sin^{-1} 0.5 = 30°$, or $\frac{\pi}{6}$ radians.

11 In previous units, you learned the values of the sine function for some special angles. Use what you know about the sine of special angles and the inverse sine function to complete the following table. Use radian measures between $-\frac{\pi}{2}$ and $\frac{\pi}{2}$ and degree measures between $-90°$ and $90°$.

u	-1	$-\frac{\sqrt{3}}{2}$		$-\frac{1}{2}$	0	$\frac{1}{2}$			1
$\sin^{-1} u$ (in radians)			$-\frac{\pi}{4}$			$\frac{\pi}{6}$		$\frac{\pi}{3}$	
$\sin^{-1} u$ (in degrees)	-90					30	45		

12 Use algebra and trigonometry to find all solutions of the following equations in both degrees and radians. Give exact solutions, if possible.

a. $2 \sin x = -\sqrt{3}$

b. $2 \sin x - 3 = 1$

c. $3 \sin x + 2 = 4$

d. $6 - 3 \sin x = 2$

SUMMARIZE THE MATHEMATICS

In this investigation, you learned about the inverse sine function and some of its properties. You also used the inverse sine function to solve linear equations involving the sine.

a The sine function defined over the set of real numbers has no inverse function. Why not?

b Explain how the inverse sine function was defined in this lesson in order to overcome the difficulties you noted in Part a.

c Describe the conditions under which $\sin^{-1}(\sin x) = x$. Explain your reasoning.

d Consider an equation of the form $a \sin x + b = c$ which is linear in $\sin x$.

   i. How is solving this equation similar to and different from solving $ax + b = c$ which is linear in x?

   ii. Under what conditions on a, b, and c does $a \sin x + b = c$ have solutions?

   iii. If you know one solution, how many solutions will there be? Describe how to find all solutions.

Be prepared to share your thinking and procedures with the class.

 CHECK YOUR UNDERSTANDING

Use your understanding of the inverse sine function to help complete these tasks.

a. Determine whether the following statements are true or false. Explain how you know without use of a calculator or computer inverse sine function.

   i. $\sin^{-1} 1 = \frac{\pi}{2}$

   ii. $\sin^{-1} 0 = 0$

   iii. $\sin^{-1}(-1) = -\frac{\pi}{2}$

   iv. $\sin^{-1}\left(\frac{1}{\sqrt{2}}\right) = 45°$

   v. $\sin^{-1}\left(\frac{\sqrt{3}}{2}\right) = \frac{2\pi}{3}$

   vi. $\sin^{-1}\left(\frac{1}{2}\right) = 150°$

b. Use algebra and trigonometry to determine whether each equation has a solution for x. If so, find all solutions in radians. Give exact solutions, if possible.

   i. $2 \sin x = \sqrt{2}$

   ii. $2 \sin x - 2 = 1$

   iii. $3 \sin x + 2 = 0$

   iv. $2 - 3 \sin x = 2$

Inverses of the Cosine and Tangent

In Investigation 1, you developed the inverse sine function and used it to find solutions of equations involving the sine. Inverses of the cosine and tangent functions can be used in the same way to answer questions about phenomena that have been modeled with those functions.

As you work on the problems of this investigation, look for answers to these questions:

> *How are the inverses of the cosine and tangent functions defined?*
>
> *How can the inverse cosine and inverse tangent functions be used to solve equations involving the cosine or tangent?*

Defining and Using the Inverse Cosine Function Suppose that an amusement park is planning a new roller coaster with part of its track shaped like the graph of $y = 12 \cos 0.1x + 6$, where x and y are in meters and $0 \le x \le 100$.

For positive values of y, the roller coaster track is above ground. For negative values of y, it is below ground in a tunnel. In order to find the horizontal distance to the point where the roller coaster enters the tunnel, you need to solve $12 \cos 0.1x + 6 = 0$ or its equivalent $\cos 0.1x = -0.5$. An inverse cosine function would help to find the required value of x.

1 Examine the following graph of $y = \cos x$.

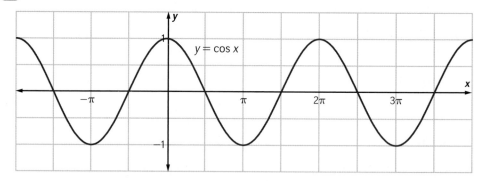

a. Why does the cosine function not have an inverse, when defined on its entire domain?

b. Would the cosine function have an inverse if it was defined on these restricted domains?

   i. $-\dfrac{\pi}{2} \le x \le \dfrac{\pi}{2}$ **ii.** $0 \le x \le \pi$

   iii. $-\pi \le x \le 0$ **iv.** $\dfrac{\pi}{2} \le x \le \dfrac{3\pi}{2}$

c. Of the feasible restricted domains for $y = \cos x$, which seems most useful for defining $\cos^{-1} x$?

The standard mathematical definition of the **inverse cosine** function is

$$\cos^{-1} k = x \text{ if } \cos x = k \text{ and } 0 \le x \le \pi.$$

The inverse cosine function is also called the *arccosine* function and written arccos x.

2 Analyze the properties of the inverse cosine function.

  a. What are the domain and range of $y = \cos^{-1} x$?

  b. What window will show a complete graph of $y = \cos^{-1} x$?

  c. Sketch a graph of $y = \cos^{-1} x$. Label the x- and y-intercepts.

  d. Label the point on the graph of $y = \cos^{-1} x$ that corresponds to a solution of $\cos x = 0.5$.

  e. Label the point on the graph of $y = \cos^{-1} x$ that shows a solution of $\cos x = -0.3$.

3 Use the inverse cosine function, a graph of cos x, and reasoning like what you applied in studying the sine and inverse sine functions to find *all solutions* for these equations.

  a. $\cos x = \dfrac{1}{\sqrt{2}}$

  b. $6 \cos x = 3$

  c. $2 \cos x + \sqrt{3} = 0$

  d. $5 \cos 0.3x = 4$

4 The roller coaster described at the start of this investigation had a track matching the graph of $y = 12 \cos 0.1x + 6$, where x and y are in meters and $0 \le x \le 100$. For positive values of y, the roller coaster track is above ground. For negative values of y, it is below ground in a tunnel.

  a. Graph the function that represents this portion of the track.

  b. At what point does the track reach its maximum height?

  c. Write and solve an equation that gives the horizontal distance traveled when the roller coaster first enters the tunnel.

  d. What are the lowest point(s) of the track? At what value(s) of x does it occur?

  e. Write and solve an inequality to find where the roller coaster is in the tunnel.

5 At the start of this lesson, the function $d(t) = 11 - 3 \cos 0.5t$ was proposed as a model of change in tidal water depth at Durdle Door.

a. If $t = 0$ indicates noon on one day, what is the water depth at that time?

b. Use the inverse cosine to find the first time after noon when the water depth is 10 feet.

c. Use a graph of $d(t) = 11 - 3 \cos 0.5t$ and the inverse cosine function to find *all times* after noon when the water depth is 10 feet.

Defining and Using the Inverse Tangent Function If you have ever traveled on a highway with steep hills, you have probably seen signs like the one on the right. It is quite likely that the 9% grade figure was not found by sighting from top to bottom of the hill with a protractor. Instead, road engineers probably used other instruments to measure the horizontal and vertical distances involved.

If the 9% figure represents the slope of the road, it is also the tangent of $\angle A$ in the following diagram.

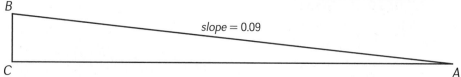

slope = 0.09

To find the measure of that angle, you need to solve the equation $\tan x = 0.09$. An inverse tangent function would provide the solution.

6 Examine the following graph of $y = \tan x$.

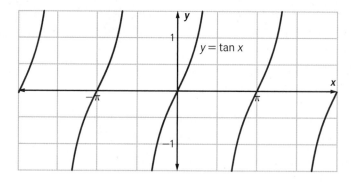

The graph of $y = \tan x$ is different from the sine and cosine functions in several ways.

a. What are the domain and range of $y = \tan x$?

b. Does $y = \tan x$ have any asymptotes? Explain.

c. What is the period of $y = \tan x$?

d. Why does the tangent function not have an inverse, when defined on its entire domain?

e. Would the tangent function have an inverse if it was defined on these restricted domains? Why or why not?

    i. $-\frac{\pi}{2} < x < \frac{\pi}{2}$

    ii. $-\pi < x < \pi$

    iii. $-\frac{3\pi}{2} < x < 0$

    iv. $\frac{\pi}{2} < x < \frac{3\pi}{2}$

f. Of the feasible restricted domains in Part d for $y = \tan x$, which seems most useful for defining $\tan^{-1} x$?

The standard mathematical definition of the **inverse tangent** function is

$$\tan^{-1} k = x \text{ if } \tan x = k \text{ and } -\frac{\pi}{2} < x < \frac{\pi}{2}.$$

The inverse tangent function is also called the *arctangent* function and written arctan x.

7 Analyze properties of the inverse tangent function.

a. What are the domain and range of $y = \tan^{-1} x$.

b. Use a calculator or computer function graphing tool in radian mode to help make a sketch of $y = \tan^{-1} x$ in a window like $-5 \le x \le 5$ and $-2 \le y \le 2$. Then add the following features to your sketch.

    i. Sketch the asymptotes.

    ii. Label the point on the graph that represents a solution of $\tan x = 1$?

    iii. Label the point on the graph that represents a solution of $\tan x = 3.5$.

    iv. Label the x- and y-intercepts.

8 With slight adjustments, the strategies that you used to solve equations involving the sine and cosine will work to solve equations involving the tangent. Find all solutions of the following equations in both radians and degrees. Give exact solutions, if possible.
After you find one solution, remember the pattern in the graph of $y = \tan x$ and its period to find all others.

a. $\tan x = 1$

b. $\sqrt{3} \tan x = 1$

c. $-2 \tan x - 3 = 5$

d. $\sqrt{3} \tan x + 4 = -1$

9 To find the angle of depression of a road with 9% grade, you need to solve the equation $\tan x = 0.09$. Find the solution to this equation that makes sense in the problem situation. Give your answer in degrees and in radians.

10 **Properties of $\sin^{-1}$, $\cos^{-1}$, and $\tan^{-1}$** In Investigation 1, you derived the following properties of the inverse sine function.

- $\sin(\sin^{-1} k) = k$ provided $-1 \le k \le 1$.
- $\sin^{-1}(\sin x) = x$ provided $-\frac{\pi}{2} \le x \le \frac{\pi}{2}$.
- If $-1 \le k \le 1$, one solution of $\sin x = k$ is $x = \sin^{-1} k$, and this solution is between $-\frac{\pi}{2}$ and $\frac{\pi}{2}$ (or between $-90°$ and $90°$), inclusive.

a. Replace "sin" with "cos" in the above statements. Make appropriate revisions to state similar properties of the inverse cosine function.

b. Replace "sin" with "tan" in the statements above. Make appropriate revisions to state similar properties of the inverse tangent function.

SUMMARIZE THE MATHEMATICS

In this investigation, you learned about the inverse cosine and inverse tangent functions, and you used those functions to help solve linear equations involving the cosine or tangent.

a What are the domain and range of $y = \cos^{-1} x$?

b Sketch the graph of $y = \cos^{-1} x$.

c Describe a strategy that uses the inverse cosine function to solve equations in the form $a \cos x + b = c$.

d What are the domain and range of $y = \tan^{-1} x$?

e Sketch the graph of $y = \tan^{-1} x$.

f Describe a strategy that uses the inverse tangent function to solve equations in the form $a \tan x + b = c$.

Be prepared to share your thinking and procedures with the class.

CHECK YOUR UNDERSTANDING

Use your understanding of the inverse cosine and tangent functions to help complete these tasks.

a. Find all solutions of each equation.

   i. $3 - 4 \cos x = 6$

   ii. $\tan x - 3 = 12$

b. The daily average temperature in a southern hemisphere city like Johannesburg, South Africa, varies throughout the year—warmest from November through February and coolest from May through September.

Suppose that the function $T(m) = 60 + 10 \cos 0.5m$ is a good model for estimating the daily temperature in degrees Fahrenheit at a time m months into the year. Write and solve equations that answer these questions.

   i. At what time(s) of the year is the daily average temperature 65°F?

   ii. At what time(s) of the year is the daily average temperature 53°F?

APPLICATIONS

1 Use the definitions of the sine and inverse sine functions to write the following equations in different equivalent forms.

a. $\sin\left(\frac{\pi}{3}\right) = \frac{\sqrt{3}}{2}$ **b.** $\sin^{-1}\left(\frac{\sqrt{2}}{2}\right) = \frac{\pi}{4}$ **c.** $\sin\left(-\frac{\pi}{6}\right) = -\frac{1}{2}$

2 Use information in the diagram to find the requested values. Be prepared to explain how you know that your answers are correct, without the use of technology.

a. $\sin^{-1} 0.259 = \underline{\quad}$ **b.** $\sin^{-1} 0.966 = \underline{\quad}$

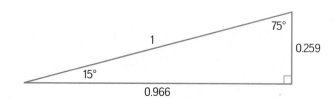

3 Use algebraic reasoning and the inverse sine function to find a solution for each of the following equations. Give exact solutions in both radians and degrees, if possible.

a. $\sin x = \frac{1}{2}$ **b.** $\sin 3x = \frac{1}{2}$ **c.** $8 \sin 0.4x - 1 = 3$

4 Use algebraic reasoning and the inverse sine function to find all solutions in radians for the following equations. Give exact solutions, if possible.

a. $2 \sin x + \sqrt{3} = 0$ **b.** $4 \sin x + 3 = 1$ **c.** $3 - \sin x = 2 \sin x$

5 Sarasota, Florida, like many cities in tropical climates, has a seasonal change in population each year. Suppose that the number of people living in Sarasota at any time of the year can be approximated by the function

$$p(t) = 50 + 25 \sin 0.5t,$$

where $p(t)$ is in thousands of people, t is in months after November 1, and $0 \le t \le 12$.

a. Graph the population function for a one-year period in an appropriate window. Start your graph using $t = 0$ to stand for November 1.

b. What is the maximum predicted number of people living in Sarasota? When does that maximum occur?

c. What is the minimum predicted number of people living in Sarasota? When does that minimum occur?

d. On what date(s) in the year is the number of people living in Sarasota about 60,000? Show how to find the answer to this question using algebraic reasoning and the inverse sine function.

6 Determine which of the following mathematical statements are true. Explain how you know without the use of technology.

a. $\cos\left(\frac{\pi}{3}\right) = \frac{1}{2}$

b. $\cos^{-1}\left(\frac{\sqrt{2}}{2}\right) = \frac{\pi}{4}$

c. $\cos^{-1}\left(\frac{\sqrt{3}}{2}\right) = \frac{\pi}{6}$

7 Use information in the diagram below to find requested values of the inverse cosine function. Be prepared to explain how you know that your answers are correct, without the use of technology.

a. $\cos^{-1} 0.259 =$ _____

b. $\cos^{-1} 0.966 =$ _____

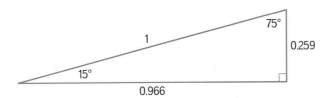

8 Use algebraic reasoning and the inverse cosine function to find a solution for each of the following equations. Give exact solutions in both radians and degrees, if possible.

a. $\cos x = \frac{1}{2}$

b. $\cos 5x = \frac{1}{2}$

c. $8 \cos 0.4x - 1 = 3$

9 Use algebraic reasoning and the inverse cosine function to find all solutions in radians for the following equations. Give exact solutions, if possible.

a. $2 \cos x - \sqrt{3} = 0$

b. $4 \cos x - 3 = 1$

c. $2 - \cos x = 3 \cos x$

10 The meandering path of a river often can be approximated by a sine or cosine curve. Suppose that two towns are located so that one is six miles directly east of the other and the path of the river between those towns is approximated by the graph of $y = 5 \cos x$ for $-2 \le x \le 4$.

The x-coordinate represents location from west to east in miles and y represents location north or south of the line connecting the two towns in miles.

a. Graph the path of the river over the six-mile interval between the two towns.

b. Answer parts i and ii relative to the origin of this graph.

i. What is the northernmost point of the river between the two towns?

ii. What is the river's southernmost point between the two towns?

c. Suppose a west-east road lies on the line $y = 1.5$ and a north-south road on the line $x = 0$ on the graph.

   i. How many bridges are needed for these straight roads?

   ii. At what points on the map coordinate system are the bridges located?

d. If it was decided that the two towns build direct roads connecting the bridge on the north-south road to the bridges on the west-east road, how long will each of those roads be?

11 Determine which of the following mathematical statements are true. Explain how you know, without the use of technology.

   a. $\tan 45° = \frac{1}{2}$

   b. $\tan^{-1} 1 = 45°$

   c. $\tan^{-1} \sqrt{3} = \frac{\pi}{3}$

12 Use information in the diagram below to find requested values of the inverse tangent function. Explain how you know that your answers are correct, without using the **tan** or **tan**$^{-1}$ function of technology.

   a. $\tan^{-1} (\underline{\quad}) = 75°$

   b. $\tan^{-1} (\underline{\quad}) = 15°$

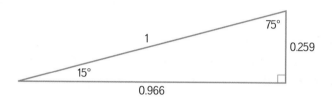

75°

1

0.259

15°

0.966

13 Use algebraic reasoning and the inverse tangent function to find all solutions for the following equations. Give exact solutions, if possible.

   a. $\tan x = 1$

   b. $5 \tan x = 1$

   c. $5 \tan x - 1 = 19$

14 The center of the Ferris wheel at a county fair is 7 meters above the ground. The Ferris wheel itself is 12 meters in diameter. The angular velocity of the wheel is 18° per second.

   a. Write an equation involving the sine function that gives the distance y in meters above the ground of a seat on the wheel t seconds from the time that it is at the 3:00 position on the wheel. (*Hint:* The seat will return to its starting point in 20 seconds.)

   b. Using the equation that you wrote in Part a, write an equation in t whose solutions are the times at which the seat is 10 meters above the ground. Find all solutions between 0 and 30 seconds.

Acclaim Images

CONNECTIONS

15 One tool for determining missing parts of a triangle $\triangle ABC$ is the Law of Sines:

$$\frac{a}{\sin A} = \frac{b}{\sin B} = \frac{c}{\sin C}.$$

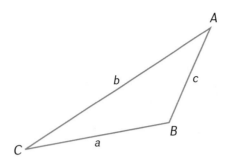

a. Write and solve a trigonometric equation whose solution is m∠A if $a = 2.4$ feet, $b = 3.7$ feet, and m∠$B = 112°$. Is there just one solution? Explain.

b. Solve the general equation, $\frac{a}{\sin A} = \frac{b}{\sin B}$, for m∠$A$.

16 Compare the use of inverse functions in solving exponential and trigonometric equations by solving each pair of equations using radians for the sine functions. Note similarities and differences in strategies used.

a. $20(10^x) = 5$ and $20 \sin x = 5$

b. $20(10^{3x}) = 5$ and $20 \sin 3x = 5$

c. $20(10^{3x}) - 3 = 5$ and $20 \sin 3x - 3 = 5$

17 An ice cream store in Indiana is a seasonal business. Stores that remain open all year have a fluctuation in daily income (in thousands of dollars) similar to that shown in the following table.

Date	Daily Income	Date	Daily Income
January 1	0.73	July 1	5.89
February 1	0.66	August 1	6.35
March 1	0.85	September 1	5.77
April 1	1.35	October 1	4.52
May 1	2.78	November 1	2.83
June 1	4.67	December 1	1.44

a. On which of the tabled dates does the store make maximum income? Minimum income?

b. Plot the income data y against x, time in months after January 1. Use January 1 as $x = 0$.

c. Given the shape of the plot in Part b, what kind of function might provide a good fit for these data?

d. Use your calculator or computer software to fit these data to a sine function. On some calculators, there is a statistics function called **sinreg**. What function rule best fits the data?

e. Calculate the average daily income of the 12 values in the table. What average daily income is predicted by your equation in Part d?

18 The Law of Cosines can be used to find unknown angle or side measurements in triangles. In $\triangle ABC$,

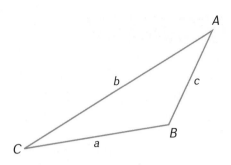

$$c^2 = a^2 + b^2 - 2ab \cos C.$$

a. Find the measure of $\angle C$ in case $a = 5$, $b = 7$, and $c = 3.25$.

b. Find the measure of $\angle A$ in case $a = 9.6$, $b = 3$, and $c = 11$.

19 The daily attendance of California's scenic Yosemite National Park varies throughout the year with a maximum of about 20,000 visitors on August 1 and a minimum of about 2,000 visitors on February 1.

a. Show that if the function

$$A(m) = 9 \cos 0.5m + 11$$

is used to predict daily attendance in thousands at a time m months after August 1, the function has maximum and minimum values that agree with the given data.

b. About what daily attendance does your model predict for January 1?

c. On what other day is the predicted daily attendance the same as that of January 1?

d. Members of the Stanford University Hiking Club dislike crowds, but they love to hike in Yosemite. They have decided to avoid the park on days when the predicted attendance is 16,000 or greater. On what days should they avoid Yosemite?

20 Use of the Law of Sines to find side lengths of triangles generally requires a computation like $a = (\sin A)\left(\dfrac{b}{\sin B}\right)$. Before electronic calculators were invented, that calculation was tedious. It involved multiplying and dividing by decimals accurate to four or more places. To avoid such messy computation, people used logarithms to change the operations to sums or differences that are easier to compute by hand. The logs of trigonometric functions could be found in published tables.

a. Write an equivalent expression for $\log\left((\sin A)\left(\dfrac{b}{\sin B}\right)\right)$ that requires only addition and subtraction of logarithms. Explain why the result is equal to $\log a$.

b. Suppose $b = 4.5$, $m\angle A = 42°$, and $m\angle B = 103°$.

i. Use the expression derived in Part a to find $\log a$.

ii. Use the value of $\log a$ to determine a.

21 The graph of each equation in the form $y = kx, x \geq 0$ meets the positive x-axis at the origin to form an angle with measure between $-90°$ and $90°$. The graph of the equation $y = -2x, x \geq 0$ meets the positive x-axis at an angle of approximately $-63°$.

a. Complete a copy of the following table to explore the relationship between angle and slope.

Slope	-4	-2	-1.5	-1	-0.5	0	0.5	1	1.5	2	4
Angle		-63									

b. Plot the (*slope, angle*) data.

c. Identify a function type that seems likely to model well the dependence of *angle measurement* on *slope*. Use a curve-fitting tool to find the best-fit model of that type.

d. Identify the intercepts, local maximum and minimum points, and asymptotes of the function model from Part c.

22 Use the coordinate definition of the trigonometric functions and properties of their inverses to determine the values of these expressions.

a. $\sin\left(\tan^{-1}\left(\frac{4}{3}\right)\right)$

b. $\tan\left(\arcsin\left(-\frac{3}{4}\right)\right)$

c. $\sin\left(\cos^{-1}\left(-\frac{1}{5}\right)\right)$

d. $\sin^{-1}\left(\sin\left(-300°\right)\right)$

e. $\arccos\left(\sin\left(-\frac{\pi}{4}\right)\right)$

f. $\tan\left(\cos^{-1} x\right), 0 < x < 1$

23 Use your calculator to complete a copy of the following table. Describe interesting patterns you see in the table. Explain your results.

	\multicolumn{7}{c|}{Radian Measure (x)}						
	1	2	3	4	5	6	7
$\sin^{-1}(\sin x)$	1						
$\cos^{-1}(\cos x)$				2.2832			
$\tan^{-1}(\tan x)$							0.7168

REFLECTIONS

24 Sketch the graph $y = \sin x$ for $-\frac{\pi}{2} \leq x \leq \frac{\pi}{2}$ and the line $y = x$ on the same coordinate system. Sketch the image of the graph of $y = \sin x$ reflected across the line $y = x$. What is the equation for this reflection image? Explain.

25 To remember the meaning of the inverse sine function, many students say to themselves:

The notation $\sin^{-1} x$ *means to find the angle whose sine is x. For example,* $\sin^{-1} 0.5 = 30°$, *or* $\frac{\pi}{6}$, *because in a* $30°$-$60°$ *right triangle, the side opposite the* $30°$ *angle is one-half the hypotenuse.*

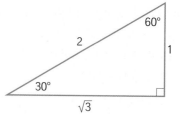

a. Do you agree that this is a helpful and correct way to think about the inverse sine function?

b. What does this way of thinking and the diagram above suggest is the value of $\sin^{-1} \frac{\sqrt{3}}{2}$?

c. What happens when you try to apply this way of thinking to find $\sin^{-1}(-0.5)$?

d. What somewhat similar way of thinking about inverse functions can be used when you are asked to find log 1,000 or log 0.01?

26 Explain the meaning of each of the following expressions. Evaluate them for $t = 0.5$.

a. $\cos^{-1} t$

b. $\cos(t^{-1})$

c. $(\cos t)^{-1}$

d. $\arccos t$

27 Each of the following equations is true for some values of t and false for other values of t.

• For each equation, give one value of t for which the equation is true and one value for which it is false.

• Describe the set of all t for which each equation is true.

a. $\sin^{-1}(\sin t) = t$ (radians)

b. $\cos^{-1}(\cos t) = t$ (degrees)

c. $\tan^{-1}(\tan t) = t$ (radians)

EXTENSIONS

28 Simple sound waves can be represented by functions of the form $y = a \sin bx$. Certain sound waves, when played simultaneously, will produce a particularly pleasant sensation to the human ear. Such sound waves form the basis of music. For example, two sound waves are said to be separated by an *octave* if the ratio of their frequencies is 2:1.

a. The frequency of a middle C note is about 264 Hz, or 264 cycles per second. What is the period of this sound wave in fractions of a second?

b. If the note has amplitude 70, how can its fluctuating air pressure y be written as a function of time x in seconds?

c. Suppose that second note with equal amplitude and higher frequency is separated by an octave from middle C. Write a function for the sound wave associated with this note.

d. To begin to make music, these two notes can be played together. Write and graph a third function that is the sum of these two for $0 \leq x \leq 0.01$.

e. Suppose another note is separated from middle C by a third. That is, the frequencies of the two notes are in the ratio of 5:4. What are the two possible frequencies of this note?

29 Verify that each equation is true for all x where $0 \leq x \leq 1$. You will be showing that each equation is an *identity*, that is, a statement that is true for all replacements of the variable for which the statement is defined.

(*Hint*: In Part a, the rotational symmetry of the sine function about the origin means that $\sin(-x) = -\sin x$ for all x.)

a. $\sin^{-1}(-x) = -\sin^{-1} x$

(*Hint*: In Parts b–d, make use of the fact that a point on the terminal side of $\sin^{-1} x$ at distance 1 from the origin has coordinates $\left(\sqrt{1-x^2}, x\right)$ and a point on the terminal side of $\cos^{-1} x$ at distance 1 from the origin has coordinates $\left(x, \sqrt{1-x^2}\right)$.)

b. $\sin^{-1} x + \cos^{-1} x = \dfrac{\pi}{2}$

c. $\cos(\sin^{-1} x) = \sqrt{1-x^2}$

d. $\tan(\cos^{-1} x) = \dfrac{\sqrt{1-x^2}}{x}$

30 More complicated trigonometric equations and other equations containing both trigonometric functions and other functions can be solved using symbolic reasoning. Use symbolic reasoning to solve the following equations in radians.

a. $\cos^2 x + 2\cos x + 1 = 0$

b. $2\sin x \cos x + \sin x = 0$

c. $\tan^2 3x = 4\tan 3x$

d. $\sin^2 2x = 5\sin 2x - 6$

e. $\log(\cos x) + 2.5 = 1.9$

f. $5(10^{\tan x}) + 15 = 64$

31 At one time in the history of scientific computer programming, some languages, like BASIC and FORTRAN, had only one inverse trigonometric function: the inverse tangent. The inverse sine and inverse cosine were both computed in these languages using expressions that involved the inverse tangent. Complete the following tasks to see how this can be done.

a. Let $y = \sin^{-1} x$. The goal is to write y in terms of the inverse tangent of a function of x. Start by writing x in terms of y.

b. Consider the angle with measure y in standard position. Use your equation for x in Part a to write an expression for $\tan y$ in terms of x.

c. Transform this equation to one in which y is written in terms of the inverse tangent of a function of x.

d. Show that your equation gives correct values of $\sin^{-1} 0.5$ and $\sin^{-1} (-0.8)$.

e. Repeat the steps in Parts a–d to show how to write $\cos^{-1} x$ in terms of the inverse tangent of a function of x. You will need to take care that the equation you arrive at gives the correct value of $\cos^{-1} x$ if $x < 0$.

32 The graph of $x = \sin y$ is given at the right. This graph is clearly not the graph of a function of x. For that reason, your calculator in function mode will not draw this graph. Rather, you will need to set your calculator to parametric mode. With parametric equations, a third variable t is introduced, and the x- and y-coordinates are written as separate functions of t.

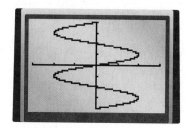

a. With your calculator set to parametric mode, press ⬚ Y= . You will then enter a function of t for X_T (or X_{1T}). Enter **sin(T)**. If the resulting pair of equations, X_T and Y_T, is to be equivalent to $x = \sin y$, what should you enter for Y_T?

b. Experiment with window settings to get a graph that looks like this one. A good choice for Tstep is 0.1. What settings for x, y, and t worked for you?

c. Without changing the window settings for x and y, how could you change the settings for t so that the resulting graph is that of $y = \sin^{-1} x$? Explain.

d. Repeat Parts a–c with the graph of $x = \cos y$.

33 Earth travels around the Sun in an elliptical orbit. The distance d in kilometers between Earth and the Sun at time t in days since January 1 is approximated by the following formula.

$$d = \frac{1.5 \times 10^8}{1 - 0.0166 \cos\left(\frac{2\pi(t - 185)}{365}\right)}$$

a. What is the minimum distance from Earth to the Sun? On what date does the minimum distance occur?

b. What is the maximum distance from Earth to the Sun? On what date does the maximum distance occur?

c. Write an equation whose solutions are the two days during the year at which Earth is 1.48×10^8 km from the Sun. Use symbol manipulation methods to solve the equation. Check your solutions by substitution.

34 The center of a water wheel with radius 4 meters is 2.5 meters above the surface of the water. A point P is located at $(4, 0)$ on the circumference of the wheel when the wheel begins to rotate counterclockwise through an angle θ in radians.

  a. Explain why point P is below the water's surface between consecutive solutions of $4 \sin \theta + 2.5 = 0$.

  b. Find the interval of θ between 0 and 2π for which point P is below the water's surface by solving this equation.

35 Use symbolic reasoning and properties of the inverse trigonometric functions (in radian mode) to solve the following equations for x.

  a. $2 \sin^{-1} x = 1$

  b. $\cos^{-1} 2x = 0.3$

  c. $3 \tan^{-1} \pi x = -3$

REVIEW

36 Solve each equation or inequality. Graph the solution to each inequality on a number line.

  a. $4 + 2x(x - 5) = (x + 3)(2x - 8)$ **b.** $2x^2 + 6x = 8$

  c. $x + 3 = \dfrac{1}{x}$ **d.** $(x^2 + 4x)(x - 3) = 0$

  e. $10x - 21 \leq x^2$ **f.** $12 - 8x \geq 4(x + 9)$

  g. $|x + 7| = 10$ **h.** $|x - 3| > 5$

37 In the diagram below, $AC = AB$, $AD = AE$, and $DC = EF$.

  a. Prove that $\triangle EBF$ is isosceles.

  b. Prove that quadrilateral $DEFC$ is a parallelogram.

  c. Is $\triangle EBF \sim \triangle ABC$? Provide reasoning to support your answer.

  d. If $m\angle C = 80°$, find the following angle measures.

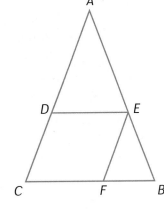

  i. $m\angle B$ **ii.** $m\angle ADE$

  iii. $m\angle A$ **iv.** $m\angle EFC$

38 Antonio has an account balance of $1,500 on his credit card. His credit card company charges a monthly interest rate of 1.5%. He can afford to pay $50 a month to pay off his debt and has made a pledge to himself not to purchase anything more with his credit card.

    a. How much will his credit card balance be after 1 month? After 2 months?

    b. Write a recursive formula for the account balance from one month to the next.

    c. How long will it take him to reach a zero balance?

    d. How much interest will he have paid when his account reaches a zero balance?

39 Use relationships between angle measures and arc lengths to determine each indicated value.

    a. In the circle with center O, $m\angle BOA = 5x + 16°$ and $m\angle BCA = 3x - 2°$.

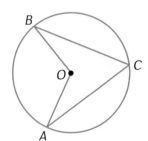

       i. $m\angle BCA$

       ii. $m\overparen{BCA}$

    b. In the circle at the right, $AB = BC = CD$ and $m\angle C = 125°$.

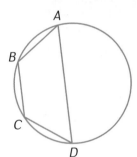

       i. $m\overparen{BAD}$

       ii. $m\overparen{BCD}$

       iii. $m\overparen{BC}$

       iv. $m\angle BAD$

    c. In the diagram at the right, $\overrightarrow{AB}$ and $\overrightarrow{AC}$ are tangent to the circle at points B and C, $m\angle A = 68°$, and $AC = 8$ in.

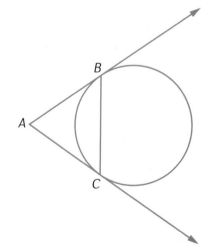

       i. AB

       ii. BC

       iii. Radius of the circle

40 For each rational function below, use algebraic reasoning to:

- find coordinates of all x-intercepts and the y-intercept of the graph.

- describe the domain of the function.

- find equations of all vertical and horizontal asymptotes.

- sketch a graph on which you label intercepts and asymptotes.

a. $f(x) = \dfrac{5}{x - 2}$

b. $g(x) = \dfrac{3x - 4}{2x + 1}$

41 A national survey of 13–17-year-olds indicated that 43% of teens say they have experienced some form of cyberbullying. Suppose you randomly select 120 teens to survey about cyberbullying. (**Source:** www.ncpc.org/resources/files/pdf/bullying/cyberbullying.pdf)

a. How many of them would you expect to say they have experienced some form of cyberbullying?

b. Is the binomial distribution of the number of teens who say they have experienced cyberbullying approximately normal?

c. What is the standard deviation of this binomial distribution?

d. Would you be surprised to find that only 42 teens in your sample reported having been cyberbullied? Explain.

42 Consider the graphs of $f(x)$ and $g(x)$ shown below. The two graphs intersect at the points $(-2, 0)$ and $(5, 7)$.

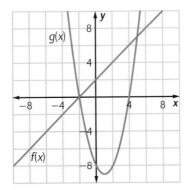

a. For what values of x is $f(x) - g(x) < 0$?

b. Find a rule for $f(x)$.

c. If $g(x)$ is a quadratic function with vertex $(1, -9)$, write the vertex form and standard polynomial form for $g(x)$.

d. What is the degree of each of the following functions? Explain your reasoning.

- $f(x) \cdot g(x)$

- $f(x) + g(x)$

43 Without using technology, plot 4 points on the graph of each function below. Then sketch the graphs of each function and label its intercept(s).

a. $f(x) = \log x$

b. $g(x) = \log_2 x$

Looking Back

The lessons of this unit extended your knowledge and skill in work with algebraic and circular functions and expressions in three ways. First, you learned what it means for a function to have an inverse, how to find inverse functions when they exist, and how to use inverse functions to solve a variety of problems. Second, you learned the definition and key properties of an important inverse function that gives base-10 logarithms, and you developed strategies for using that function to solve exponential equations. Third, you extended your understanding of the basic sine, cosine, and tangent functions. Then you learned the definitions, properties, and applications of three inverse circular functions.

The next several tasks give you an opportunity to review your knowledge of inverse functions and apply that knowledge to some new contexts.

1 In which of these situations will the indicated function have an inverse? For those functions that do have inverses, explain what information the inverse would provide.

a. In most states, there is a rule that shows how to calculate sales tax as a function of the price of any purchase.

b. The price for a package of cheese sold in a market is a function of the weight of the package.

c. The time that it takes the sound of thunder to reach the ears of a person is a function of the distance from the lightning strike producing the thunder to the listener's ears.

d. The depth of water at a particular location in an ocean harbor varies over time.

2 Sketch graphs of these functions. Determine which have inverses. Then find rules and sketch graphs for those that do.

a. $f(x) = 0.5x$

b. $g(x) = 1.5x + 4$

c. $r(t) = \frac{500}{t}$

d. $s(x) = \log x$

3 Use properties of logarithms to write each expression in an equivalent form.

a. $4 \log x$

b. $\log \left(\frac{2x}{5y} \right)$

c. $\log \left(\frac{1}{a} \right)$

d. $\log y - \log 4x$

e. $\log 6x^2y$

4 Light, radio and television signals, x-rays, and the waves produced by alternating electric current are forms of electromagnetic radiation. The next table gives frequencies for a sample of such electromagnetic waves.

a. Write each frequency in scientific notation.

b. Suppose that you had a table that gave logarithms only for numbers between 0 and 10. Show how you could use those table values, the results of your work in Part a, and properties of logarithms to estimate the logarithms of the given frequencies.

Type of Electromagnetic Wave	Frequency in Cycles per Second
Alternating electrical current	60
AM radio	from 540,000 to 1,600,000
Television and FM radio	from 54,000,000 to 216,000,000
Light	from infrared 390,000,000,000,000 to ultraviolet 770,000,000,000,000
x-rays	from 30,000,000,000,000,000 to 100,000,000,000,000,000,000

5 Use what you know about logarithms to solve these equations without rewriting the exponential expressions in equivalent base-10 form. Give both exact expressions and approximate solutions.

a. $3^x = 100,000$

b. $\log 10^8 = 4x$

c. $5(2)^{x+1} = 3,000$

6 If you buy a new car for $25,000, its resale value decreases by about 20% each year. Write and solve equations representing these questions.

a. When will the value of the car be only $10,000?

b. How long will it take for the value of the car to be reduced to only $5,000?

c. When is the car worth half the purchase price?

7 Determine which of the following mathematical statements are true. Explain how you know without the use of technology. If the statement is false, provide a correction.

a. $\tan^{-1} \sqrt{3} = 60°$

b. $\tan^{-1}(-1) = 45°$

c. $\sin^{-1} 0 = \pi$

d. $\cos^{-1} \left(\frac{\sqrt{2}}{2} \right) = \frac{\pi}{4}$

8 Use algebraic reasoning and the inverse trigonometric functions to find all solutions in radians for the following equations. Give exact solutions, if possible.

a. $6 \sin x = 3$

b. $2 \tan x - 1 = 7$

c. $\cos 2x = 0$

9 In wildlife preserves, like national parks, the populations of many animal species vary periodically in relationship to the populations of other species that are either predators or prey.

For example, rabbits are prey for foxes, so when the fox population is high, it causes the rabbit population to fall, and vice versa.

a. Suppose that the fox population in a study region is given by the function $f(t) = 20 \sin t + 100$, where t is the time in years after the first census in the region.

   i. What are the period and amplitude of cycles in the fox population?

   ii. At what time(s) was the fox population a maximum and when a minimum during the first cycle of change?

   iii. At what time(s) will the fox population be about 115? When about 90?

b. Suppose that the rabbit population in the same study region is given by the function $r(t) = 300 \cos t + 2{,}500$.

   i. What are the period and amplitude of cycles in the rabbit population?

   ii. At what time(s) was the rabbit population a maximum and when a minimum during the first cycle of change?

   iii. At what time(s) will the rabbit population be about 2,750? When about 2,300?

SUMMARIZE THE MATHEMATICS

In this unit, you explored inverses of functions with particular attention to inverses of exponential functions and of trigonometric/circular functions.

a What does it mean to say that two functions f and g are inverses of each other?

b How can you determine whether a function $f(x)$ has an inverse by studying:

  i. a table of $(x, f(x))$ values?

  ii. a graph of the function?

  iii. a definition in words or a symbolic rule for the function?

c Among the families of functions that you have studied most closely and used most often:

  i. which almost always have inverses?

  ii. which do not generally have inverses?

d How can the statement "$\log_{10} a = b$" be expressed in equivalent form using exponents?

e How can properties of logarithms be used to write these algebraic expressions in equivalent form?

  i. $\log mn = \ldots$ **ii.** $\log m^n = \ldots$ **iii.** $\log \dfrac{m}{n} = \ldots$

f How can properties of logarithms be used to solve exponential equations like $a(b^x) = c$?

g What does it mean to say that $\sin^{-1} x = k$?

h What are the domain and range of $\sin^{-1} x$? Of $\cos^{-1} x$? Of $\tan^{-1} x$?

i How can inverse trigonometric functions be used to solve trigonometric equations?

Be prepared to share your examples and descriptions with the class.

 CHECK YOUR UNDERSTANDING

Write, in outline form, a summary of the important mathematical concepts and methods developed in this unit. Organize your summary so that it can be used as a quick reference in future units and courses.

GLOSSARY/GLOSARIO

English	**Español**

A

Alternate exterior angles (p. 36) In the diagram, transversal *t* cuts lines ℓ and *m*, forming angles 1 through 8. Pairs of alternate exterior angles are $\angle 1$ and $\angle 8$, and $\angle 2$ and $\angle 7$. The angles in each pair are not between the lines and are on opposite sides of the transversal.

Ángulos alternos externos (pág. 36) En el diagrama, la transversal *t* corta las líneas ℓ y *m*. Los pares de ángulos alternos externos son $\angle 1$ y $\angle 8$, y $\angle 2$ y $\angle 7$. Los ángulos en cada par no están entre las líneas y están en lados opuestos de la transversal.

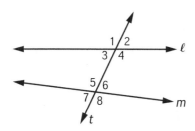

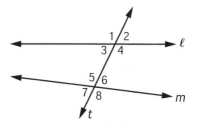

Alternate interior angles (p. 36) In the diagram, transversal *t* cuts lines ℓ and *m*, forming angles 1 through 8. Pairs of alternate interior angles are $\angle 3$ and $\angle 6$, and $\angle 4$ and $\angle 5$. The angles in each pair are between the two lines and on opposite sides of the transversal.

Ángulos alternos internos (pág. 36) En el diagrama, la transversal *t* corta las líneas ℓ y *m*, formando los ángulos 1 a 8. Los pares de ángulos alternos internos son $\angle 3$ y $\angle 6$, y $\angle 4$ y $\angle 5$. Los ángulos en cada par no están entre las rectas y están en lados opuestos de la transversal.

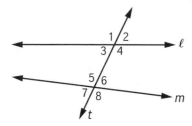

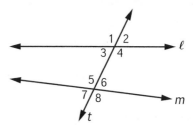

Amplitude (p. 434) Half the difference *maximum value − minimum value*, if they exist, in one cycle of a periodic graph.

Amplitud (pág. 434) Diferencia media entre *valor máximo − valor mínimo*, si existen, en un ciclo de una gráfica periódica.

Angular velocity (p. 421) The rate at which a rotating object such as a pulley, sprocket, or drive shaft turns.

Velocidad angular (pág. 421) La tasa a la que un objeto que rota, como una polea, rueda de engranaje o eje motor, gira.

Annual percentage rate or **APR** (p. 465) A standard way to state the effective annual interest rate on a loan.

Tasa anual porcentual o **TAP** (pág. 465) Forma estándar de enunciar la tasa de interés anual de un préstamo.

Annual percentage yield or **APY** (p. 473) The effective annual interest rate on an investment, taking into account the effect of *compounding interest*.

Renta anual porcentual o **RAP** (pág. 473) Tasa anual de interés de una inversión teniendo en cuenta el efecto del *interés compuesto*.

English

Arc intercepted by an angle (p. 405) For an angle whose sides cut a circle at points A and B, the portion of the circle that lies in the interior of the angle, together with points A and B (the *endpoints of the arc*).

Arithmetic growth (p. 483) Growth that is modeled by an *arithmetic sequence*.

Arithmetic sequence (p. 483) A sequence of numbers in which the difference between any two consecutive terms is a fixed nonzero constant. Symbolically, $a_n = a_{n-1} + d$.

Arithmetic series (p. 493) A *series* whose terms are those of an *arithmetic sequence*.

Attracting fixed point (p. 521) A *fixed point* such that function iteration sequences that start close to it converge to that point.

Average run length or **ARL** (p. 308) When using control charts, the expected number of items tested until a test gives a false alarm.

Español

Arco interceptado por un ángulo (pág. 405) Para un ángulo cuyos lados cortan un círculo en los puntos A y B, la porción del círculo que se encuentra en el interior del ángulo, junto con los puntos A y B (*los extremos del arco*).

Crecimiento aritmético (pág. 483) Crecimiento basado en una *sucesión aritmética*.

Sucesión aritmética (pág. 483) Secuencia de números en los que la diferencia entre dos números consecutivos cualesquiera es una constante fija diferente a cero. Simbólicamente, $a_n = a_{n-1} + d$.

Serie aritmética (pág. 493) *Serie* cuyos términos son los de una *sucesión aritmética*.

Punto fijo de atracción (pág. 521) Un *punto fijo* tal que las sucesiones de función de iteración que comienzan cercanas al mismo convergen en ese punto.

Longitud media de recorrido o **LMR** (pág. 308) En el uso de diagramas de control, el número esperado de artículos examinados hasta que la prueba da como resultado una falsa alarma.

B

Balance (p. 406) In the case of a loan, the money still owed.

Binomial situation (p. 260) A probabilistic situation with a fixed number of independent trials, each with two possible outcomes, and the same probability of a success on each trial.

Saldo (pág. 406) En el caso de un préstamo, el dinero que todavía se adeuda.

Situación binomial (pág. 260) Situación probabilística con un número fijo de pruebas independientes, cada una con dos resultados posibles y la misma probabilidad de éxito en cada prueba.

C

Central angle of a circle (p. 401) An angle of measure less than 180° whose vertex is at the center of the circle and whose sides contain radii of the circle.

Centroid of a Triangle (p. 203) The point of concurrency of the medians of a triangle.

Chord of a circle (p. 397) A segment that joins two distinct points on a circle.

Circle (p. 397) The set of all points in a plane that are equidistant from a given point O, called the *center* of the circle.

Ángulo central de un círculo (pág. 401) Ángulo que mide menos de 180° cuyo vértice está en el centro del círculo y cuyos lados son radios del círculo.

Centroide de un triángulo (pág. 203) Punto de concurrencia de las medianas de un triángulo.

Cuerda de un círculo (pág. 397) Segmento que une dos puntos de un círculo.

Círculo (pág. 397) Conjunto de todos los puntos de un plano que son equidistantes de un punto dado O, llamado *centro* del círculo.

English

Circular functions (p. 430) Functions (sine and cosine) pairing points on the real number line with coordinates of points on a unit circle by "mapping" the real number line around the unit circle—non-negative real axis is wrapped *counterclockwise* starting at $(1, 0)$; negative real axis is wrapped in *clockwise* direction. For any real number r, if the wrapping procedure pairs r with (x, y), then $\cos r = x$ and $\sin r = y$.

Circumcenter (p. 201) The point of concurrency of the perpendicular bisectors of the sides of a triangle; this is the center of the *circumscribed circle* (*circumcircle*) of the triangle.

Closed-form formula for a sequence See *Function formula for a sequence*.

Combined recursive formula (p. 489) A *recursive formula* of the form $A_n = rA_{n-1} + b$. (Also called an *affine recurrence relation* or a *nonhomogeneous first-order linear difference equation*.)

Common difference (p. 485) The constant difference between any two consecutive terms in an *arithmetic sequence*.

Common logarithm (p. 560) If $10^x = y$, then x is called the base 10 logarithm of y; it is often denoted $x = \log y$.

Common ratio (p. 487) The constant ratio of any two consecutive terms in a *geometric sequence*.

Completing the square (p. 350) The process by which a quadratic expression is rewritten in the form $a(x - h)^2 + k$, called a *complete square* or *vertex form*.

Complex number (p. 355) Any complex number can be expressed in the form $a + bi$ where a and b are real numbers and $i = \sqrt{-1}$.

Composition of transformations (p. 209) The process of applying two transformations in succession. The transformation that maps the *original preimage* to the *final image* is called the *composite transformation*.

Compound interest (p. 465) Interest that is applied to previous interest as well as to the original amount of money borrowed or invested.

Español

Funciones circulares (pág. 430) Puntos de apareamiento de las funciones (seno y coseno) en una recta numérica real con coordenadas en el círculo unitario mediante "aplicación"—el eje real no negativo se envuelve en *dirección contraria a las manecillas del reloj* comenzando en $(1, 0)$; el eje real negativo se envuelve en *dirección de las manecillas del reloj*. Para cualquier número real r, si el procedimiento de envolver empareja r con (x, y), entonces $\cos r = x$ y $\sin r = y$.

Circuncentro (pág. 201) Punto de concurrencia de las bisectrices perpendiculares de los lados de un triángulo; este es el centro del *círculo circunscrito* (*circuncírculo*) del triángulo.

Fórmula cerrada de una sucesión Ver *Fórmula de la función de una sucesión*.

Fórmula recurrente combinada (pág. 489) *Fórmula recurrente* del tipo $A_n = rA_{n-1} + b$. (También llamada *relación recurrente afín* o *ecuación diferencial lineal de primer orden no homogénea*).

Diferencia común (pág. 485) Diferencia constante entre dos términos consecutivos cualesquiera de una *sucesión aritmética*.

Logaritmo común (pág. 560) Si $10^x = y$, entonces x se llama logaritmo en base 10 de y; con frecuencia se indica $x = \log y$.

Razón común (pág. 487) Razón constante de dos términos consecutivos cualesquiera de una *sucesión geométrica*.

Completar el cuadrado (pág. 350) El proceso por el cual una expresión cuadrática es reescrita en la forma $a(x - h)^2 + k$, llamada *cuadrado completo* o *forma del vértice*.

Número complejo (pág. 355) Todo número complejo puede expresarse con la forma $a + bi$ donde a y b son números reales y $i = \sqrt{-1}$.

Composición de transformaciones (pág. 209) Proceso de aplicar dos transformaciones sucesivas. La transformación que relaciona la *preimagen original* con la *imagen final* se llama *transformación compuesta*.

Interés compuesto (pág. 465) Interés aplicado al interés anterior, así como también al monto inicial de dinero prestado o invertido.

English

Concentric circles (p. 397) Two or more circles in the same plane that have the same center.

Conclusion (p. 10) In an "if-then" statement, the condition that follows "then." Symbolically, in the statement $p \Rightarrow q$, the conclusion is q.

Concurrent lines (p. 200) Three or more lines that intersect at a common point.

Conditional statement See *If-then statement*.

Congruent figures (p. 162) Two figures are congruent if and only if one figure is the image of the other under a rigid transformation (line reflection, translation, rotation, glide reflection) or a composite of rigid transformations. Congruent figures have the same shape and size, regardless of position or orientation.

Consecutive integers (p. 16) Adjacent integers on a number line. These can be expressed symbolically as n and $n + 1$.

Constraint (p. 132) A limitation on values that variables may assume in a problem situation. For example, the linear constraint $3x + 5y < 21$ expresses a condition for acceptable combinations of values for the variables x and y.

Contrapositive of an if-then statement (p. 23) Reverses the order and negates both parts of the if-then statement. In symbols, the contrapositive of $p \Rightarrow q$ is *not q* $\Rightarrow$ *not p*.

Control chart (or, **run chart**) (p. 284) A type of plot over time where observations from an industrial process are plotted in order of occurrence and checked for patterns that indicate that the process has gone out of control.

Control group (or, **comparison group**) (p. 77) In an experiment, a randomly selected group of subjects that gets no treatment, gets a placebo, or gets an established or standard treatment. Otherwise, the control group is treated the same as the group or groups that get the experimental treatment or treatments.

Converse of an if-then statement (p. 10) Reverses the order of the two parts of the if-then statement. In symbols, given the original statement $p \Rightarrow q$, its converse statement is $q \Rightarrow p$.

Español

Círculos concéntricos (pág. 397) Dos o más círculos en el mismo plano que tienen el mismo centro.

Conclusión (pág. 10) En una oración condicional, la condición que sigue al término "entonces". Simbólicamente, en una oración $p \Rightarrow q$, la conclusion es q.

Rectas concurrentes (pág. 200) Tres o más rectas que se intersecan en un punto en común.

Enunciado condicional Ver *Oración condicional*.

Figuras congruentes (pág. 162) Dos figuras son congruentes si y solo si una de las figuras es la imagen de la otra bajo una transformación rígida (reflexión lineal, traslación, rotación, reflexión de deslizamiento) o una combinación de transformaciones rígidas. Las figuras congruentes tienen la misma forma y tamaño, sin importar su posición u orientación.

Números enteros consecutivos (pág. 16) Números enteros adyacentes en una recta numérica. Pueden expresarse simbólicamente como n y $n + 1$.

Restricción (pág. 132) Una limitación en los valores que una variable puede asumir de una situación problemática. Por ejemplo, la restricción lineal $3x + 5y < 21$ expresa una condición sobre las combinaciones de valores aceptables para las variables x e y.

Contrapositivo de una oración condicional (pág. 23) Invierte el orden y niega ambas partes de una oración condicional. En símbolos, el contrapositivo de $p \Rightarrow q$ es *no q* $\Rightarrow$ *no p*.

Diagrama de control (pág. 284) Tipo de diagrama en el tiempo donde se registran las observaciones de un proceso industrial y se buscan patrones que indiquen que el proceso está fuera de control.

Grupo de control (o **grupo de comparación**) (pág. 77) En un experimento, un grupo de sujetos seleccionados al azar que no recibe tratamiento, recibe un placebo o recibe el tratamiento establecido o estándar. De otra manera, el grupo de control se trata de la misma manera que el grupo o los grupos que reciben el tratamiento experimental.

Converso de un enunciado condicional (pág. 10) Invierte el orden de las dos partes de una oración condicional. En símbolos, dado el enunciado original $p \Rightarrow q$, su enunciado converso es $q \Rightarrow p$.

English	Español

Corresponding angles (p. 36) In the diagram, transversal t cuts lines ℓ and m, forming angles 1 through 8. Pairs of corresponding angles are $\angle 1$ and $\angle 5$, $\angle 2$ and $\angle 6$, $\angle 3$ and $\angle 7$, and $\angle 4$ and $\angle 8$. Angles in each pair are in the same relative position with respect to each line and the transversal.

Ángulos correspondientes (pág. 36) En el diagrama, la transversal t corta las líneas ℓ y m, formando los ángulos 1 a 8. Los pares de ángulos correspondientes son $\angle 1$ y $\angle 5$, $\angle 2$ y $\angle 6$, $\angle 3$ y $\angle 7$, y $\angle 4$ y $\angle 8$. Los ángulos en cada par están en la misma posición relativa con respecto a cada recta y a la transversal.

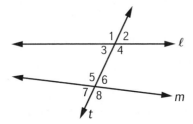

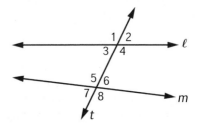

Cycle—function The graph of one period of a *periodic function*. In the case of circular motion, a cycle corresponds to one full revolution.

Ciclo—función La gráfica de un periodo de una *función periódica*. En el caso del movimiento circular, un ciclo corresponde a una revolución completa.

Cycle—iteration (p. 521) A sequence of numbers that repeats over and over when *iterating a function*.

Ciclo—iteración (pág. 521) Secuencia de números que se repite una y otra vez en la *iteración de una función*.

D

Deductive reasoning (p. 2) Reasoning strategy that involves reasoning *from* facts, definitions, and accepted properties *to* conclusions using principles of logic.

Razonamiento deductivo (pág. 2) Estrategia de razonamiento que involucra el razonamiento a partir de hechos, definiciones y propiedades aceptadas para obtener conclusiones que usan principios de lógica.

Degree measure of a circular arc (p. 401) The degree measure of a minor arc and the degree measure of its corresponding central angle are the same. The degree measure of a major arc is 360° minus the degree measure of its corresponding central angle.

Medida gradual de un arco circular (pág. 401) La medida gradual de un arco menor y la de su ángulo central correspondiente son iguales. La medida gradual de un arco mayor es 360° menos la medida gradual de su ángulo central correspondiente.

Degree of a polynomial (p. 324) The highest exponent of the variable that appears in the polynomial expression.

Grado de un polinomio (pág. 324) El mayor exponente de la variable que aparece en la expresión polinomial.

Difference equation See *Recursive formula*.

Ecuación de diferencia Ver *Fórmula recursiva*.

Discrete dynamical system (p. 462) A situation (system) involving change (dynamical) in which the nature of the change is step-by-step (discrete).

Sistema dinámico discreto (pág. 462) Situación (sistema) que involucra un cambio (dinámico) en la que la naturaleza del cambio se da paso a paso (discreto).

Distance from a point to a line (p. 202) The length of the perpendicular segment from the point to the line.

Distancia desde un punto a una recta (pág. 202) Longitud del segmento perpendicular desde el punto a la recta.

English	Español

Double blind (p. 78) An experiment where neither the subjects nor the person who evaluates how well the treatment works knows which treatment the subject received.

Doble ciego (pág. 78) Experimento en el cual ni el sujeto ni la persona que evalúa la eficacia del tratamiento sabe qué tratamiento recibió el sujeto.

— **E** —

Equilic quadrilateral (p. 226) A quadrilateral with a pair of congruent opposite sides that, when extended, meet to form a 60° angle. The other two sides are called bases.

Cuadrilátero equílico (pág. 226) Cuadrilátero con un par de lados opuestos congruentes que, al extenderse, se encuentran para formar un ángulo de 60°. Los otros dos lados se llaman bases.

Expected value (or, **expected number**) (p. 261) In probability, the long-run average or mean value of a probability distribution.

Valor esperado (o **número esperado**) (pág. 261) En probabilidad, el promedio a largo plazo o valor medio de una distribución de probabilidad.

Experiment (p. 77) A research study in which available subjects are randomly assigned to two or more different treatments in order to compare how responses to the treatments differ.

Experimento (pág. 77) Tipo de investigación en el que los sujetos son asignados en forma aleatoria a dos o más tratamientos distintos para comparar como difieren las respuestas a los tratamientos.

Explicit formula for a sequence See *Function formula for a sequence.*

Fórmula explícita de una sucesión Ver *Fórmula de la función de una sucesión.*

Exponential growth (p. 486) Growth that is modeled by a *geometric sequence* or an exponential function.

Crecimiento exponencial (pág. 486) Crecimiento basado en una *sucesión geométrica* o una función exponencial.

Exterior angle of a triangle (p. 46) The angle formed by the side of a triangle and the extension of an adjacent side.

Ángulo exterior de un triángulo (pág. 46) Ángulo que se forma por un lado del triángulo y la extensión de un lado adyacente.

Exterior angles on the same side of the transversal (p. 36) In the diagram, transversal *t* cuts lines ℓ and *m*, forming angles 1 through 8. Pairs of exterior angles on the same side of the transversal are $\angle 1$ and $\angle 7$, and $\angle 2$ and $\angle 8$.

Ángulos exteriores en el mismo lado de la transversal (pág. 36) En el diagrama, la transversal *t* corta las rectas ℓ y *m*, formando los ángulos 1 a 8. Los pares de ángulos exteriores en el mismo lado de la transversal son $\angle 1$ y $\angle 7$, y $\angle 2$ y $\angle 8$.

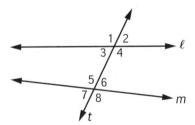

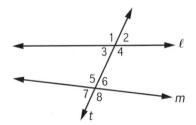

Exterior of a circle (p. 397) The set of all points in the plane of a circle whose distance from the center is greater than the circle's radius.

Exterior de un círculo (pág. 397) Conjunto de todos los puntos del plano de un círculo cuya distancia desde el centro es mayor que el radio del círculo.

English	**Español**

F

Factorial notation (p. 21) A compact way of writing the product of consecutive positive whole numbers. Symbolically, $n! = n \cdot (n-1) \cdot (n-2) \cdot \cdots \cdot 2 \cdot 1$.

Notación factorial (pág. 21) Forma compacta de escribir el producto de números enteros positivos consecutivos. En símbolos, $n! = n \cdot (n-1) \cdot (n-2) \cdot \cdots \cdot 2 \cdot 1$.

False alarm (p. 294) In statistical process control, this is when a test signals that a process may be out of control when it is in control.

Falsa alarma (pág. 294) En control de proceso estadístico, ésta ocurre cuando una prueba señala que el proceso puede estar fuera de control cuando está bajo control.

Feasible region (p. 136) In a linear programming problem, the feasible region is the set of all points whose coordinates satisfy all given constraints.

Zona factible (pág. 136) En problemas de programación lineal, la zona factible es el conjunto de todos los puntos cuyas coordenadas satisfacen todas las limitaciones dadas.

Finite differences table (p. 495) A table corresponding to a numerical sequence in which the first column of the table is a consecutive list of indices for the sequence, the second column is the corresponding terms of the sequence, the third column contains the differences of consecutive terms from the second column (called first differences), the fourth column contains the differences of consecutive terms in the third column (called second differences), and so on.

Tabla de diferencias finitas (pág. 495) Tabla correspondiente a una sucesión numérica en la que la primera columna de la tabla es una lista consecutiva de los índices de la sucesión, la segunda columna son los términos correspondientes de la secuencia, la tercera columna contiene las diferencias de los términos consecutivos de la segunda columna (llamada primera diferencia), la cuarta columna contiene las diferencias de los números consecutivos de la tercera columna (llamada segunda diferencia) y así sucesivamente.

Fixed point of a function (p. 520) For the function f, a value x such that $f(x) = x$. When *iterating a function* if you reach that value, you never leave it.

Punto fijo de una función (pág. 520) Para la función f, una valor x tal que $f(x) = x$. En el proceso de *iteración de una función*, cuando se alcanza ese valor, nunca se deja.

Function formula for a sequence (pp. 484 and 489) A non-recursive formula that expresses the nth term in a sequence as a function of n. (Also called an *explicit formula* or a *closed-form formula*.)

Fórmula de la función de una sucesión (págs. 484 y 489) Fórmula no recurrente que expresa el término nth en una sucesión como una función de n. (También llamada *fórmula explícita* o *fórmula cerrada*).

G

Geometric growth (p. 486) Growth that is modeled by a *geometric sequence*.

Crecimiento geométrico (pág. 486) Crecimiento basado en una *sucesión geométrica*.

Geometric mean (p. 186) The geometric mean of two positive integers a and b is the positive number x such that $\frac{a}{x} = \frac{x}{b}$ or $x = \sqrt{ab}$.

Media geométrica (pág. 186) La media geométrica de dos enteros positivos a y b es el número positivo x tal que $\frac{a}{x} = \frac{x}{b}$ ó $x = \sqrt{ab}$.

Geometric sequence (p. 485) A sequence of numbers in which the ratio of any two consecutive terms is a fixed constant. Symbolically, $a_{n+1} = r \cdot a_n$.

Sucesión geométrica (pág. 485) Sucesión de números en la que la razón de cualquier par de números consecutivos es una constante fija. Simbólicamente, se presenta así: $a_{n+1} = r \cdot a_n$.

English	Español

Geometric series (p. 493) A *series* whose terms are those of a *geometric sequence*.

Serie geométrica (pág. 493) *Serie* cuyos términos son los de una *sucesión geométrica*.

Glide-reflection (p. 215) A rigid transformation that is the composition of a reflection across a line and a translation in a direction parallel to the line.

Reflexión del deslizamiento (pág. 215) Transformación rígida que es la composición de una reflexión en una recta y una traslación en dirección paralela a esa recta.

Global maximum or minimum point (p. 343) A global maximum point for a function $f(x)$ is a pair $(a, f(a))$ with the property that $f(a) \geq f(x)$ for all x. The pair $(b, f(b))$ is a global minimum point if $f(b) \leq f(x)$ for all x.

Punto global máximo o mínimo (pág. 343) El punto global máximo de una función $f(x)$ es un par $(a, f(a))$ con la propiedad que $f(a) \geq f(x)$ para toda x. El par $(b, f(b))$ es el punto global mínimo si $f(b) \leq f(x)$ para toda x.

Graphical iteration (p. 519) A graphical representation of *iterating a function*, in which the graph of $y = x$ is drawn on the same set of coordinate axes as the graph of the function being iterated and the process of iteration is shown graphically by moving vertically to the graph of the function, then horizontally to the graph of $y = x$, then vertically to the graph of the function, then horizontally to the graph of $y = x$, and so on.

Iteración gráfica (pág. 519) Representación gráfica de la *iteración de una función*, en la que la gráfica de $y = x$ se dibuja en el mismo conjunto de ejes de coordenadas que la de la función que se quiere iterar, y el proceso de iteración se muestra gráficamente desplazándose verticalmente a la gráfica de la función, luego en forma horizontal a la gráfica de $y = x$, luego verticalmente a la gráfica de la función, luego horizontalmente a la gráfica de $y = x$ y así sucesivamente.

Great circle (p. 48) A circle on the surface of a sphere formed by a plane passing through the center of the sphere.

Gran círculo (pág. 48) Círculo en la superficie de una esfera formado por un plano que pasa por el centro de la esfera.

H

Horizontal asymptote (p. 367) A line with equation $y = k$ is a horizontal asymptote for the graph of a function $f(x)$ whenever the values of $f(x)$ approach k as a limit as $x \rightarrow +\infty$ or $x \rightarrow -\infty$.

Asíntota horizontal (pág. 367) Una recta con ecuación $y = k$ es una asíntota horizontal para la gráfica de una función $f(x)$ cuando los valores de $f(x)$ se aproximan a k como $x \rightarrow +\infty$ o $x \rightarrow -\infty$.

Hypothesis of an if-then statement (p. 10) The condition that follows "if." Symbolically, in an if-then statement $p \Rightarrow q$, p is the hypothesis.

Hipótesis de una oración condicional (pág. 10) La condición que sigue al "Si". Simbólicamente, en una oración condicional $p \Rightarrow q$, p es la hipótesis.

I

If and only if statement (p. 22) A combination of an if-then statement and its converse. In symbols, "p if and only if q" is written as $p \Leftrightarrow q$, and is understood to mean $p \Rightarrow q$ and $q \Rightarrow p$.

Oración Si y sólo si (pág. 22) Combinación de oraciones condicionales y su inverso. En símbolos, "p si y solo si q" se escribe $p \Leftrightarrow q$, y se supone que significa $p \Rightarrow q$ y $q \Rightarrow p$.

If-then statement (p. 10) Frequently used in deductive arguments because if the hypothesis is satisfied then the conclusion follows. If-then statements can be represented symbolically as $p \Rightarrow q$ (read "if p, then q" or "p implies q"), where p represents the hypothesis and q represents the conclusion.

Oración condicional (pág. 10) Frecuentemente usada en los argumentos deductivos porque si se satisface la hipótesis, entonces sigue la conclusión. Las oraciones condicionales pueden representarse simbólicamente como $p \Rightarrow q$ (se lee "si p, entonces q" o "p implica q"), donde p representa la hipótesis y q representa la conclusión.

English

Incenter (p. 202) The point of concurrency of the bisectors of the angles of a triangle; this is the center of the *incircle* (*inscribed circle*) of the triangle.

Independent events (p. 308) In probability, two events A and B are said to be independent if the probability that B occurs does not change depending on whether or not A occurred.

Inductive reasoning (p. 2) Reasoning strategy used to discover general patterns or principles based on evidence from experiments or several cases.

Inequality (p. 109) A statement like $3x + 5y < 9$ or $t^2 + 2 \geq 4$ composed of numbers or algebraic expressions connected by an inequality symbol $(<, \leq, >, \geq)$. The solution of an inequality is all values of the variable(s) for which the statement is true.

Inscribed angle in a circle (p. 404) An angle whose vertex is on the circle and whose sides contain chords of the circle.

Interior angles on the same side of the transversal (p. 36) In the diagram, transversal t cuts lines ℓ and m, forming angles 1 through 8. Pairs of interior angles on the same side of the transversal are $\angle 4$ and $\angle 6$, and $\angle 3$ and $\angle 5$.

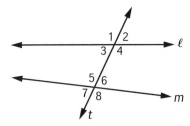

Interior of a circle (p. 397) The set of all points in the plane of a circle whose distance from the center is less than the circle's radius.

Interval (p. 116) All numbers on a number line between two specified endpoints. The *closed* interval $[a, b]$ is all numbers x such that $a \leq x \leq b$. The *open* interval (a, b) is all numbers x such that $a < x < b$.

Inverse cosine function (p. 584) The inverse cosine or arccos function is defined as $\cos^{-1} k = x$ if $\cos x = k$ and $0 \leq x \leq \pi$.

Español

Incentro (pág. 202) Punto donde se unen las bisectrices de los ángulos de un triángulo; éste es el centro del *incentro* (*círculo inscrito*) del triángulo.

Eventos independientes (pág. 308) En probabilidad, se dice que dos eventos A y B son independientes si la probabilidad de que B ocurra no cambia si A ocurre o no.

Razonamiento inductivo (pág. 2) Estrategia de razonamiento que se usa para descubrir patrones o principios generales, basándose en pruebas de experimentos o de otros casos.

Desigualdad (pág. 109) Oración como $3x + 5y < 9$ o $t^2 + 2 \geq 4$ compuesta por números o expresiones algebraicas relacionadas por un símbolo de desigualdad $(<, \leq, >, \geq)$. La solución de una desigualdad son todos los valores para la o las variables que hacen verdadera la oración.

Ángulo inscrito de un círculo (pág. 404) Ángulo cuyo vértice está en el círculo y cuyos lados contienen cuerdas del círculo.

Ángulos interiores en el mismo lado de la transversal (pág. 36) En el diagrama, la transversal t corta las rectas ℓ y m, formando los ángulos 1 a 8. Los pares de ángulos interiores en el mismo lado de la transversal son $\angle 4$ y $\angle 6$, y $\angle 3$ y $\angle 5$.

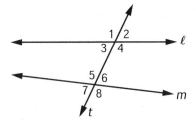

Interior de un círculo (pág. 397) Conjunto de todos los puntos del plano de un círculo cuya distancia desde el centro es menor que el radio del círculo.

Intervalo (pág. 116) Todos los números de una recta de numeros entre dos extremos especificados. El intervalo *cerrado* $[a, b]$ son todos los números x tal que $a \leq x \leq b$. El intervalo *abierto* (a, b) son todos los números x tal que $a < x < b$.

Función coseno inverso (pág. 584) La función coseno inverso o arcocoseno se define como $\cos^{-1} k = x$ si $\cos x = k$ y $0 \leq x \leq \pi$.

English

Inverse function (p. 543) For a given function f, if the function g has domain equal to the range of f, range equal to the domain of f, and if the composite, g following f is x for all x, then g is the inverse of f, denoted f^{-1}.

Inverse sine function (p. 579) The inverse sine or arcsin function is defined as $\sin^{-1} k = x$ if $\sin x = k$ and $-\frac{\pi}{2} \le x \le \frac{\pi}{2}$.

Inverse of an if-then statement (p. 22) The negation of the hypothesis and conclusion in an if-then statement. In symbols, the inverse of $p \Rightarrow q$ is not $p \Rightarrow$ not q.

Inverse tangent function (p. 587) The inverse tangent or arctan function is defined as $\tan^{-1} k = x$ if $\tan x = k$ and $-\frac{\pi}{2} < x < \frac{\pi}{2}$.

Isosceles trapezoid (p. 217) A quadrilateral with exactly one pair of parallel sides and the nonparallel sides congruent.

Iterating a function (p. 515) The process of sequentially feeding the outputs of a function back into itself as inputs.

Iteration (p. 462) The process of repeating the same procedure or computation over and over.

Español

Función inversa (pág. 543) Para una función dada f, si la función g tiene un dominio igual al rango de f, rango igual al dominio de f y si el compuesto, g seguido de f es x para toda x, entonces g es la inversa de f y se denota f^{-1}.

Función seno inverso (pág. 579) La función seno inverso o arcoseno se define como $\sin^{-1} k = x$ si $\sin x = k$ y $-\frac{\pi}{2} \le x \le \frac{\pi}{2}$.

ElInverso de una oración Si y sólo si (pág. 22) Negación de la hipótesis y la conclusión en una oración condicional. En símbolos, el inverso de $p \Rightarrow q$ es no $p \Rightarrow$ no q.

Función tangente inversa (pág. 587) La función tangente inversa o arcotangente se define como $\tan^{-1} k = x$ si $\tan x = k$ y $-\frac{\pi}{2} < x < \frac{\pi}{2}$.

Trapecio isósceles (pág. 217) Cuadrilátero con exactamente un par de lados paralelos y lados no paralelos congruentes.

Iteración de una función (pág. 515) Proceso de reingresar secuencialmente los resultados de una función en sí misma como valor de entrada.

Iteración (pág. 462) Proceso de repetir el mismo procedimiento o cálculo una y otra vez.

L

Law of Cosines (p. 168) In any triangle ABC with sides of lengths a, b, and c opposite $\angle A$, $\angle B$, and $\angle C$, respectively: $c^2 = a^2 + b^2 - 2ab \cos C$.

Law of Sines (p. 168) In any triangle ABC with sides of lengths a, b, and c opposite $\angle A$, $\angle B$, and $\angle C$, respectively: $\frac{\sin A}{a} = \frac{\sin B}{b} = \frac{\sin C}{c}$.

Line reflection (p. 13) A transformation that maps each point P of the plane onto an image point P' as follows: If point P is not on line ℓ, then ℓ is the perpendicular bisector of $\overline{PP'}$. If point P is on ℓ, then $P' = P$. That is, P is its own image.

Linear pair (p. 30) A pair of adjacent angles whose noncommon sides are opposite rays.

Ley del coseno (pág. 168) En todo triángulo ABC con lados de longitud a, b y c opuestos a $\angle A$, $\angle B$ y $\angle C$, respectivamente: $c^2 = a^2 + b^2 - 2ab \cos C$.

Ley del seno (pág. 168) En todo triángulo ABC con lados de longitud a, b y c opuestos a $\angle A$, $\angle B$ y $\angle C$, respectivamente: $\frac{\sin A}{a} = \frac{\sin B}{b} = \frac{\sin C}{c}$.

Recta de reflexión (pág. 13) Transformación que representa cada punto P del plano en un punto imagen P' de la siguiente manera: Si el punto P no está en la recta ℓ, entonces ℓ es la bisectriz perpendicular de $\overline{PP'}$. Si el punto P está en ℓ, entonces $P' = P$. Es decir, P es su propia imagen.

Par lineal (pág. 30) Par de ángulos adyacentes cuyos lados no comunes son rayos opuestos.

English

Linear programming (p. 128) A mathematical procedure to find values of variables that satisfy a set of linear *constraints* and optimize the value of a linear *objective function*.

Linear velocity (p. 421) The distance that a point on a revolving circle moves in a unit of time.

Local maximum or minimum point (p. 324) A local maximum point for a function $f(x)$ is a pair $(a, f(a))$ with the property that $f(a) \geq f(x)$ for all x in an open interval containing a. The pair $(b, f(b))$ is a local minimum point if $f(b) \leq f(x)$ for all x in an open interval containing a.

Logically equivalent statements (p. 23) Two statements p and q are logically equivalent if each implies the other. In symbols, $p \Rightarrow q$ and $q \Rightarrow p$.

Logistic equation (or logistic map) (p. 518) An equation of the form $f(x) = rx(1 - x)$.

Lurking variable (p. 79) A variable that helps to explain the association between the conditions or treatments and the response, but is not the explanation that the study was designed to test.

Español

Programación lineal (pág. 128) Un procedimiento matemático en el que la tarea es encontrar los valores de variables que satisfacen un grupo de *restricciones* lineales y optimizan el valor de una *función objetiva* lineal.

Velocidad lineal (pág. 421) Distancia que un punto recorre en una unidad de tiempo en un círculo rotativo.

Punto local máximo o mínimo (pág. 324) El punto local máximo para una función $f(x)$ es un par $(a, f(a))$ con la propiedad que $f(a) \geq f(x)$ para toda x en un intervalo abierto que contiene a. El par $(b, f(b))$ es el punto local mínimo si $f(b) \leq f(x)$ para toda x en un intervalo abierto que contiene a.

Enunciados lógicamente equivalentes (pág. 23) Dos oraciones p y q son equivalentes desde el punto de vista lógico si cada una implica la otra. Simbólicamente, se representa así: $p \Rightarrow q$ y $q \Rightarrow p$.

Ecuación logística (pág. 518) Ecuación con la forma $f(x) = rx(1 - x)$.

Variable escondida (pág. 79) Variable que ayuda a explicar la asociación entre las condiciones o tratamientos y la respuesta, pero que no es la explicación que el estudio debía probar.

M

Major arc of a circle (p. 401) An arc with degree measure greater than 180°.

Minor arc of a circle (p. 401) An arc with degree measure less than 180°.

Modus ponens (Latin: mode that affirms) (p. 10) A fundamental principle of logic. If an if-then statement is true in general and the hypothesis is known to be true in a particular case, then the conclusion is also true in that case. In symbols, $p \Rightarrow q$, given p, conclude q. Also called Affirming the Hypothesis.

Mutually exclusive events (or, disjoint events) (p. 305) Two events are said to be mutually exclusive if it is impossible for both of them to occur on the same trial.

Arco mayor de un círculo (pág. 401) Arco con una medida mayor de 180°.

Arco menor de un círculo (pág. 401) Arco con una medida menor de 180°.

Modus ponens (del Latín "modo que afirma") (pág. 10) Principio fundamental de la lógica. Si una oración condicional es verdadera en general y la hipótesis es verdadera en un caso en particular, entonces la conclusión es verdadera en ese caso. En símbolos, $p \Rightarrow q$, dado p, se concluyes q. También se conoce como "afirmar la hipótesis".

Eventos mutuamente excluyentes (o eventos desunidos) (pág. 305) Se dice que dos eventos son mutuamente excluyentes si es imposible que ambos ocurran en el mismo experimento.

English	Español

N

Negation of a statement (p. 22) A statement that negates the given statement. In symbols, the negation of statement p is *not p*.

Negación de una oración (pág. 22) Oración que niega una oración dada. En símbolos, la negación de la oración p *no* es p.

Normal distribution (p. 237) A theoretical probability distribution that is bell-shaped, symmetric, continuous, and defined for all real numbers.

Distribución normal (pág. 237) Distribución de probabilidad teórica que es tiene la forma de una campana, y es simétrica, continua y definida para todos los números reales.

O

Objective function (p. 137) In a linear programming problem, an algebraic expression like $7x - 3y$ whose value is to be optimized within the constraints of the problem.

Función objetiva (pág. 137) En un problema de programación lineal, es una expresión algebraic como $7x - 3y$ cuyo valor debe ser optimizado dentro de las restricciones del problema.

Observational study (p. 89) A statistical study in which the conditions to be compared are already present in the subjects that are observed.

Estudio observacional (pág. 89) Estudio estadístico en el que las condiciones que se comparan están presentes, en ese momento, en los sujetos que serán observados.

Odds (p. 278) An alternative way of expressing a probability. When outcomes are equally likely, the odds of an event are the number of favorable outcomes to the number of unfavorable outcomes. For example, if the odds that an event occurs are 3 to 5, then the probability the event occurs is $\frac{3}{(3+5)}$, or $\frac{3}{8}$.

Posibilidades (pág. 278) Forma alternativa de expresar una probabilidad. Cuando los resultados son igualmente posibles, las posibilidades de un evento son el número de resultados favorables al número de resultados desfavorables. Por ejemplo, si las posibilidades de que un evento ocurra son 3 a 5, entonces la probabilidad de que ocurra el evento es $\frac{3}{(3+5)}$, ó $\frac{3}{8}$.

Orientation of a figure (p. 15) Determined by the clockwise or counterclockwise cyclic labeling of at least three points on a figure.

Orientación de una figura (pág. 15) Está determinada por la clasificación del ciclo horario o antihorario de al menos tres puntos de una figura.

P

Parallelogram (p. 205) A quadrilateral that has two pairs of opposite sides the same length (congruent), or equivalently, a quadrilateral that has two pairs of parallel sides.

Paralelogramo (pág. 205) Cuadrilátero que tiene dos pares de lados opuestos de la misma longitude (congruentes), o de manera equivalente, cuadrilátero que tiene dos pares de lados paralelos.

Percentile (p. 241) A value x in a distribution lies at the pth percentile if $p\%$ of the values in the distribution are less than or equal to x.

Percentil (pág. 241) Un valor x de una distribución se encuentra en el percentil $p.°$ si $p\%$ de los valores de la distribución son menores que o iguales a x.

Period (p. 434) The length of a smallest interval (in the domain) that contains a cycle of a periodic graph.

Período (pág. 434) Longitud del intervalo más corto (en el dominio) que contiene un ciclo en una gráfica periódica.

Periodic graph (p. 434) A graph of a pattern of change that repeats itself over and over again.

Gráfica periódica (pág. 434) Gráfica de un patron de cambio que se repite una y otra vez.

617

GLOSSARY/GLOSARIO

English

Perpendicular bisector of a segment (p. 13) A line or line segment perpendicular to the segment at its midpoint.

Placebo (p. 77) A sham or empty treatment that to subjects of an experiment appears to be the real treatment.

Placebo effect (p. 78) The phenomenon that people tend to do better when given a treatment, even if it is a placebo.

Polynomial expression (p. 324) An algebraic expression in the general form $a_n x^n + a_{n-1} x^{n-1} + \cdots + a_1 x + a_0$, where $a_n, a_{n-1}, \ldots, a_1, a_0$ are constants.

Polynomial function (p. 323) A function whose rule can be expressed in the form $f(x) = a_n x^n + a_{n-1} x^{n-1} + \cdots + a_1 x + a_0$.

Postulates (p. 31) Mathematical statements that are accepted as true without proof. Also called axioms.

Present value (p. 476) The amount of money that, if invested today at a fixed *compound interest* rate, would give the same yield as a number of regular payments in the future such that at exactly the end of the payment period the *balance* is zero.

Prime number (p. 10) An integer greater than 1 that has exactly two factors, 1 and itself.

Principal (p. 465) In finance, the amount of money borrowed or invested.

Proportion (p. 66) A statement of equality between ratios.

Español

Bisectriz perpendicular de un segmento (pág. 13) Recta o segmento de recta perpendicular del segmento en su punto medio.

Placebo (pág. 77) Tratamiento simulado o falso para los sujetos de un experimento que parece ser el tratamiento real.

Efecto placebo (pág. 78) Fenómeno por el que las personas tienden a sentirse mejor con un tratamiento, incluso si es un placebo.

Expresión polinomial (pág. 324) Una expresión polinomial es cualquier expresión algebraica con la forma general $a_n x^n + a_{n-1} x^{n-1} + \cdots + a_1 x + a_0$, donde $a_n, a_{n-1}, \ldots, a_1, a_0$ son constantes.

Función polinomial (pág. 323) Función cuya regla puede ser expresada en la forma $f(x) = a_n x^n + a_{n-1} x^{n-1} + \cdots + a_1 x + a_0$.

Postulados (pág. 31) Enunciados matemáticos que son aceptados como verdaderos sin evidencia. También se les llama axiomas.

Valor actual (pág. 476) La cantidad de dinero que, si se invierte hoy a una tasa fija de *interés compuesto*, daría el mismo rendimiento en cuanto al número dado de pagos regulares en el futuro, de forma tal que al final del período de pago el *saldo* sería cero.

Número primo (pág. 10) Un entero mayor que 1 que tiene exactamente dos factores, 1 y sí mismo.

Capital (pág. 465) En finanzas, la cantidad de dinero que se pide prestado o se invierte.

Proporción (pág. 66) Un enunciado de igualdad entre razones.

Q

Quadratic formula (p. 353) The formula $x = \dfrac{-b \pm \sqrt{b^2 - 4ac}}{2a}$ that gives the solutions of any quadratic equation in the form $ax^2 + bx + c = 0$, where a, b, and c are constants and $a \neq 0$.

Fórmula cuadrática (pág. 353) La fórmula $x = \dfrac{-b \pm \sqrt{b^2 - 4ac}}{2a}$ que da las soluciones de cualquier ecuación cuadrática en la forma $ax^2 + bx + c = 0$, donde a, b y c son constantes y $a \neq 0$.

R

Radian (p. 427) The measure of a central angle of a circle that intercepts an arc equal in length to the radius of the circle. One radian equals $\dfrac{180}{\pi}$ degrees, which is approximately 57.2958°.

Radian (pág. 427) Medida del ángulo central de un círculo que intercepta un arco de igual longitud que el radio del círculo. Un radian equivale a $\dfrac{180}{\pi}$ grados, lo cual es aproximadamente 57.2958°.

English

Random sample of size *n* (p. 89) A sample selected by a method equivalent to writing the name of every member of the population on a card, mixing the cards well, and drawing *n* cards.

Randomization distribution (p. 84) In an experiment, a distribution of possible differences between the mean responses from two treatments, generated by assuming that each response would have been the same had each subject been assigned to the other treatment.

Randomization test (or, **permutation test**) (p. 85) A statistical test that can be used to determine whether the difference in the mean response from two treatments in an experiment is statistically significant.

Rare event (p. 267) An event that lies in the outer 5% of a distribution. In a waiting-time distribution, rare events typically are in the upper 5% of the distribution. In a binomial distribution, rare events may be those in the upper 2.5% and lower 2.5%.

Rational expression (p. 366) A quotient of two polynomial expressions.

Rational function (p. 364) A rational function is a function with rule that can be expressed as the quotient of two polynomials.

Recursion (p. 462) A sequential process in which a step is described in terms of previous steps.

Recursive formula (p. 467) A formula involving *recursion*. (Also called a recurrence relation or a difference equation.)

Recursive formula for a sequence (p. 483) A formula that expresses a given term in a sequence as a function of previous terms.

Regular polygon (p. 4) A polygon in which all sides are congruent and all angles are congruent.

Remote interior angles of a triangle (p. 46) The two angles of a triangle that are not adjacent to a given exterior angle.

Español

Muestra aleatoria de tamaño *n* (pág. 89) Muestra que se obtiene por un método equivalente a escribir el nombre de cada miembro de la población en una tarjeta, mezclarlas bien y elegir *n* tarjetas.

Distribución aleatoria (pág. 84) En un experimento, la distribución de diferencias posibles entre las respuestas medias de dos tratamientos, generadas al suponer que cada respuesta hubiera sido la misma si a cada uno de los sujetos se le hubiera asignado el otro tratamiento.

Prueba aleatoria (pág. 85) Prueba estadística que se puede usar para determinar si la diferencia en la respuesta media de dos tratamientos de un experimento es estadísticamente significativa.

Evento casual (pág. 267) Evento que ocurre en el 5% externo de una distribución. En una distribución de tiempo de espera, los eventos casuales normalmente se hallan en el 5% superior de la distribución. En una distribución binomial, los eventos casuales pueden ser aquellos que se encuentran en el 2.5% superior y 2.5% inferior.

Expresión racional (pág. 366) El cociente de dos expresiones polinomiales.

Función racional (pág. 364) Una función racional es una función con una regla que se puede expresar como el cociente de dos polinomios.

Recursión (pág. 462) Proceso secuencial en el cual un paso es descrito en términos de los pasos previos.

Fórmula recurrente (pág. 467) Fórmula que involucra *recursión*. (También llamada relación recurrente o ecuación de diferencia).

Fórmula recurrente de una sucesión (pág. 483) Fórmula que expresa un término dado en una sucesión como una función de los términos previos.

Polígono regular (pág. 4) Polígono con todos los lados y los ángulos congruentes.

Ángulos interiores remotos de un triángulo (pág. 46) Los dos ángulos de un triángulo que no son adyacentes a un ángulo exterior dado.

English

Español

Repelling fixed point (p. 521) A *fixed point* such that function iteration sequences move away from it (except for a sequence that begins at the fixed point).

Punto fijo de repulsión (pág. 521) *Punto fijo* tal que las sucesiones de iteración de la función se alejan de él (a excepción de la sucesión que comienza en el punto fijo).

Response variable (p. 76) In an experiment, the outcome to be measured.

Variable de respuesta (pág. 76) En un experimento, el resultado que se debe medir.

Rigid transformation (pp. 208–214) A transformation of points in the plane that preserves all distances. Such a transformation repositions a figure in a plane without changing its shape or size.

Transformación rígida (págs. 208–214) Transformación de los puntos en un plano que conserva todas las distancias. Tal transformación reposiciona una figura en un plano sin cambiar el tamaño o la forma de la figura.

Rotation (p. 210) A transformation that "turns" all points in a plane through a specified angle about a fixed point called the rotation center. That is, if points P' and Q' are the images of points P and Q under a counterclockwise rotation about point C, then $CP = CP'$, $CQ = CQ'$, and $m\angle PCP' = m\angle QCQ'$.

Rotación (pág. 210) Transformación que "gira" todos los puntos en un plano a través de un ángulo específico con relación a un punto fijo llamado centro de rotación. Es decir, si los puntos P' y Q' son las imágenes de los puntos P y Q en una rotación antihoraria alrededor del punto C, entonces $CP = CP'$, $CQ = CQ'$, y $m\angle PCP' = m\angle QCQ'$.

S

Sample survey (or, **poll**) (p. 89) Observation of a random sample in order to estimate a characteristic of the larger population from which the sample was taken.

Encuesta por muestreo (pág. 89) Observación de una muestra aleatoria para estimar una característica de la población total a partir de la muestra tomada.

Scale factor of a size transformation (p. 165) A positive constant that scales (multiplies) all lengths (or distances) in the plane. A scale factor greater than 1 produces enlarged figures, and a scale factor less than 1 produces reduced figures; in both cases, the figures transformed by the size transformation are similar to the original.

Factor de escala de una transformación de (pág. 165) Constante positiva que aumenta (multiplica) todas las longitudes (o distancias) en el plano. Una factor de escala mayor que 1 produce figuras ampliadas, y un factor de escala menor que 1 produce figuras reducidas; en ambos caso, las figuras transformadas son semejantes a la original.

Scientific notation (p. 571) Expression of a number in the form $a \cdot 10^k$, where a has absolute value between 1 and 10 and k is an integer. For example, 3.245×10^3 or 3.245×10^{-3} are expressed in scientific notation.

Notación científica (pág. 571) Expresión de un número en la forma $a \cdot 10^k$, donde a tiene un valor Vabsoluto entre 1 y 10 y k es un número entero. Por ejemplo, 3.245×10^3 ó 3.245×10^{-3}, están expresados en notación científica.

Sequential change (p. 462) Change that occurs step-by-step.

Cambio secuencial (pág. 462) Cambio que ocurre paso a paso.

Series (p. 493) An expression in which the terms of a sequence are added together.

Serie (pág. 493) Expresión en la que los términos de una sucesión son sumados juntos.

English	**Español**

Sierpinski Triangle (p. 189) A fractal that begins at the initial stage as an equilateral triangle. At each iterative step, in each remaining triangle, the middle triangle(s) formed by connecting the midpoints of each side are removed.

Triángulo de Sierpinski (pág. 189) Fractal que comienza en la etapa inicial como un triángulo equilátero. En cada paso de la iteración, en cada triángulo restante, se quitan el/los triángulo(s) del medio formado(s) al conectar los puntos medios de cada lado.

Similar figures Figures that are related by a size transformation or by a similarity transformation. Such figures have the same shape, regardless of position or orientation.

Figuras semejantes Figuras relacionadas por una transformación de tamaño o por una transformación de semejanza. Tales figuras tienen la firma forma, sin importar la posición u orientación.

Similar polygons (p. 165) A special case of similar figures. Corresponding angles have the same measure and the ratios of lengths of corresponding sides is a constant.

Polígonos semejantes (pág. 165) Caso especial de figuras semejantes. Los ángulos correspondientes tiene la misma medida y la longitud de los radios de los lados correspondientes es una constante.

Similarity transformation (p. 234) Composite of a size transformation and a rigid transformation. Such a transformation resizes a figure in a plane without changing its shape.

Transformación de semejanza (pág. 234) Combinación de una transformación de tamaño y una transformación rígida. Tal transformación cambia las medidas de una figura en el plano sin cambiar su forma.

Single blind (p. 78) An experiment in which the subjects do not know which treatment they are getting.

Simple ciego (pág. 78) Experimento en el cual los sujetos no saben qué tratamiento reciben.

Size transformation (p. 177) A transformation with center C and magnitude $k > 0$ that maps each point P of the plane onto an image point P' as follows: Point C is its own image. For $P \neq C$, the image point P' is on $\overline{CP}$ and $CP' = k \cdot CP$.

Transformación de tamaño (pág. 177) Una transformación con centro C y magnitud $k > 0$ que representa cada punto P en un plano sobre un punto de imagen P' de la siguiente forma: el punto C es su propia imagen. Para $P \neq C$, la imagen del punto P' está en $\overline{CP}$, y $CP' = k \cdot CP$.

Spherical geometry (p. 48) The geometry on a sphere in which all "lines" (shortest paths on the surface) are great circles.

Geometría esférica (pág. 48) Geometría en una esfera en la que todas las "rectas" (los caminos más cortos de la superficie) son grandes círculos.

Standardized value (or, *z*-score) (p. 243) The (positive or negative) number of standard deviations a given value lies from the mean in a distribution.

Valor estandarizado (o **valor *z***) (pág. 243) En una distribución, el número (positivo o negativo) de desviaciones estándar de un número dado a partir de la media.

Statistically significant (p. 85) In an experiment, the conclusion that it is unreasonable to attribute the difference in mean response from the treatments solely to the particular random assignment of treatments to subjects. The researcher concludes that one treatment causes a larger mean response than the other.

Estadísticamente significativo (pág. 85) En un experimento, la conclusión de que no es razonable atribuir la diferencia de la respuesta media al tratamiento sólo a la asignación aleatoria particular de los tratamientos a los pacientes. El investigador puede concluir que el experimento establece que un tratamiento causa una respuesta media mayor que el otro.

English	Español

Subjects (p. 77) The available group of people (or animals, plants, or objects) to which treatments are applied in an experiment.

Supplementary angles (p. 35) A pair of angles whose measures add to 180°.

Sujetos (pág. 77) Grupo de personas disponibles (o animales, plantas u objetos) a los que se somete a un tratamiento en un experimento.

Ángulos suplementarios (pág. 35) Par de ángulos cuyas medidas suman 180°.

T

Tangent line to a circle (p. 397) A line that intersects a circle in only one point.

Theorem (p. 32) In mathematics, a statement that has been proved true.

Transformation (p. 26) A correspondence between all points in a plane and themselves such that (1) each point in the plane has a unique image point in the plane, and (2) for each point in the plane, there is a point that is the preimage of the given point.

Translation (p. 209) A transformation that "slides" all points in the plane the same distance (magnitude) and same direction. That is, if points P' and Q' are the images of points P and Q under a translation, then $PP' = QQ'$ and $\overline{PP'} \parallel \overline{QQ'}$.

Transmission factor (p. 443) The number by which the speed of the driver in a pulley system is multiplied to get the speed of the follower.

Transversal (p. 35) A line that intersects two coplanar lines in two distinct points.

Trapezoid (p. 7) A quadrilateral with two opposite sides parallel.

Treatments (p. 76) The conditions to be compared in an experiment. Treatments should be randomly assigned to subjects.

Trial (p. 260) In probability, one repetition of a random process.

Rectas tangentes de un círculo (pág. 397) Recta que interseca un círculo sólo en un punto.

Teorema (pág. 32) En matemáticas, enunciado que se ha probado como verdadero.

Transformación (pág. 26) Correspondencia entre todos los puntos en un plano y entre estos mismos, tal que (1) cada punto en el plano tiene un punto imagen único en el plano y (2) para cada punto en el plano existe un punto que es la preimagen del punto dado.

Traslación (pág. 209) Transformación que "desliza" todos los puntos de un plano la misma distancia (magnitud) y en la misma dirección. Es decir, si los puntos P' y Q' son imágenes de los puntos P y Q en una traslación, entonces $PP' = QQ'$ y $\overline{PP'} \parallel \overline{QQ'}$.

Factor de transmisión (pág. 443) Número por el que la velocidad del conductor en un sistema de poleas se multiplica para obtener la velocidad del seguidor.

Transversal (pág. 35) Recta que interseca dos rectas coplanares en dos puntos distintos.

Trapecio (pág. 7) Cuadrilátero con dos lados opuestos paralelos.

Tratamientos (pág. 76) Condiciones que se deben comparar en un experimento. Los tratamientos deben aplicarse a los sujetos en forma aleatoria.

Prueba (pág. 260) En probabilidad, una repetición de un proceso aleatorio.

English	**Español**

V

Valid argument (p. 3) An argument that uses correct rules of logic. When the premises (hypotheses) are true, and the argument is valid, then the conclusion must be true.

Argumento válido (pág. 3) Argumento que utiliza reglas correctas de la lógica. Cuando las premisas (hipótesis) son verdaderas, y el argumento es válido, entonces la conclusión debe ser verdadera.

Venn diagram (p. 21) A diagram where mutually exclusive events are represented by non-overlapping circles and events that are not mutually exclusive are represented by overlapping circles.

Diagrama de Venn (pág. 21) Diagrama en el que los eventos mutuamente excluyentes se representan mediante círculos que no se superponen y los eventos que no son mutuamente excluyentes se representan mediante círculos superpuestos.

Vertex form of a quadratic See *Completing the square.*

Forma de vértice de una expresión cuadrática Ver *Completar el cuadrado.*

Vertical angles (p. 30) Two angles whose sides form two pairs of opposite rays. When two lines or two segments intersect, vertical angles are formed.

Ángulos opuestos por el vértice (pág. 30) Dos ángulos cuyos lados forman dos pares de rayos opuestos. Cuando dos líneas o dos segmentos se intersecan, se forman ángulos verticales.

Vertical asymptote (p. 367) A line with equation $x = k$ is a vertical asymptote for the graph of a function $f(x)$ whenever values of $f(x)$ approach $+\infty$ or $-\infty$ as $x \to k$ (from one side only, or from both sides).

Asíntota vertical (pág. 367) Una recta con ecuación $x = k$ es una asíntota vertical para el gráfico de una función $f(x)$ cuando los valores de $f(x)$ tiendan a $+\infty$ ó $-\infty$, en tanto $x \to k$ (de un lado solamente, o de ambos lados).

X

x-bar chart (p. 300) A control chart where the means of samples of measurements, rather than individual measurements, are plotted.

Gráfica de barras x (pág. 300) Tipo de diagrama de control en el que las medias de las muestras de mediciones, en vez de ser mediciones individuales, son combinadas.

Z

Zeroes of polynomials (p. 329) A number k is a zero of the polynomial $f(x) = a_n x^n + a_{n-1} x^{n-1} + \cdots + a_1 x + a_0$ if $f(x) = 0$. The number k will be an x-intercept of the graph of f; k is also called a root of the polynomial.

Cero de un polinomio (pág. 329) Un número k es cero de un polinomio $a_n x^n + a_{n-1} x^{n-1} + \cdots + a_1 x + a_0$ si $f(x) = 0$. El número k serán una intercepción en x de gráfica de f; k es también llamado raíz del polinomio.

Math Online A mathematics multilingual glossary is available at www.glencoe.com/apps/eGlossary612/grade.php. The Glossary includes the following languages:

Arabic	English	Korean	Tagalog
Bengali	Haitian Creole	Russian	Urdu
Cantonese	Hmong	Spanish	Vietnamese

INDEX OF MATHEMATICAL TOPICS

─────── **T** ───────

INDEX OF CONTEXTS

INDEX OF CONTEXTS

Alignment of *Core-Plus Mathematics*: Course 3 with the CCSS

In the following charts, CCSS indicators in bold font are those "focused on" in the Investigation. Indicators not in bold are "connected to" in the Investigation. The Mathematical Practices permeate every lesson. The *CCSS Guide to Core-Plus Mathematics* provides, for each high school mathematics standard, corresponding page references in Courses 1–4.

Course 3	CCSS Content Standards
Unit 1 – Reasoning and Proof	
Lesson 1 Reasoning Strategies	
Investigation 1 Reasoned Arguments	**Modeling Conceptual Category (CCSS p. 72),** A-SSE.3, G-MG.1
Investigation 2 If-Then Statements	**G-CO.2, G-CO.4, G-CO.6, G-CO.9, G-CO.12,** A-SSE.4, G-CO.10
Lesson 2 Geometric Reasoning and Proof	
Investigation 1 Reasoning About Intersecting Lines and Angles	**G-CO.1, G-CO.9, G-CO.12,** A-CED.4
Investigation 2 Reasoning About Parallel Lines and Angles	**G-CO.1, G-CO.9, G-CO.12,** A-CED.4, A-REI.10, G-CO.10
Lesson 3 Algebraic Reasoning and Proof	
Investigation 1 Reasoning With Algebraic Expressions	**A-SSE.1,** A-SSE.4, F-BF.2, F-LE.1, F-LE.4
Investigation 2 Reasoning With Algebraic Equations	**A-SSE.1, A-REI.1,** F-TF.8, G-SRT.10
Lesson 4 Statistical Reasoning	
Investigation 1 Design of Experiments	**S-IC.3,** S-IC.6
Investigation 2 By Chance or from Cause?	**S-ID.1, S-ID.2, S-ID.3, S-IC.5,** A-SSE.1, S-IC.4
Investigation 3 Statistical Studies	**S-ID.1, S-IC.3,** S-IC.6
Unit 2 – Inequalities and Linear Programming	
Lesson 1 Inequalities in One Variable	
Investigation 1 Getting the Picture	**A-SSE.1, A-CED.1, A-REI.10, F-IF.1, F-IF.2, F-IF.4, F-IF.5, F-IF.7,** A-SSE.3, A-REI.4, F-TF.2
Investigation 2 Quadratic Inequalities	**A-SSE.1, A-SSE.3, A-REI.4, A-REI.10, F-IF.7,** A-CED.1
Investigation 3 Complex Inequalities	**A-SSE.3, A-REI.2, A-REI.4, A-REI.7, A-REI.10, A-REI.11, F-IF.7,** A-CED.1, F-TF.2, G-SRT.11
Lesson 2 Inequalities in Two Variables	
Investigation 1 Solving Inequalities	**A-CED.1, A-REI.12, F-IF.4, F-IF.7, F-LE.5,** A-SSE.1, A-REI.10, F-IF.1, F-LE.1
Investigation 2 Linear Programming—A Graphic Approach	**A-CED.1, A-CED.2, A-CED.3, A-REI.6, A-REI.12, F-IF.4,** A-SSE.1, F-IF.1, F-LE.1
Investigation 3 Linear Programming—Algebraic Methods	**A-SSE.1, A-CED.1, A-CED.2, A-CED.3, A-REI.6, A-REI.11, AREI.12, F-IF.4, F-IF.7, F-LE.5,** F-IF.1, F-IF.5, F-LE.1

Course 3	CCSS Content Standards
Unit 3 – Similarity and Congruence	
Lesson 1 Reasoning about Similar Figures	
Investigation 1 When are Two Polygons Similar?	**G-SRT.2, G-SRT.3, G-MG.3,** G-CO.12
Investigation 2 Sufficient Conditions for Similarity of Triangles	**G-SRT.2, G-SRT.3, G-SRT.4, G-SRT.5, G-MG.3,** G-SRT.8, G-SRT.10, G-SRT.11
Investigation 3 Reasoning with Similarity Conditions	**G-SRT.1, G-SRT.2, G-SRT.3, G-SRT.4, G-SRT.5, G-C.1, G-MG.3,** A-SSE.2, G-GPE.4
Lesson 2 Reasoning about Congruent Figures	
Investigation 1 Congruence of Triangles Revisited	**G-CO.9,** G-CO.10
Investigation 2 Congruence in Triangles	**G-CO.10, G-CO.12, G-CO.13,** G-GPE.4
Investigation 3 Congruence of Quadrilaterals	**G-CO.8, G-CO.11,** G-GPE.4.
Investigation 4 Congruence and Similarity: A Transformation Approach	**G-CO.2, G-CO.4, G-CO.5, G-CO.6, G-CO.7, G-CO.8, G-SRT.2, GSRT.3**
Unit 4 – Samples and Variation	
Lesson 1 Normal Distributions	
Investigation 1 Characteristics of a Normal Distribution	**S-ID.1, S-ID.2, S-ID.4, S-IC.1,** N-Q.3
Investigation 2 Standardized Values	**S-ID.2, S-ID.3, S-ID.4,** F-IF.7
Investigation 3 Using Standardized Values to Find Percentiles	**S-ID.2, S-ID.4,** S-ID.8
Lesson 2 Binomial Distributions	
Investigation 1 Shape, Center, and Spread	**S-ID.1, S-ID.3, S-ID.4, S-IC.1, S-MD.2, S-MD.3, S-MD.4,** F-IF.7, S-CP.1, S-CP.2
Investigation 2 Binomial Distributions and Making Decisions	**S-ID.1, S-ID.3, S-ID.4, S-IC.1, S-MD.2, S-MD.3, S-MD.4,** S-CP.1, S-CP.2
Lesson 3 Statistical Process Control	
Investigation 1 Out of Control Signals	**S-ID.1, S-ID.2, S-ID.4, S-IC.1, S-MD.7**
Investigation 2 False Alarms	**S-ID.4, S-IC.1, S-CP.7, S-CP.8, S-MD.7,** S-CP.2, S-CP.3
Investigation 3 The Center Limit Theorem	**S-ID.1, S-ID.2, S-ID.4, S-MD.7**

Course 3	CCSS Content Standards
Unit 5 – Polynomial and Rational Functions	
Lesson 1 Polynomial Expressions and Functions	
Investigation 1 Modeling with Polynomial Functions	**A-SSE.1, A-APR.3, A-CED.2, F-IF.1, F-IF.2, F-IF.4, F-IF.7, F-BF.1,** A-CED.1, F-BF.3, S-ID.6
Investigation 2 Addition, Subtraction, and Zeroes	**A-SSE.1, A-APR.1, A-APR.3, A-CED.2, F-IF.1, F-IF.2, F-IF.4, F-IF.7, F-BF.1,** A-SSE.2, A-CED.1
Investigation 3 Zeroes and Products of Polynomials	**A-SSE.1, A-SSE.2, A-APR.1, A-APR.2, A-APR.3, F-IF.1, F-IF.2, FIF.4, F-IF.7, F-BF.1,** A-CED.1, A-CED.2, F-IF.5, F-BF.3, G-MG.3
Lesson 2 Quadratic Polynomials	
Investigation 1 Completing the Square	**A-SSE.1, A-SSE.3, A-REI.4, A-REI.10, G-GPE.1,** F-IF.7
Investigation 2 The Quadratic Formula and Complex Numbers	**N-CN.1, N-CN.2, N-CN.7, A-SSE.1, A-REI.1, A-REI.4, A-REI.10,** N-CN.4, N-CN.9, N-VM.8, F-IF.7
Lesson 3 Rational Expressions and Functions	
Investigation 1 Domains and Graphs of Rational Functions	**A-SSE.1, A-CED.2, F-IF.2, F-IF.4, F-IF.5, F-IF.7,** A-CED1, F-IF.1, G-SRT.6
Investigation 2 Simplifying Rational Expressions	**A-SSE.1, A-SSE.3, A-CED.2, F-IF.2, F-IF.4, F-IF.5, F-IF.7, F-IF.8,** A-CED.1, F-IF.1
Investigation 3 Adding and Subtracting Rational Expressions	**A-SSE.1, A-SSE.3, A-APR.6, A-APR.7, A-CED.2, F-IF.2, F-IF.4, FIF.5, F-IF.7, F-IF.8, F-BF.1,** A-CED.1, F-IF.1
Investigation 4 Multiplying and Dividing Rational Expressions	**A-SSE.1, A-APR.6, A-APR-7, F-IF.2, F-IF.5, F-IF.7, F-IF.8, F-FB.1,** G-SRT.5
Unit 6 – Circles and Circular Functions	
Lesson 1 Circles and Their Properties	
Investigation 1 Tangents to a Circle	**G-CO.1, G-CO.12, G-C.2, G-C.4,** F-BF.3, G-SRT.5, G-GPE.4, G-MG.3
Investigation 2 Chords, Arcs, and Central Angles	**G-CO.12, G-C.2,** G-SRT.5, G-GPE.4, S-MD.7
Investigation 3 Angles Inscribed in a Circle	**G-CO.13, G-C.2, G-C.3, G-C.5,** G-SRT.5, G-GPE.4, G-MG.3
Lesson 2 Circular Motion and Periodic Functions	
Investigation 1 Angular and Linear Velocity	**G-MG.1, G-MG.3,** N-VM.12, G-CO.6, G-SRT.7
Investigation 2 Modeling Circular Motion	**F-TF.2, F-TF.5,** N-VM.11
Investigation 3 Revolutions, Degrees, and Radians	**F-BF.3, F-TF.1, F-TF.2, F-TF.3, G-C.5,** G-SRT.7, G-MG.1, G-MG.3
Investigation 4 Patterns of Periodic Change	**F-BF.3, F-TF.5,** N-Q.1, F-IF.7, G-CO.6

Course 3	CCSS Content Standards
Unit 7 – Recursion and Iteration	
Lesson 1 Modeling Sequential Change	
Investigation 1 Modeling Population Change	**A-SSE.1, A-CED.2, F-BF.2**, A-REI.2, F-IF.1
Investigation 2 The Power of Notation and Technology	**A-SSE.1, A-REI.2, A-REI.10, F-IF.2, F-IF.3, F-BF.2**, N-VM.8, A-REI.1, F-IF.1
Lesson 2 A Recursive View of Functions	
Investigation 1 Arithmetic and Geometric Sequences	**A-SSE.1, A-CED.2, F-IF.3, F-BF.1, F-LE.2**, S-ID.6, S-MD.1, S-MD.2
Investigation 2 Some Sums	**A-SSE.1, A-SSE.4, A-CED.2, F-IF.3, F-BF.1**, A-REI.1, F-IF.2, F-LE.5, S-ID.6
Investigation 3 Finite Differences	**A-SSE.1, A-CED.2, F-IF.3, F-BF.1**, N-VM.9, A-REI.8, A-REI.9, F-IF.2, S-ID.6
Lesson 3 Iterating Functions	
Investigation 1 Play It Again … and Again	**F-IF.1, F-IF.2, F-IF.3**, A-CED.1
Investigation 2 Iterating Linear Functions	**F-IF.1, F-IF.2, F-IF.3**, A-CED.1, A-REI.6
Unit 8 – Inverse Functions	
Lesson 1 What is an Inverse Function?	
Investigation 1 Coding and Decoding Messages	**F-IF.1, F-BF.4**, N-VM.8
Investigation 2 Finding and Using Inverse Functions	**F-IF.1, F-IF.5, F-BF.1, F-BF.4**, N-VM.8, A-REI.10
Lesson 2 Common Logarithms and Their Properties	
Investigation 1 Common Logarithms Revisited	**A-CED.1, A-REI.1, F-IF.7, F-IF.8, F-BF.5, F-LE.4**, A-SSE.1, A-SSE.3
Investigation 2 Covering All the Bases	**A-CED.1, A-REI.1, F-BF.5, F-LE.4**, A-SSE.1
Investigation 3 Properties of Logarithms	**A-REI.1, F-BF.5, F-LE.4**, A-SSE.1, A-CED.2
Lesson 3 Inverse Trigonometric Functions	
Investigation 1 The Ups and Downs of the Sine	**F-IF.4, F-IF.5, F-BF.4, F-TF.5, F-TF.6, F-TF.7**, A-CED.1, A-REI.1, S-ID.6
Investigation 2 Inverses of the Cosine and Tangent	**F-IF.4, F-IF.5, F-BF.4, F-TF.5, F-TF.6, F-TF.7**, A-CED.1, A-REI.1, F-IF.7